AFRICA

ELEVENTH EDITION

Dr. Wayne Edge

OTHER BOOKS IN THE GLOBAL STUDIES SERIES

- China
- Europe
- India and South Asia
- Japan and the Pacific Rim
- Latin America
- The Middle East
- Russia, the Eurasian Republics, and Central/Eastern Europe

McGraw-Hill/Dushkin Company
2460 Kerper Boulevard, Dubuque, Iowa 52001
Visit us on the Internet—http://www.dushkin.com

Staff

Larry Loeppke	*Managing Editor*
Nichole Altman	*Developmental Editor*
Lori Church	*Permissions Assistant*
Maggie Lytle	*Cover*
Tara McDermott	*Designer*
Kari Voss	*Typesetting Supervisor/Co-designer*
Jean Smith	*Typesetter*
Sandy Wille	*Typesetter*
Karen Spring	*Typesetter*

Sources for Statistical Reports

U.S. State Department *Background Notes* (2003)
C.I.A. *World Factbook* (2002)
World Bank *World Development Reports* (2002/2003)
UN *Population and Vital Statistics Reports* (2002/2003)
World Statistics in Brief (2002)
The Statesman's Yearbook (2003)
Population Reference Bureau *World Population Data Sheet* (2002)
The World Almanac (2003)
The Economist Intelligence Unit (2003)

Copyright

Cataloging in Publication Data
Main entry under title: Global Studies: Africa. 11th ed.
1. Africa—History—1960–.
I. Title:Africa. II. Edge, Wayne, *comp*.
ISBN 0–07–319535–9 960.3 91–71258 ISSN 1098-3880

Eleventh Edition
Printed in the United States of America 1234567890BAHBAH54 Printed on Recycled Paper

AFRICA

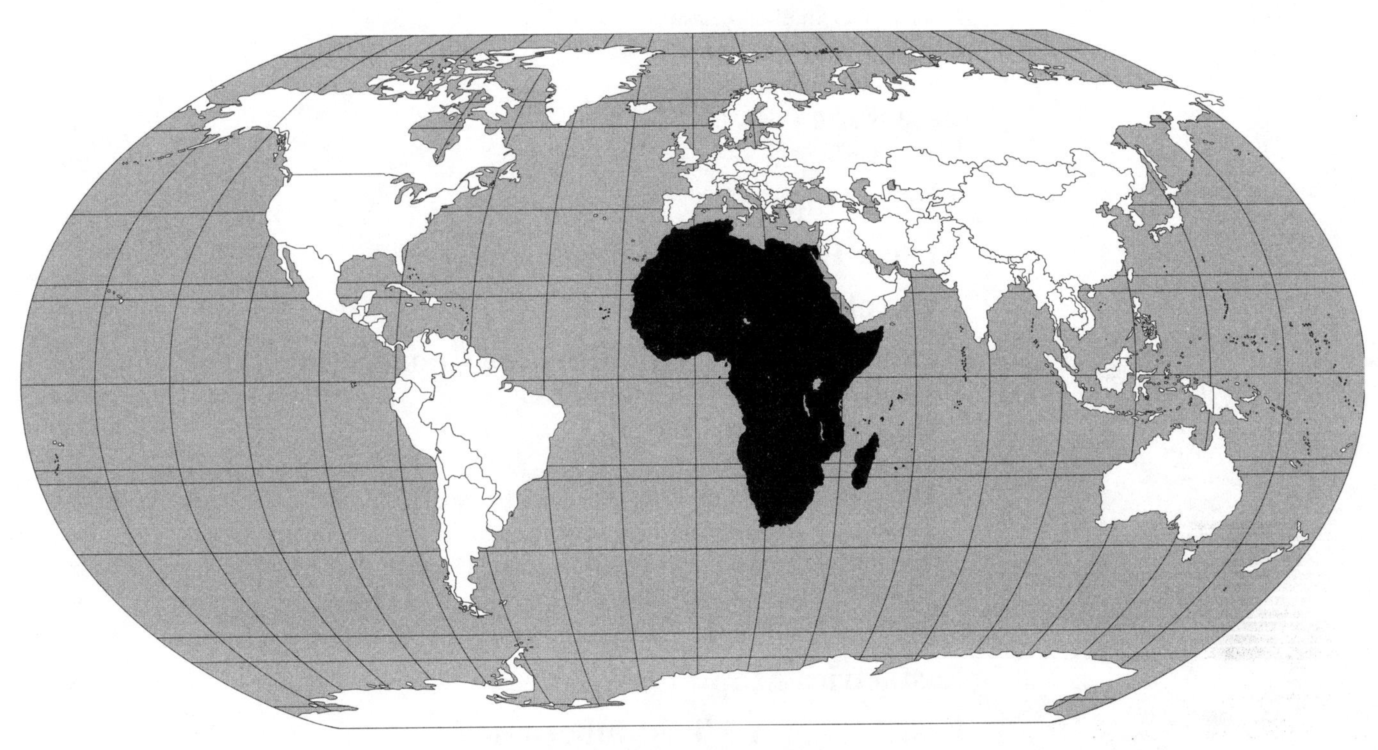

AUTHOR/EDITORS

Dr. Wayne Edge

Dr. Wayne Edge is a longtime resident of Botswana. Having obtained his Ph.D. in political science at the University of Delaware, he was until recently a lecturer at the University of Botswana. He has also taught at the University of the Virgin Islands, Lincoln University, the University of Delaware, and the School for International Training, Botswana Program. Dr. Edge's other publications include authoring *The Autobiography of Motsamai Mpho* (Lepopo Publishers, 1996); and coediting, with M. Lekorewe, as well as contributing to *Botswana: Politics and Society* (J. L. Van Schaik Publishers, 1998).

Contents

Using Global Studies: Africa

THE GLOBAL STUDIES SERIES

The Global Studies series was created to help readers acquire a basic knowledge and understanding of the regions and countries in the world. Each volume provides a foundation of information—geographic, cultural, economic, political, historical, artistic, and religious—that will allow readers to better assess the current and future problems within these countries and regions and to comprehend how events there might affect their own well-being. In short, these volumes present the background information necessary to respond to the realities of our global age.

Each of the volumes in the Global Studies series is crafted under the careful direction of an author/editor—an expert in the area under study. The author/editors teach and conduct research and have traveled extensively through the regions about which they are writing.

In this *Global Studies: Africa* edition, the author/editor has written an introductory essay on the continent as a whole, several regional essays, and country reports for each of the countries included.

MAJOR FEATURES OF THE GLOBAL STUDIES SERIES

The Global Studies volumes are organized to provide concise information on the regions and countries within those areas under study. The major sections and features of the books are described here.

Regional Essays

For *Global Studies: Africa,* the author/editor has written several essays focusing on the religious, cultural, sociopolitical, and economic differences and similarities of the countries and peoples in the various regions of Africa. Regional maps accompany the essays.

Country Reports

Concise reports are written for each of the countries within the region under study. These reports are the heart of each Global Studies volume. *Global Studies: Africa, Eleventh Edition,* contains 54 country reports.

The country reports are composed of five standard elements. Each report contains a detailed map visually positioning the country among its neighboring states; a summary of statistical information; a current essay providing important historical, geographical, political, cultural, and economic information; a historical timeline, offering a convenient visual survey of a few key historical events; and four "graphic indicators," with summary statements about the country in terms of development, freedom, health/welfare, and achievements.

A Note on the Statistical Reports

The statistical information provided for each country has been drawn from a wide range of sources. (The most frequently referenced are listed on page ii.) Every effort has been made to provide the most current and accurate information available. However, sometimes the information cited by these sources differs to some extent; and, all too often, the most current information available for some countries is somewhat dated. Aside from these occasional difficulties, the statistical summary of each country is generally quite complete and up to date. Care should be taken, however, in using these statistics (or, for that matter, any published statistics) in making hard comparisons among countries. We have also provided comparable statistics for the United States and Canada, which can be found on pages x and xi.

World Press Articles

Within each Global Studies volume is reprinted a number of articles carefully selected by our editorial staff and the author/editor from a broad range of international periodicals and newspapers. The articles have been chosen for currency, interest, and their differing perspectives on the subject countries. There are 13 articles in *Global Studies: Africa, Eleventh Edition.*

WWW Sites

An extensive annotated list of selected World Wide Web sites can be found on the facing page viii in this edition of *Global Studies: Africa.* In addition, the URL addresses for country-specific Web sites are provided on the statistics page of most countries. All of the Web site addresses were correct and operational at press time. Instructors and students alike are urged to refer to those sites often to enhance their understanding of the region and to keep up with current events.

Glossary, Bibliography, Index

At the back of each Global Studies volume, readers will find a glossary of terms and abbreviations, which provides a quick reference to the specialized vocabulary of the area under study and to the standard abbreviations used throughout the volume.

Following the glossary is a bibliography, which lists general works, national histories, and current-events publications and periodicals that provide regular coverage on Africa.

The index at the end of the volume is an accurate reference to the contents of the volume. Readers seeking specific information and citations should consult this standard index.

Currency and Usefulness

Global Studies: Africa, like the other Global Studies volumes, is intended to provide the most current and useful information available necessary to understand the events that are shaping the cultures of the region today.

This volume is revised on a regular basis. The statistics are updated, regional essays and country reports revised, and world press articles replaced. In order to accomplish this task, we turn to our author/editor, our advisory boards, and—hopefully—to you, the users of this volume. Your comments are more than welcome. If you have an idea that you think will make the next edition more useful, an article or bit of information that will make it more current, or a general comment on its organization, content, or features that you would like to share with us, please send it in for serious consideration.

Selected World Wide Web Sites for Africa

(Some Web sites continually change their structure and content, so the information listed here may not always be available. Check our Web site at: http://www.dushkin.com/online/ —Ed.)

GENERAL SITES

BBC World Service

http://www.bbc.co.uk/worldservice/index.htm

The BBC, one of the world's most successful radio networks, provides the latest news from around the world, including news from almost all of the African countries.

C-SPAN ONLINE

http://www.c-span.org

Access C-SPAN International on the Web for International Programming Highlights and archived C-SPAN programs.

International Network Information Center at University of Texas

http://inic.utexas.edu

This gateway has pointers to international sites, including Africa, as well as African Studies Resources.

I-Trade International Trade Resources & Data Exchange

http://www.i-trade.com

Monthly exchange-rate data, U.S. Global Trade Outlook, and recent World Fact Book statistical demographic and geographic data for 180-plus countries can be found on this Web site.

Penn Library: Resources by Subject

http://www.library.upenn.edu/resources/subject/subject.html

This vast site is rich in links to information about African studies, including demography and population.

ReliefWeb

http://www.reliefweb.int

The UN's Department of Humanitarian Affairs clearinghouse for international humanitarian emergencies presents daily news updates, including Reuters, VOA, PANA.

Social Science Information Gateway (SOSIG)

http://sosig.esrc.bris.ac.uk

The Economic and Social Research Council [ESRC] project catalogs 22 subjects and lists developing countries' URL addresses.

United Nations System

http://www.unsystem.org

The official Web site for the United Nations system of organizations can be found here. Everything is listed alphabetically, and examples include UNICC and the Food and Agriculture Organization.

UN Development Programme (UNDP)

http://www.undp.org

Publications and current information on world poverty, Mission Statement, UN Development Fund for Women, and more are available on this Web site. Be sure to see the Poverty Clock.

UN Environmental Programme (UNEP)

http://www.unep.org

This UNEP official site provides information on UN environmental programs, products, services, and events. A search engine is also available.

U.S. Agency for International Development (USAID)

http://www.usaid.gov/regions/afr/

The U.S. policy regarding assistance to African countries is presented at this site.

U.S. Central Intelligence Agency (CIA)

http://www.odci.gov

This site includes publications of the CIA, such as the World Fact Book, Factbook on Intelligence, Handbook of International Economic Statistics, and CIA maps.

U.S. Department of State

http://www.state.gov/countries/

Organized alphabetically, data on human rights issues, international organizations, and country reports as well as other data are available here.

World Bank Group

http://www.worldbank.org/html/extdr/regions.htm

News (press releases, summary of new projects, speeches), publications, topics in development, and reports on countries and regions can be accessed on this Web site. Links to other financial organizations are also provided.

World Health Organization (WHO)

http://www.who.int/en

Maintained by WHO's headquarters in Geneva, Switzerland, this site uses the Excite search engine to conduct keyword searches.

World Trade Organization (WTO)

http://www.wto.org

WTO's Web site topics include information on world trade systems, data on textiles, intellectual property rights, legal frameworks, trade and environmental policies, recent agreements, and other issues.

GENERAL AFRICA SITES

Africa News Web Site: Crisis in the Great Lakes Region

http://www.africanews.org/specials/greatlakes.html

African News Web Site on Great Lakes (Rwanda, Burundi, Zaire, and Kenya, Tanzania, Uganda) can be found here with frequent updates and good links to other sites. It is possible to order e-mail crisis updates here.

African Policy Information Center (APIC)

http://www.africapolicy.org

Developed by the Washington Office on Africa to widen policy debate in the United States on African issues, this Web site includes special topic briefs, regular reports, and documents on African politics.

Africa: South of the Sahara

http://www-sul.stanford.edu/depts/ssrg/africa/guide.html

On this site, Topics and Regions link headings will lead to a wealth of information.

African Studies WWW (U.Penn)

http://www.sas.upenn.edu/African_Studies/AS.html

This Web site provides facts about each African country, which includes news, statistics, and links to other Web sites.

AllAfrica Global Media

http://allafrica.com

From this page, explore African news by region or country. Topics covered include conflict and security; economy, business, and finance; environment and sustainable development; health; and human rights, plus many more.

Library of Congress Country Studies

http://lcweb2.loc.gov/frd/cs/cshome.html#toc

Of the 71 countries that are covered in the continuing series of books available at this Web site, at least a dozen of them are in Africa.

South African Government Index

http://www.polity.org.za/gnuindex.html

This official site includes links to government agencies, data on structures of government, and links to detailed documents.

Weekly Mail & Guardian [Johannesburg]

http://www.mg.co.za

This free electronic daily South African newspaper includes archived back issues as well as links to other related sites on Africa.

World History Archives

http://www.hartford-hwp.com/archives/

Hartford Web Publishing offers historical archives for the continent of Africa as a whole as well as for all the regions and countries in Africa.See individual country report pages for additional Web sites.

See individual country report pages for additional Web sites.

The United States (United States of America)

GEOGRAPHY

Area in Square Miles (Kilometers): 3,717,792 (9,629,091) (about 1/2 the size of Russia)
Capital (Population): Washington, DC (3,997,000)
Environmental Concerns: air and water pollution; limited freshwater resources, desertification; loss of habitat; waste disposal; acid rain
Geographical Features: vast central plain, mountains in the west, hills and low mountains in the east; rugged mountains and broad river valleys in Alaska; volcanic topography in Hawaii
Climate: mostly temperate, but ranging from tropical to arctic

PEOPLE

Population

Total: 280,563,000
Annual Growth Rate: 0.89%
Rural/Urban Population Ratio: 24/76
Major Languages: predominantly English; a sizable Spanish-speaking minority; many others
Ethnic Makeup: 77% white; 13% black; 4% Asian; 6% Amerindian and others
Religions: 56% Protestant; 28% Roman Catholic; 2% Jewish; 4% others; 10% none or unaffiliated

Health

Life Expectancy at Birth: 74 years (male); 80 years (female)
Infant Mortality: 6.69/1,000 live births
Physicians Available: 1/365 people
HIV/AIDS Rate in Adults: 0.61%

Education

Adult Literacy Rate: 97% (official)
Compulsory (Ages): 7–16; free

COMMUNICATION

Telephones: 194,000,000 main lines
Daily Newspaper Circulation: 238/1,000 people
Televisions: 776/1,000 people
Internet Users: 165,750,000 (2002)

TRANSPORTATION

Highways in Miles (Kilometers): 3,906,960 (6,261,154)
Railroads in Miles (Kilometers): 149,161 (240,000)
Usable Airfields: 14,695
Motor Vehicles in Use: 206,000,000

GOVERNMENT

Type: federal republic
Independence Date: July 4, 1776
Head of State/Government: President George W. Bush is both head of state and head of government
Political Parties: Democratic Party; Republican Party; others of relatively minor political significance
Suffrage: universal at 18

MILITARY

Military Expenditures (% of GDP): 3.2%
Current Disputes: various boundary and territorial disputes; "war on terrorism"

ECONOMY

Per Capita Income/GDP: $36,300/$10.082 trillion
GDP Growth Rate: 0%
Inflation Rate: 3%
Unemployment Rate: 5.8%
Population Below Poverty Line: 13%
Natural Resources: many minerals and metals; petroleum; natural gas; timber; arable land
Agriculture: food grains; feed crops; fruits and vegetables; oil-bearing crops; livestock; dairy products
Industry: diversified in both capital and consumer-goods industries
Exports: $723 billion (primary partners Canada, Mexico, Japan)
Imports: $1.148 trillion (primary partners Canada, Mexico, Japan)

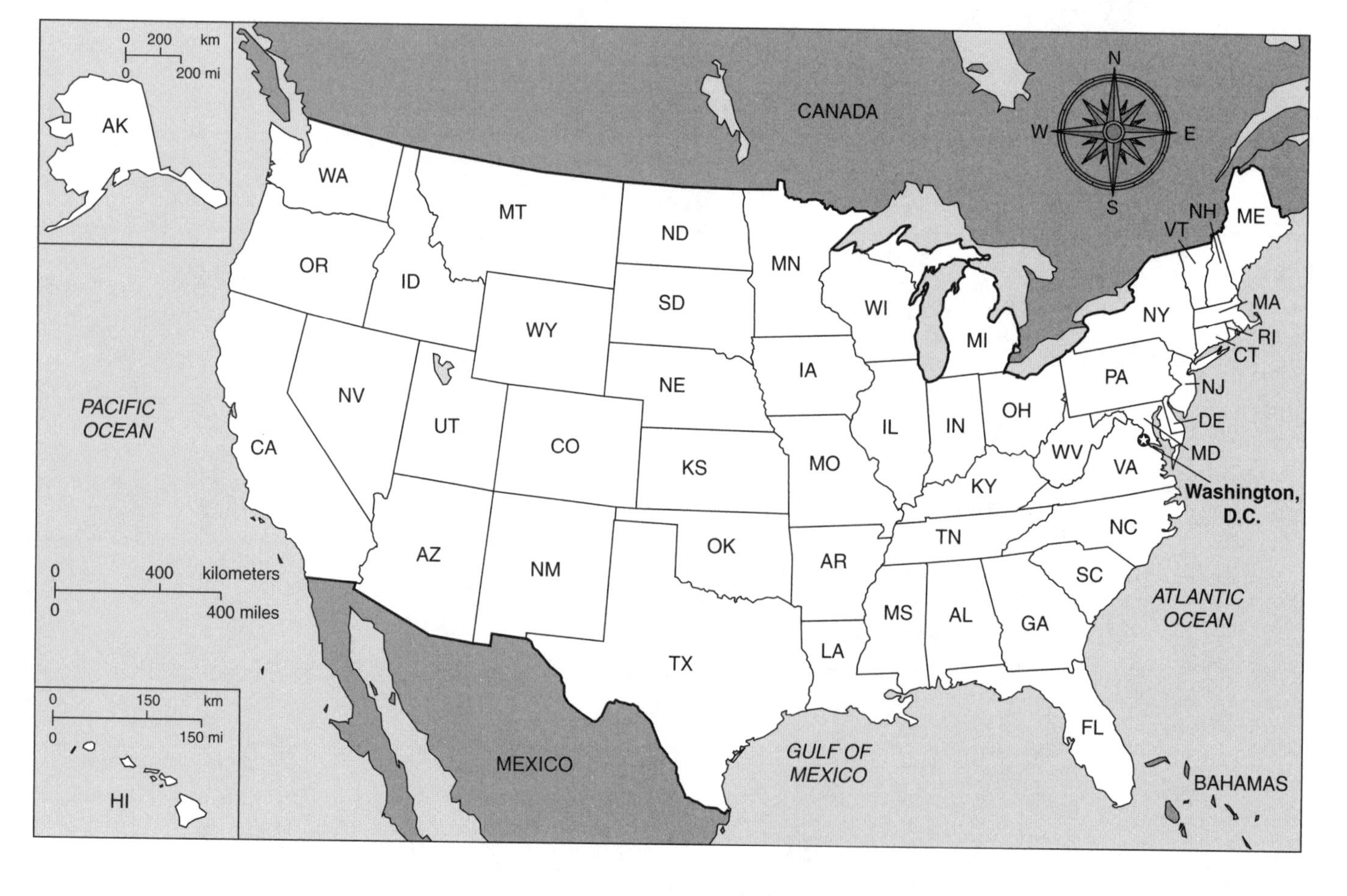

Canada

GEOGRAPHY

Area in Square Miles (Kilometers): 3,850,790 (9,976,140) (slightly larger than the United States)
Capital (Population): Ottawa (1,094,000)
Environmental Concerns: air and water pollution; acid rain; industrial damage to agriculture and forest productivity
Geographical Features: permafrost in the north; mountains in the west; central plains; lowlands in the southeast
Climate: varies from temperate to arctic

PEOPLE

Population

Total: 31,903,000
Annual Growth Rate: 0.96%
Rural/Urban Population Ratio: 23/77
Major Languages: both English and French are official
Ethnic Makeup: 28% British Isles origin; 23% French origin; 15% other European; 6% others; 2% indigenous; 26% mixed
Religions: 46% Roman Catholic; 36% Protestant; 18% others

Health

Life Expectancy at Birth: 76 years (male); 83 years (female)
Infant Mortality: 4.95/1,000 live births
Physicians Available: 1/534 people
HIV/AIDS Rate in Adults: 0.3%

Education

Adult Literacy Rate: 97%
Compulsory (Ages): primary school

COMMUNICATION

Telephones: 20,803,000 main lines
Daily Newspaper Circulation: 215/1,000 people
Televisions: 647/1,000 people
Internet Users: 16,840,000 (2002)

TRANSPORTATION

Highways in Miles (Kilometers): 559,240 (902,000)
Railroads in Miles (Kilometers): 22,320 (36,000)
Usable Airfields: 1,419
Motor Vehicles in Use: 16,800,000

GOVERNMENT

Type: confederation with parliamentary democracy
Independence Date: July 1, 1867
Head of State/Government: Queen Elizabeth II; Prime Minister Jean Chrétien
Political Parties: Progressive Conservative Party; Liberal Party; New Democratic Party; Bloc Québécois; Canadian Alliance
Suffrage: universal at 18

MILITARY

Military Expenditures (% of GDP): 1.1%
Current Disputes: maritime boundary disputes with the United States

ECONOMY

Currency ($U.S. equivalent): 1.46 Canadian dollars = $1
Per Capita Income/GDP: $27,700/$875 billion
GDP Growth Rate: 2%
Inflation Rate: 3%
Unemployment Rate: 7%
Labor Force by Occupation: 74% services; 15% manufacturing; 6% agriculture and others
Natural Resources: petroleum; natural gas; fish; minerals; cement; forestry products; wildlife; hydropower
Agriculture: grains; livestock; dairy products; potatoes; hogs; poultry and eggs; tobacco; fruits and vegetables
Industry: oil production and refining; natural-gas development; fish products; wood and paper products; chemicals; transportation equipment
Exports: $273.8 billion (primary partners United States, Japan, United Kingdom)
Imports: $238.3 billion (primary partners United States, European Union, Japan)

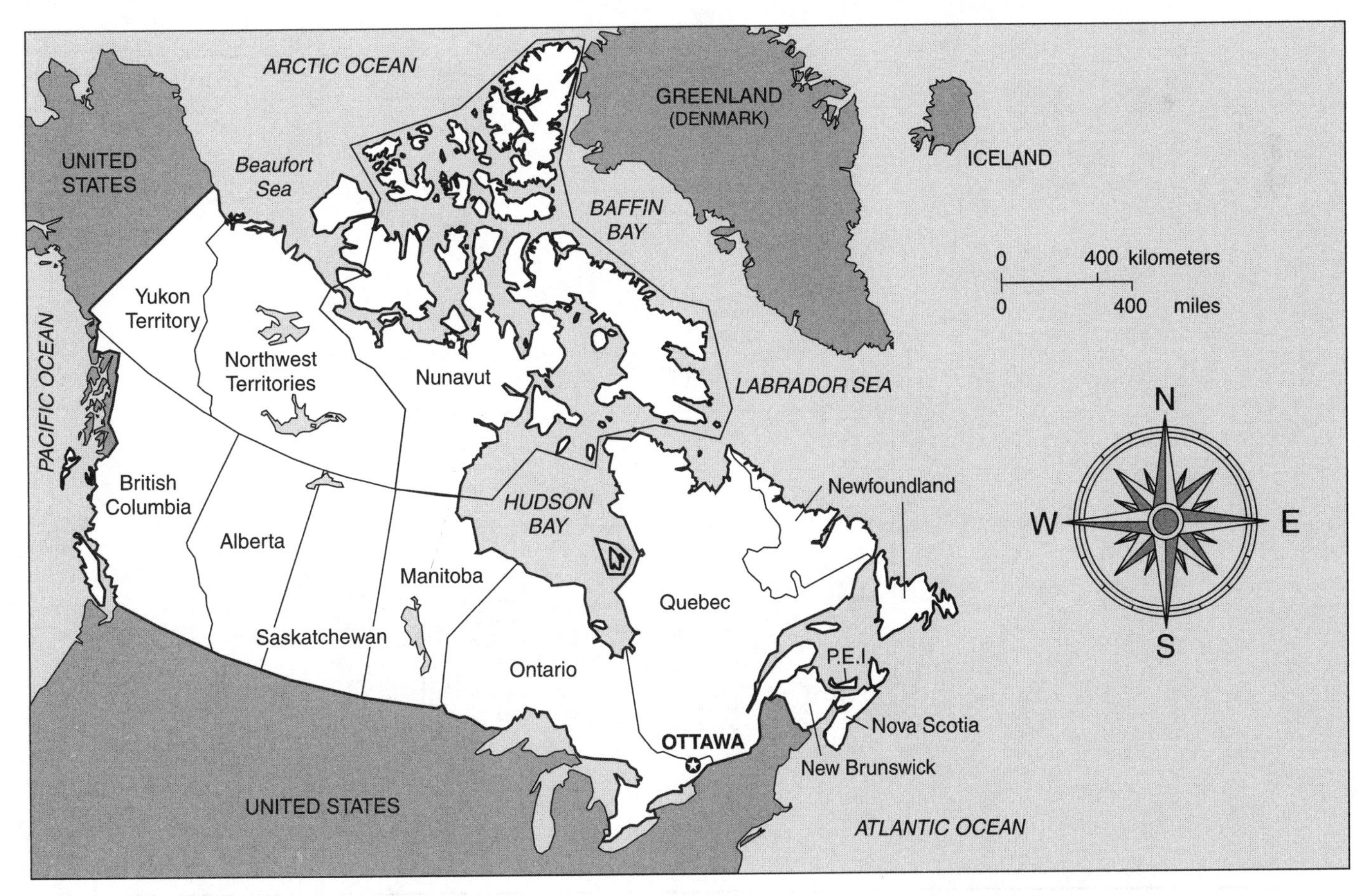

GLOBAL STUDIES

This map is provided to give you a graphic picture of where the countries of the world are located, the relationship they have with their region and neighbors, and their positions relative to major economic and political power blocs. We have focused on certain areas to illustrate these crowded regions more clearly. Africa is shaded for emphasis.

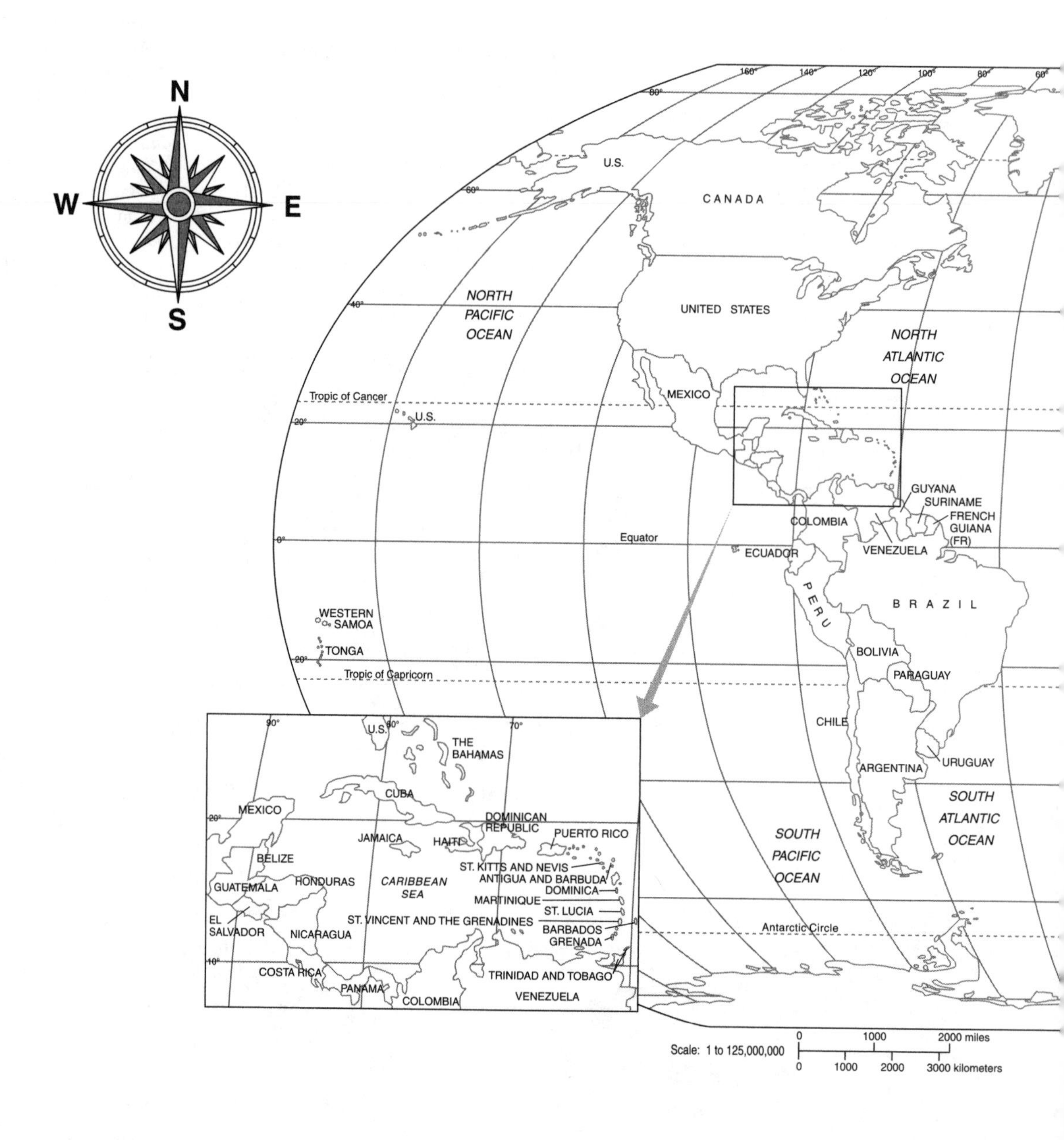

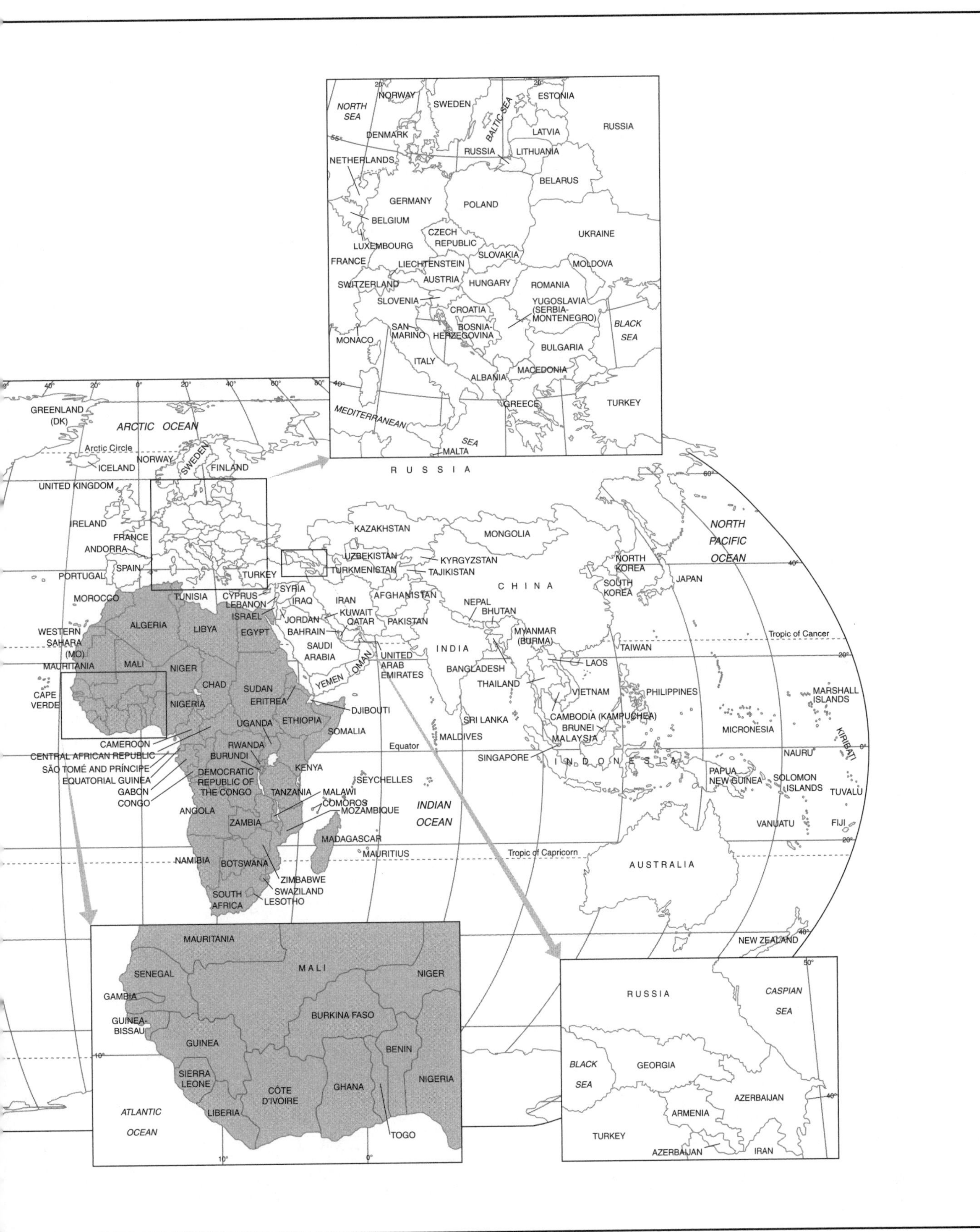
NORTH SEA
NORWAY
SWEDEN
BALTIC SEA
ESTONIA
RUSSIA
DENMARK
LATVIA
LITHUANIA
NETHERLANDS
BELARUS
GERMANY
POLAND
BELGIUM
CZECH REPUBLIC
UKRAINE
LUXEMBOURG
SLOVAKIA
FRANCE
LIECHTENSTEIN
MOLDOVA
SWITZERLAND
AUSTRIA
HUNGARY
ROMANIA
SLOVENIA
CROATIA
YUGOSLAVIA (SERBIA-MONTENEGRO)
BLACK SEA
SAN MARINO
BOSNIA-HERZEGOVINA
MONACO
BULGARIA
ITALY
ALBANIA
MACEDONIA
GREECE
TURKEY
MEDITERRANEAN SEA
MALTA
GREENLAND (DK)
ARCTIC OCEAN
Arctic Circle
ICELAND
FINLAND
UNITED KINGDOM
R U S S I A
IRELAND
KAZAKHSTAN
MONGOLIA
NORTH PACIFIC OCEAN
ANDORRA
UZBEKISTAN
KYRGYZSTAN
NORTH KOREA
PORTUGAL
SPAIN
TURKMENISTAN
TAJIKISTAN
JAPAN
SOUTH KOREA
MOROCCO
TUNISIA
CYPRUS
SYRIA
LEBANON
IRAQ
IRAN
AFGHANISTAN
C H I N A
ISRAEL
JORDAN
KUWAIT
NEPAL
BHUTAN
WESTERN SAHARA (MO)
ALGERIA
LIBYA
EGYPT
BAHRAIN
QATAR
PAKISTAN
MYANMAR (BURMA)
Tropic of Cancer
SAUDI ARABIA
I N D I A
TAIWAN
MAURITANIA
MALI
NIGER
UNITED ARAB EMIRATES
LAOS
OMAN
BANGLADESH
CHAD
YEMEN
CAPE VERDE
SUDAN
THAILAND
VIETNAM
PHILIPPINES
MARSHALL ISLANDS
NIGERIA
ERITREA
DJIBOUTI
UGANDA
ETHIOPIA
SRI LANKA
CAMBODIA (KAMPUCHEA)
MICRONESIA
SOMALIA
BRUNEI
KIRIBATI
CAMEROON
RWANDA
MALDIVES
MALAYSIA
CENTRAL AFRICAN REPUBLIC
Equator
BURUNDI
SINGAPORE
NAURU
I N D O N E S I A
SÃO TOMÉ AND PRÍNCIPE
KENYA
EQUATORIAL GUINEA
DEMOCRATIC REPUBLIC OF THE CONGO
SEYCHELLES
PAPUA NEW GUINEA
SOLOMON ISLANDS
GABON
TUVALU
TANZANIA
MALAWI
CONGO
COMOROS
INDIAN OCEAN
ANGOLA
MOZAMBIQUE
ZAMBIA
VANUATU
FIJI
MADAGASCAR
MAURITIUS
Tropic of Capricorn
NAMIBIA
BOTSWANA
AUSTRALIA
ZIMBABWE
SOUTH AFRICA
SWAZILAND
LESOTHO
NEW ZEALAND
MAURITANIA
SENEGAL
MALI
NIGER
GAMBIA
RUSSIA
CASPIAN SEA
BURKINA FASO
GUINEA-BISSAU
GUINEA
BENIN
BLACK SEA
GEORGIA
SIERRA LEONE
NIGERIA
GHANA
CÔTE D'IVOIRE
AZERBAIJAN
ATLANTIC OCEAN
LIBERIA
ARMENIA
TOGO
TURKEY
AZERBAIJAN
IRAN

Africa

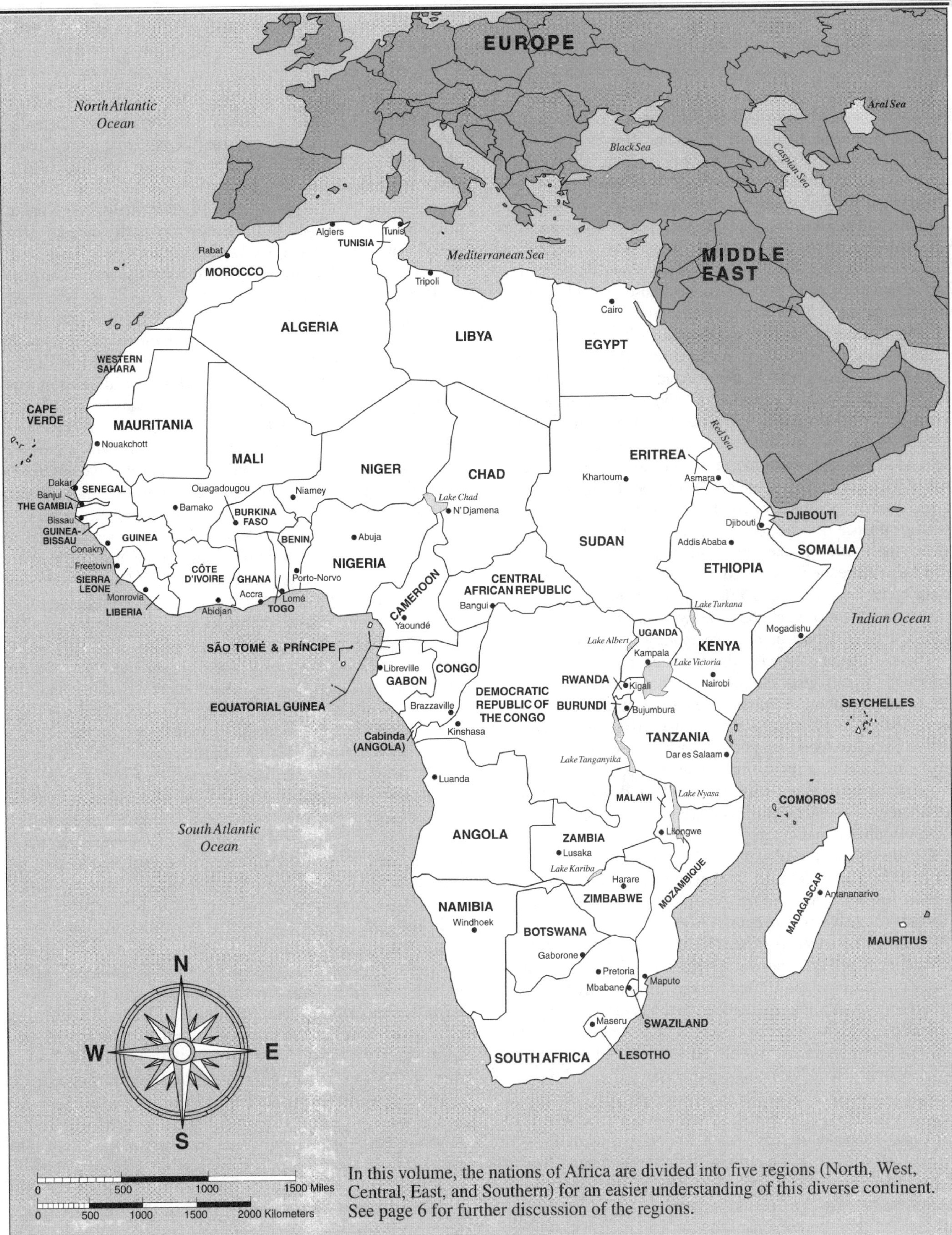

In this volume, the nations of Africa are divided into five regions (North, West, Central, East, and Southern) for an easier understanding of this diverse continent. See page 6 for further discussion of the regions.

Africa: Looking for a Renaissance

At first glance the study of Africa and its people seems to be a monumental task, because the complexity and diversity of the cultural and geographic regions are astounding. However, within the myriad challenges of studying the continent there are wonderful stories to be told about each nation and the variety of people contained within it. What stands out about Africa and its heritage is its antiquity, and how this antiquity might rest peacefully inside a rapidly changing world. From Egypt to South Africa, the African countries present the oldest known civilizations and the point of origin of humanity itself, but each nation is also uniquely modern, facing the struggles of today and searching for a way to bring about development and improve the quality of life for the people.

All nations, including those in Africa, are built to bring about a better life for the population, and to avoid a government that actively kills their people in pursuit of nebulous, unattainable, and contradictory goals stated by corrupt leaders who are motivated by personal greed, that is disguised as religious or ethnic zealousy. Thus while the mainstream media continue to bring forward a vision of the continent that is full of tales of woe, its important to remember that the vast majority of the African population holds out some semblance of hope for a better day. Contrary to many accounts, most African states are developing and building in their own way. There are wars, based on a variety of issues; religion, ethnicity, money, and simple differences of opinion, but for every country in violent conflict, there are two who strive to gain their niche against great odds in the contemporary global village.

The question of course is: what has happened in the last hundred years to this great continent—a continent that holds the key to the evolution of humanity, and a continent that developed the first language skills that allow us to communicate and evolve complex social structures? This text examines the history of the continent in an effort to understand how Africa has evolved and how it continues to struggle to develop. Because Africa, regardless of the glories of its past, is a continent hungry for development, and looking for a Renaissance.

During the early 1960s, most of the African continent was liberated from colonial rule. Seventeen nations gained their independence in the great *Uhuru* ("Freedom") year of 1960 alone. The times were electric. In country after country, the banners of new states, whose leaders offered idealistic promises to remake the continent and thus the world, replaced the flags of various European states and the United Nations. Hopes were high, and even the most ambitious of goals seemed obtainable. Non-Africans also spoke of the resource-rich continent as being on the verge of a developmental takeoff. Some of the old racist myths about Africa were at last being questioned.

Yet four decades after the great freedom year, conditions throughout Africa are sobering rather than euphoric. For most Africans, independence has been a desperate struggle for survival rather than an exhilarating path to development. Nowadays Africa is often described in the global media as a "continent in crisis," a "region in turmoil," "on a precipice," and "suffering"—phrases that echo the sensationalist writings of nineteenth-century missionaries eager to convince others of the continent's need for "salvation." Unfortunately, the modern headlines are too often accurate, and certainly far more accurate than the mission tracts of yesteryear. Today, millions of Africans are indeed seeking some form of salvation from the grinding poverty, pestilence, and in many areas, wars that afflict their lives. Added to these miseries is the ongoing HIV/AIDS pandemic, which is devastating much of the continent; the majority of AIDS sufferers and fatalities have been African.

Perhaps the scale of Africa's maladies helps to explain why contemporary evangelists are so much more successful in swelling their congregations than were their counterparts in the past. It is certainly not for lack of competition; Africa is a continent of many, often overlapping faiths. In addition to Islam, Christianity, and other spiritual paths, Africans have embraced a myriad of secular ideologies: Marxism, African socialism, people's capitalism, structural adjustment, pan-Africanism, authenticity, nonracialism, the one-party state, and the multiparty state. The list is endless, but salvation seems ever more distant.

Africa's current circumstances are indeed difficult, yet it is also true that the postcolonial era has brought progress as well as problems. The goals so optimistically pronounced at independence have, for the most part, not been abandoned. Even when the states have faltered, the societies that they encompass have remained dynamic and adaptable to shifting opportunities. The support of strong families continues to allow most Africans to overcome enormous adversity. There are starving children in Africa today, but there are also many more in school uniforms studying to make their dreams a reality. Africa as a whole remains a dynamic, ever-changing continent that in recent years has seen much progress as well as instances of regression. For example, the use of mobile wireless telephones (cell phones) has spread across the continent like wildfire, bringing mass-communications capacity to millions for the first time. While most Africans remain on the other side of the "digital divide" in terms of their capacity to participate in the "global information age," the introduction of new information technologies is bringing change. Textile businesses in such diverse places as Mauritius, Lesotho, and Ghana use the new technologies to monitor their niche markets in Europe and North America on a daily basis. This trend will probably grow if trade access between the developed and developing world continues to be liberalized through the World Trade Organization (WTO) and other multilateral agreements such as the African Growth and Opportunity Act (AGOA) and the Cotonou Convention, which, respectively, have helped to open up U.S. and European Union markets to African goods.

It is also worth noting that most of what appears in the global media about Africa is but a snapshot of the continent in moments of catastrophic deprivation, war, and degradation. The global flow of information is dominated by a small number of media companies that must for the most part capture the eyes and ears of consumers in the world's wealthier societies to capture market share and sell advertising. Incremental progress in

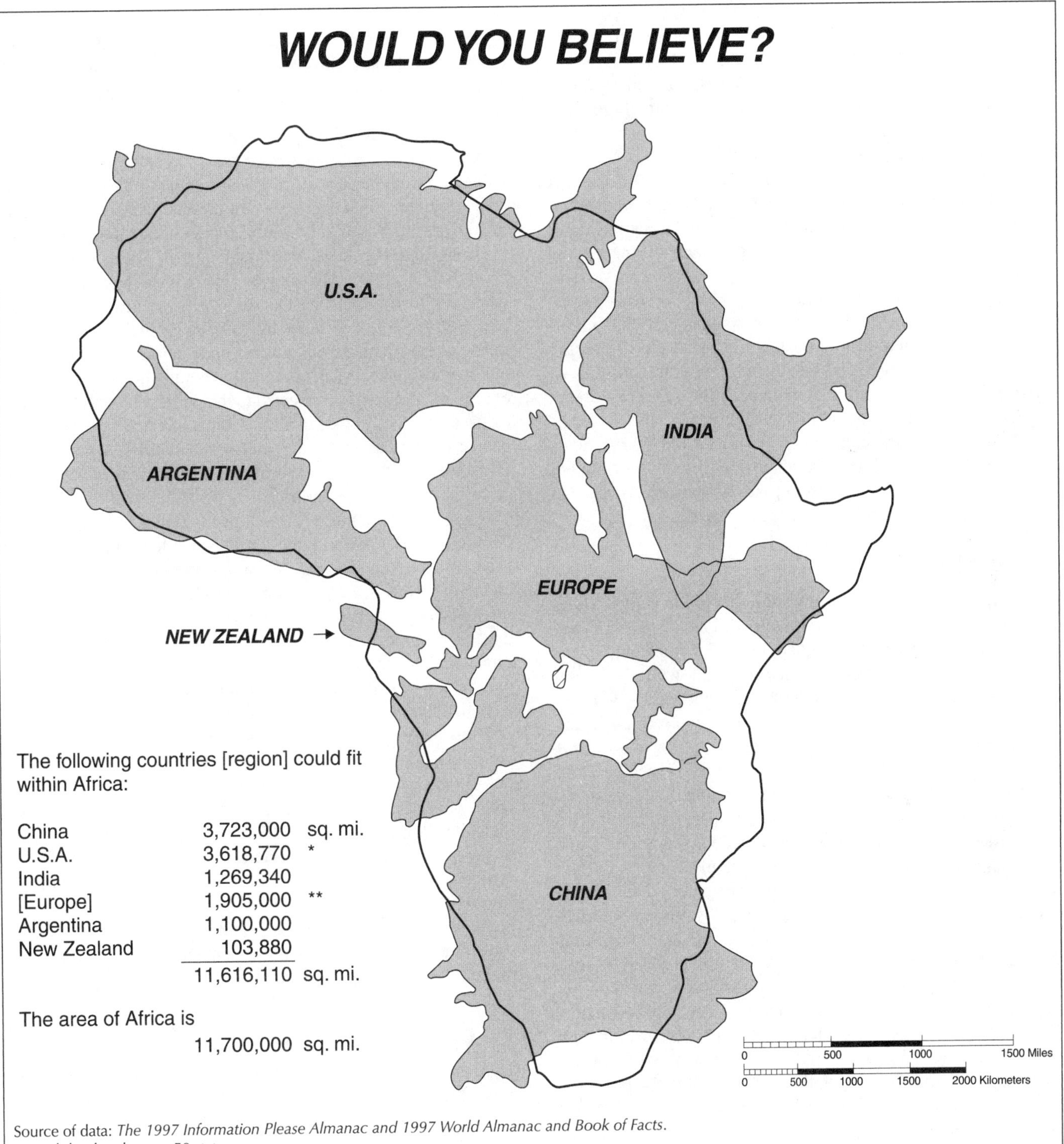

Source of data: *The 1997 Information Please Almanac and 1997 World Almanac and Book of Facts.*
* Total, land and water, 50 states
** *1997 Information Please Almanac.* Includes Iceland. Excludes the former European Soviet Union and European Turkey.

improving the quality of life for ordinary Africans is rarely perceived by these media companies as newsworthy. Thus, events such as the building of a new school, road, or water system that can dramatically transform the lives of people in a particular African locality are rarely of interest to the international broadcasters that increasingly influence what appears on the air waves of Africa itself as well as the rest of the world. Outsiders also often simply lose sight of the fact that Africa is a vast continent and not a country.

It would be naïve to downplay the serious challenges that face what is, in per capita economic terms, the world's least-developed continent. According to the 2002 United Nations Human Development Report, 39 of the world's 50 least developed countries are in Africa. But it is also worth appreciating that, with the passing of time, many of the ills that were inherited from colonialism have been remedied. In the process, new perspectives on how to move forward are being tested. The corruption and lack of capacity that characterize many African governments are mitigated by the resilience of local communities. Africans not only seek salvation from deprivation but also actively work for the delivery of a better tomorrow in which their children can enjoy the material benefits of being participants, rather than onlookers, in their continent's integration into the process of globalization.

A prerequisite for the realization of such dreams is the maintenance of peace. Without peace, there can be no prospect of freedom and development. In this context, recent progress in bringing an end to the armed conflicts in Angola, Sierra Leone, the Horn of Africa, and, perhaps most significantly, Congo, along with the consolidation of stability in such other areas as Mozambique and South Africa, have become sources of renewed hope.

If the momentum for peace can be maintained by the continent's growing ranks of legitimately elected leaders, ongoing talk among statespeople and intellectuals about the onset of an era of rebirth for the continent will appear less as a pipedream and more as a realistic possibility. Such a renaissance is predicated upon faith in the commitment of post–Cold War Africa's capacity to resolve its internal problems in partnership with, but without undue interference from, nations outside Africa that have in the past contributed to the continent's instability.

A DIVERSE CONTINENT

Africa, which is almost four times the size of the United States (excluding Alaska), ranks just below Asia as the world's biggest continent. Well over one quarter of the membership of the United Nations consists of African states—53 in all. Such facts are worth noting, for even well-educated outsiders often lose sight of Africa's continental scope when they discuss its problems and prospects.

Not only is the African continent vast but, archaeology tells us, it was also the cradle of human civilization. (New evidence suggests that today's non-Africans are the descendants of Africans who moved out of the continent as recently as 80,000 years ago.) It should therefore not be surprising that the ways of life as well as genetic makeup of the 800 million or so contemporary inhabitants of Africa are characterized by extraordinary diversity. In southern Africa's Kgalahari Desert (shared by South Africa, Namibia, and Botswana), a group of individuals, known to the outside world as Bushmen, continue to practice their ancient way of life as hunters and gatherers. Geneticists consider Bushmen to be the oldest surviving human beings—within their genetic structure, Bushman carry the original genes of the human race. To add majesty to their heritage, contemporary linguists claim that the click language spoken by the Bushmen is also the oldest known language in the world. Contemporary Africans speak more than 1,000 languages and live their lives according to a rich variety of household arrangements, kinship systems, and religious beliefs. The art and music styles of the continent are as varied as the people.

Given its diversity, it is not easy to generalize about Africa. For each statement, there is an exception. However, one aspect that is constant to all African societies is that they have always been changing, albeit in modern times at an ever-increasing rate. Cities have grown and people have moved back and forth between village and town, giving rise to new social groups, institutions, occupations, religions, and forms of communication that have made their mark in the countryside as well as in the urban centers. All Africans, whether they be urban computer programmers or hunter-gatherers living in the remote corners of the Kalahari Desert, have taken on new practices, interests, and burdens, yet they have retained their African identity. Uniquely African institutions, values, and histories underlie contemporary lifestyles throughout the continent.

Memories of past civilizations are a source of pride and community. The medieval Mali and Ghana empires, the glory of Pharaonic Egypt, the Fulani caliphate of northern Nigeria, the knights of Kanem and Bornu, the Great Zimbabwe, and the Kingdom of the Kongo, among others, are all remembered. Africa is a paradox. Twenty-first century folks may be able to go to the moon, but nowhere are there monuments to challenge the greatest of the African past. Yet, even these wonders are under siege. In Sudan, the largest nation of the continent, cultures are under attack, and many of the oldest archeological sites, including pyramids, are threatened with extinction. The past is connected to the present through the generations and by ties to the land. In a continent where the majority of people are still farmers, land remains "the mother that never dies." It is valued for its fruits and because it is the place to which the ancestors came and were buried.

The art of personal relationships continues to be important. People typically live in large families. Children are considered precious, and large families are still desired for social as well as economic reasons. Elders are an important part of a household; nursing homes and retirement communities generally do not exist. People are not supposed to be loners. "I am because we are" remains a valued precept. In this age of nation-states, the "we" may refer to one's ethnic community, while obligations to one's extended family often take precedence over other loyalties.

Most Africans believe in a spiritual as well as a material world. The continent contains a rich variety of indigenous belief systems, which often coexist with the larger religions of Islam and various Christian sects. Many families believe that their lives are influenced by their ancestors. Africans from all walks of life will seek the services of professional "traditional" healers to explain an illness or suggest remedies for such things as sterility or bad fortune. But this pattern of behavior does not preclude one from turning to scientific medicine; all African governments face strong popular demands for better access to modern health-care facilities.

Islam has long been a strong force in Africa. Today that religion rivals Christianity as the fastest-growing faith on the continent. The followers of both religions often adapt their faiths to accommodate local traditions and values. Some people also join new religious movements and churches, such as the Brotherhood of the Cross and Star in Nigeria or the Church of Simon Kimbangu in Congo, that link Christian and indigenous beliefs with new ideas and rituals. Like other institutions in the towns and cities, the churches and mosques provide their followers with social networks.

Local art, like local religion, often reflects the influence of the changing world. An airplane is featured on a Nigerian gelede mask, the Apollo space mission inspires a Burkinabe carver, and an Ndebele dance wand is a beaded electric pole.

THE TROUBLED PRESENT

Some of the crises in Africa today threaten its peoples' traditional resiliency. The facts are grim: In material terms, the average African is poorer today than at independence, and it is predicted that poverty will only increase in the immediate future. Drought conditions in recent decades have led to food shortages across the continent. In the 1980s, widespread famine occurred in 22 African nations; the Food and Agriculture Organization (FAO) of the United Nations estimated that 70 percent of all Africans did not have enough to eat. An outpouring of assistance and relief efforts at the time saved as many as 35 million lives. Overall per capita food production in Africa dropped by 12 percent between 1961 and 1995. One factor in the decline has been the tendency of agricultural planners to ignore the fact that up to 70 percent of Africa's food crops are grown by women. It has also been estimated that up to 40 percent of the continent's food crops go uneaten as a result of inadequate transport and storage facilities. Although agricultural production rose modestly in the late 1990s, the food crisis continues. Since 1990, large parts of East and Southern Africa, in particular, have faced the prospect of renewed hunger. Other areas have become dependent on outside food aid. Declining commodity prices on world markets, explosive population growth, and recurring drought and locust infestations have often counterbalanced marginal advances in agricultural production through better incentives to farmers. Problems of climate irregularity, and obtaining and transporting needed goods and supplies require continued assistance and long-range planning. Wood, the average person's source of energy, grows scarcer every year, and most governments have had to contend with the rising cost of imported fuels.

Armed conflicts have devastated portions of Africa. Recent carnage in Angola, Djibouti, Sierra Leone, Somalia, and the Democratic Republic of the Congo (D.R.C., formerly Zaire), due to internal strife encouraged to greater or lesser degrees by outside forces, places them in a distinct class of suffering—a class that has also included (and that may yet again) Chad, Eritrea, Ethiopia, Mozambique, and Rwanda. By the end of 2002, civil war persisted in Burundi, Liberia, Sudan, Central African Republic, and northern Uganda, while threatening to ignite in long-stable Côte d'Ivoire. More than 3 million people have died in these countries over the past decade, while millions more have become refugees. Except for scattered enclaves, normal economic activities have been greatly disrupted or have ceased altogether.

Sudan today, in 2005, brings the whole issue of governance into sharp focus. Governance of whom, for what ends? Fifty thousand Sudanese have been killed in six months, while an additional million people have been displaced from their homes, and there is no end in sight.

The Sudan is a test case that seems to be headed for continued disaster, and there are many throughout all of Africa who see no reason for it to remain unified as a country. What is clear is that there is no justification for Sudan's national unity, if that unity means the death and destruction of its people and their culture. The entire continent of Africa is unified by the need to develop its human and physical resources, and in order to do that the wars must stop.

Almost all African governments are deeply in debt. In 1991, the foreign debt owed by all the sub-Saharan African countries except South Africa already stood at about $175 billion. Although it is smaller in its absolute amount than that of Latin America, as a percentage of its economic output the continent's debt is the highest in the world and is rising swiftly. The combined gross national product (GNP) for the same countries, whose total populations are in excess of 500 million, was less than $150 billion, a figure that represents only 1.2 percent of the global GNP and is about equal to the GNP of Belgium, a country of 10 million people. In Zambia, an extreme example, the per capita foreign debt theoretically owed by each of its citizens is nearly $1,000, more than twice its annual per capita income.

In order to obtain money to meet debts and pay for their running expenses, many African governments have been obliged to accept the stringent terms of global lending agencies, most notably the World Bank and the International Monetary Fund (IMF). These lending terms have led to great hardship, especially in urban areas, through austerity measures such as the abandonment of price controls on basic foodstuffs and the freezing of wages. Many African governments and experts are questioning both the justice

MEASURING MISERY

United Nations Human Development Index, 2002*: Ranking of African States Among World Nations. Total Number of States Ranked: 173. African States Not Ranked: Liberia, Somalia

Rank	Country	Rank	Country
47	Seychelles	149	Djibouti
64	Libya	150	Uganda
67	Mauritius	151	Tanzania
100	Cape Verde	152	Mauritania
106	Algeria	153	Zambia
107	South Africa	154	Senegal
111	Equatorial Guinea	155	Democratic Republic of the Congo
115	Egypt	156	Côte d'Ivoire
117	Gabon	157	Eritrea
119	Sao Tome and Principe	158	Benin
122	Namibia	159	Guinea
123	Morocco	160	The Gambia
125	Swaziland	161	Angola
126	Botswana	162	Rwanda
128	Zimbabwe	163	Malawi
129	Ghana	164	Mali
132	Lesotho	165	Central African Republic
134	Kenya	166	Chad
135	Cameroon	167	Guinea-Bissau
136	Republic of Congo	168	Ethiopia
137	Comoros	169	Burkina Faso
139	Sudan	170	Mozambique
141	Togo	171	Burundi
147	Madagascar	172	Niger
148	Nigeria	173	Sierra Leone

*Standings among 173 countries, with the ranking 173 indicating the lowest development.

Source: UNDP, Human Development Report 2002.

THE AIDS PANDEMIC

Perhaps the greatest challenge currently facing Africa is the spread of HIV/AIDS. The statistics are chilling. According to the United Nations, of the 36 million people worldwide living with the HIV virus, some 24 million live in sub-Saharan Africa. Of the 5.3 million new infections estimated for the year 2000, 3.8 million were in Africa. AIDS is already the leading cause of death on the continent. In all, 2.4 million AIDS-related deaths were recorded for Africa in 2000, representing about 80 percent of the worldwide total.

AIDS-related fatalities have also resulted in a rapidly growing number of "AIDS orphans." According to Kingsley Amoako, executive secretary of the Economic Commission for Africa, more than 12 million children have been orphaned in Africa due to AIDS (out of the global estimate of just over 13 million). Speaking at a gathering of African leaders in November 2000, Amoako noted that, "Within the next 10 years, it is projected that there will be 40 million AIDS orphans in Africa.... The AIDS pandemic is undermining social and economic structures and reversing the fragile gains made since independence ... in parts of Africa, AIDS is killing one in every three adults, making orphans out of every tenth child and decimating entire communities."

The worst-hit parts of the continent in recent years have been East and Southern Africa, with some countries having infection rates of more than a fifth of their adult populations. According to published figures, the most affected countries are Botswana, South Africa, and Zimbabwe, where it is currently estimated that one in every two people under age 15 could die from the disease.

Inevitably, the spread of HIV/AIDS is having a devastating impact on economic and social development. For example, it is estimated that in the next decade, South Africa's gross domestic product will be 17 percent lower than it would have been without the pandemic.

Amid the gloom there is, nonetheless, grounds for hope that the scourge can ultimately be overcome. According to a UN report issued at the end of 2000, some parts of the continent are finally seeing a decrease in new HIV cases. The report notes that this has resulted in a modest overall decrease in the total number of new HIV cases in Africa as a whole. The decrease has been partially attributed to the gradual success of prevention programs, especially in the East African countries.

Africa's ability to fight HIV/AIDS is compromised by its debt burden and the high cost of HIV/AIDS treatment drugs. One of the most outspoken figures on the relationship between disease and debt on the continent has been Botswana's president, Festus Mogae. In a direct appeal to the wealthier nations, he observed:

> Your wealth in recent years increased by trillions and therefore what we owe is peanuts. It will not affect anybody, not the balance sheets of banks or anybody. It's just a matter of principle. You are insisting on repayment as a matter of principle, but it has no financial consequences for anybody else except the debtor. For him it's a lot of money.... Pharmaceutical companies have come forward and offered us discounts. Some of these discounts are very generous but are still more than our faint means can allow us to afford, and therefore we are still not able to take full advantage of the offer.... We are saying the rest of the world, including and especially the United States and the rest of the G-7, at the governmental level should do something to make it possible for us to access these treatments that are currently available.

and practicality of these terms. In response there have been recent, though many would argue insufficient, moves by donor countries to arrange for debt relief in the poorest countries.

Another factor that helps to account for Africa's relative poverty is the low level of industrial output of all but a few of its countries. The decline of many commodity prices on the world market has further reduced national incomes. As a result, the foreign exchange needed to import food, machinery, fuel, and other goods is very limited in most African countries. In 1987, the continent's economy grew by only 0.8 percent, far below its population growth rate of about 3.2 percent. In the same year, cereal production declined 8 percent and overall agricultural production grew by only 0.5 percent. There has been some modest improvement in subsequent years. Recent estimates put the continent's economic growth rate at 1.5 percent—still the world's lowest, and far below that of the population growth rate. Perhaps more significant has been the high growth that has been recorded in those states, like Mozambique, that have managed to move away from civil war to governance based on democratic consensus.

But perhaps the greatest challenge facing Africa today is in the area of health. The HIV/AIDS pandemic has spread at an alarming rate over the past two decades, already claiming millions of lives and threatening millions more. Southern Africa has been especially hard hit in recent years, with more than one in five adults said to be HIV-positive in a number of countries. As a result, estimates of life expectancy have been declining dramatically. Botswana is a notable example. By the early 1990s, average overall life expectancy in the nation had risen to about 65 years, due to sustained public investment in providing universal access to primary health care. But the most recent estimate presented at the 2002 United Nations AIDS Conference in Spain suggest that average life expectancy in Botswana could drop to as little as 27 years by 2010. Put another way, unless the pandemic can be brought under control, it has been estimated that AIDS will eventually claim the lives of one out of every two Batswana (as the people of Botswana are known) born in the new millennium. The battle against AIDS in Africa has been complicated by the existence of multiple strains of the HIV virus, of which the most virulent is currently concentrated in Southern Africa. The spread of HIV/AIDS has contributed to the resurgence of diseases such as tuberculosis. Mortality due to malaria—Africa's traditional scourge—has also been rising in many areas, as has cholera. Diseases afflicting livestock, such as rinderpest and foot-and-mouth disease, have also been making a comeback in certain regions.

THE EVOLUTION OF AFRICA'S ECONOMIES

Africa has seldom been rich, although it has vast resources, and some rulers and other elites have become very wealthy. In earlier centuries, the slave trade greatly contributed to limiting economic development in many African regions. During the period of European exploration and colonialism, Africa's involvement in the world economy greatly increased with the emergence of new forms of "legitimate" commerce. But colonial-era policies and practices assured that this development was of little long-term benefit to most of the continent's peoples.

During the 70 or so years of European colonial rule over most of Africa, its nations' economies were shaped to the advantage of the imperialists. Cash crops such as cocoa, coffee, and rubber began to be grown for the European market. Some African farmers benefited from these crops, but the cash-crop economy also involved large foreign-run plantations. It also encouraged the trends toward use of migrant labor and the decline in food production. Many people became dependent for their livelihood on the forces of the world market, which, like the weather, were beyond their immediate control.

Mining also increased during colonial times, again for the benefit of the colonial rulers. The ores were extracted from African soil by European companies. African labor was employed, but the machinery came from abroad. The copper, diamonds, gold, iron ore, and uranium were shipped overseas to be processed and marketed in the Western economies. Upon independence, African governments received a varying percentage of the take through taxation and consortium agreements. But mining remained an enclave industry, sometimes described as a "state within a state" because such industries were run by outsiders who established communities that used imported machinery and technicians and exported the products to industrialized countries.

Inflationary conditions in other parts of the world have had adverse effects on Africa. The raw materials that Africans export today often receive low prices on the world market, while the manufactured goods that African countries import are expensive. Local African industries lack spare parts and machinery, and farmers frequently cannot afford to transport crops to market. As a result, the whole economy slows down. Thus, Africa, because of the policies of former colonial powers and current independent governments, is tied into the world economy in ways that do not always serve its peoples' best interests.

THE PROBLEMS OF GOVERNANCE

Outside forces are not the only cause of Africa's current crises. Too often, Africa has been a misgoverned continent. After independence, the idealism that characterized various nationalist movements, with their promises of popular self-determination, gave way in most states to cynical authoritarian regimes. By 1989, only Botswana, Mauritius, soon-to-be-independent Namibia, and, arguably, The Gambia and Senegal could reasonably claim that their governments were elected in genuinely free and fair elections.

During the 1980s, the government of Robert Mugabe in Zimbabwe, in Southern Africa, undoubtedly enjoyed majority support, but political life in the country was already seriously marred by violence and intimidation aimed at the Mugabe regime's potential opposition. Past multiparty contests in the North African nations of Egypt, Morocco, and Tunisia, as well as in the West African state of Liberia, had been manipulated to assure that the ruling establishments remained unchallenged. Elsewhere, the continent was divided between military and/or one-party regimes, which often combined the seemingly contradictory characteristics of weakness and absolutism at the top. While a few of the one-party states, most notably Tanzania, then offered people genuine, if limited, choices of leadership, most were, to a greater or lesser degree, simply vehicles of personal rule.

But since 1990 there has been a democratic reawakening in Africa, which has toppled the political status quo in some areas and threatened its survival throughout the continent. Whereas in 1989 some 35 nations were governed as single-party states, by 1994 there were none, though Swaziland and Uganda were experimenting with no-party systems. In a number of countries—Benin, Cape Verde, Central African Republic, Congo, Madagascar, Mali, Malawi, Niger, São Tomé and Príncipe, Senegal, South Africa, and Zambia—ruling parties were decisively rejected in multiparty elections, while elections in other areas led to a greater sharing of power between the old regimes and their formerly suppressed oppositions.

In many countries, the democratic transformation is still ongoing and remains fragile. There have been accusations of manipulation and voting fraud in a growing number of countries in the past decade; while in Algeria, The Gambia, Niger, Nigeria, and Sierra Leone, the seeming will of the electorates has been overridden by military coups.

A very fragile democracy has since been restored to Nigeria, Africa's most populous state, while military rule in Sierra Leone has given way to an ongoing attempt to restore a democratic consensus through UN–sanctioned intervention by international peacekeeping forces. In The Gambia, where three decades of multiparty democracy were ended through a military coup, there have also been elections. But their legitimacy has been questioned.

Events in Benin have most closely paralleled the recent changes of Central/Eastern Europe. Benin's military-based, Marxist-Leninist regime of Mathieu Kérékou was pressured into relinquishing power to a transitional civilian government made up of technocrats and former dissidents. (Television broadcasts of this "civilian coup" enjoyed large audiences in neighboring countries.) In several other countries, such as Equatorial Guinea, Gabon, and Togo, mounting opposition has resulted in the semblance without the substance of free elections by long-ruling military autocrats. In the Democratic Republic of the Congo, attempts to establish a framework for reform through a multiparty consultative conference were overshadowed by the almost complete collapse of state structures. A victory by externally backed rebels in 1997 was followed by renewed civil war and foreign intervention. Continued conflict in the country has contributed to the further destabilization of neighboring states. Many people fear that these countries may soon experience turmoil similar to that which has engulfed Ethiopia, Liberia, Rwanda, Somalia, and Uganda, where military autocrats have been overthrown by armed rebels.

Why did most postcolonial African governments, until recently, take on autocratic forms? And why are these forms now being so widely challenged? There are no definitive answers to either of these questions. One common explanation for authoritarianism in Africa has been the weakness of the states themselves. Most African governments have faced the difficult task of maintaining national unity with diverse, ethnically divided citizenries. Although the states of Africa may overlay and overlap historic kingdoms, most are products of colonialism. Their boundaries were fashioned during the late-nineteenth-century European partition of the continent, which divided and joined ethnic groups by lines drawn in Europe. The successful leaders of African independence movements worked within the colonial boundaries; and when they joined together in the Organization of African Unity (OAU), they agreed to respect the territorial status quo.

While the need to stem interethnic and regional conflict has been one justification for placing limits on popular self-determination, another explanation can be found in the administrative systems that the nationalist leaderships inherited. All the European colonies in Africa functioned essentially as police states. Not only were various forms of opposition curtailed, but intrusive security establishments were created to watch over and control the indigenous populations. Although headed by Europeans, most colonial security services employed local staff members who were prepared to assume leadership roles at independence. A wave of military coups swept across West Africa during the 1960s; elsewhere, aspiring dictators like "Life President" Ngwazi Hastings Banda of Malawi were quick to appreciate the value of the inherited instruments of control.

Africa's economic difficulties have also frequently been cited as contributing to its political underdevelopment. On one hand, Nigeria's last civilian government, for example, was certainly undermined in part by the economic crisis that engulfed it due to falling oil revenues. On the other hand, in a pattern reminiscent of recent changes in Latin America, economic difficulties resulting in high rates of inflation and indebtedness seem to be tempting some African militaries, such as Benin's, to return to the barracks and allow civilian politicians to assume responsibility for the implementation of inevitably harsh austerity programs.

External powers have long sustained African dictatorships through their grants of military and economic aid—and, on occasion, direct intervention. For example, a local attempt in 1964 to restore constitutional rule in Gabon was thwarted by French paratroopers, while Joseph Desiré Mobutu's kleptocratic hold over Zaire (now the D.R.C.) relied from the very beginning on overt and covert assistance from the United States and other Western states. The former Soviet bloc and China also helped in the past to support their share of unsavory African allies, in places like Ethiopia, Equatorial Guinea, and Burundi. But the end of the Cold War has led to a reduced desire on the part of outside powers to prop up their unpopular African allies. At the same time, the major international lending agencies have increasingly concerned themselves with the perceived need to adjust the political as well as economic structures of debtor nations. This new emphasis is justified in part by the alleged linkage between political unaccountability and economic corruption and mismanagement.

The ongoing decline of socialism on the continent is also having a significant political effect. Some regimes have professed a Marxist orientation, while others have felt that a special African socialism could be built on the communal and cooperative traditions of their societies. In countries such as Guinea-Bissau and Mozambique, a revolutionary socialist orientation was introduced at the grassroots level during the struggles for independence, within areas liberated from colonialism. The various socialist governments have not been free of personality cults, nor from corruption and oppressive measures. And many governments that have eschewed the socialist label have, nonetheless, developed public corporations and central-planning methods similar to those governments that openly profess Marxism. In recent years, virtually all of Africa's governments, partly in line with IMF and World Bank requirements but also because of the inefficiency and losses of many of their public corporations, have placed greater emphasis on private-sector development.

REASONS FOR OPTIMISM

Although the problems facing African countries have grown since independence, so have the continent's collective achievements. The number of people who can read and write in local languages as well as in English, French, or Portuguese has increased enormously. More people can peruse newspapers, follow instructions for fertilizers, and read the labels on medicine bottles. Professionals trained in modern technology who, for example, plan electrification schemes, organize large office staffs, or develop medical facilities are more available because of the large number of African universities that have developed since the end of colonialism. Health care has also expanded and improved in most areas. Outside of the areas that have been ravaged by war, life expectancy has generally increased and infant mortality rates have declined.

The problems besetting Africa have caused deep pessimism in some quarters, with a few observers going so far as to question whether the postcolonial division of the continent into multiethnic states is viable. But the states themselves have proved to be surprisingly resilient. Central authority has reemerged in such traumatized, once seemingly ungovernable countries as Uganda, Mozambique, and most recently Sierra Leone. Despite the terrible wars that are still being waged in a few nations, mostly in the form of civil wars, postwar African governments have been notably successful in avoiding armed conflict with one another (although this fact has been severely tested by external interventions in the Democratic Republic of the Congo). Of special significance is South Africa's recent transformation into a nonracial democracy, which has been accompanied by its emergence as the leading member of the Southern African Development Community and has brought an end to its previous policy of regional destabilization.

Another positive development is the increasing attention that African governments and intra-African agencies are giving to women, as was exemplified in a global population summit held in Cairo, Egypt, in 1994. The pivotal role of women in agriculture and other activities is increasingly being recognized and supported. In many countries, prenatal and hospital care for mothers and their babies have increased, conditions for women workers in factories have improved, and new cooperatives for women's activities have been developed. Women are also playing a more prominent role in the political life of many African countries.

The advances that have been made in Africa are important ones, but they could be undercut by continued economic decline. Africa needs debt relief and outside aid just to maintain the gains that have been made. Yet as an African proverb observes, "Someone else's legs will do you no good in traveling." Africa, as the individual country reports in this volume observe, is a continent of many and varied resources. There are mineral riches and a vast agricultural potential. However, the continent's people, the youths who make up more than half the population and the elders whose wisdom is revered, are its greatest resource. The rest of the world, which has benefited from the continent's material resources, can also learn from the social strengths of African families and communities.

Central Africa

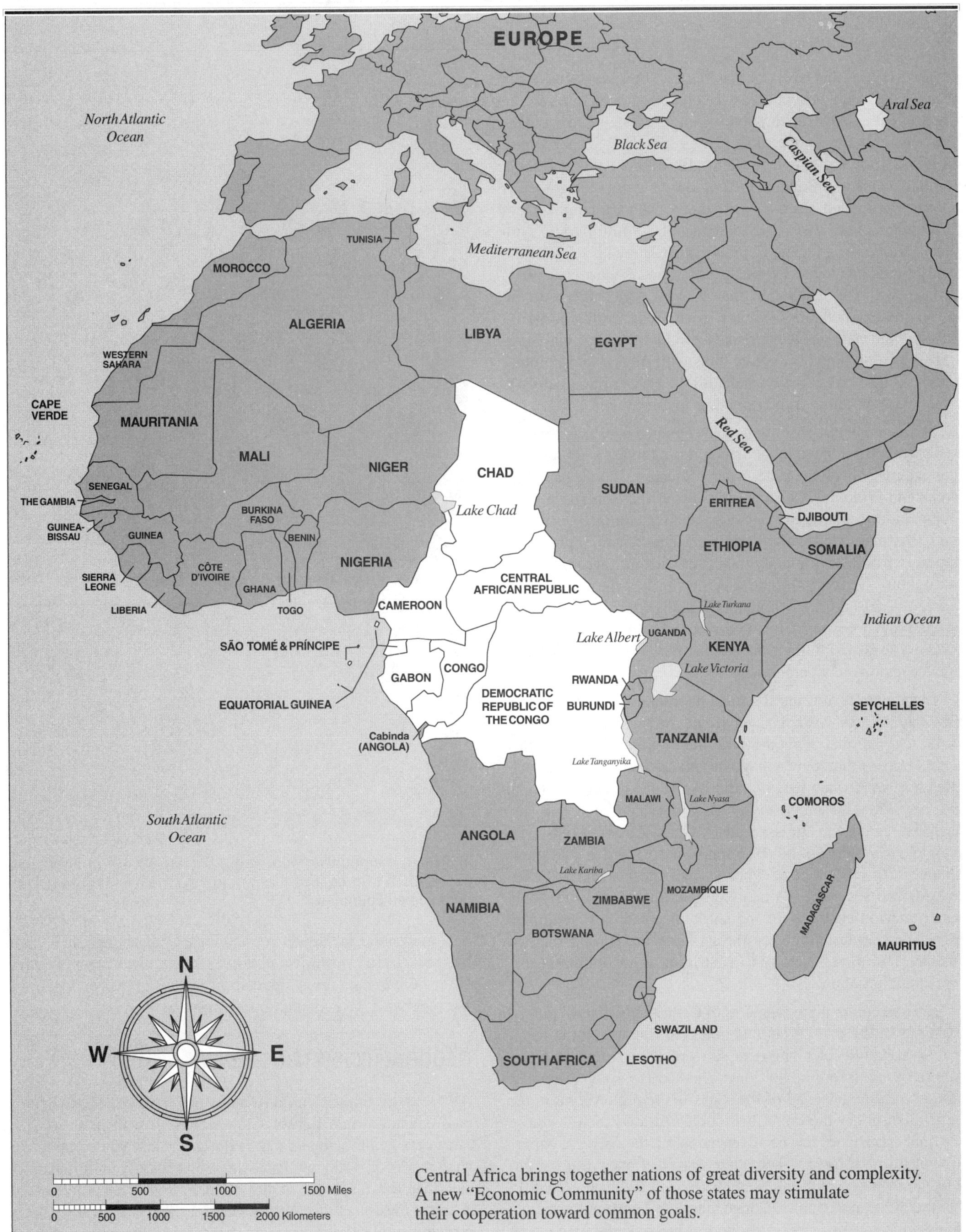

Central Africa brings together nations of great diversity and complexity. A new "Economic Community" of those states may stimulate their cooperation toward common goals.

Central Africa: Possibilities for Cooperation

The Central African region, as defined in this book, brings together countries that have not shared a common past; nor do they necessarily seem destined for a common future. Cameroon, Chad, Central African Republic, Congo (Congo-Brazzaville), the Democratic Republic of the Congo (the D.R.C., or Congo-Kinshasa, formerly known as Zaire), Equatorial Guinea, Gabon, and São Tomé and Príncipe are not always grouped together as one region. Indeed, users of this volume who are familiar with the continent may also associate the label "Central Africa" with such states as Angola and Zambia rather than with some of the states mentioned here. Geographically, Chad is more closely associated with the Sahelian nations of West Africa than with the heavily forested regions of Central Africa to its south. Similarly, the southern part of Democratic Republic of Congo has long-standing cultural and economic links with Angola and Zambia, which in this text are associated with the states of Southern Africa, largely because of their political involvements.

Yet the eight countries that are designated here as belonging to Central Africa have much in common. French is a predominant language in all the states except Equatorial Guinea and São Tomé and Príncipe. All except São Tomé and Príncipe and the Democratic Republic of the Congo share a common currency, the CFA franc. And while Chad's current economic prospects appear to be exceptionally poor, the natural wealth found throughout the rest of Central Africa makes the region as a whole one of enormous potential. Finally, in the postcolonial era, all the Central African governments have made some progress in realizing their developmental possibilities through greater regional cooperation.

The countries of Central Africa incorporate a variety of peoples and cultures, resources, environments, systems of government, and national goals. Most of the modern nations overlay both societies that were village-based and localized, and societies that were once part of extensive state formations. Islam has had little influence in the region, except in Chad and northern Cameroon. In most areas, Christianity coexists with indigenous systems of belief. Sophisticated wooden sculptures are one of the cultural achievements associated with most Central African societies. To many people, the carvings are only material manifestations of the spiritual potential of complex local cosmologies. However, the art forms are myriad and distinctive, and their diversity is as striking as the common features that they share.

The postcolonial governments of Central Africa have ranged from outwardly conservative regimes (in Gabon and the Democratic Republic of Congo) to self-proclaimed revolutionary Marxist-Leninist orders (in Congo and São Tomé and Príncipe). More fundamentally, all of the states in the region have in the past fallen under the control of unelected autocracies, whose continued existence has been dependent on the coercive capacities of military forces—sometimes external ones. But the authoritarian status quo has been challenged in recent years. In Central African Republic, Congo, and São Tomé and Príncipe, democratic openings have resulted in the peaceful election of new governments. However, the elected government in Congo has since been overthrown by forces loyal to its former dictator. Elsewhere in the region, opposition parties have been legalized but otherwise have made limited progress.

(United Nations photo 117,717)

In Africa, cooperative work groups such as the one pictured above often take on jobs that would be done by machinery in industrialized countries.

GEOGRAPHIC DISTINCTIVENESS

All the states of the Central African region except Chad encompass equatorial rain forests. Citizens who live in these regions must cope with a climate that is hot and moist while facing the challenges of utilizing (and in some cases, unfortunately, clearing) the resources of the great forests. The problems of living in these areas account, in part, for the relatively low, albeit growing, population densities of most of the states. The difficulty

(United Nations photo)

Regional cooperation will be necessary to utilize the natural resources of Central African countries without damaging their precious but fragile environment. Environmentalists warn that the Central African forests are rapidly being destroyed.

of establishing roads and railroads impedes communication and thus economic development. The peoples of the rain-forest areas tend to cluster along riverbanks and existing rail lines. In modern times, largely because of the extensive development of minerals, many inhabitants have moved to the cities, accounting for a comparatively high urban population in all the states.

Central Africa's rivers have long been its lifelines. The watershed in Cameroon between the Niger and Congo (or Zaire) Rivers provides a natural divide between the West and Central African regions. The Congo River is the largest in the region, but the Oubangi, Chari, Ogooue, and other rivers are important also for the communication and trading opportunities they offer. The rivers flow to the Atlantic Ocean, a fact that has encouraged the orientation of Central Africa's external trade toward Europe and the Americas.

Many of the countries of the region have similar sources of wealth. The rivers are capable of generating enormous amounts of hydropower. The rain forests are also rich in lumber, which is a major export of most of the countries. Other forest products, such as rubber and palm oil, are widely marketed.

Extensive lumbering and clearing activities for agriculture, have created worldwide concern about the depletion of the rain forests. As a result, in recent years there have been some organized boycotts in Europe of the region's hardwood exports, although far more trees are felled to process plywood.

(United Nations photo 71673)

Refugees from the conflict in the Republic of the Congo are seen here in the make-shift living quarters that they erected in the area allotted to them outside Elizabethville.

As one might expect, Central Africa as a whole is one of the areas least affected by the drought conditions that periodically plague Africa. Nevertheless, serious drought is a well-known visitor in Chad, Central African Republic, and the northern regions of Cameroon, where it contributes to local food shortages. Savanna lands are found in some parts lying to the north and the south of the forests. Whereas rain forests have often inhibited travel, the savannas have long been transitional areas, great avenues of migration linking the regions of Africa, while providing agricultural and pastoral opportunities for their residents.

The Central African countries share other resources besides the products of the rain forest. Cameroon, Congo, and Gabon derive considerable revenues from their petroleum reserves. Other important minerals include diamonds, copper, gold, manganese, and uranium. The processes involved in the exploitation of these commodities, as well as the demand for them in the world market, are issues of common concern among the producing nations. Many of the states also share an interest in exported cash crops such as coffee, cocoa, and cotton, whose international prices are subject to sharp fluctuations. The similarity of their environments and products provides an economic incentive for Central African cooperation.

LINKS TO FRANCE

Many of the different ethnic groups in Central Africa overlap national boundaries. Examples include the Fang, who are found in Cameroon, Equatorial Guinea, and Gabon; the Bateke of Congo and Gabon; and the BaKongo, who are concentrated in Angola as well as in Congo and the Democratic Republic of the Congo. Such cross-border ethnic ties are less important as sources of regional unity than the European colonial systems that the countries inherited. While Equatorial Guinea was controlled by Spain, São Tomé and Príncipe by Portugal, and the D.R.C. by Belgium, the predominant external power in the region remains France. Central African Republic, Chad, Congo, and Gabon were all once part of French Equatorial Africa. Most of Cameroon was also governed by the French, who were awarded the bulk of the former German colony of the Kamerun as a "trust territory" in the aftermath of World War I. French administration provided the five states with similar colonial experiences.

Early colonial development in the former French colonies and the Democratic Republic of the Congo was affected by European "concessions" companies, institutions that were sold extensive rights (often 99-year leases granting political as well as economic powers) to exploit such local products as ivory and

(United Nations photo 96708)

Here, in the Democratic Republic of the Congo, a bulldozer levels an access road to the Kamnyola Bridge over the Livumvi River.

rubber. At the beginning of the twentieth century, just 41 companies controlled 70 percent of all of the territory of contemporary Central African Republic, Congo, and Gabon. Mining operations as well as large plantations were established that often relied on forced labor. Individual production by Africans was also encouraged, often through coercion rather than economic incentives. While the colonial companies encouraged production and trade, they did little to aid the growth of infrastructure or long-term development. Only in the D.R.C. was industry promoted to any great extent.

In general, French colonial rule, along with that of the Belgians, Portuguese, and Spanish and the activities of the companies, offered few opportunities for Africans to gain training and education. There was also little encouragement of local entrepreneurship. An important exception to this pattern was the policies pursued by Felix Eboue, a black man from French Guiana (in South America) who served as a senior administrator in the Free French administration of French Equatorial Africa during the 1940s. Eboue increased opportunities for the urban elite in Central African Republic, Congo, and Gabon. He also played an important role in the Brazzaville Conference of 1944, which, recognizing the part that the people of the French colonies had played in World War II, abolished forced labor and granted citizenship to all. Yet political progress toward self-government was uneven. Because of the lack of local labor development, there were too few people at independence who were qualified to shoulder the bureaucratic and administrative tasks of the regimes that took power. People who could handle the economic institutions for the countries' benefit were equally scarce. And in any case, the nations' economies remained for the most part securely in outside—largely French—hands.

The Spanish on the Equatorial Guinea island of Fernando Po, and the Portuguese of São Tomé and Príncipe, also profited from their exploitation of forced labor. Political opportunities in these territories were even more limited than on the African mainland. Neither country gained independence until fairly recently: Equatorial Guinea in 1968, São Tomé and Príncipe in 1975.

In the years since independence, most of the countries of Central Africa have been influenced, pressured, and supported by France and the other former colonial powers. French firms in Central African Republic, Congo, and Gabon continue to dominate the exploitation of local resources. Most of these companies are only slightly encumbered by the regulations of the independent states in which they operate, and all are geared toward European markets and needs. Financial institutions are generally branches of French institutions, and all the former French colonies as well as Equatorial Guinea are members of the Central African Franc (CFA) Zone. French expatriates occupy senior positions in local civil-service establishments and in companies; many more of them are resident in the region today than was true 30 years ago. In addition, French troops are stationed in Central African Republic, Chad, and Gabon, regimes that owe their very existence to past French military interventions. Besides being a major trading partner, France has contributed significantly to the budgets of its former possessions, especially the poorer states of Central African Republic and Chad.

Despite having been under Belgian rule, the Democratic Republic of the Congo is an active member of the Francophonic bloc in Africa. In 1977, French troops put down a rebellion in the southeastern part of Zaire (as the D.R.C. was then known). Zaire, in turn, sent its troops to serve beside those of France in Chad and Togo. Since playing a role in the 1979 coup that brought the current regime to power, France has also had a predominant influence in Equatorial Guinea.

REGIONAL COOPERATION AND CONFLICT

Although many Africans in Central Africa recognize that closer links among their countries would be beneficial, there have been fewer initiatives toward political unity or economic integration in this region than in East, West, or Southern Africa. In the years before independence, Barthelemy Boganda, of what is now Central African Republic, espoused and publicized the idea of a "United States of Latin Africa," which was to include Angola and Zaire as well as the territories of French Equatorial Africa, but he was frustrated by Paris as well as by local politicians. When France offered independence to its colonies in 1960, soon after Boganda's death, the possibility of forming a federation was discussed. But Gabon, which was wealthier than the other countries, declined to participate. Central African Republic, Chad, and Congo drafted an agreement that would have created a federal legislature and executive branch governing all three countries, but local jealousies defeated this plan.

There have been some formal efforts at economic integration among the former French states. The Customs and Economic Union of the Central African States (UDEAC) was established in 1964, but its membership has been unstable. Chad and Central African Republic withdrew to join Zaire in an alternate organization. (Central African Republic later returned, bringing the number of members to six.) The East and Central African states together planned an "Economic Community" in 1967, but it never materialized.

In the 1980s there were efforts to make greater progress toward economic cooperation. Urged on by the United Nations Economic Commission for Africa, and with the stimulus of the Lagos Plan of Action, representatives of Central African states met in 1982 to prepare for a new economic grouping. In 1983, all the Central African states as well as Rwanda and Burundi in East Africa signed a treaty establishing the Economic Community of Central African States (ECCA). ECCA's goals were broader than those of UDEAC. Members hoped that the union would stimulate industrial activity, increase markets, and reduce the dependence on France and other countries for trade and capital. But with dues often unpaid and meetings postponed, ECCA has so far failed to meet its potential.

Hopes that the Central African states could work collectively toward a brighter future have been further compromised by the spread in recent years of extreme political instability and brutal and far-reaching armed conflicts both within and between states in the region. This is particularly true of the ongoing civil war in the region's biggest state, the Democratic Republic of the Congo; this war has not only divided the region but has involved states as far afield as Libya, Nigeria, and South Africa.

In the mid-1990s, there was a growing optimism that democratization might enable the region to become a center of a continental renaissance. This view was greatly boosted by the fall of the corrupt, authoritarian regime of Mobutu Sese Seko, which led to the nominal transformation of Zaire into the Democratic Republic of the Congo. But such hopes suffered a serious setback with the resurgence of dictatorship and kleptocracy under Mobutu's successor, Laurent Kabila, who remained in power through external backers and through implicit appeals for genocide against ethnic groups perceived as his enemies. With such continued divide-and-misrule, the chances of Boganda's vision becoming a reality appeared more remote. But since the assassination of Laurent Kabila and the rise of his son Joseph Kabila in the D.R.C., prospects for peace in Central Africa's largest state seem realistic. After years of autocracy, there is an opportunity for a long-term cease-fire and cessation of hostilities among the warring factions within the region. Most notably, by the end of 2002, foreign state military forces had largely pulled out of the country. Regional initiatives may have failed in the past, but this is not to say that there will not be more effective cooperation in the future. The key element for the creation of new regional organizations rests in the ability of the Central African states to sit at a negotiating table and recognize the existence of their common interest. In this respect, peace in the D.R.C. represents the best hope for stability and development in Central Africa.

Cameroon (Republic of Cameroon)

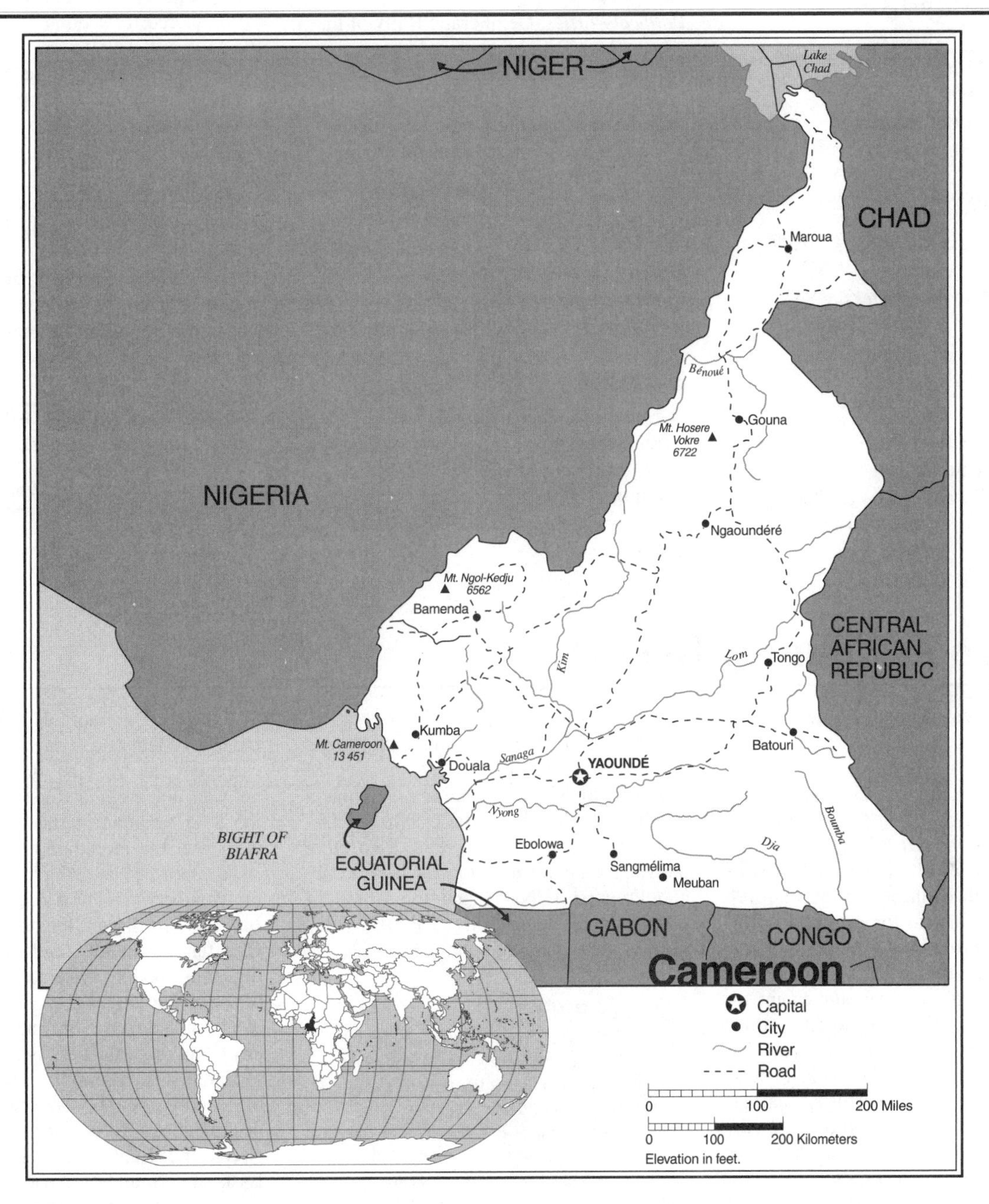

Cameroon Statistics

GEOGRAPHY

Area in Square Miles (Kilometers): 183,568 (475,400) (about the size of California)
Capital (Population): Yaoundé (1,119,000)
Environmental Concerns: deforestation; overgrazing; desertification; poaching; overfishing; water-borne disease
Geographical Features: diverse, with coastal plain in southwest, dissected plain in center, mountains in west, and plains in north
Climate: from tropical to semiarid

PEOPLE

Population

Total: 16,185,000
Annual Growth Rate: 1.97%
Rural/Urban Population Ratio: 54/46
Major Languages: English; French; Fulde; Ewondo; Duala; Bamelke; Bassa; Bali; others
Ethnic Makeup: 31% Cameroonian Highlander; 19% Equatorial Bantu; 11% Kirdi; 10% Fulani; 29% others
Religions: 40% indigenous beliefs; 40% Christian; 20% Muslim

Health

Life Expectancy at Birth: 54 years (male); 55 years (female)
Infant Mortality: 68.8/1,000 live births
Physicians Available: 1/11,848 people
HIV/AIDS Rate in Adults: 7.73%

Education

Adult Literacy Rate: 79%
Compulsory (Ages): 6–12; free

COMMUNICATION

Telephones: 95,000 main lines
Televisions: 72/1,000 people
Internet Users: 60,000 (2002)

TRANSPORTATION

Highways in Miles (Kilometers): 20,580 (34,300)
Railroads in Miles (Kilometers): 693 (1,111)
Usable Airfields: 49
Motor Vehicles in Use: 153,000

GOVERNMENT

Type: unitary republic
Independence Date: January 1, 1960 (from UN trusteeship under French administration)
Head of State/Government: President Paul Biya; Prime Minister Peter Mafany Musonge
Political Parties: Democratic Rally of the Cameroon People; National Union for Democracy and Progress; Social Democratic Front; Cameroonian Democratic Union; Union of Cameroonian Populations; others
Suffrage: universal at 20

MILITARY

Military Expenditures (% of GDP): 1.4%
Current Disputes: various border conflicts, especially with Nigeria

ECONOMY

Currency ($ U.S. equivalent): 529.43 CFA francs =$1
Per Capita Income/GDP: $1,700/$26.4 billion
GDP Growth Rate: 4.2%
Inflation Rate: 2%
Unemployment Rate: 30%
Labor Force by Occupation: 70% agriculture; 13% industry and commerce; 17% other
Population Below Poverty Line: 48%
Natural Resources: petroleum; timber; bauxite; iron ore; hydropower
Agriculture: coffee; cocoa; cotton; rubber; bananas; oilseed; grain; roots; livestock; timber
Industry: petroleum production and refining; food processing; light consumer goods; textiles; lumber
Exports: $2.1 billion (primary partners Italy, France, Netherlands)
Imports: $1.5 billion (primary partners France, Germany, United States, Japan)

SUGGESTED WEB SITES

http://www.sas.upenn.edu/African_Studies/Country_Specific/Cameroon.html
http://www.cameroon.net
http://www.telp.com/cameroon/home.htm

Cameroon Country Report

Over the past decade, Cameroonians have been united by at least two things: support for their world-class national soccer team, the Indomitable Lions, and condemnation of neighboring Nigeria about a long-simmering border dispute over the Bakassi Peninsula. The latter conflict was referred by both countries to the International Court of Justice for resolution. But when the Court ruled in Cameroon's favor in October 2002, the government of Nigeria reneged on its previous agreement to accept the verdict. In August 2003, after talks in Cameroon, Nigeria said that it would not hand over the Bakassi Peninsula for at least three years. Yet in December 2003, Nigeria did in fact hand over 32 villages to Cameroon as part of the 2002 International Court of Justice's ruling. In January 2004, Nigeria agreed to have joint border patrols with Cameroon.

Although it claims to operate a multiparty democracy, freedom of expression is severely limited in Cameroon. Politically, Cameroonians are deeply divided. During the October 2004 general elections, incumbent president Paul Biye was once again reelected by a large majority, even with many of the opposition parties boycotting. The poll, which was boycotted by the largest opposition parties, was a followup to the controversial elections of March 1992, which ended a quarter-century of one-party rule by Biya's Cameroon People's Democratic Party (CPDM). Although the CPDM relinquished its monopoly of power, it retained control of the government. With a plurality of 88 out of 180 seats, the CPDM was able to form a coalition government with the Movement for the Defense of Democracy, which won six seats. Two other parties, the National Union for Democracy and Progress (UNDP) and the Union of Cameroonian Populations (UPC), divided most of the remaining seats.

DEVELOPMENT

The Cameroon Development Corporation coordinates more than half of the agricultural exports and, after the government, employs the most people. Cocoa and coffee comprise more than 50% of Cameroon's exports. Lower prices for these commodities in recent years have reduced the country's income.

Like the 1998 elections, the legitimacy of the 1992 poll was compromised by the boycott of some opposition parties—most notably the Social Democratic Front (SDF), the Democratic Union (CDU), and a faction of the UPC—allowing the CPDM to win numerous constituencies by default. The situation was later aggravated, in October 1992, when the CPDM's Paul Biya was declared the victor in a snap election accompanied by opposition allegations of vote-rigging.

Cameroon's fractious politics is partially a reflection of its diversity. In geographical terms, the land is divided between the tropical forests in the south, the drier savanna of the north-central region, and the mountainous country along its western border, which forms a natural division between West and Central Africa. In terms of religion, the country has many Christians, Muslims, and followers of indigenous belief systems. More than a dozen major languages, with numerous dialects, are spoken. The languages of southern Cameroon are linguistically classified as Bantu. The "Bantu line" that runs across the country, roughly following the course of the Sanaga River, forms a boundary between the Bantu languages of Central, East, and Southern Africa and the non-Bantu tongues of North and West Africa. Many scholars believe that the roots of the Bantu language tree are buried in Cameroonian soil. Cameroon is also unique among the continental African states in sharing two European languages, English and French, as its official mediums. Relations between Anglophone and Francophone Cameroon have been troubled in recent years. In October 2001, violence flared between government security forces and protesters favoring the separation of English-speaking Cameroon.

Cameroon's use of both English and French is a product of its unique colonial heritage. Three European powers have ruled

(United Nations photo 152,295 by Shaw McCutcheon)

Cameroon has experienced political unrest in recent years as various factions have moved to establish a stable form of government. At the heart of the political turmoil is the need to raise the living standards of the population through an increase in agricultural production. These farmers with their cattle herds are one part of this movement.

over Cameroon. The Germans were the first. From 1884 to 1916, they laid the foundation of much of the country's communications infrastructure and, primarily through the establishment of European-run plantations, export agriculture. During World War I, the area was divided between the British and French, who subsequently ruled their respective zones as League of Nations (later the United Nations) mandates. French "Cameroun" included the eastern four fifths of the former German colony, while British "Cameroon" consisted of two narrow strips of territory that were administered as part of its Nigerian territory.

In the 1950s, Cameroonians in both the British and French zones began to agitate for unity and independence. At the core of their nationalist vision was the "Kamerun Idea," a belief that the period of German rule had given rise to a pan-Cameroonian identity. The largest and most radical of the nationalist movements in the French zone was the Union of the Cameroonian People, which turned to armed struggle. Between 1955 and 1963, when most of the UPC guerrillas were defeated, some 10,000 to 15,000 people were killed. Most of the victims belonged to the Bamileke and Bassa ethnic groups of southwestern Cameroon, which continues to be the core area of UPC support. (Some sources refer to the UPC uprising as the Bamileke Rebellion.)

FREEDOM

While Cameroon's human-rights record has improved since its return to multipartyism, political detentions and harassment continue. Amnesty International has drawn attention to the alleged starvation of detainees at the notorious Tchollire prison. Furthermore, the nation's vibrant free press has become a prime target of repression, with several editors arrested in 1998.

To counter the UPC revolt, the French adopted a dual policy of repression against the guerrillas' supporters and the devolution of political power to local non-UPC politicians. Most of these "moderate" leaders, who enjoyed core followings in both the heavily Christianized southeast and the Muslim north, coalesced as the Cameroonian Union, whose leader was Ahmadou Ahidjo, a northerner. In pre-independence elections, Ahidjo's party won just 51 out of the 100 seats. Ahidjo thus led a divided, war-torn state to independence in 1960.

In 1961, the southern section of British Cameroon voted to join Ahidjo's republic. The northern section opted to remain part of Nigeria. The principal party in the south was the Kamerun National Democratic Party, whose leader, John Foncha, became the vice president of the Cameroon republic, while Ahidjo served as president. The former British and French zones initially maintained their separate local parliaments, but the increasingly authoritarian Ahidjo pushed for a unified form of government. In 1966, all of Cameroon's legal political groups were dissolved into Ahidjo's new Cameroon National Union (CNU), creating a de facto one-party state. Trade unions and other mass organizations were also brought under CNU control. In 1972, Ahidjo proposed the abolition of the federation and the creation of a constitution for a unified Cameroon. This was approved by a suspiciously lopsided vote of 3,217,058 to 158.

In 1982, Ahidjo, believing that his health was graver than was actually the case, suddenly resigned. His handpicked successor was Paul Biya. To the surprise of many, the heretofore self-effacing Biya quickly proved to be his own man. He brought young technocrats into the ministries and initially called for a more open and democratic society. But as he pressed forward, Biya came into increasing conflict with Ahidjo, who tried to reassert his authority as CNU chairman. The ensuing power struggle took on overtones of an ethnic conflict between Biya's largely southern Christian supporters and Ahidjo's core following of northern Muslims. In 1983, Ahidjo lost and went into exile. The next year, he was tried and convicted, in absentia, for allegedly plotting Biya's overthrow.

HEALTH/WELFARE

The overall literacy rate in Cameroon, about 63%, is among the highest in Africa. There exists, however, great disparity in regional figures as well as between males and females. In addition to public schools, the government devotes a large proportion of its budget to subsidizing private schools.

In April 1984, only two months after the conviction, Ahidjo's supporters in the Presidential Guard attempted to overthrow Biya. The revolt was put down, but up to 1,000 people were killed. In the coup's aftermath, Biya combined repression with attempts to restructure the ruling apparatus. In 1985, the CNU was overhauled as the Cameroon People's Democratic Movement. However, President Biya became increasingly reliant on the support of his own Beti group.

An upsurge of prodemocracy agitation began in 1990. In March, the Social Democratic Front was formed in Bamenda, the main town of the Anglophonic west, over government objections. In May, as many as 40,000 people from the vicinity of Bamenda, out of a total population of about 100,000, attended an SDF rally. Government troops opened fire on school children returning from the demonstration. This action led to a wave of unrest, which spread to the capital city of Yaoundé. The government media tried to portray the SDF as a subversive movement of "English speakers," but it attracted significant support in Francophonic areas. Dozens of additional opposition groups, including the UNDP (which is loyal to the now-deceased Ahidjo's legacy) and the long-underground UPC, joined forces with the SDF in calling for a transition government, a new constitution, and multiparty elections.

Throughout much of 1991, Cameroon's already depressed economy was further crippled by opposition mass action, dubbed the "Ghost Town Campaign." A series of concessions by Biya culminated in a November agreement between Biya and most of the opposition (the SDF being among the holdouts) to formulate a new constitution and prepare for elections.

ACHIEVEMENTS

The strong showing by Cameroon's national soccer team, the Indomitable Lions, in the 1990 and 1994 World Cup competitions is a source of pride for sports fans throughout Africa. Their success, along with the record numbers of medals won by African athletes in the 1988 and 1992 Olympics, is symbolic of the continent's coming of age in international sports competitions.

One unrealized hope has been that democratic reform would help move Cameroon away from its consistent Transparency International rating as one of the world's most corrupt countries. Endemic corruption has become associated with environmental degradation. In recent years conservationists have been especially concerned about the construction of an oil pipeline, funded by the World Bank, without an environmental-impact study, and the allocation of about 80 percent of the country's forest for logging.

Timeline: PAST

1884
The establishment of the German Kamerun Protectorate

1916
The partition of Cameroon; separate British and French mandates are established under the League of Nations

1955
The UPC (formed in 1948) is outlawed for launching revolts in the cities

1960
The Independent Cameroon Republic is established, with Ahmadou Ahidjo as the first president

1961
The Cameroon Federal Republic reunites French Cameroon with British Cameroon after a UN-supervised referendum

1972
The new Constitution creates a unitary state

1980s
Ahidjo resigns and is replaced by Paul Biya; Lake Nyos releases lethal volcanic gases, killing an estimated 2,000 people

1990s
Nationwide agitation for a restoration of multiparty democracy; Biya retains the presidency in disputed elections

PRESENT

2000s
New clashes over the Bakassi Peninsula

Central African Republic

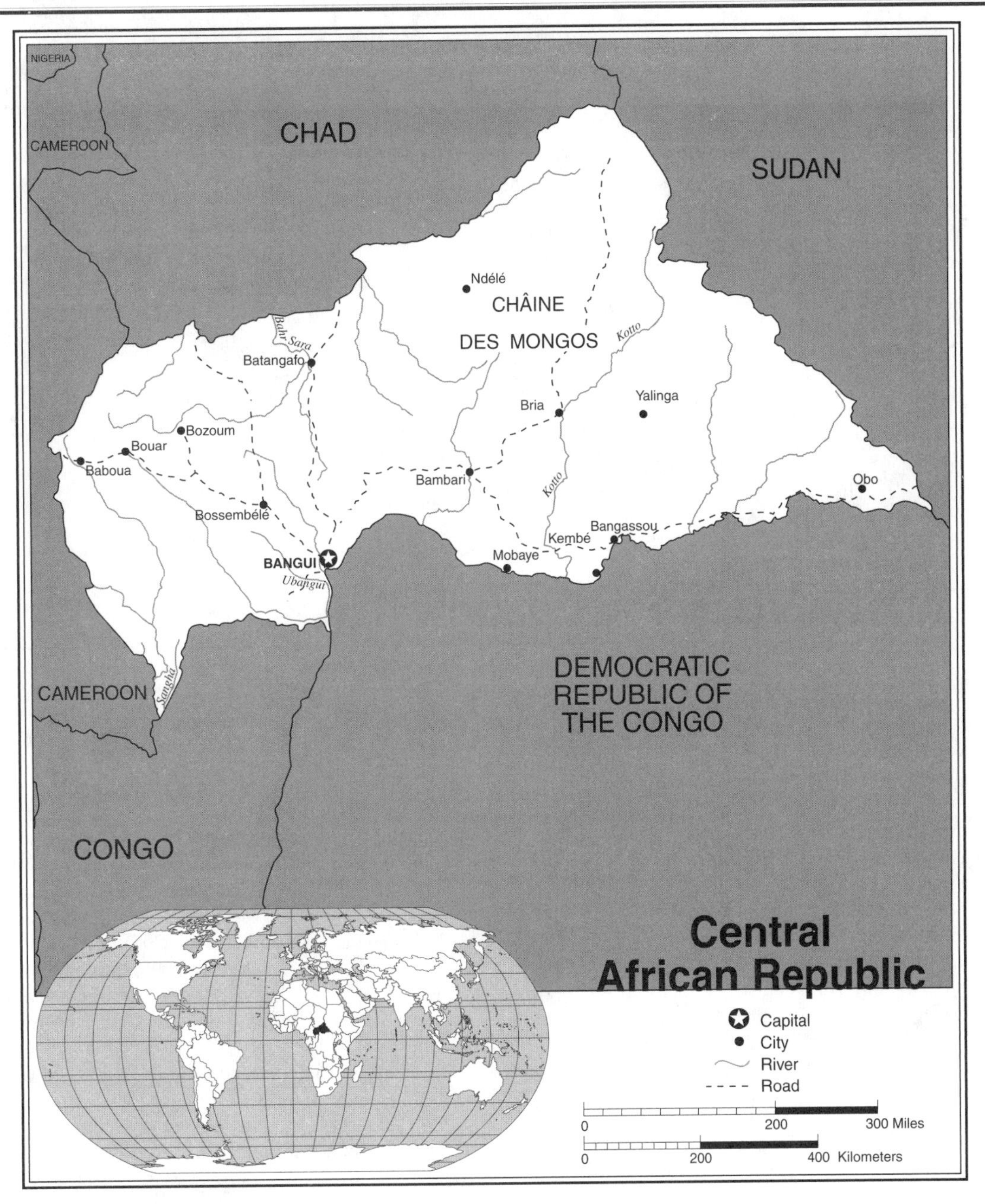

Central African Republic Statistics

GEOGRAPHY

Area in Square Miles (Kilometers): 240,324 (662,436) (about the size of Texas)

Capital (Population): Bangui (666,000)

Environmental Concerns: poaching; desertification; deforestation; potable water

Geographical Features: vast, flat to rolling, monotonous plateau; scattered hills in the northeast and southwest; landlocked

Climate: tropical

PEOPLE

Population

Total: 3,742,482

Annual Growth Rate: 1.86%

Rural/Urban Population Ratio: 60/40

Major Languages: French; Songo; Arabic; Hunsa; Swahili

Ethnic Makeup: 34% Baya; 27% Banda; 21% Mandja; 10% Sara; 8% others

Religions: 35% indigenous beliefs; 25% Protestant; 25% Roman Catholic; 15% Muslim

Health

Life Expectancy at Birth: 42 years (male); 45 years (female)

Infant Mortality: 103.8/1,000 live births
Physicians Available: 1/18,660 people
HIV/AIDS Rate in Adults: 13.84%

Education

Adult Literacy Rate: 60%
Compulsory (Ages): 6–14

COMMUNICATION

Telephones: 10,000 main lines
Televisions: 5/1,000 people
Internet Users: 5,000 (2002)

TRANSPORTATION

Highways in Miles (Kilometers): 14,286 (23,810)
Railroads in Miles (Kilometers): none
Usable Airfields: 51
Motor Vehicles in Use: 20,000

GOVERNMENT

Type: republic
Independence Date: August 13, 1960 (from France)
Head of State/Government: President Frańcois Bozize; Prime Minister Celestin Gaombalet
Political Parties: Movement for the Liberation of the Central African People; Central African Democratic Assembly; Movement for Democracy and Development; others
Suffrage: universal at 21

MILITARY

Military Expenditures (% of GDP): 1.1%
Current Disputes: internal strife

ECONOMY

Currency ($U.S. Equivalent): 581.2 CFA francs = $1
Per Capita Income/GDP: $1,300/$4.6 billion
GDP Growth Rate: 1.8%
Inflation Rate: 3.6%
Unemployment Rate: 8%
Natural Resources: diamonds; uranium; timber; gold; petroleum; hydropower
Agriculture: cotton; coffee; tobacco; manioc; millet; corn; bananas; timber
Industry: diamond mining; sawmills; breweries; textiles; footwear; assembly of bicycles and motorcycles
Exports: $166 million (primary partners Benelux, Côte d'Ivoire, Spain)
Imports: $154 million (primary partners France, Cameroon, Benelux)

SUGGESTED WEB SITES

http://www.sas.upenn.edu/African_Studies/Country_Specific/Cent-A-R.html
http://www.emulateme.com/car.htm

Central African Republic Country Report

In March 2003, the government of the Central African Republic (CAR) was overthrown when rebels loyal to former Army chief-of-staff General François Bozize seized control of the capital, Bangui. The country's president, Ange-Félix Patassé, was out of the country at the time. For many years Patassé, elected in 1993 and reelected in 1998, had struggled to bring political and financial stability to the poor, landlocked country, following three decades of military dictatorship. He failed.

In 1996, the French military rescued the government from an army mutiny. The intervention was resented by many as a sign of France's continuing control. A national-reconciliation agreement was signed in March 1998 allowing UN peacekeepers to oversee elections in 1999 and the training of a new army.

DEVELOPMENT

C.A.R.'s timber industry has suffered from corruption and environmentally destructive forms of exploitation. However, the nation has considerable forestry potential, with dozens of commercially viable and renewable species of trees.

But in 2000, further unrest was sparked by government's failure to pay civil servants their back wages, resulting in a general strike organized by 15 opposition groups. In May 2001, Patassé survived an attempted coup, with the help of Libyan, Chadian, and Congolese rebel forces. Former president André Kolingba and Army chief-of-staff Bozize led the coup. Thereafter a curfew was instituted, which was lifted a year later. The lifting of the curfew was meant to signal the return of "security and peace." But in October 2002, rebels loyal to Bozize seized control of much of Bangui before being driven out by Libyan and progovernment forces.

Since gaining independence in 1960, the political, economic, and military presence of France has remained pervasive in C.A.R. At the same time, the country's natural resources, as well as French largess, have been dissipated. Yet with diamonds, timber, and a resilient peasantry, the country is better endowed than many of its neighbors. In recent years, Libya has taken a special interest in the country's mostly still untapped natural wealth.

C.A.R.'s population has traditionally been divided between the so-called river peoples and savanna peoples, but most are united by the Songo language. What the country has lacked is a leadership committed to national development rather than to internationally sanctioned waste.

BOGANDA'S VISION

In 1959, as Central African Republic moved toward independence, Barthelemy Boganda, a former priest and the leader of the territory's nationalist movement, did not share the euphoria exhibited by many of his colleagues. To him, the French path to independence was a trap. Where there once had been a united French Equatorial Africa (A.E.F.), there were now five separate states, each struggling toward its own nationhood. Boganda had led the struggle to transform the territory into a true Central African Republic. But in 1958, French president Charles de Gaulle overruled all objections in forcing the breakup of the A.E.F. Boganda believed that, thus balkanized, the Central African states would each be too weak to achieve true independence, but he still hoped that A.E.F. reunification might prove possible after independence.

FREEDOM

C.A.R.'s human-rights record remains poor. Its security forces are linked to summary executions and torture. Other human-rights abuses include harsh prison conditions, arbitrary arrest, detention without trial, and restrictions on freedom of assembly. President Patassé granted amnesty to former senior officials of the Kolingba regime and mutineers.

In 1941, Boganda had founded the Popular Movement for the Social Evolution of Black Africa (MESAN). While Boganda was a pragmatist willing to use moderate means in his struggle, his vision was radical, for he hoped to unite French, Belgian, and Portuguese territories into an independent republic. His movement succeeded in gaining a local following among the peasantry as well as intellectuals. In 1958, Boganda led the territory to self-government, but he died in a mysterious plane crash just before independence.

Boganda's successors have failed to live up to his stature. At independence, David Dacko, a nephew of Boganda's who succeeded to the leadership of MESAN but also cultivated the political support of local French settlers who had seen Boganda as an agitator, led the country. Dacko's MESAN became the vehicle of the wealthy elite.

HEALTH/WELFARE

The literacy rate is low in Central African Republic—63%. Teacher training is currently being emphasized, especially for primary-school teachers. Poaching has diminished C.A.R.'s reputation as one of the world's last great wildlife refuges.

A general strike in December 1965 was followed by a military coup on New Year's Eve, which put Dacko's cousin, Army Commander Jean-Bedel Bokassa, in power. Dacko's overthrow was justified by the need to launch political and economic reforms. But more likely motives for the coup were French concern about Dacko's growing ties with China and Bokassa's own budding megalomania.

The country suffered greatly under Bokassa's eccentric rule. During the 1970s, he was often portrayed, alongside Idi Amin of Uganda, as an archetype of African leadership at its worst. It was more the sensational nature of his brutality—such as public torture and dismemberment of prisoners—rather than its scale that captured headlines. In 1972, he made himself "president-for-life." Unsatisfied with this position, in 1976 he proclaimed himself emperor, in the image of his hero Napoleon Bonaparte. The French government underwrote the $22 million spent on his coronation ceremony, which attracted widespread coverage in the global media.

ACHIEVEMENTS

Despite recurrent drought, a poor infrastructure, and inefficient official marketing, the farmers of Central African Republic have generally been able to meet most of the nation's basic food needs.

In 1979, reports surfaced that Bokassa himself had participated in the beating deaths of schoolchildren who had protested his decree that they purchase new uniforms bearing his portrait. The French government finally decided that its ally had become a liability. While Bokassa was away on a state visit to Libya, French paratroopers returned Dacko to power. In 1981, Dacko was once more toppled, in a coup that installed Prime Minister (General) André Kolingba. In 1985, Kolingba's provisional military regime was transformed into a one-party state. But in 1991, under a combination of local and French pressure, he agreed to the legalization of opposition parties.

Pressure for multiparty politics had increased as the government sank deeper into debt, despite financial intervention on the part of France, the World Bank, and the International Monetary Fund. Landlocked C.A.R.'s economy has long been constrained by high transport costs. But a perhaps greater burden has been the smuggling of its diamonds and other resources, including poached ivory, by officials who are high up in the government.

Timeline: PAST

1904
Separate French administration of the Oubangui-Chàri colony is established

1912–1913
Gold and diamonds are discovered

1949
Barthelemy Boganda sets up MESAN, which gains wide support

1960
Boganda dies; David Dacko, his successor, becomes president at independence

1966
Jean-Bedel Bokassa takes power after the first general strike

1976
Bokassa declares himself emperor

1979
Bokassa is involved in the massacre of schoolchildren; Dacko is returned as head of state

1981
André Kolingba takes power from Dacko

1990s
Ange-Félix Patassé wins the presidency; Patassé requires French intervention to overcome an army mutiny

PRESENT

2000s
The government takes steps to reduce street crime

General François Bozize overthrows the Patassé government

Chad (Republic of Chad)

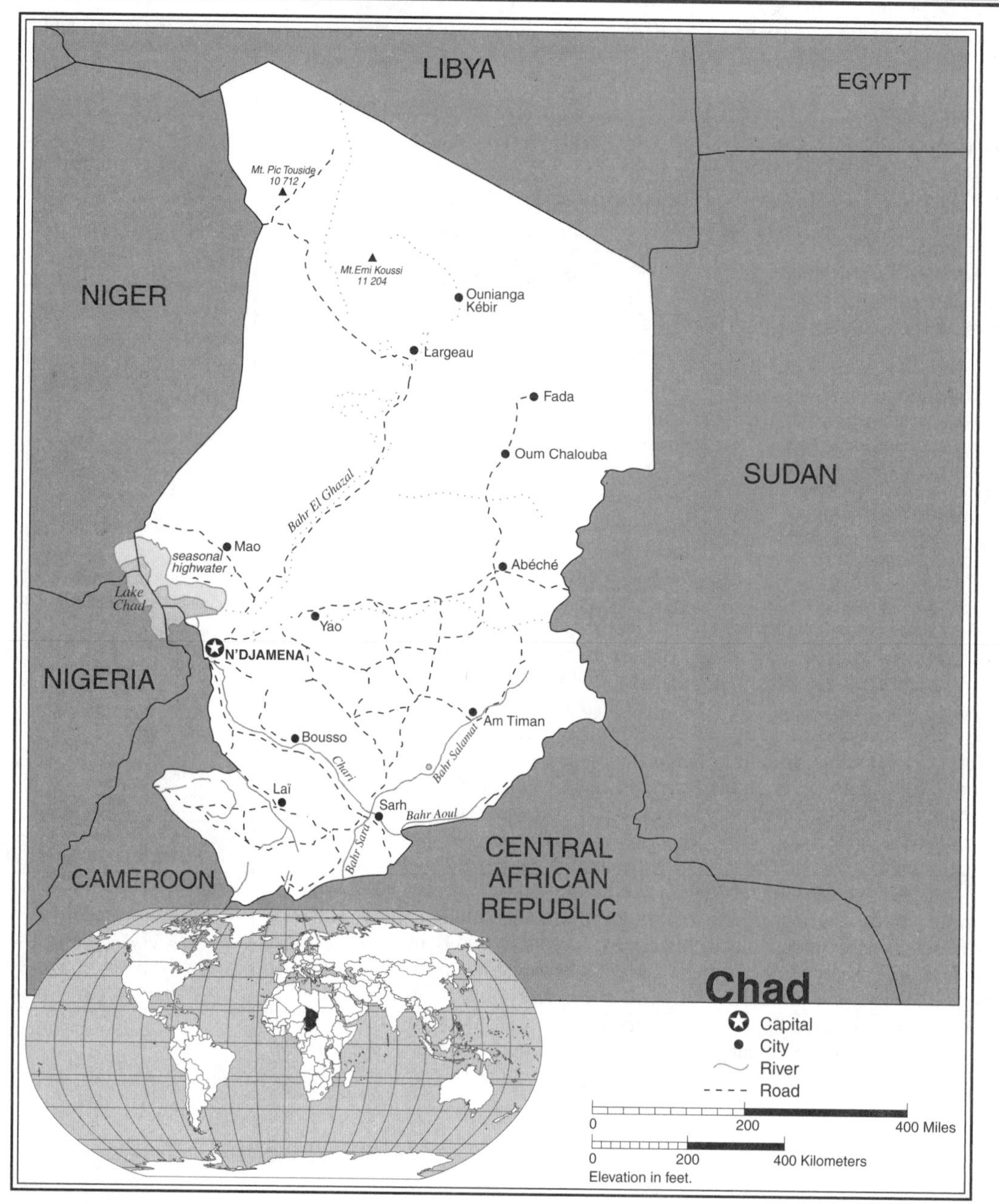

Chad Statistics

GEOGRAPHY

Area in Square Miles (Kilometers): 496,000 (1,284,634) (about 3 times the size of California)

Capital (Population): N'Djamena (826,000)

Environmental Concerns: soil and water pollution; desertification; insufficient potable water; waste disposal

Geographical Features: broad, arid plains in the center; desert in the north; mountains in the northwest; lowlands in the south; landlocked

Climate: tropical in the south; desert in the north

PEOPLE

Population

Total: 9,538,544

Annual Growth Rate: 3.27%

Rural/Urban Population Ratio: 77/23

Major Languages: French; Arabic; Sara; Sango; others

Ethnic Makeup: 200 distinct groups

Religions: 51% Muslim; 35% Christian; 7% animist; 7% others

Health

Life Expectancy at Birth: 49 years (male); 53 years (female)

Infant Mortality: 93.4/1,000 live births

Physicians Available: 1/27,765 people

HIV/AIDS Rate in Adults: 5%–7%

Education

Adult Literacy Rate: 40%
Compulsory (Ages): 6–14

COMMUNICATION

Telephones: 10,300 main lines
Televisions: 8/1,000 people
Internet Users: 15,000 (2002)

TRANSPORTATION

Highways in Miles (Kilometers): 19,620 (32,700)
Railroads in Miles (Kilometers): none
Usable Airfields: 49
Motor Vehicles in Use: 24,000

GOVERNMENT

Type: republic
Independence Date: August 11, 1960 (from France)
Head of State/Government: President Idriss Déby; Prime Minister Moussa Faki Mahamat
Political Parties: Patriotic Salvation Movement; National Union for Development and Renewal; many others
Suffrage: universal at 18

MILITARY

Military Expenditures (% of GDP): 2.1%
Current Disputes: civil war; border conflicts over Lake Chad area

ECONOMY

Currency ($ U.S. equivalent): 581 CFA francs = $1
Per Capita Income/GDP: $1,030/$8.9 billion
GDP Growth Rate: 8%
Inflation Rate: 3%
Labor Force by Occupation: 80%+ agriculture
Population Below Poverty Line: 80%
Natural Resources: petroleum; uranium; natron; kaolin; fish (Lake Chad)
Agriculture: subsistence crops; cotton; peanuts; fish; livestock
Industry: livestock products; breweries; natron; soap; textiles; cigarettes; construction materials
Exports: $172 million (primary partners Portugal, Germany, Thailand)
Imports: $223 million (primary partners France, Nigeria, Cameroon)

SUGGESTED WEB SITES

http://www.chadembassy.org/site/index.cfm
http://www.sas.upenn.edu/African_Studies/Country_Specific/Chad.html

Chad Country Report

After decades of civil war between northern- and southern-based armed movements, in 1997 Chad completed its transition to civilian rule, under the firm guidance of its president, Idriss Déby. A former northern warlord who seized power in 1990, in 1996 Déby achieved a second-round victory in the country's first genuinely contested presidential elections since its independence in 1960. The result was seen as an endorsement of his government's gradual progress in rebuilding Chad's state structures, which had all but collapsed by the 1980s, when the fractured country was referred to as the "Lebanon of Africa."

DEVELOPMENT

Chad has potential petroleum and mineral wealth that would greatly help the economy if stable central government can be created. Deposits of chromium, tungsten, titanium, gold, uranium, and tin as well as oil are known to exist. Roads are in poor condition and are dangerous.

In June 2001, Chad's highest court confirmed Déby's reelection, after a controversial poll in which the results of about one quarter of the polling stations were cancelled due to alleged irregularities. Six unsuccessful candidates were briefly picked up for questioning by police after the poll. While Déby's success—through both the ballot and bullet—in defeating, marginalizing, and/or reconciling rival factions has restored a semblance of statehood to Chad over the past five years, he continues to preside over a bankrupt government whose control over much of the countryside is tenuous.

In 1998, a new armed insurgency broke out in the north, led by former defense chief Youssouf Togimi. In January 2002, a Libyan-brokered peace deal was agreed to by the government and Togoimi's rebels (known as the Movement for Democracy and Justice in Chad, or MDJT). But further clashes have put implementation of the peace agreement into jeopardy. The accord provides for a ceasefire, release of prisoners, integration of the rebels into the national army, and government positions for MDJT leaders.

Déby also faces continuing challenges to his control over the south, where there have been calls for Chad to become a federal, rather than unitary, state. Many in the south still resent the central government in the capital city of N'Djamena, believing it to represent predominantly northern interests. On-going fighting within the Central African Republic, Chad's southern neighbor, could further destabilize the situation.

CIVIL WAR

Chad's conflicts are partially rooted in the country's ethnic and religious divisions. It has been common for outsiders to portray the struggle as being between Arab-oriented Muslim northerners and black Christian southerners, but Chad's regional and ethnic allegiances are much more complex. Geographically, the country is better divided into three zones: the northern Sahara, a middle Sahel region, and the southern savanna. Within each of these ecological areas live peoples who speak different languages and engage in a variety of economic activities. Wider ethno-regional and religious loyalties have emerged as a result of the Civil War, but such aggregates have tended to be fragile and their allegiances shifting.

FREEDOM

Despite some modest improvement, Chad's human-rights record remains poor. Its security forces are linked to torture, extra-judicial killings, beatings, disappearances, and rape. A recent Amnesty International report on Chad was entitled "Hope Betrayed." Antigovernment rebel forces are also accused of atrocities. The judiciary is not independent.

At Chad's independence, France turned over power to François Tombalbaye, a Sara-speaking Christian southerner. Tombalbaye ruled with a combination of repression, ethnic favoritism, and incompetence, which quickly alienated his regime from broad sectors of the population. A northern-based coalition of armed groups, the National Liberation Front, or Frolinat, launched an increasingly successful insurgency. The intervention of French troops on Tombalbaye's behalf failed to stem the rebellion. In 1975, the army, tired of the war and upset by the president's increasingly conspicuous brutality, overthrew

Tombalbaye and established a military regime, headed by Felix Malloum.

Malloum's government was also unable to defeat Frolinat; so, in 1978, it agreed to share power with the largest of the Frolinat groups, the Armed Forces of the North (FAN), led by Hissène Habré. This agreement broke down in 1979, resulting in fighting in N'Djamena. FAN came out ahead, while Malloum's men withdrew to the south. The triumph of the "northerners" immediately led to further fighting among various factions—some allied to Habré, others loyal to his main rival within the Frolinat, Goukkouni Oueddie. Earlier Habré had split from Oueddie, whom he accused of indifference toward Libya's unilateral annexation in 1976 of the Aouzou Strip, along Chad's northern frontier. At the time, Libya was the principal foreign backer of Frolinat.

HEALTH/WELFARE

In 1992, there were reports of catastrophic famine in the countryside. Limited human services were provided by external aid agencies. Medicines are in short supply or completely unavailable.

In 1980, shortly after the last French forces withdrew from Chad, the Libyan Army invaded the country, at the invitation of Oueddie. Oueddie was then proclaimed the leader in a "Transitional Government of National Unity" (GUNT), which was established in N'Djamena. Nigeria and other neighboring states, joined by France and the United States, pressed for the withdrawal of the Libyan forces. This pressure grew in 1981 after Libyan leader Muammar al-Qadhafi announced the merger of Chad and Libya. Following a period of intense multinational negotiations, the Libyan military presence was reduced at Oueddie's request.

The removal of the Libyan forces from most of Chad was accompanied by revived fighting between GUNT and FAN, with the latter receiving substantial U.S. support, via Egypt and Sudan. A peacekeeping force assembled by the Organization of African Unity proved ineffectual. The collapse of GUNT in 1982 led to a second major Libyan invasion. The Libyan offensive was countered by the return of French forces, assisted by Zairian troops and by smaller contingents from several other Francophonic African countries. Between 1983 and 1987, the country was virtually partitioned along the 16th Parallel, with Habré's French-backed, FAN–led coalition in the south and the Libyan-backed remnants of GUNT in the north.

A political and military breakthrough occurred in 1987. Habré's efforts to unite the country led to a reconciliation with Malloum's followers and with elements within GUNT. Oueddie himself was apparently placed under house arrest in Libya. Emboldened, Habré launched a major offensive north of the 16th Parallel that rolled back the better-equipped Libyan forces, who by now included a substantial number of Lebanese mercenaries. A factor in the Libyan defeat was U.S.–supplied Stinger missiles, which allowed Habré's forces to neutralize Libya's powerful air force (Habré's government lacked significant air power of its own). A cease-fire was declared after the Libyans had been driven out of all of northern Chad with the exception of a portion of the disputed Aouzou Strip.

In 1988, Qadhafi announced that he would recognize the Habré government and pay compensation to Chad. The announcement was welcomed—with some skepticism—by Chadian and other African leaders, although no mention was made of the conflicting claims to the Aouzou Strip.

ACHIEVEMENTS

In precolonial times, the town of Kanem was a leading regional center of commerce and culture. Since independence in 1960, perhaps Chad's major achievement has been the resiliency of its people under the harshest of circumstances. The holding of truly contested elections is also a significant accomplishment.

The long-running struggle for Chad took another turn in November 1990, with the sudden collapse of Habré's regime in the face of a three-week offensive by guerrillas loyal to his former army commander, Idriss Déby. Despite substantial Libyan (and Sudanese) backing for his seizure of power, Déby had the support of France, Nigeria, and the United States (Habré had supported Iraq's annexation of Kuwait). A 1,200-man French force began assisting Déby against rebels loyal to Habré and other faction leaders.

Between January and April 1993, Déby's hand was strengthened by the successful holding of a "National Convention," in which a number of formerly hostile groups agreed to cooperate with the government in drawing up a new constitution. In April 1994, his government was further boosted by Qadhafi's unexpected decision to withdraw his troops from the Aouzou Strip, leaving Chad in undisputed control of the territory. The move followed an International Court of Justice ruling in Chad's favor.

Timeline: PAST

1960
Independence is achieved under President François Tombalbaye

1965–1966
Revolt breaks out among peasant groups; FROLINAT is formed

1978
Establishment of a Transitional Government of National Unity (GUNT) with Hissène Habré and Goukkouni Oueddie

1980s
Habré seizes power and reunites the country in a U.S.–supported war against Libya

1990s
Habré is overthrown by Idriss Déby; Déby promises to create a multiparty democracy, but conditions remain anarchic

PRESENT

2000s
Chad's northern provinces bordering Libya remain heavily landmined

Persistent armed insurgency in the north

Dèby is confirmed as reelected, after a controversial poll

A BETTER FUTURE?

After a three-year-long civil war between various factions and the government, Chad seemed in 2003 to be on the verge of peace within the country after signing a set of peace accords with the National Resistance Army (ANR) rebel movement (signed in January 2003), which was active in the east, and with the Movement for Democracy and Justice (signed in January 2002), who were operational in the north. Throughout 2004, violence broke out in Sudan's Darfur region, and Chad was faced with a major influx of more than 300,000 people who crossed its borders as refugees. By March 2004, the violence had reached such a high level that the United Nations Council for Refugees said that it was moving Sudanese refugees deeper into Chad to avoid attacks from Sudanese militias. Throughout April and May 2004, Chadian troops were occasionally fighting pro-Sudanese government militias, as fighting in Sudan's Darfur region spilled over the border. The good news from Sudan during this period of 2002 to 2004 was that in October 2003 Chad became an oil exporter with the opening of a pipeline connecting its oil fields with Cameroon.

The long, drawn-out conflict in Chad has led to immense suffering. Up to a half a million people—the equivalent of 10 percent of the total population—have been killed in the fighting.

Even if peace could be restored, the overall prospects for national development are bleak. The country has potential mineral wealth, but its geographic isolation and current world prices are disincentives to investors. Local food self-sufficiency should be obtainable despite the possibility of recurrent drought, but geography limits the potential of export crops. Chad thus appears to be an extreme case of the more general African need for a radical transformation of prevailing regional and global economic interrelationships. Had outside powers devoted half the resources to Chad's development over the past decades as they have provided to its civil conflicts, perhaps the country's future would appear brighter.

Congo (Republic of the Congo; Congo-Brazzaville)

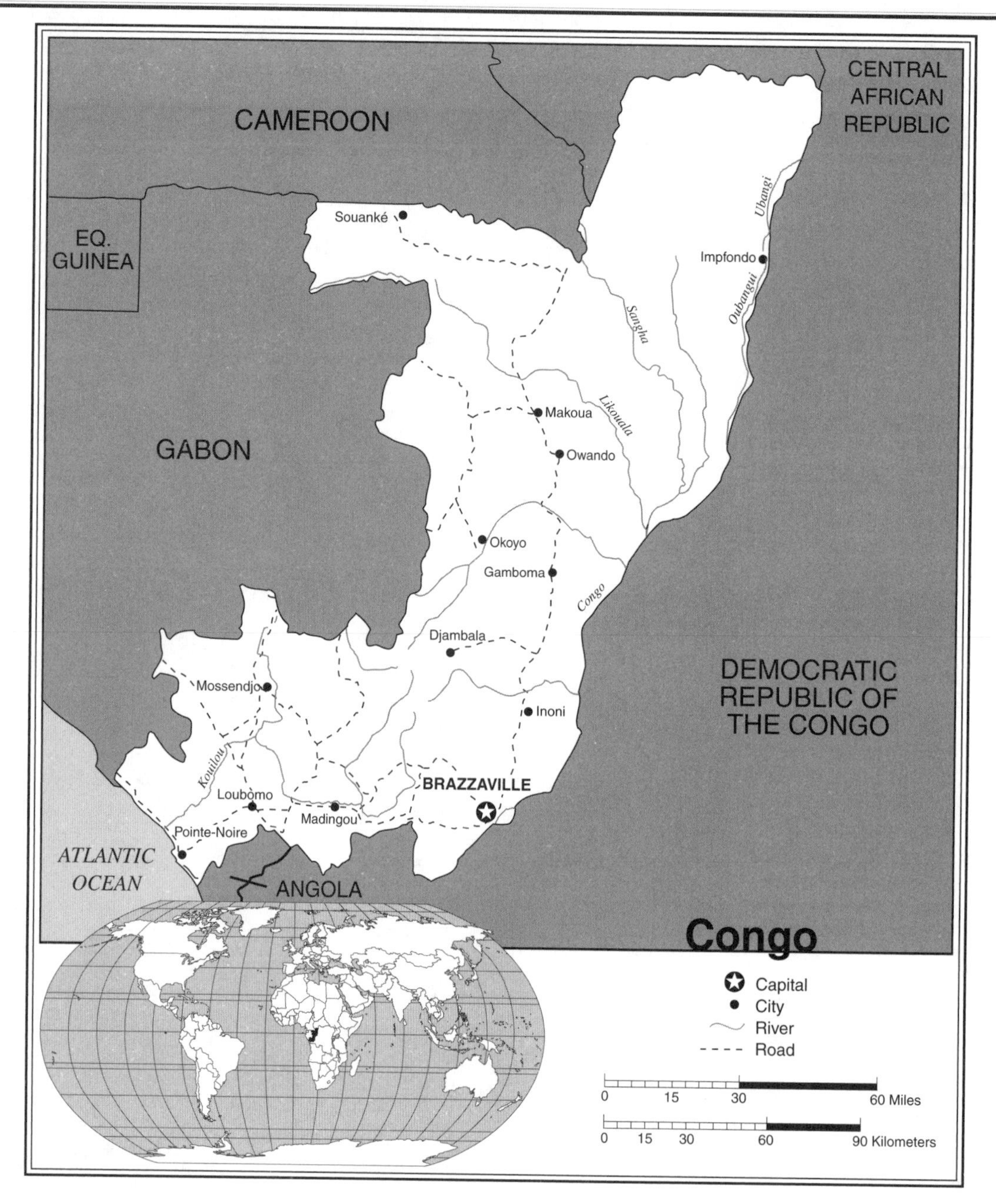

Congo Statistics

GEOGRAPHY

Area in Square Miles (Kilometers): 132,000 (342,000) (about the size of Montana)

Capital (Population): Brazzaville (1,360,000)

Environmental Concerns: air and water pollution; deforestation

Geographical Features: coastal plain; southern basin; central plateau; northern basin

Climate: tropical; particularly enervating climate astride the equator

PEOPLE

Population

Total: 2,959,000

Annual Growth Rate: 1.42%

Rural/Urban Population Ratio: 38/62

Major Languages: French; Lingala; Kikongo; Teke; Sangha; M'Bochi; others

Ethnic Makeup: 48% BaKongo; 20% Sangha; 17% Teke; 15% others

Religions: 50% Christian; 48% indigenous beliefs; 2% Muslim

Health

Life Expectancy at Birth: 44 years (male); 51 years (female)

Infant Mortality: 97/1,000 live births
Physicians Available: 1/3,873 people
HIV/AIDS Rate in Adults: 6.43%

Education

Adult Literacy Rate: 75%
Compulsory (Ages): 6–16

COMMUNICATION

Telephones: 22,000 main lines
Televisions: 17/1,000 people
Internet Users: 15,000 (2002)

TRANSPORTATION

Highways in Miles (Kilometers): 7,680 (12,800)
Railroads in Miles (Kilometers): 494 (797)
Usable Airfields: 33
Motor Vehicles in Use: 47,000

GOVERNMENT

Type: republic
Independence Date: August 15, 1960 (from France)
Head of State/Government: President Denis Sassou-Nguesso is both head of state and head of government
Political Parties: Democratic and Patriotic Forces; Congolese Movement for Democracy and Integral Development; many others
Suffrage: universal at 18

MILITARY

Military Expenditures (% of GDP): 2.8%
Current Disputes: civil conflicts; boundary issue with Democratic Republic of Congo

ECONOMY

Currency ($ U.S. Equivalent): 581.2 CFA francs = $1
Per Capita Income/GDP: $900/$2.5 billion
GDP Growth Rate: 1.3%
Inflation Rate: 2.4%

Natural Resources: timber; potash; lead; zinc; uranium; petroleum; natural gas; copper; phosphates

Agriculture: cassava; cocoa; coffee; sugarcane; rice; peanuts; vegetables; forest products

Industry: processing of agricultural and forestry goods; cement; brewing; petroleum

Exports: $2.6 billion (primary partners United States, South Korea, China)

Imports: $725 million (primary partners France, United States, Italy)

SUGGESTED WEB SITES

http://www.sas.upenn.edu/African_Studies/Country_Specific/Congo.html

Congo Country Report

Once considered one of Africa's most promising economies, over the past decade the Republic of the Congo—not to be confused with its larger neighbor the Democratic Republic of the Congo (D.R.C., the former Zaire)—has been afflicted by civil strife that has killed thousands while displacing up to one third of the population. In March 2002, the current president, General Denis Sassou-Nguesso, claimed 89 percent of the vote in presidential elections in which his two main political rivals, former president Pascal Lissouba and prime minister Bernard Kolelas, were barred from the contest. The poll was supposed to be the culmination of a two-year peace process that returned the country to democracy under a new Constitution. But in the same month, renewed fighting broke out between government forces and "Ninja" rebels loyal to Kolelas. In June, battles between government troops and the Ninja spread to Brazzaville, killing about 100 people.

The current round of political conflict in Congo began in 1997, when forces loyal to Sassou-Nguesso, who had previously ruled as a virtual dictator between 1979 and 1992, launched a rebellion. They seized control of Brazzaville in October 1997. The return of Sassou-Nguesso, who had garnered only 17 percent of the vote while losing the presidency to Lissouba just two years earlier, would not have been possible without the intervention of Angolan government forces, whose overt violation of Congolese sovereignty attracted little international comment. The Angolans were apparently motivated by allegations that Lissouba was supportive of UNITA rebels in their own country.

DEVELOPMENT

Congo's Niari Valley has become the nation's leading agricultural area, due to its rich alluvial soils. The government has been encouraging food-processing plants to locate in the region.

The fall of the capital did not end the fighting, which resulted in the Congolese Army as well as the country as a whole becoming largely split along north–south regional lines. Opposition to Sassou-Nguesso has been concentrated among the BaKongo people of the south, who make up about half of the total population. With continued Angolan backing, Sassou-Nguesso's forces were able to drive back those of his rivals.

At the end of 1999, many of the rebels agreed to a cease-fire accord. This was followed by a peace agreement that provided for a national dialogue; demilitarization of political parties; and the reorganization of the army, including the re-admission of rebel units into the security forces. By September 2001, some 15,000 rebels had been disarmed in a cash-for-arms scheme. This was rewarded by the International Monetary Fund, which cancelled some $4 billion in debt. But other rebels have not as yet been bought off.

In December 2001, Lissouba was convicted in absentia by the High Court of treason and corruption charges, and sentenced to 30 years hard labor. Although Sassou-Nguesso's forces have maintained the upper hand, political stability and economic growth are unlikely to return to Congo in the absence of a truly inclusive process of reconciliation.

FREEDOM

Until 1990, political opposition groups, along with Jehovah's Witnesses and certain other religious sects, were vigorously suppressed. The new Constitution provides for basic freedoms of association, belief, and speech.

The overthrow of Lissouba was all the more unfortunate in that his government had seemed to have been making progress in negotiating an end to the violence among the political factions that has plagued the country for more than a decade. A 1995 agreement was supposed to lead to the disarmament of party militias. But instead, efforts to assure the disarmament of Sassou-Nguesso's men set off the revolt in June 1997. Previously, in November 1993, large sections of Brazzaville had been a battleground between troops loyal to the elected Pan-African Union of Social

(United Nations photo 154086 by Sean Sprague)

Since Congo achieved its independence in 1960, it has made enormous educational strides. Almost all children in the country now attend school, which has had a tremendous effect on helping to realize the potential of the country's natural resources.

Democracy (UPADS) government of President Lissouba and the Ninjas—armed supporters of opposition leader Bernard Kolelas's Union for the Renewal of Democracy (URD). After several weeks of fighting, peace was finally restored with the intervention of an Organization of African Unity mediator. The crisis underscored the continuing fragility of Congo's difficult political transition into a multiparty democracy.

Sassou-Nguesso had previously ruled Congo as the head of a self-proclaimed, Marxist-Leninist one-party state. But in 1990, the ruling Congolese Workers Party (PCT) agreed to abandon both its past ideology and its monopoly of power. In 1991, a four-month-long "National Conference" met to pave the way for a new constitutional order. An interim government headed by Andre Milongo was appointed, pending elections, while Sassou-Nguesso was stripped of all but ceremonial power. In the face of coup attempts by elements of the old order and a deteriorating economy, legislative and executive elections were finally held in the second half of 1992, resulting in Lissouba's election and a divided National Assembly.

A new National Assembly election in 1993 resulted in a decisive UPADS victory over a URD–PCT alliance, but the losers rejected the results for five months. With the economy experiencing a prolonged depression, political tensions remained high.

CONGO REPUBLIC

The Republic of the Congo takes its name from the river that forms its southeastern border with the Democratic Republic of Congo. Because Zaire prior to 1971 also called itself the Congo, the two countries are sometimes confused. Close historical and ethnic ties do in fact exist between the nations. The BaKongo are the largest ethnolinguistic group in Congo and western former Zaire as well as in northern Angola. During the fifteenth and sixteenth centuries, this group was united under the powerful Kingdom of the Kongo, which ruled over much of Central Africa while establishing commercial and diplomatic ties with Europe. But the kingdom had virtually disappeared by the late nineteenth century, when the territory along the northwest bank of the Congo River—the modern republic—was annexed by France, while the southeast bank—Zaire—was placed under the brutal rule of King Leopold of Belgium.

Despite the establishment of this political division, cultural ties between Congo and the former Zaire, the former French and Belgian Congos, remained strong. Brazzaville sits across the river from the D.R.C. capital of Kinshasha. The metropolitan region formed by these two centers has, through such figures as the late Congolese artist Franco, given rise to *soukous,* a musical style that is now popular in such places as Tokyo and Paris as well as throughout much of Africa.

ECONOMIC DEVELOPMENT

Brazzaville, which today houses well over a third of Congo's population, was established during the colonial era as the administrative headquarters of French Equatorial Africa, a vast territory that included the present-day states of Chad, Central African Republic, Gabon, and Congo Republic. As a result, the city expanded, and the area around it developed as an imperial crossroads. The Congolese paid a heavy price for this growth. Thousands died while working under brutal conditions to build the Congo-Ocean Railroad, which linked Brazzaville with Pointe-Noire on the coast. Many more suffered as forced laborers for foreign concessionaires during the early decades of the twentieth century.

HEALTH/WELFARE

Almost all Congolese between ages 6 and 16 currently attend school. Adult-literacy programs have also proved successful, giving the country one of the highest literacy rates in Africa. However, 30% of Congolese children under age 5 are reported to suffer from chronic malnutrition.

While the economies of many African states stagnated or declined during the 1970s and 1980s, Congo generally experienced growth, a result of its oil wealth. Hydrocarbons account for 90 percent of the total value of the nation's exports. But the danger of this dependence has been apparent since 1986, when falling oil prices led to a sharp decline in gross domestic product. An even greater threat to the nation's economic health is its mounting debt. As a result of heavy borrowing during the oil-boom years, by 1989 the total debt was estimated to be 50 percent greater than the value of the country's annual economic output. The annual cost of servicing the debt was almost equal to domestic expenditure.

The debt led to International Monetary Fund pressure on Congo's rulers to introduce austerity measures as part of a Structural Adjustment Program (SAP). The PCT regime and its interim successor were willing to move away from the country's emphasis on central planning toward a greater reliance on market economics. But after an initial round of severe budgetary cutbacks, both administrations found it difficult to reduce their spending further on such things as food subsidies and state-sector employment.

With nearly two thirds of Congo's population now urbanized, there has been deep concern about the social and political consequences of introducing harsher austerity measures. Many urban-dwellers are already either unemployed or underemployed; even those with steady formal-sector jobs have been squeezed by wages that fail to keep up with inflation. The country's powerful trade unions, which are hostile to SAP, have been in the forefront of the democratization process.

ACHIEVEMENTS

There are a number of Congolese poets and novelists who combine their creative efforts with teaching and public service. Tchicaya U'Tam'si, who died in 1988, wrote poetry and novels and worked for many years for UNESCO.

Although most Congolese are facing tough times in the short run, the economy's long-term prospects remain promising. Besides oil, the country is endowed with a wide variety of mineral reserves. Timber has long been a major industry. And after years of neglect, the agricultural sector is growing. The goal of a return to food self-sufficiency appears achievable. Cocoa, coffee, tobacco, and sugarcane are major cash crops, while palm-oil estates are being rehabilitated.

The small but well-established Congolese manufacturing sector also has much potential. Congo's urbanized population is relatively skilled, thanks to the enormous educational strides that have been made since independence. Almost all children in Congo now attend school. Prior to the devastation that occurred during the 1997 revolt, the infrastructure serving Brazzaville and Pointe-Noire, coupled with the previous government's emphasis on private-sector growth, was potentially attractive to outside investors.

LOOKING AHEAD

Civil wars and militia conflicts have plagued the Republic of Congo throughout its recent past, but hopes for stability were raised in 2003 when the government signed a peace accord with rebels in the south. After three relatively peaceful but coup-ridden decades of independence, the former French colony experienced the first of two destructive bouts of fighting in 1993 when disputed parliamentary elections led to bloody, ethnically based fighting between pro-government forces and the opposition. The finding and exploitation of oil has been a curse on the Congo, as the soldiers have moved out of the barrages and into the streets in a more or less continual struggle for political power during the last decade. However, in 2004 a ceasefire agreement followed by the inclusion of some opposition members in the government helped to restore peace. The government forces, who were backed by Angolan troops, held the upper hand until a peace deal was signed in March 2003 with remnants of the civil war militias, known as Ninjas, who had continued to fight government forces in the southern Pool region.

Timeline: PAST

1910
Middle Congo becomes part of French Equatorial Africa

1944
Conference establishes French Union; Felix Eboue establishes positive policies for African advancement

1960
Independence is achieved, with Abbe Fulbert Youlou as the first president

1963
A general strike brings the army and a more radical government (National Revolutionary Movement) to power

1968–1969
A new military government under Marien Ngouabi takes over; the Congolese Workers' Party is formed

1977
Ngouabi is assassinated; Colonel Yhombi-Opango rules

1979–1992
Denis Sassou-Nguesso is president

1990s
Pascal Lissouba is elected president; former dictator Sassou-Nguesso again seizes power

PRESENT

2000s
Congo tries to recover from the civil conflict of the late 1990s

Security problems remain despite the peace process

Democratic Republic of the Congo
(Congo-Kinshasa; formerly Zaire)

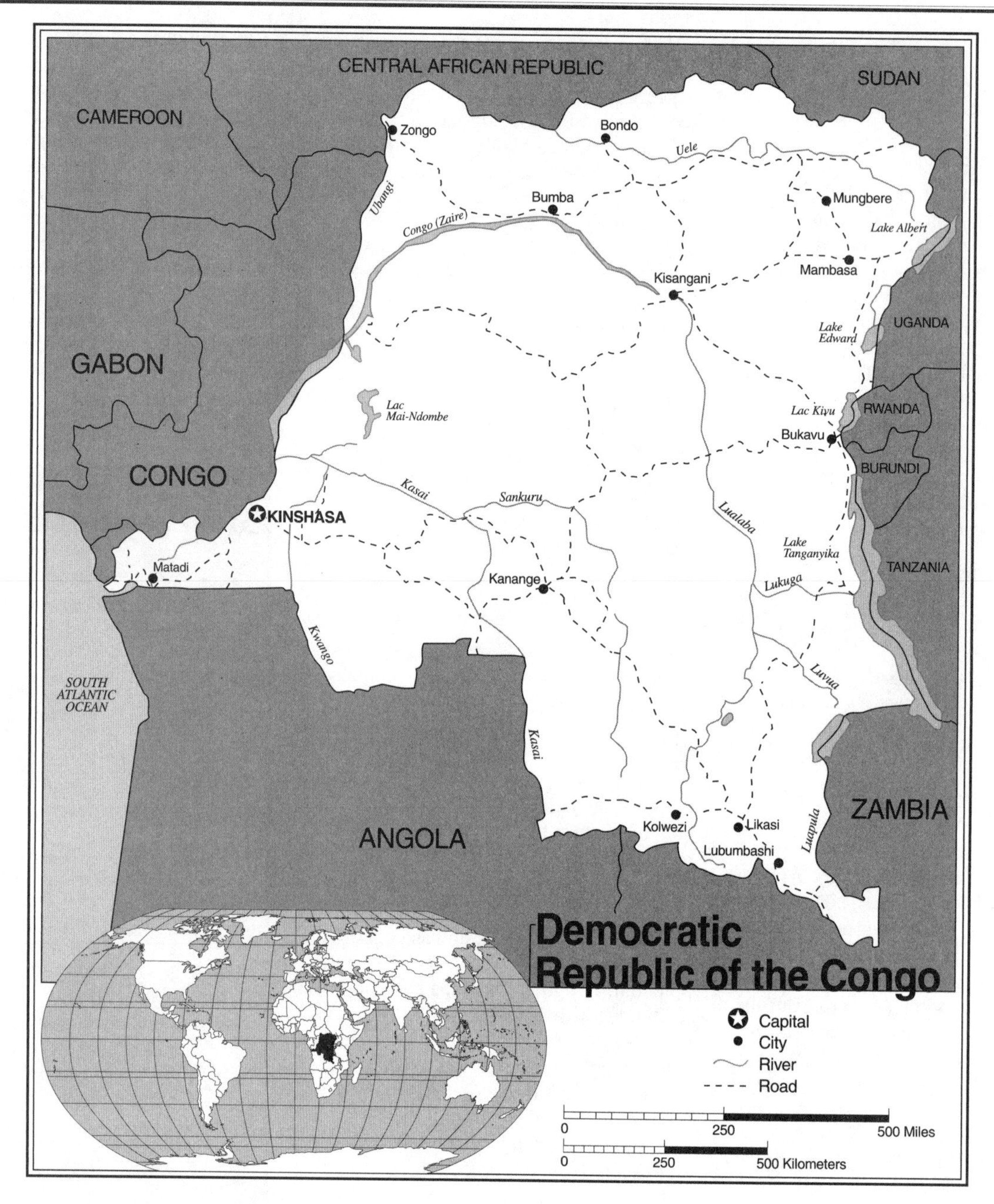

Democratic Republic of the Congo Statistics

GEOGRAPHY

Area in Square Miles (Kilometers): 905,063 (2,300,000) (1/4 the size of the United States)

Capital (Population): Kinshasa (5,064,000)

Environmental Concerns: poaching; water pollution; deforestation; soil erosion; mining

Geographical Features: the vast central basin is a low-lying plateau; mountains in the east; dense tropical rain forest

Climate: tropical equatorial

PEOPLE

Population

Total: 58,317,930
Annual Growth Rate: 2.79%
Rural/Urban Population Ratio: 70/30
Major Languages: French; Lingala; Kingwana; others

Ethnic Makeup: Bantu majority; more than 200 African groups
Religions: 70% Christian; 20% indigenous beliefs; 10% Muslim

Health

Life Expectancy at Birth: 47 years (male); 51 years (female)
Infant Mortality: 98/1,000 live births
Physicians Available: 1/15,584 people
HIV/AIDS Rate in Adults: 5%

Education

Adult Literacy Rate: 77.3%
Compulsory (Ages): 6–12

COMMUNICATION

Telephones: 21,000 main lines
Daily Newspaper Circulation: 3/1,000 people
Internet Users: 50,000 (2002)

TRANSPORTATION

Highways in Miles (Kilometers): 97,340 (157,000)
Railroads in Miles (Kilometers): 3,206 (5,138)
Usable Airfields: 232
Motor Vehicles in Use: 530,000

GOVERNMENT

Type: dictatorship; presumably in transition to representative government
Independence Date: June 30, 1960 (from Belgium)
Head of State/Government: President Joseph Kabila is both head of state and head of government
Political Parties: Popular Movement of the Revolution; Democratic Social Christian Party; others
Suffrage: universal and compulsory at 18

MILITARY

Military Expenditures (% of GDP): 1.4%
Current Disputes: civil war; boundary issues with surrounding states; military conflicts with neighboring states

ECONOMY

Currency ($ U.S. equivalent): 3,275 new Zaires = $1
Per Capita Income/GDP: $590/$32 billion
GDP Growth Rate: –4%
Labor Force by Occupation: 65% agriculture; 19% services; 16% industry
Natural Resources: cobalt; copper; cadmium; petroleum; zinc; diamonds; manganese; tin; gold; silver; bauxite; iron ore; coal; hydropower; timber; others
Agriculture: coffee; palm oil; rubber; tea; manioc; root crops; corn; fruits; sugarcane; wood products
Industry: mineral mining and processing; consumer products; cement; diamonds
Exports: $750 million (primary partners Belgium, United States, South Africa)
Imports: $1.02 billion (primary partners South Africa, Benelux, Nigeria)

SUGGESTED WEB SITE

`http://www.sas.upenn.edu/African_Studies/Country_Specific/DR_Congo.html`

Democratic Republic of the Congo Country Report

Due to its geographical location and staggering mineral resources, the Democratic Republic of Congo (D.R.C.), also formerly known as Zaire, is perhaps the most strategically important nation in Africa, with the potential to become one of its richest. Unfortunately, after years of almost unbelievable levels of corruption, external interference, and greed, it suffers from the lack of a strong central government with credible structures. At times in its history, it has been difficult to imagine a nation in more distress than the D.R.C. After being ruled from 1965 to 1997 by the notoriously corrupt dictator Joseph Mobutu, in recent years the D.R.C. has all but fallen apart as armed internal and external forces have divided the country.

A series of agreements in 2002 led to the withdrawal of most external-state forces from Congolese soil and the promise of a "government of national unity." With an on-again, off-again peace process now bearing some fruit, there may be some grounds for cautious optimism.

The D.R.C.'s latest conflict began in August 1998, when rebels of the Congolese Assembly for Democracy, heavily backed by the Rwandan and Ugandan militaries, advanced rapidly from the east. This occurred shortly after the 15-month-old D.R.C. regime headed by Laurent Kabila, who had replaced Mobutu, called for the withdrawal of all foreign forces from Congolese soil. Early rebel hopes of a quick victory were frustrated by the intervention of Angolan, Namibian, and Zimbabwean forces. The fighting quickly became deadlocked, leading to mediation efforts under the chairmanship of Botswana's former president Sir Ketumile Masire, and backed by the United Nations and the Organization of African Unity. These negotiations appeared to be getting nowhere until January 2001, when Laurent Kabila was assassinated and replaced as president by his son Joseph. Meanwhile, the rebel forces had become divided, with their Rwandan and Ugandan mentors at times turning on each other. By May 2001, the International Rescue Committee, a New York–based refugee agency, had estimated that the war had directly and indirectly killed some 2.5 million people.

Laurent Kabila had come to power as the leader of the Alliance of Democratic Forces for the Liberation of Congo-Zaire, a rebel movement that was itself heavily dependent on external support from Rwanda, Uganda, and Zimbabwe. In the final months of 1996, Kabila's forces seized control of eastern Zaire, and they arrived in Kinshasa in May 1997. Once in power, Kabila changed the name of the country from Zaire, an identity that had been imposed by Mobutu in 1971, to its former title: Democratic Republic of the Congo.

DEVELOPMENT

Western aid and development assistance were drastically reduced in 1992, but new aid was pledged in 1994 as a reward for Mobutu's cooperation in dealing with the Rwandan conflict. An agreement was signed with Egypt for the long-term development of Zaire's hydroelectric power.

Hopes that the new government would bring an end to the chronic corruption and mismanagement that have long plagued the country, or lead to an opening for democracy and improved human rights, soon proved misplaced. A leading Mobutu opponent, Etienne Tshisekedi wa Mulumba, was assaulted by Kabila's men within days of their victory. Tshisekedi remains the head of a nonviolent opposition coalition, the Union for Democracy and Social Progress, that is calling for a transition to democracy through a negotiated "government of national unity." So far, this plea has been all but ignored by outside powers as well as the warring factions.

Meanwhile, the country's already decayed social and economic infrastructure continues to disintegrate. The informal sector now dominates the local economy. In the process, national unity is being challenged by the reemergence of secessionist tendencies in the mineral-rich provinces of Katanga (Shaba) and Kasai. Western and eastern former Zaire has been inundated by an influx of some 2 million refugees from the killing fields of Rwanda and Burundi. With its vast size and wealth, D.R.C.'s fate will ultimately affect the future of neighboring states as well as its own citizens. Since independence, its instability has been a destabilizing influence for its neighbors in Central, Eastern, and Southern Africa. At peace, it could become the hub of an African renaissance.

Geographically, the country is located at Africa's center. It encompasses the entire Congo River Basin, whose waters are the potential source of 13 percent of the world's hydroelectric power. This immense area, about one quarter the size of the United States, encompasses a variety of land forms. It contains good agricultural possibilities and a wide range of natural resources, some of which have been intensively exploited for decades.

The D.R.C. links Africa from west to east. Its very narrow coastline faces the Atlantic. Forces from the East African coast have long influenced the landlocked eastern D.R.C. In the mid-nineteenth century, Swahili, Arab, and Nyamwezi traders from Tanzania established their hegemony over much of southeastern Zaire, pillaging the countryside for ivory and slaves. While the slave trade has left bitter memories, the Swahili language has spread to become a lingua franca throughout the eastern third of the country.

The 55 million people of the Democratic Republic of the Congo belong to more than 200 different ethnic groups, speak some 700 languages and dialects, and have varied lifestyles. Boundaries established in the late nineteenth century hemmed in portions of the Azande, Konga, Chokwe, and Songye peoples, yet they maintain contact with their kin in other countries.

Many important precolonial states were centered here, including the Luba, Kuba, and Lunda kingdoms, the latter of which, in earlier centuries, exploited the salt and copper of southeastern Zaire. The Kingdom of the Kongo, located at the mouth of the Congo River, flourished during the fifteenth and sixteenth centuries, establishing important diplomatic and commercial relations with Portugal. The elaborate political systems of these kingdoms are an important heritage for the Democratic Republic of the Congo.

LEOPOLD'S GENOCIDE

The European impact, like the Swahili and Arab influences from the east, had deeply destructive results. The Congo Basin was explored and exploited by private individuals before it came under Belgian domination. As a private citizen, King Leopold of Belgium sponsored H. M. Stanley's expeditions to explore the basin. In 1879, Leopold used Stanley's "treaties" as a justification for setting up the "Congo Independent State" over the whole region. This state was actually a private proprietary colony. To turn a profit on his vast enterprise, Leopold acted under the assumption that the people and resources in the territory were his personal property. His commercial agents and various concessionaires, to whom he leased portions of his colony, began to brutally coerce the local African population into providing ivory, wild rubber, and other commodities. The armed militias sent out to collect quotas of rubber and other goods committed numerous atrocities against the people, including destroying whole villages.

FREEDOM

The regimes in the Democratic Republic of the Congo have shown little respect for human rights. One Amnesty International report concluded that all political prisoners were tortured, and death squads were active.

No one knows for sure how many Africans perished in Congo Independent State as a result of the brutalities of Leopold's agents. Some critics estimate that the territory's population was reduced by 10 *million* people over a period of 20 years. Many were starved to death as forced laborers. Others were massacred in order to induce survivors to produce more rubber. Women and children were suffocated in "hostage houses" while their men did their masters' bidding. Thousands fled to neighboring territories.

For years the Congo regime was able to keep information of its crimes from leaking overseas, but eventually reports from missionaries and others did emerge. Public outrage was stirred by accounts such as E. D. Morel's *Red Rubber* and Mark Twain's caustic *King Leopold's Soliloquy,* as well as gruesome pictures of men, women, and children whose hands had been severed by troops (who were expected to produce the hands for their officers as evidence of their diligence). Joseph Conrad's fictionalized account of his experiences, *The Heart of Darkness,* became a popular literary classic. Finally even the European imperialists, during an era when their racial arrogance was at its height, could no longer stomach Leopold, called by some "the king with ten million murders on his soul."

During the years of Belgian rule, 1908 to 1960, foreign domination was less genocidal, but a tradition of abuse had nevertheless been established. The colonial authorities still used armed forces for "pacification" campaigns, tax collection, and labor recruitment. Local collaborators were turned into chiefs and given arbitrary powers that they would not have had under indigenous political systems. Concessionary companies continued to use force to recruit labor for their plantations and mines. The colonial regime encouraged the work of Catholic missionaries. Health facilities as well as a paternalistic system of education were developed. A strong elementary-school system was part of the colonial program, but the Belgians never instituted a major secondary-school system, and there was no institution of higher learning. By independence, only 16 Congolese had been able to earn university degrees, all but two in non-Belgian institutions. A small group of high-school–educated Congolese, known as *évolués* ("evolved ones"), served the needs of an administration that never intended nor planned for Congo's independence.

In the 1950s, the independence movements that were emerging throughout Africa affected the Congolese, especially townspeople. The Belgians began to recognize the need to prepare for a different future. Small initiatives were allowed; in 1955, nationalist associations were first permitted, and a 30-year timetable for independence was proposed. This sparked heated debate. Some évolués agreed with the Belgians' proposal. Others, including the members of the Alliance of the Ba-Kongo (ABAKO), an ethnic association in Kinshasa, and the National Congolese Movement (MNC), led by Prime Minister Patrice Lumumba, rejected it.

A serious clash at an ABAKO demonstration in 1959 resulted in some 50 deaths. In the face of mounting unrest, further encouraged by the imminent independence of the French Congo (Republic of the Congo), the Belgians conceded a rapid transition to independence. A constitutional conference in January 1960 established a federal-government system for the future independent state. But there was no real preparation for this far-reaching political change.

THE CONGO CRISIS

Democratic Republic of the Congo became independent on June 30, 1960, under the leadership of President Joseph Kasavubu and Prime Minister Patrice Lumumba. Within a week, an army mutiny had stimulated widespread disorder. The scars of Congo's uniquely bitter colonial experience showed. Unlike in Africa's other postcolonial states, hatred of the white former masters turned to violence in Congo, resulting in the hurried flight of the majority of its large European community. Ethnic and regional bloodshed took a much greater toll among the African population. The wealthy Katanga Province (now Shaba) and South Kasai seceded.

Lumumba called upon the United Nations for assistance, and troops came from a variety of countries to serve in the UN force. Later, as a result of a dispute with President Kasavubu, Lumumba sought Soviet aid. Congo could have become a Cold War battlefield, but the army, under Lumumba's former confidant, Joseph Desiré Mobutu, intervened. Lumumba was arrested and turned over to the Katanga rebels; he was later assassinated. Western interests and, in particular, the U.S. Central Intelligence Agency (CIA) played a substantial if not fully revealed role in the downfall of the idealistic Lumumba and the rise of his cynical successor, Mobutu. Rebellions by Lumumbists in the northeast and Katanga secessionists, supported by foreign mercenaries, continued through 1967.

MOBUTUISM

Mobutu seized full power in 1965, ousting Kasavubu in a military coup. With ruthless energy, he eliminated the rival political factions within the central government and crushed the regional rebellions. Mobutu banned party politics. In 1971, he established the Second Republic as a one-party state in which all power was centralized around the "Founding President." Every citizen, at birth, was legally expected to be a disciplined member of Mobutu's Popular Revolutionary Movement (MPR). With the exception of some religious organizations, virtually all social institutions were to function as MPR organs. The official ideology of the MPR republic became "Mobutuism"—the words, deeds, and decrees of "the Guide" Mobutu. All citizens were required to sing his praises daily at the workplace, at schools, and at social gatherings. In hymns and prayers, the name Mobutu was often substituted for that of Jesus. A principal slogan of Mobutuism was "authenticity." Supposedly this meant a rejection of European values and norms for African ones.

But it was Mobutu alone who defined what was authentic. He added to his own name the title *Sese Seko* ("the All Powerful") while declaring all European personal names illegal. He also established a national dress code; ties were outlawed, men were expected to wear his abacost suit, and women were obliged to wear the *paigne,* or wrapper. (The former Zaire was perhaps the only place in the world where the necktie was a symbol of political resistance.) The name of the country was changed from Congo to *Zaire,* a word derived from the sixteenth-century Portuguese mispronunciation of the (Ki)Kongo word for "river."

HEALTH/WELFARE

In 1978, more than 5 million students were registered for primary schools and 35,000 for college. However, the level of education has since declined. Many teachers were laid off in the 1980s, though nonexistent "ghost teachers" remained on the payroll. The few innovative educational programs that do manage to exist are outside of the state system.

Outside of Zaire, some took Mobutu's protestations of authenticity at face value, while a few other African dictators, such as Togo's Gnassingbé Eyadéma, emulated aspects of his fascist methodology. But the majority of Zairians grew to loathe his "cultural revolution."

Authenticity was briefly accompanied by a program of nationalization. Union Minière and other corporations were placed under government control. In 1973 and 1974, plantations, commercial institutions, and other businesses were also taken over, in what was called a "radicalizing of the Zairian Revolution."

But the expropriated businesses simply enriched a small elite. In many cases, Mobutu gave them away to his cronies, who often sold off the assets. Consequently, the economy suffered. Industries and businesses were mismanaged or ravaged. Some individuals became extraordinarily wealthy, while the population as a whole became progressively poorer with each passing year. Mobutu allegedly became the wealthiest person in all of Africa, with a fortune estimated in excess of $5 billion (about equal to Zaire's national debt), most of which was invested and spent outside of Africa. He and his relatives owned mansions all over the world.

Until his last year in power, no opposition to Mobutu was allowed. Those critical of the regime faced imprisonment, torture, or death. The Roman Catholic Church and the Kimbanguist Church of Jesus Christ Upon This Earth were the only institutions able to speak out. Strikes were not allowed. In 1977 and 1978, new revolts in the Shaba Province were crushed by U.S.–backed Moroccan, French, and Belgian military interventions. Thus in 1997, rebels under Laurent Kabila ousted the ailing Mobutu and renamed the country the Democratic Republic of the Congo.

ECONOMIC DISASTER

The country's economic potential was developed by and for the Belgians, but by 1960, that development had gone further than in most other African colonial territories. It started with a good economic base, but the chaos of the early 1960s brought development to a standstill, and the Mobutu years were marked by regression. Development projects were initiated, but often without careful planning. World economic conditions, including falling copper and cobalt prices, contributed to Zaire's difficulties.

But the main obstacle to any sort of economic progress was the rampant corruption of Mobutu and those around him. The governing system in Zaire was a kleptocracy (rule by thieves). A well-organized system of graft transferred wealth from ordinary citizens to officials and other elites. With Mobutu stealing billions and those closest to him stealing millions, the entire society operated on an invisible tax system; for example, citizens had to bribe nurses for medical care, bureaucrats for documents, principals for school admission, and police to stay out of jail. For most civil servants, who were paid little or nothing, accepting bribes was a necessary activity. This fundamental fact also applied to most soldiers, who thus survived by living off the civilian population.

ACHIEVEMENTS

Kinshasha has been called the dance-music capital of Africa. The most popular sound is *souskous,* or "Congo rumba." The grand old man of the style is Rochereau Tabu Ley. Other artists, like Papa Wemba, Pablo Lubidika, and Sandoka, have joined him in spreading its rhythms internationally.

Ordinary people have suffered. By 1990, real wages of urban workers in the country were only 2 percent of what they were in 1960. Rural incomes had also deteriorated. The official 1990 price paid to coffee farmers, for example, was only one fifth of what it was in 1954 under the hugely exploitive Belgian regime. The situation has worsened since, due to periods of hyperinflation.

Much of the state's coffee and other cash crops have long been smuggled, more often than not through the connivance of senior government officials. Thus, although the country's agriculture has great economic potential, the returns from this sector continue to shrink. Despite its immense size and plentiful rainfall, it must import about 60 percent of its food requirements. Rural people move to the city or, for lack of employment, move back to the country and take up subsistence agriculture, rather than cash-crop farming, in order to ensure their own survival. The deterioration of roads and bridges has led to the decline of all trade.

In 1983, the government adopted International Monetary Fund austerity measures, but this only cut public expenditures. It had no effect on the endemic corruption, nor did it increase taxes on the rich. Under Mobutu's regime, more than 30 percent of former Zaire's budget went for debt servicing.

In June 1997, Kabila announced short-term economic priorities, including job creation, road and hospital rebuilding, and a national fuel-supply pipeline. But it was unclear where the money would come from to implement these plans.

U.S. SUPPORT FOR MOBUTU

Mobutu's regime was able to sidestep its financial crises and maintain power through the support of foreign powers, especially Belgium, France, Germany, and the United States. A U.S. intelligence report prepared in the mid-1950s concluded that the then–Belgian Congo was indeed the hub of Africa and thus vital to America's strategic interests. U.S. policy was thus the first to promote and then to perpetuate Mobutu as a pro-Western source of stability in the region. Mobutu himself skillfully cultivated this image.

Mobutu collaborated with the United States in opposing the Marxist-oriented Popular Movement for the Liberation of Angola. By so doing, he not only set himself up as an important Cold War ally but also was able to pursue regional objectives of his own. The National Front for the Liberation of Angola, long championed by the CIA as a counterforce to the MPLA, was led by an in-law of Mobutu, Holden Roberto. Mobutu also long coveted Angola's oil-rich enclave of Cabinda and thus sought CIA and South African assistance for the "independence" movement there. In recent years, millions of U.S. dollars were spent upgrading the airstrip at Kamina in Shaba Province, used by the CIA to supply the guerrillas of the National Union for the Total Independence of Angola, another faction opposed to the MPLA government. In 1989, Mobutu attempted to set himself up as a mediator between the government and the UNITA rebels, but even the latter grew to distrust him.

The United States had long known of Mobutu's human-rights violations and of the oppression and corruption that characterized his regime; high-level defectors as well as victims had publicized its abuses. Since 1987, Mobutu responded with heavily financed public-relations efforts aimed at lobbying U.S. legislators. U.S. support for Mobutu continued, but the eventual collapse of his authority led Washington belatedly to search for alternatives.

Mobutu also allied himself with other conservative forces in Africa and the Middle East. Moroccan troops came to his aid during the revolts in Shaba Province in 1977 and 1978. For his part, Mobutu was a leading African supporter of Morocco's stand with regard to the Western Sahara dispute. Under his rule the country was also an active member of the Francophonic African bloc. In 1983, Mobutu dispatched 2,000 Zairian troops to Chad in support of the government of Hissène Habré, then under attack from Libya, while in 1986, his men again joined French forces in propping up the Eyadéma regime in Togo. He also maintained and strengthened his ties with South Africa (today, the former Zaire imports almost half its food from that state). In 1982, he renewed the diplomatic ties with Israel that had been broken after the Arab–Israeli War of 1973. Israelis subsequently joined French and Belgians as senior advisers and trainers working within the former Zairian Army. In 1990, the outbreak of violent unrest in Kinshasha once more led to the intervention of French and Belgian troops.

Despite Mobutu's cultivation of foreign assistance to prop up his dictatorship, internal opposition grew. In 1990, he tried to head off his critics both at home and abroad by promising to set up a new Third Republic, based on multiparty democracy. Despite this step, repression intensified.

In October 1996, while Mobutu was in Europe recovering from cancer surgery, rebel troops under the leadership of Laurent Kabila seized their first major town, Uvira. Thousands of Rwandan Hutu refugees were forced to flee back to Rwanda. When Mobutu returned home in April 1997, he declared a nationwide state of emergency. Kabila's supporters then closed down Kinshasa as part of the campaign to oust Mobutu. Following negotiations with South Africa's Mandela, Mobutu left Kinshasa and went into exile.

THE STRUGGLE CONTINUES

Although the history of the Democratic Republic of Congo (D.R.C.) has been one of civil war and corruption, signs of change are on the horizon. A vast country with immense economic resources, the D.R.C. has been at the centre of what could be termed Africa's international war. But a peace deal, and the formation of a transitional government in 2003 appeared to signal the end of the five-year conflict which pitted government forces, supported by Angola, Namibia, and Zimbabwe, against rebels backed by Uganda and Rwanda. In April 2003, President Laurent Kabila signed a new constitution, under which an interim government would rule for two years, pending elections in 2005. The constitution was drawn up at talks in South Africa between D.R.C.'s warring factions. In May 2003, the last Ugandan troops left eastern D.R.C. as reports of bloody clashes between rival militias in Bunia emerged. The French arrived in Bunia to fill the vacuum left by the Ugandans as part of a UN-mandated rapid-reaction force.

D.R.C. leaders of the main former rebel groups were sworn in as vice presidents in July 2003 with an interim parliament inaugurated in August 2003.

By December 2003, former government soldiers of the Mobutu regime and those of two main rebel groups formed a united force. This led in March 2004 to gunmen attacking military bases in Kinshasa in an apparent coup attempt. In June 2004, rebel soldiers occupied the eastern border town

Timeline: PAST

1879
Leopold sets up the Congo Independent State as his private kingdom

1906
Congo becomes a Belgian colony

1960
Congo gains independence; civil war begins; a UN force is involved; Patrice Lumumba is murdered

1965
Joseph Desiré Mobutu takes command in a bloodless coup

1971
The name of the state is changed to Zaire

1990s
Central authority crumbles; millions of Rwandan and Burundian refugees flood into Zaire; Mobutu is overthrown

PRESENT

2000s
Civil war continues, but hopes of a permanent cease-fire have emerged

Laurent Kabila is assassinated and replaced by his son, Joseph Kabila of Bukavu for a week. The government accused Rwanda of supporting the rebels. Popular protests in Kinshasa over the UN's failure to prevent the town's capture turned violent. A second reported coup attempt by rebel guards was neutralized in June 2004.

The political and military situation in D.R.C. is stable. The fact that there were mass popular protests in support of the transitional government against the slow speed of the UN's response and opposing a coup, indicates that there is a certain degree of maturity amongst the population. The people of the D.R.C. will not stand for radical changes brought about by violence from any minority group. Moreover, without majority participation in some form of democratic election, no government can expect to achieve a mandate to rule the country.

The war claimed an estimated three million lives, either as a direct result of fighting or because of disease and malnutrition. It has been called possibly the worst emergency to unfold in Africa in recent decades. The war had an economic as well as a political side. Fighting was fueled by the country's vast mineral wealth, with all sides taking advantage of the anarchy to plunder its natural resources. Former rebels have joined a power-sharing government, and general elections are planned for 2005.

Equatorial Guinea (Republic of Equatorial Guinea)

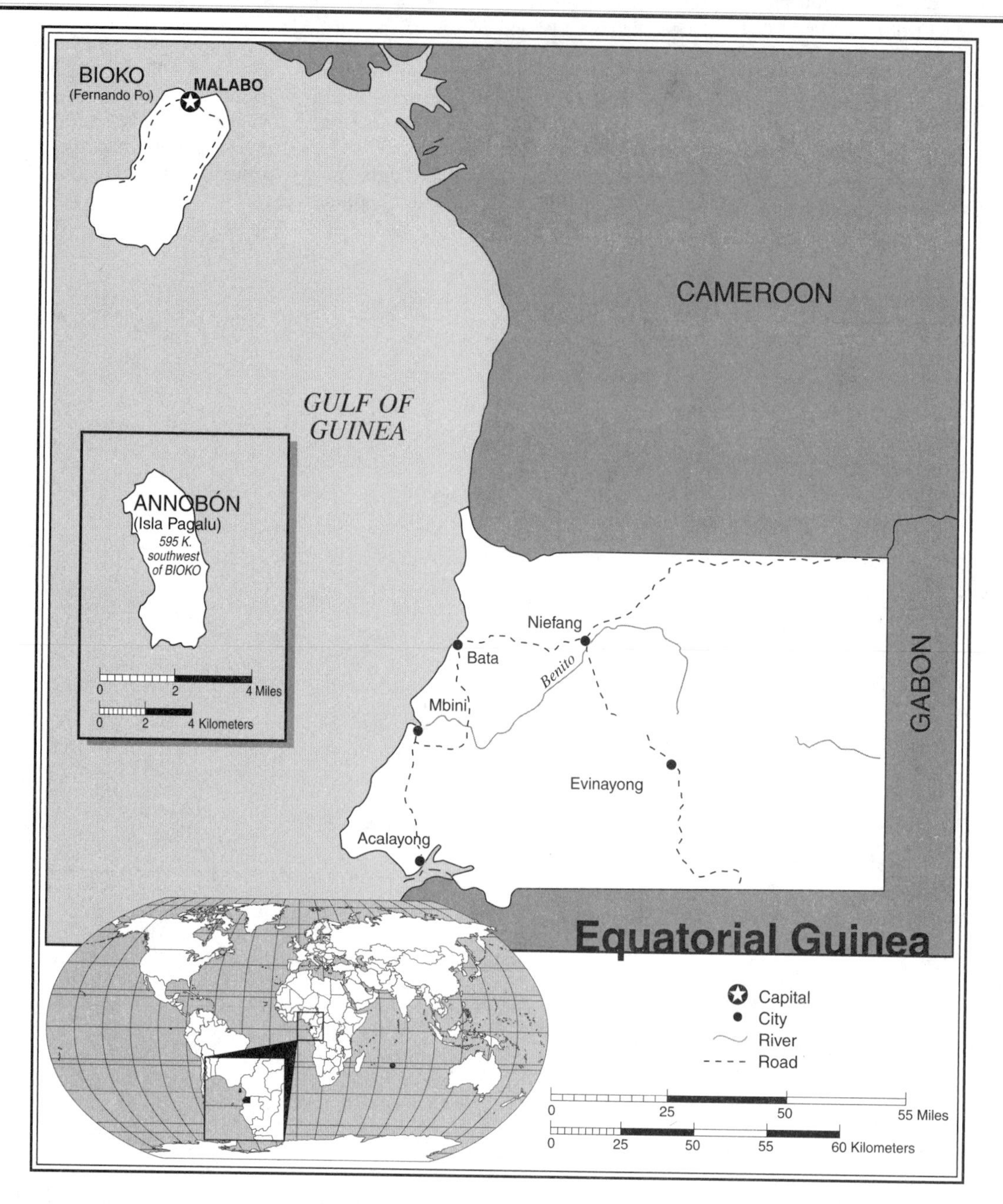

Equatorial Guinea Statistics

GEOGRAPHY

Area in Square Miles (Kilometers): 10,820 (28,023) (about the size of Maryland)
Capital (Population): Malabo (58,000)
Environmental Concerns: desertification; lack of potable water
Geographical Features: coastal plains rise to interior hills; volcanic islands; mainland portion and 5 inhabited islands
Climate: tropical

PEOPLE

Population

Total: 523,051
Annual Growth Rate: 2.47%
Rural/Urban Population Ratio: 57/43
Major Languages: Spanish; French; Bubi; Fang; Ibo
Ethnic Makeup: primarily Bioko, Rio Muni; fewer than 1,000 Europeans
Religions: nominally Christian, predominantly Roman Catholic; indigenous beliefs

Health

Life Expectancy at Birth: 52 years (male); 56 years (female)
Infant Mortality: 91/1,000 live births
Physicians Available: 1/3,532 people
HIV/AIDS Rate in Adults: 3.4%

Education

Adult Literacy Rate: 78.5%
Compulsory (Ages): 6–11; free

COMMUNICATION

Telephones: 4,000 main lines
Televisions: 88/1,000 people
Internet Users: 1,800 (2002)

TRANSPORTATION

Highways in Miles (Kilometers): 1,786 (2,880)
Railroads in Miles (Kilometers): none
Usable Airfields: 3
Motor Vehicles in Use: 7,600

GOVERNMENT

Type: republic
Independence Date: October 12, 1968 (from Spain)
Head of State/Government: President (Brigadier General) Teodoro Obiang Nguema Mbasogo; Prime Minister Miguel Abia Biteo Borico
Political Parties: Democratic Party for Equatorial Guinea; Progressive Democratic Alliance; Popular Action of Equatorial Guinea; Convergence Party for Social Democracy; others
Suffrage: universal at 18

MILITARY

Military Expenditures (% of GDP): 2.5%
Current Disputes: maritime boundary disputes with Cameroon, Gabon, and Nigeria

ECONOMY

Currency ($ U.S. Equivalent): 581.2 CFA francs = $1
Per Capita Income/GDP: $2,100/$1.04 billion
GDP Growth Rate: 6%
Inflation Rate: 6%
Unemployment Rate: 30%
Natural Resources: timber; petroleum; gold; manganese; uranium
Agriculture: cocoa; coffee; timber; rice; yams; cassava; bananas; palm oil; livestock
Industry: fishing; sawmilling; petroleum; natural gas
Exports: $2.1 billion (primary partners China, Japan, United States)
Imports: $736 million (primary partners United States, France, Spain)

SUGGESTED WEB SITE

`http://www.sas.upenn.edu/African_Studies/E-Guinea.html`

Equatorial Guinea Country Report

Few countries have been more consistently misruled than Equatorial Guinea. Having been traumatized during its first decade of independence by the sadistic Macias Nguema (1968–1979), the country continues to decay under his nephew and former security chief, Obiang Nguema Mbasogo.

In June 2002, the country's High Court handed 68 people prison sentences of up to 20 years for an alleged coup plot against President Obiang Nguema. Those jailed included a main opposition leader, Placido Mico Abogo. The European Union noted with concern that alleged "confessions" from among the accused seemed to have been obtained under duress.

DEVELOPMENT

The exploitation of oil and gas by U.S., French, and Spanish companies should greatly increase government revenues. The U.S. company Walter International recently finished work on a gas-separation plant.

Many observers look upon the trial as another betrayal of repeated promises of political reform. Since the last (1999) elections, which were characterized by widespread intimidation, including arrest and torture against the opposition, Obiang has sought to encourage exiled opponents to return to Equatorial Guinea. A few have returned to register their parties for elections, scheduled for 2003, and eight of the principal opposition groups have formed a coalition to try to remove Obiang from power. In response, the president appears to be trying to distance himself from his own government. In 2001, he replaced his cabinet for failing to "respect the majority opinion of the people," immediately prior to the formation of the opposition's coalition of eight parties.

Obiang officially transformed his regime into a multiparty democracy back in 1992. But this gesture is now dismissed as a thinly disguised sham for the benefit of the French, Spanish, and Americans who provide assistance to his regime. Outside interest in the country is focused on its newfound oil wealth, which in recent years has fueled economic growth that has so far been of little benefit to ordinary people.

As a result of the government's failure to honor its commitments, all the significant opposition groups boycotted the 1993 elections, describing them as a farce. Subsequent opposition attempts to come to an accommodation with the government were set back in 1995, when a prominent opposition leader was arrested. In 1996, Obiang claimed 97.85 percent of the vote in a new presidential poll.

In December 2002, President Obiang Nguema was unanimously re-elected, with 100 percent of the vote. The opposition leaders had pulled out of the poll, citing fraud and irregularities. However, by August 2003, exiled opposition leaders had formed a self-proclaimed government-in-exile with their base in Madrid, Spain. This led to the government releasing opposition leader Placido Mico Abogo and 17 other political prisoners. In March 2004, President Obiang said that 15 mercenaries (linked to the suspected mercenaries detained in Zimbabwe at the same time) were responsible for an alleged coup attempt. The trial of the 14 accused coup plotters from South Africa and Armenia took place in April 2004. A crackdown on immigrants followed the alleged coup attempt, with hundreds of foreigners deported.

FREEDOM

In September 1998, Amnesty International cautiously welcomed a decree by President Obiang Nguema Mbasogo commuting the death sentences of 15 political opponents, including 4 exiles judged in absentia, who had been convicted in a summary trial the previous June. The reprieves were considered a vindication of those arguing for the continued need to put international pressure on the regime.

In April 2004, during the parliamentary and municipal elections, President Obiang's ruling Democratic Party of Equatorial Guinea (PDGE) and allied parties took 98 of 100 seats in parliament, and all but seven of 244 municipal posts. Foreign independent election observers criticized both the poll and election result. Indeed, the ruling party and its leader have been able to garner some of the highest voting figures in the history of world democratic politics. One certainly must look closely at election results of 100 percent for the president and 98 percent for parliamentary and municipal candidates from the ruling party—these results show either an amazing degree of popularity for the man and his party, or else something is obviously desperately wrong in Equatorial Guinea. In most countries throughout the world, even announcing such electoral results would bring the population protesting into the streets, with mass riots and police killings resulting.

Equatorial Guinea's current suffering contrasts with the mood of optimism that characterized the country when it gained its independence from Spain in 1968. Confidence was then buoyed by a strong and growing gross domestic product, potential mineral riches, and exceptionally good soil.

HEALTH/WELFARE

At independence, Equatorial Guinea had one of the best doctor-to-population ratios in Africa, but Macias's rule left it with one of the lowest. Health care is gradually reviving, however, with major assistance coming from public and private sources.

The republic is comprised of two small islands, Fernando Po (now officially known as Bioko) and Annobón, and the larger and more populous coastal enclave of Rio Muni. Before the two islands and the enclave were united, during the 1800s, as Spain's only colony in sub-Saharan Africa, all three areas were victimized by their intense involvement in the slave trade.

Spain's major colonial concern was the prosperity of the large cocoa and coffee plantations that were established on the islands, particularly on Fernando Po. Because of resistance from the local Bubi, labor for these estates was imported from elsewhere in West Africa. Coercive recruitment and poor working conditions led to frequent charges of slavery.

Despite early evidence of its potential riches, Rio Muni was largely neglected by the Spanish, who did not occupy its interior until 1926. In the 1930s and 1940s, much of the enclave was under the political control of the Elar-ayong, a nationalist movement that sought to unite the Fang, Rio Muni's principal ethnic group, against both the Spanish and the French rulers in neighboring Cameroon and Gabon. The territory has remained one of the world's least developed areas.

In 1968, then–Fascist-ruled Spain entrusted local power to Macias Nguema, who had risen through the ranks of the security service. Under his increasingly deranged misrule, virtually all public and private enterprise collapsed; indeed, between 1974 and 1979, the country had no budget. One third of the nation's population went into exile; tens of thousands of others were either murdered or allowed to die of disease and starvation. Many of the survivors were put to forced labor, and the rest were left to subsist off the land. Killings were carried out by boys conscripted between the ages of seven and 14.

Although no community in Equatorial Guinea was left unscarred by Macias's tyranny, the greatest disruption occurred on the islands. By 1976, the entire resident-alien population had left, along with most surviving members of the educated class. On Annobón, the government blocked all international efforts to stem a severe cholera epidemic in 1973. The near-total depopulation of the island was completed in 1976, when all able-bodied men on Annobón, along with another 20,000 from Rio Muni, were drafted for forced labor on Fernando Po.

ACHIEVEMENTS

At independence, 90% of all children attended school, but the schools were closed under Macias. Since 1979, primary education has revived and now incorporates most children.

If Equatorial Guinea's first decade of independence was hell, the years since have at best been purgatory. No sector of the economy is free of corruption. Uncontrolled—and in theory illegal—logging is destroying Rio Muni's environment, while in Malabo, the police routinely engage in theft. Food is imported and malnutrition commonplace. It has been reported that the remaining population of Annobón is being systematically starved while Obiang Nguema Mbasogo collects huge payments from international companies that use the island as a toxic-waste dump.

At least one fifth of the Equato-Guinean population continue to live in exile, mostly in Cameroon and Gabon. This community has fostered a number of opposition groups. The government relies financially on French and Spanish aid. But Madrid's commitment has been strained by criticism from the Spanish press, which has been virtually alone in publicizing the continued suffering of Equatorial Guinea's people.

Timeline: PAST

1500s
Europeans explore modern Equatorial Guinea

1641
The Dutch establish slave-trading stations

1778
Spain claims the area of Equatorial Guinea; de facto control is not completed until 1926

1930
The League of Nations investigates charges of slavery on Fernando Po

1958
The murder of nationalist leader Acacio Mane leads to the founding of political parties

1963
Local autonomy is granted

1968
Independence; Macias Nguema begins his reign

1979
A coup ends the dictatorial regime of Macias Nguema; Obiang Nguema Mbasogo becomes the new ruler

1990s
A shift to multipartyism is accompanied by wave of political detentions; Obiang Nguema Mbasogo claims electoral victory

PRESENT

2000s
The exploitation of large oil reserves boosts the economy, but ordinary people still suffer

Gabon (Gabonese Republic)

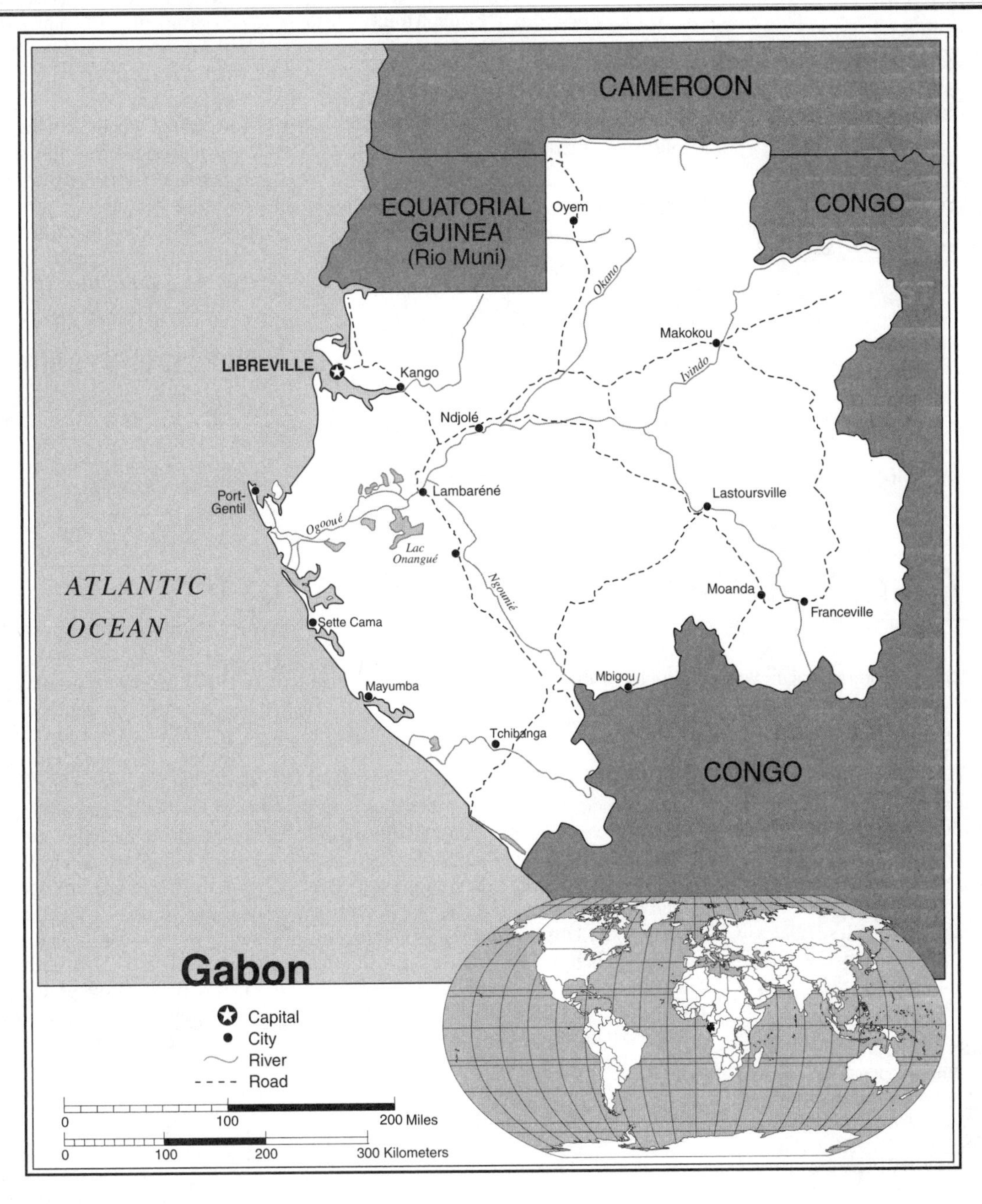

Gabon Statistics

GEOGRAPHY

Area in Square Miles (Kilometers): 102,317 (264,180) (about the size of Colorado)

Capital (Population): Libreville (573,000)

Environmental Concerns: deforestation; poaching

Geographical Features: narrow coastal plain; hilly interior; savanna in the east and south

Climate: tropical

PEOPLE

Population

Total: 1,355,246

Annual Growth Rate: 2.5%

Rural/Urban Population Ratio: 19/81

Major Languages: French; Fang; Myene; Eshira; Bopounou; Bateke; Bandjabi

Ethnic Makeup: about 95% African, including Eshira, Fang, Bapounou, and Bateke; 5% European

Religions: 55%–75% Christian; less than 1% Muslim; remainder indigenous beliefs

Health

Life Expectancy at Birth: 48 years (male); 50 years (female)

Infant Mortality: 93.5/1,000 live births

Physicians Available: 1/2,337 people

HIV/AIDS Rate in Adults: 8.1%

Education

Adult Literacy Rate: 63.2%
Compulsory (Ages): 6–16

COMMUNICATION

Telephones: 39,000 main lines
Televisions: 35/1,000 people
Internet Users: 35,000 (2002)

TRANSPORTATION

Highways in Miles (Kilometers): 4,650 (7,500)
Railroads in Miles (Kilometers): 402 (649)
Usable Airfields: 59
Motor Vehicles in Use: 33,000

GOVERNMENT

Type: republic; multiparty presidential regime
Independence Date: August 17, 1960 (from France)
Head of State/Government: President El Hadj Omar Bongo; Prime Minister Jean-Francois Ntoutoume-Emane
Political Parties: Gabonese Democratic Party; Gabonese Party for Progress; National Woodcutters Rally; others
Suffrage: universal at 21

MILITARY

Military Expenditures (% of GDP): 2%
Current Disputes: maritime boundary dispute with Equatorial Guinea

ECONOMY

Currency ($ U.S. equivalent): 529.43 CFA francs = $1
Per Capita Income/GDP: $5,500/$6.7 billion
GDP Growth Rate: 2.5%
Inflation Rate: 1.5%
Unemployment Rate: 21%
Labor Force by Occupation: 60% agriculture; 25% services and government; 15% industry and commerce
Natural Resources: petroleum; iron ore; manganese; uranium; gold; timber; hydropower
Agriculture: cocoa; coffee; palm oil
Industry: petroleum; lumber; mining; chemicals; ship repair; food processing; cement; textiles
Exports: 2.5 billion (primary partners United States, France, China)
Imports: $921 million (primary partners France, Côte d'Ivoire, United States)

SUGGESTED WEB SITES

http://www.gabonnews.com
http://www.sas.upenn.edu/African_Studies/Country_Specific/Gabon.html
http://www.presidence-gabon.com/index-a.html

Gabon Country Report

Since independence, Gabon has achieved one of the highest per capita gross domestic products in Africa, due to exploitation of the country's natural riches, especially its oil. But there is a wide gap between such statistical wealth and the real poverty that still shapes the lives of most Gabonese. Disparities in income have helped fuel crime. In a controversial response, in July 2002 the government razed four village suburbs of the capital city Libreville, saying that the areas had become havens for foreign criminal gangs.

DEVELOPMENT

The Trans-Gabonais Railway is one of the largest construction projects in Africa. Work began in 1974 and, after some delays, most of the line is now complete. The railway has opened up much of Gabon's interior to commercial development. The Chinese have become very important foreign investors in Gabon in recent years. In April 2004 the French oil firm Total signed a contract to export Gabonese oil to China. In September 2004 yet another agreement was signed with a Chinese company, this time to exploit iron ore. Both contracts represent a major boost for the Gabonese economy as they will generate thousands of jobs.

Guided by only two presidents since independence from France in 1960, Gabon has proven to be one of the most stable nations in Africa. With more than 40 tribal groups in the nation the country has managed to avoid the ethnic violence that has frequently wrecked havok on many of the other African states. At the top of the local governing elite is President Omar Bongo, whose main palace, built a decade ago at a reported cost of $300 million, symbolizes his penchant for grandeur. Shortly after taking office in 1967, Bongo institutionalized his personal rule as the head of a one-party state. Until recently, his Democratic Party of Gabon (PDG) held a legal monopoly of power. But, although the PDG's Constitution restricted the presidency to the "Founder President," for many years it was Gabon's former colonial master, France, not the ruling party's by-laws, that upheld the Bongo regime. After a landslide victory for the Gabonese Democratic Party in January 2002, parliment changed the constitution to allow President Bongo to run for office as many times as he wishes.

The French colonial presence in Gabon dates back to 1843. Between 1898 and 1930, many Gabonese were subject to long periods of forced labor, cutting timber for French concessions companies. World War II coincided with a period of political liberalization in the territory under the Free French government of Felix Emboue, a black man born in French Guiana. Educated Gabonese were promoted for the first time to important positions in the local administration. In the 1950s, two major political parties emerged to compete in local politics: the Social Democratic Union of Gabon (UDSG), led by Jean-Hilaire Aubame; and the Gabonese Democratic Bloc (BDG) of Indjenjet Gondjout and Leon M'ba.

In the 1957 elections, the UDSG received 60 percent of the popular vote but gained only 19 seats in the 40-seat Assembly. Leon M'ba, who had the support of French logging interests, was elected leader by 21 BDG and independent deputies. As a result, it was M'ba who was at the helm when Gabon gained its independence, in 1960. This birth coincided with M'ba's declaration giving himself emergency powers, provoking a period of prolonged constitutional crisis.

FREEDOM

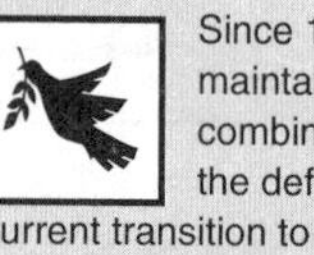

Since 1967 Bongo has maintained power through a combination of repression and the deft use of patronage. The current transition to a multiparty process, however, has led to an improvement in human rights.

In January 1964, M'ba dissolved the Assembly over its members' continued refusal to accept a one-party state under his leadership. In February, the president himself was forced to resign by a group of army officers. Power was transferred to a civilian "Provisional Government," headed by Aubame, which also included BDG politicians such as Gondjout and several prominent, unaffiliated citizens. However, no sooner had the Provisional Government been installed than Gabon was invaded by French troops. Local military units were massacred in the surprise attack, which returned M'ba to office. Upon his death, M'ba was succeeded by his hand-picked successor, Omar Bongo.

HEALTH/WELFARE

The government claims to have instituted universal, compulsory education for Gabonese up to age 16. Independent observers doubt the government's claim but concur that major progress has been made in education. Health services have also expanded greatly.

It has been suggested that France's 1964 invasion was motivated primarily by a desire to maintain absolute control over Gabon's uranium deposits, which were then vital to France's nuclear-weapons program. Many Gabonese have since believed that their country has remained a de facto French possession. France has maintained its military presence, and the Presidential Guard, mainly officered by Moroccan and French mercenaries, outguns the Gabonese Army. France dominates Gabon's resource-rich economy.

In recent decades, Gabon's status quo has been challenged by its increasingly urbanized population. Although Bongo was able to co-opt or exile many of the figures who had once opposed M'ba, a new generation of opposition has emerged both at home and in exile. During the 1980s, the underground Movement for National Recovery (MORENA) emerged as the leading opposition group. In 1989, Bongo began talks with some elements within MORENA, which led to a division within its ranks. But the breakup of MORENA failed to stem the emergence of new groups calling for a return to multiparty democracy.

Demonstrations and strikes at the beginning of 1990 led to the legalization of opposition parties. But the murder of a prominent opposition leader in May led to serious rioting at Port-Gentil, Gabon's second city. In response, France sent troops to the area. Multiparty elections for the National Assembly, in September–October 1990, resulted in a narrow victory for the PDG, amid allegations of widespread fraud. In 1992, most opposition groups united as the Coordination of Democratic Opposition. Bongo's victory claim in the December 1993 presidential elections was widely disbelieved. In September 1994, he agreed to the formation of a coalition "Transitional Government" and the drafting of a new Constitution, which was approved by 96 percent of the voters in July 1995.

ACHIEVEMENTS

Gabon will soon have a second private television station, funded by a French cable station. Profits will be used to fund films that will be shown on other African stations. Gabon's first private station is funded by Swiss and Gabonese capital.

In 1996, the PDG won a sweeping, but controversial, victory in parliamentary elections. This was followed up in 1998 by a landslide reelection victory for Bongo. The PDG's continuing hold on power was reconfirmed in the 2002 parliamentary elections. Following the poll, the National Woodcutters Rally, a party that has strong support in Libreville, agreed to take up Bongo's offer to serve as junior partners with the PDG in Gabon's first coalition government. Although opposition politicians claimed that the results had been rigged, independent observers have credited the PDG with success in retaining support of the rural base while co-opting opponents into its fold.

Timeline: PAST

1849
Libreville is founded by the French as a settlement for freed slaves

1910
Gabon becomes a colony within French Equatorial Africa

1940
The Free French in Brazzaville seize Gabon from the pro-Vichy government

1960
Independence is gained; Leon M'ba becomes president

1967
Omar Bongo becomes Gabon's second president after M'ba's death

1968
The Gabonese Democratic Party (PDG) becomes the only party of the state

1990s
Bongo agrees to multiparty elections but seeks to put limits on the opposition; riots in Port-Gentil

PRESENT

2000s
The PDG retains power

São Tomé and Príncipe (Democratic Republic of São Tomé and Príncipe)

São Tomé and Príncipe Statistics

GEOGRAPHY

Area in Square Miles (Kilometers): 387 (1,001) (about 5 times the size of Washington, D.C.)

Capital (Population): São Tomé (67,000)

Environmental Concerns: deforestation; soil erosion; soil exhaustion

Geographical Features: volcanic; mountainous; islands in the Gulf of Guinea; the smallest country in Africa

Climate: tropical

PEOPLE

Population

Total: 181,565

Annual Growth Rate: 3.1%

Rural/Urban Population Ratio: 54/46

Major Languages: Portuguese; Fang; Kriolu

Ethnic Makeup: Portuguese-African mixture; African minority

Religions: 80% Christian; 20% others

Health

Life Expectancy at Birth: 64 years (male); 67 years (female)
Infant Mortality: 47.5/1,000 live births
Physicians Available: 1/1,881 people

Education

Adult Literacy Rate: 73%
Compulsory (Ages): for 4 years between ages 7–14

COMMUNICATION

Telephones: 3,000 main lines
Televisions: 154/1,000 people
Internet Users: 15,000 (2002)

TRANSPORTATION

Highways in Miles (Kilometers): 198 (320)
Railroads in Miles (Kilometers): none
Usable Airfields: 2

GOVERNMENT

Type: republic
Independence Date: July 12, 1975 (from Portugal)
Head of State/Government: President Fradique de Menezes; Prime Minister Damiao Vaz De Almeida
Political Parties: Party for Democratic Convergence; Movement for the Liberation of São Tomé and Príncipe–Social Democratic Party; others
Suffrage: universal at 18

MILITARY

Military Expenditures (% of GDP): 0.8%
Current Disputes: none

ECONOMY

Currency ($ U.S. equivalent): 9,347 dobras = $1
Per Capita Income/GDP: $1,200/$189 million
GDP Growth Rate: 4%
Inflation Rate: 7%
Labor Force by Occupation: mainly engaged in subsistence agriculture and fishing
Natural Resources: fish; hydropower
Agriculture: cacao; coconut palms; coffee; bananas; palm kernels; copra
Industry: light construction; textiles; soap; beer; fish processing; timber
Exports: $4.1 million (primary partners Netherlands, Portugal, Spain)
Imports: $40 million (primary partners Portugal, France, United Kingdom)

SUGGESTED WEB SITES

http://www.sao-tome.com
http://www.state.gov/www.background_notes/sao_tome_0397_bgn.html
http://www.emulateme.com/saotome.htm
http://www.sas.upenn.edu/African_Studies/Country_Specific/Sao_Tome.html

São Tomé and Príncipe Country Report

In August 1995, soldiers in the small island-nation of São Tomé and Príncipe briefly deposed Miguel Trovoada, the country's first democratically elected president. The coup quickly collapsed, however, in the face of domestic and international opposition. While the country's new democracy survived, it was vulnerable to a weak economy, which showed little prospect of significant improvement anytime soon.

DEVELOPMENT

Local food production has been significantly boosted by a French-funded plan. Japan is assisting in fishery development. There is concern that tourist fishermen may adversely affect the local fishing industry.

The islands held their first multiparty elections in January 1991. The elections resulted in the defeat of the former ruling party, the Liberation Movement of São Tomé and Príncipe–Social Democratic Party (MLSTP–PSD), by Trovoada's Party for Democratic Convergence–Group of Reflection (PDC–GR). Subsequent elections in December 1992, however, reversed the PDC–GR advantage in Parliament, leading to an uneasy division of power. This division was reinforced with Trovoada's reelection in 1996, followed by an even greater MLSTP–PSD parliamentary victory in 1998.

In the July 2001 presidential elections businessman Fradique de Menezes was declared the winner. He was sworn into office in early September. However, the victory of the opposition MLSTP–PSD party in the March 2002 parliamentary elections led de Menezes to appoint Gabriel da Costa as prime minister, and both main political parties agreed to share power and form a broad-based government. de Menezes is the country's third president after Miguel Trovoada who served two five-year terms, the maximum permitted by the constitution.

The government was confronted with a massive civil-servants strike in 2001 to press for higher pay. Officials said the country's external debt in 1998 amounted to U.S. $270 million, far more than the country's annual gross domestic product.

São Tomé and Príncipe gained its independence in 1975, after a half-millennium of Portuguese rule. During the colonial era, economic life centered around the interests of a few thousand Portuguese settlers, particularly a handful of large-plantation owners who controlled more than 80 percent of the land. After independence, most of the Portuguese fled, taking their skills and capital and leaving the economy in disarray. But production on the plantations has since been revived.

The Portuguese began the first permanent settlement of São Tomé and Príncipe in the late 1400s. Through slave labor, the islands developed rapidly as one of the world's leading exporters of sugar. Only a small fraction of the profits from this boom were consumed locally; and high mortality rates, caused by brutal working conditions, led to an almost insatiable demand for more slaves. Profits from sugar declined after the mid-1500s due to competition from Brazil and the Caribbean. A period of prolonged depression set in.

FREEDOM

Before 1987, human rights were circumscribed in São Tomé and Príncipe. Gradual liberalization has now given way to a commitment to political pluralism. The current government has a good record of respect for human rights. Major problems are an inefficient judicial system, harsh prison conditions, and acts of police brutality. Outdated labor practices on the plantations limit worker rights.

In the early 1800s, a second economic boom swept the islands, when they became leading exporters of coffee, and, more important, cocoa. (São Tomé and Príncipe's position in the world market has since declined, yet these two cash crops, along with copra, have continued to be economic mainstays.) Although slavery was officially abolished during the nineteenth century, forced labor was maintained by the

Portuguese into modern times. Involuntary contract workers, known as *servicais,* were imported to labor on the islands' plantations, which had notoriously high mortality rates. Sporadic labor unrest and occasional incidents of international outrage led to some improvement in working conditions, but fundamental reforms came about only after independence. A historical turning point for the islands was the Batepa Massacre in 1953, when several hundred African laborers were killed following local resistance to labor conditions.

HEALTH/WELFARE

Since independence, the government has had enormous progress in expanding health care and education. The Sãotoméan infant mortality rate is now among the lowest in Africa, and average life expectancy is among the highest. About 65% of the population between 6 and 19 years of age now attend school.

Between 1975 and 1991, São Tomé and Príncipe was ruled by the MLSTP–PSD, which had emerged in exile as the island's leading anticolonial movement, as a one-party state initially committed to Marxist-Leninism. But in 1990, a new policy of *abertura,* or political and economic "opening," resulted in the legalization of opposition parties and the introduction of direct elections with secret balloting. Press restrictions were also lifted, and the nation's security police were purged. The democratization process was welcomed by previously exiled opposition groups, most of which united as the PDC–GR. The changed political climate was also reflected in the establishment of an independent labor movement. Previously, strikes were forbidden.

The move toward multiparty politics was accompanied by an evolution to a market economy. Since 1985, a "Free Trade Zone" was established, state farms were privatized, and private capital was attracted to build up a tourist industry. These moves were accompanied by a major expansion of Western loans and assistance to the islands—an inflow of capital that now accounts for nearly half of the gross domestic product.

The government also focused its development efforts on fishing. In 1978, a 200-mile maritime zone was declared over the tuna-rich waters around the islands. The state-owned fishing company, Empesca, began upgrading the local fleet, which still consists mostly of canoes using old-fashioned nets. The influx of aid and investment has resulted in several years of sustained economic growth.

ACHIEVEMENTS

São Tomé and Príncipe shares in a rich Luso-African artistic tradition. The country is particularly renowned for poets such as Jose de Almeida and Francisco Tenreiro, who were among the first to express in the Portuguese language the experiences and pride of Africans.

The current inhabitants of São Tomé and Príncipe are primarily of mixed African and European descent. During the colonial period, the society was stratified along racial lines. At the top were the Europeans —mostly Portuguese. Just below them were the mesticos or *filhos da terra,* the mixed-blood descendants of slaves. Descendants of slaves who arrived later were known as *forros.* Contract workers were labeled as *servicais,* while their children became known as *tongas.* Still another group was the *angolares,* who reportedly were the descendants of shipwrecked slaves. All of these colonial categories were used to divide and rule the local population; the distinctions have begun to diminish, however, as an important sociological factor on the islands.

THE FUTURE

The nation of São Tomé and Principe is precariously balanced between financial stability and social chaos. The stability arises from the recent discovery and exploitation of large offshore oil fields. In August 2002 President Fraque de Menezes announced plans for a U.S. naval base in the country, which would aim to protect São Tomé's oil interests. Yet the social chaos still lingers and the possibility of it rising up from the ranks of the army still exists.

On July 16, 2003 a military coup toppled the government. The coup occurred largely as a result of the yet to be realized oil wealth and its future distribution. President de Menezes, who was in Nigeria at the time, returned to São Tomé a week later, after an agreement had been reached with the junta. Although a general amnesty was given to the coup leaders, conditions have not improved so the threat of yet another coup still lingers.

By October 2003 oil companies bid for offshore oil blocs controlled by São Tomé and Nigeria, which are expected to generate hundreds of millions of dollars in licensing money for São Tomé. During March 2004 the government was on the verge of collapse as a major conflict arose between the president and prime minister over control of the oil deals. Four cabinet ministers were replaced. In September 2004, President de Menezes replaced Prime Minister Gabriel Arcanjo Ferreira da Costa and changed the government Cabinet after a series of corruption scandals. A new Prime Minister, Damiao Vaz De Almeida, was immediately sworn in.

Timeline: PAST

1500s
The Portuguese settle São Tomé and Príncipe

1876
Slavery is abolished, but forced labor continues

1953
The Portuguese massacre hundreds of islanders

1972
Factions within the liberation movement unite to form the MLSTP in Gabon

1975
Independence

1979
Manuel Pinto da Costa deposes and exiles Miguel Trovoada, the premier and former number-two man in the MLSTP

1990s
Economic and political liberalization; multiparty elections

PRESENT

2000s
Fradique de Menezes wins the presidency

East Africa

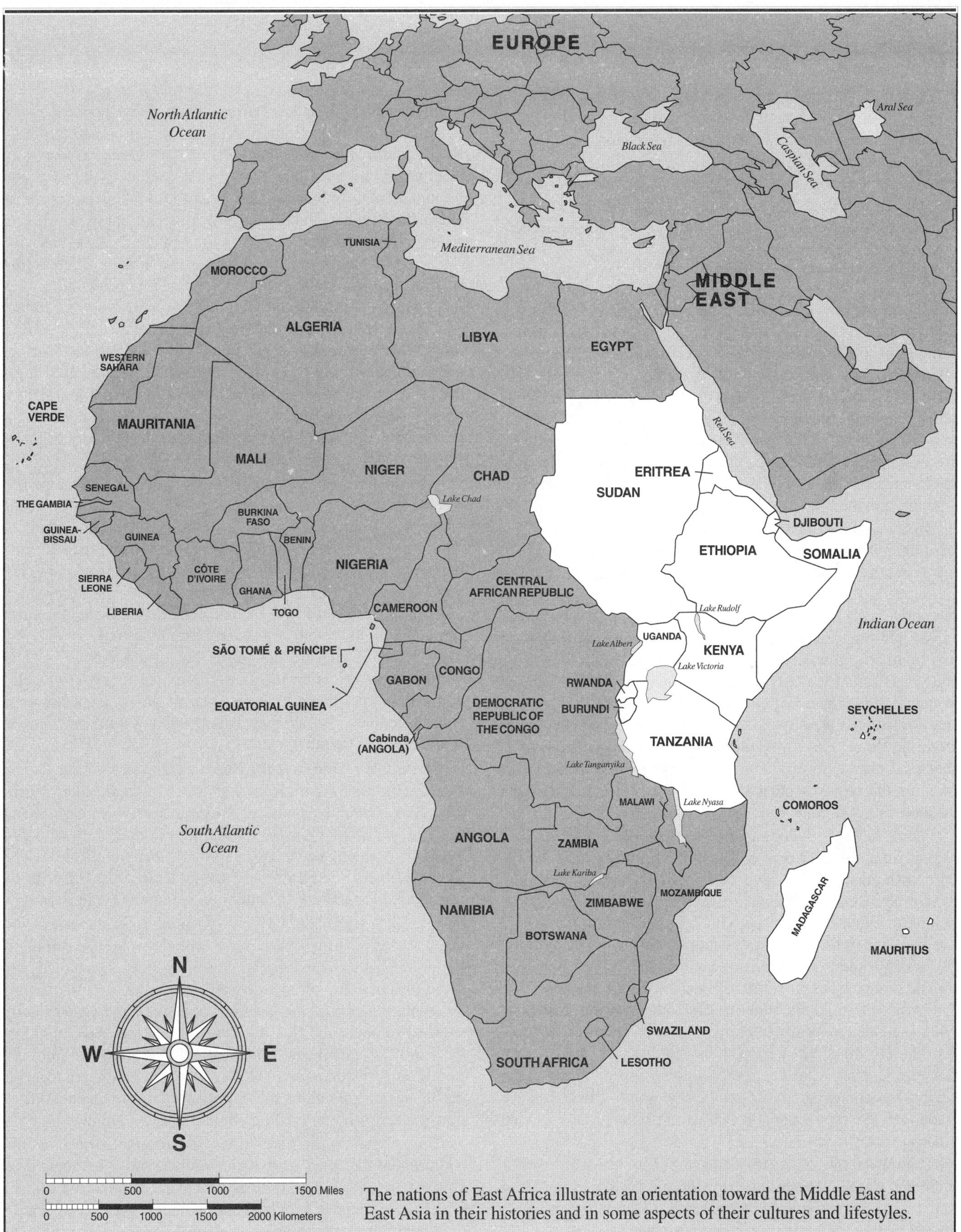

The nations of East Africa illustrate an orientation toward the Middle East and East Asia in their histories and in some aspects of their cultures and lifestyles.

East Africa: A Mixed Inheritance

The vast East African region, ranging from Sudan in the north to Tanzania and the Indian Ocean islands in the south, is an area of great diversity. Although the islands are the homes of distinctive civilizations with ties to Asia, their interactions with the African mainland give their inclusion here validity. Ecological features such as the Great Rift Valley, the prevalence of cattle-herding lifestyles, and long-standing participation in the Indian Ocean trading networks are some of the region's unifying aspects.

CATTLE-HERDING SOCIETIES

A long-horned cow would be an appropriate symbol for East Africa. Most of the region's rural inhabitants, who make up the majority of people from the Horn, to Lake Malawi, to Madagascar, value cattle for their social as well as economic importance. The Nuer of Sudan, the Somalis near the Red Sea (who, like many other peoples of the Horn, herd camels as well as cattle, goats, and sheep), and the Maasai of Tanzania and Kenya are among the pastoral peoples whose herds provide their livelihoods. Farming communities such as the Kikuyu of Kenya, the Baganda of Uganda, and the Malagasy of Madagascar also prize cattle.

Much of the East African landmass is well suited for herding. Whereas the rain forests of West and Central Africa are generally infested with tsetse flies, whose bite is fatal to livestock, most of East Africa is made up of belts of tropical and temperate savanna, which are ideal for grazing. Thus pastoralism has long been predominant in the savanna zones of West and Southern, as well as East, Africa. Tropical rain forests are found in East Africa only on the east coast of Madagascar and scattered along the mainland's coast. Much of the East African interior is dominated by the Great Rift Valley, which stretches from the Red Sea as far south as Malawi. This geological formation is characterized by mountains as well as valleys, and it features the region's great lakes, such as Lake Albert, Lake Tanganyika, and Lake Malawi.

People have been moving into and through the East African region since the existence of humankind; indeed, most of the earliest human fossils have been unearthed in this region. Today, almost all the mainland inhabitants speak languages that belong to either the Bantu or Nilotic linguistic families. There has been much historical speculation about the past migration of these peoples, but recent research indicates that both linguistic groups have probably been established in the area for a long time, although oral traditions and other forms of historical evidence indicate locally important shifts in settlement patterns into the contemporary period. Iron working and, in at least a few cases, small-scale steel production have been a part of the regional economy for more than 2,000 years. Long-distance trade and the production of various crafts have also existed since ancient times.

The inhabitants of the region have had to deal with insufficient and unreliable rainfall. Drought and famine in the Horn and in areas of Kenya and Tanzania have in recent years changed lifestyles and dislocated many people.

ISLAMIC INFLUENCE

Many of the areas of East Africa have been influenced—since at least as far back as Roman times and perhaps much further—by the Middle East and other parts of Asia. Over the past thousand years, most parts of East Africa, including the Christian highlands of Ethiopia and the inland interlake states such as Buganda, Burundi, and Rwanda, became familiar to the Muslim Arab traders of the Swahili and Red Sea coasts and the Sudanese interior. Somalia, Djibouti, and Sudan, which border the Red Sea and are close to the Arabian Peninsula, have been the countries most influenced by Arab Islamic culture. Mogadishu, the capital of Somalia, began as an Islamic trading post in the tenth century A.D. The Islamic faith, its various sects or brotherhoods, the Koran, and the Shari'a (the Islamic legal code) are predominant throughout the Horn, except in the Ethiopian and Eritrean highlands and southern Sudan. In recent years, many Somalis, Sudanese, and others have migrated to the oil-rich states of Arabia to work.

Farther south, in the communities and cultures on the perimeters of the east coast, Arabs and local Bantu-speaking Africans combined, from as early as the ninth century but especially during the 1200s to 1400s, to form the culture and the language that we now call Swahili. In the first half of the nineteenth century, Seyyid Said, the sultan of Oman, transferred his capital to Zanzibar, in recognition of the outpost's economic importance. Motivated by the rapid expansion of trade in ivory and slaves, many Arab–Swahili traders began to establish themselves and build settlements as far inland as the forests of eastern Democratic Republic of the Congo. As a result, some of the noncoastal peoples also adopted Islam, while Swahili developed into a regional lingua franca.

The whole region from the Horn to Tanzania continued to be affected by the slave trade through much of the nineteenth century. Slaves were sent north from Uganda and southern Sudan to Egypt and the Middle East, and from Ethiopia across the Red Sea. Others were taken to the coast by Arab, Swahili, or African traders, either to work on the plantations in Zanzibar or to be transported to the Persian Gulf and the Indian Ocean islands.

In the late 1800s and early 1900s, South Asian laborers from what was then British India were brought in by the British to build the East African railroad. South Asian traders already resided in Zanzibar; others now came and settled in Kenya and Tanzania, becoming shopkeepers and bankers in inland centers, such as Kampala and Nairobi, as well as on the coast, in Mombasa and Dar es Salaam or in smaller stops along the railroad. South Asian laborers were also sent in large numbers to work on the sugar plantations of Mauritius; their descendants there now make up about two thirds of that island's population.

The subregions of East Africa include the following: the countries of the Horn, East Africa proper, and the islands. The Horn includes Djibouti, Ethiopia, Eritrea, Somalia, and Sudan, which are associated here with one another not so much because of a common heritage or on account of any compatibility of

(United Nations photo 131,312 by Ray Witlin)

In the drought-affected areas of East Africa, people must devote considerable time and energy to the search for water.

their governments (indeed, they are often hostile to one another), but because of the movements of peoples across borders in recent times. *East Africa proper* is comprised of Kenya, Tanzania, and Uganda, which do have underlying cultural ties and a history of economic relations, in which Rwanda and Burundi have also shared. The *Indian Ocean islands* include the Comoros, Madagascar, Mauritius, and Seychelles, which, notwithstanding the expanses of ocean that separate them, have certain cultural aspects and current interests in common.

THE HORN

Ethiopia traditionally has had a distinct, semi-isolated history that has separated the nation from its neighbors. This early Christian civilization, which was periodically united by a strong dynasty but at other times was disunited, was centered in the highlands of the interior, surrounded by often hostile lowland peoples. Before the nineteenth century, it was in infrequent contact with other Christian societies. In the 1800s, however, a series of strong rulers reunified the highlands and went on to conquer surrounding peoples such as the Afar, Oromo, and Somali. In the process, the state expanded to its current boundaries. While the empire's expansion helped it to preserve its independence during Africa's colonial partition, sectarian and ethnic divisions—a legacy of the imperial state-building process—now threaten to tear the polity apart.

Ethiopia and the rest of the Horn have been influenced by outside powers, whose interests in the region have been primarily rooted in its strategic location. In the nineteenth century, both Britain and France became interested in the Horn, because the Red Sea was the link between their countries and the markets of Asia. This was especially true after the completion of the Suez Canal in 1869. Both of the imperial powers occupied ports on the Red Sea at the time. They then began to compete over the upper Nile in modern Sudan. In the 1890s, French forces, led by Captain Jean Baptiste Marchand, literally raced from the present-day area of Congo to reach the center of Sudan before the arrival of a larger British expeditionary force, which had invaded the region from Egypt. Ultimately, the British were able to consolidate their control over the entire Sudan.

Italian ambitions in the Horn were initially encouraged by the British, in order to counter the French. Italy's defeat by the Ethiopians at the Battle of Adowa in 1896 did not deter its efforts to dominate the coastal areas of Eritrea and southeastern Somalia. Later, under Benito Mussolini, Italy briefly (1936–1942) occupied Ethiopia itself.

During the Cold War, great-power competition for control of the Red Sea and the Gulf of Aden, with their strategic locations near the oil fields of the Middle East as well as along the Suez shipping routes, continued between the United States and the Soviet Union. Local events sometimes led to shifts in alignments. Before 1977, for instance, the United States was closely allied with Ethiopia, and the Soviet Union with Somalia. However, in 1977–1978, Ethiopia, having come under a self-proclaimed Marxist-Leninist government, allied itself with the

Soviet Union, receiving in return the support of Cuban troops and billions of dollars' worth of Socialist-bloc military aid, on loan, for use in its battles against Eritrean and Somali rebels. The latter group, living in Ethiopia's Ogaden region, were seeking to become part of a greater Somalia. In this irredentist adventure, they had the direct support of invading Somalia troops. Although the United States refused to counter the Soviets by in turn backing the irredentists, it subsequently established relations with the Somali government at a level that allowed it virtually to take over the former Soviet military facility at Berbera.

Discord and Drought

The countries of the Horn, unlike the other states in the region, are politically alienated from one another. There is thus little prospect of an effective regional community emerging among them in the foreseeable future. Although the end of the Cold War has reduced the interest of external powers, local animosities continue to wreak havoc in the region. The Horn continues to be bound together and torn apart by millions of refugees fleeing armed conflicts in all of the states. Ethiopia, Somalia, and Sudan have suffered under especially vicious authoritarian regimes that resorted to the mass murder of dissident segments of their populations. Although the old regimes have been overthrown in Ethiopia and Somalia, peace has yet to come to either society. Having gained its independence only in 1993, Eritrea, Africa's newest nation, has struggled to overcome the devastating legacy of its 30-year liberation struggle against Ethiopia. Eritrea's well-being has been further compromised by reverses in a border war with Ethiopia. Recent battlefield victories against the Eritreans have revived the passions of some Ethiopians who have never fully accepted Eritrea's succession. The stability of neighboring Djibouti, once a regional enclave of calm, has also been compromised in recent years by sometimes violent internal political conflicts.

The horrible effects of these wars have been magnified by recurrent droughts and famines. Hundreds of thousands of people have starved to death in the past decade, while many more have survived only because of international aid efforts.

Ethiopians leave their homes for Djibouti, Somalia, and Sudan for relief from war and famine. Sudanese and Somalis flee to Ethiopia for the same reasons. Today, every country harbors not only refugees but also dissidents from neighboring lands and has a citizenry related to those who live in adjoining countries. Peoples such as the Afar minority in Djibouti often seek support from their kin in Eritrea and Ethiopia. Many Somali guerrilla groups have used Ethiopia as a base, while Somali factions have continued to give aid and comfort to Ethiopia's rebellious Ogaden population. Ethiopian factions allegedly continue to assist southern rebels against the government of Sudan, which had long supported the Tigray and Eritrean rebel movements of northern Ethiopia.

At times, the states of the region have reached agreements among themselves to curb their interference in one another's affairs. But they have made almost no progress in the more fundamental task of establishing internal peace, thus assuring that the region's violent downward spiral continues.

THE SOUTHERN STATES OF EAST AFRICA

The peoples of Kenya, Tanzania, and Uganda as well as Burundi and Rwanda have underlying connections rooted in the past. The kingdoms of the Lakes Region of Uganda, Rwanda, and Burundi, though they have been politically superseded in the postcolonial era, have left their legacies. For example, myths about a heroic dynasty of rulers, the Chwezi, who ruled over an early Ugandan-based kingdom, are widespread. Archaeological evidence attests to the actual existence of the Chwezi, probably in the sixteenth century. Peoples in western Kenya and Tanzania, who have lived under less centralized systems of governance but nonetheless have rituals similar to those of the Ugandan kingdoms, also share the traditions of the Chwezi dynasty, which have become associated with a spirit cult.

The precolonial kingdoms of Rwanda and Burundi, both of which came under German and, later, Belgian control during the colonial era, were socially divided between a ruling warrior class, the Tutsis, and a much larger peasant class, the Hutus. Although both states are now independent republics, their societies remain bitterly divided along these ethnoclass lines. In Rwanda, the feudal hegemony of the Tutsis was overthrown in a bloody civil conflict in 1959, which led to the flight of many Tutsis. But in 1994, the sons of these Tutsi exiles came to power, after elements in the former Hutu-dominated regime organized a genocidal campaign against all Tutsis. In the belief that the Tutsis were back on top, millions of Hutus then fled the country. In Burundi, Tutsi rule was maintained for decades through a repressive police state, which in 1972 and 1988 resorted to the mass murder of Hutus. Elections in 1993 resulted in the country's first Hutu president at the head of a government that included members of both groups, but he was murdered by the predominantly Tutsi army. Since then, the country has been teetering on the brink of yet another catastrophe, as some of its politicians try to promote reconciliation.

Kenya and Uganda were taken over by the British in the late nineteenth century, while Tanzania, originally conquered by Germany, became a British colony after World War I. In Kenya, the British encouraged the growth of a settler community. Although never much more than 1 percent of the colony's resident population, the British settlers were given the best agricultural lands in the rich highlands region around Nairobi; and throughout most of the colonial era, they were allowed to exert a political and economic hegemony over the local Africans. The settler populations in Tanzania and Uganda were smaller and less powerful. While the settler presence in Kenya led to land alienation and consequent immiseration for many Africans, it also fostered a fair amount of colonial investment in infrastructure. As a result, Kenya had a relatively sophisticated economy at the time of its independence, a fact that was to complicate proposals for its economic integration with Tanzania and Uganda.

In the 1950s, the British established the East African Common Services Organization to promote greater economic cooperation among its Kenyan, Tanganyikan (Tanzanian), and Ugandan territories. By the early 1960s, the links among the states were so close that President Julius Nyerere of Tanzania proposed that his country delay its independence until Kenya also gained its freedom, in hopes that the two countries would then join together. This did not occur.

In 1967, the Common Services Organization was transformed by its three (now independent) members into a full-fledged "common market," known as the East African Community (EAC). The EAC collectively managed the railway system, development of harbors, and international air, postal, and telecommunication facilities. It also maintained a common currency, development bank, and other economic, cultural, and scientific services. Peoples moved freely across the borders of the three EAC states. However, the EAC soon began to unravel, as conflicts over its operations grew. It finally collapsed in 1977. The countries disputed the benefits of the association, which seemed to have been garnered primarily by Kenya. The ideologies and personalities of its leaders at the time—Nyerere, Jomo Kenyatta of Kenya, and Idi Amin of Uganda—differed greatly. Relations between Kenya and Tanzania deteriorated to the point that the border between them was closed for several years.

In 1984, Kenya, Tanzania, and Uganda signed an "East African Mediation Agreement," which allowed for the division of the EAC's assets and liabilities, along with the reopening of the Kenya–Tanzania border. This final chapter of the old Community laid the groundwork for renewed cooperation, which ultimately, in late 2001, led to the EAC's reestablishment in a lavish ceremony at Arusha, Tanzania.

By the end of the 1980s, the value of the Community to the three economies has become clear. But political factors continued to complicate the quest for integration. In 1986, Kenya and Tanzania, along with Rwanda, Burundi, Sudan, and then Zaire (today the Democratic Republic of the Congo) pledged to prevent their territories from being used by exiles seeking to destabilize their neighbors. While this broader agreement went unenforced, political relations among Kenya, Tanzania, and Uganda (which in 1986 came under the control of Yoweri Museveni's National Resistance Movement, after years of suffering under the brutal regimes of Amin and Milton Obote) began to improve.

From 1981, the three states were also linked in a loose nineteen-member state "Preferential Trade Area" for southern and eastern Africa. This body laid the basis for further cooperation in the areas of security, trade, and joint hydroelectric projects. Although members of the Economic Community of Central African States, Rwanda and Burundi were also linked with Kenya, Tanzania, and Uganda as a subregion of the UN Commission for Africa.

In 1993, the three states established a "Permanent Tripartite Commission" to look into reviving the East African Community. By then the leaders of all three countries—Daniel Mkapa of Tanzania, Daniel Arap Moi of Kenya, and Museveni—had been implementing confidence-building measures. The 1993 agreement had the goal of establishing a common market and currency zone for the region. But both Tanzania and Uganda were reluctant to move forward due to Kenya's continued industrial advantages. The 2001 treaty has allayed these concerns by dropping a strict time frame for the removal of trade restrictions.

While full economic and political union for the EAC members (who are likely to be expanded to include Rwanda, Burundi, and perhaps Ethiopia) remains a long-term goal, some important structures have already been put in place: the East African Court of Justice, the East African Legislative Assembly, and the Secretariat. In 1998, a common East African passport was introduced that allows citizens of the three nations to cross one another's borders freely. Progress is also reportedly being made toward free currency convertibility, reduced tariffs, and in the areas of defense and foreign policy. All of these steps have generally been greeted with popular support. As the Tanzanian statesman Salim Salim noted: "You can choose a friend but you cannot choose a brother.... In this case Kenyans and Ugandans are our brothers."

THE ISLANDS

The Comoros, Madagascar, Mauritius, and Seychelles each have their own unique characteristics. They all have some important traits in common. All four island nations have been strongly influenced historically by contacts with Asia as well as with mainland Africa and Europe. Madagascar and the Comoros have populations that originated in Indonesia and the Middle East as well as in Africa; the Malagasy language is related to Indonesian Malay. The citizens of Mauritius and Seychelles are of European as well as African and Asian origin.

All four island groups have also been influenced by France. Mauritius and Seychelles were not permanently inhabited until the 1770s, when French settlers arrived with their African slaves. The British subsequently took control of these two island groups and, during the 1830s, abolished slavery. Thereafter the British encouraged migration from South Asia and, to a lesser extent, from China to make up for labor shortages on the islands' plantations. Local French-based creoles remain the major languages on the islands.

In 1978, all the islands, along with opposition groups from the French possession of Réunion, formed the Indian Ocean Commission. Originally a body with a socialist orientation, the commission campaigned for the independence of Réunion and the return of the island of Diego Garcia by Britain to Mauritius, as well as the dismantling of the U.S. naval base located there. By the end of the 1980s, however, the export-oriented growth of Mauritius and the continuing prosperity of Seychelles' tourist-based economy were helping to push all nations toward a greater emphasis on market economics in their multilateral, as well as internal, policy initiatives. Madagascar and the Comoros have recently offered investment incentives for Mauritius-based private firms. Mauritians have also played prominent roles in the development of tourism in the Comoros.

In addition to their growing economic ties, the Comoros and Mauritius, and to a somewhat lesser extent, Madagascar and Seychelles, have created linkages with South Africa. In 1995, Mauritius followed South Africa's lead to become the 12th member of the Southern African Development Community (SADC).

Burundi (Republic of Burundi)

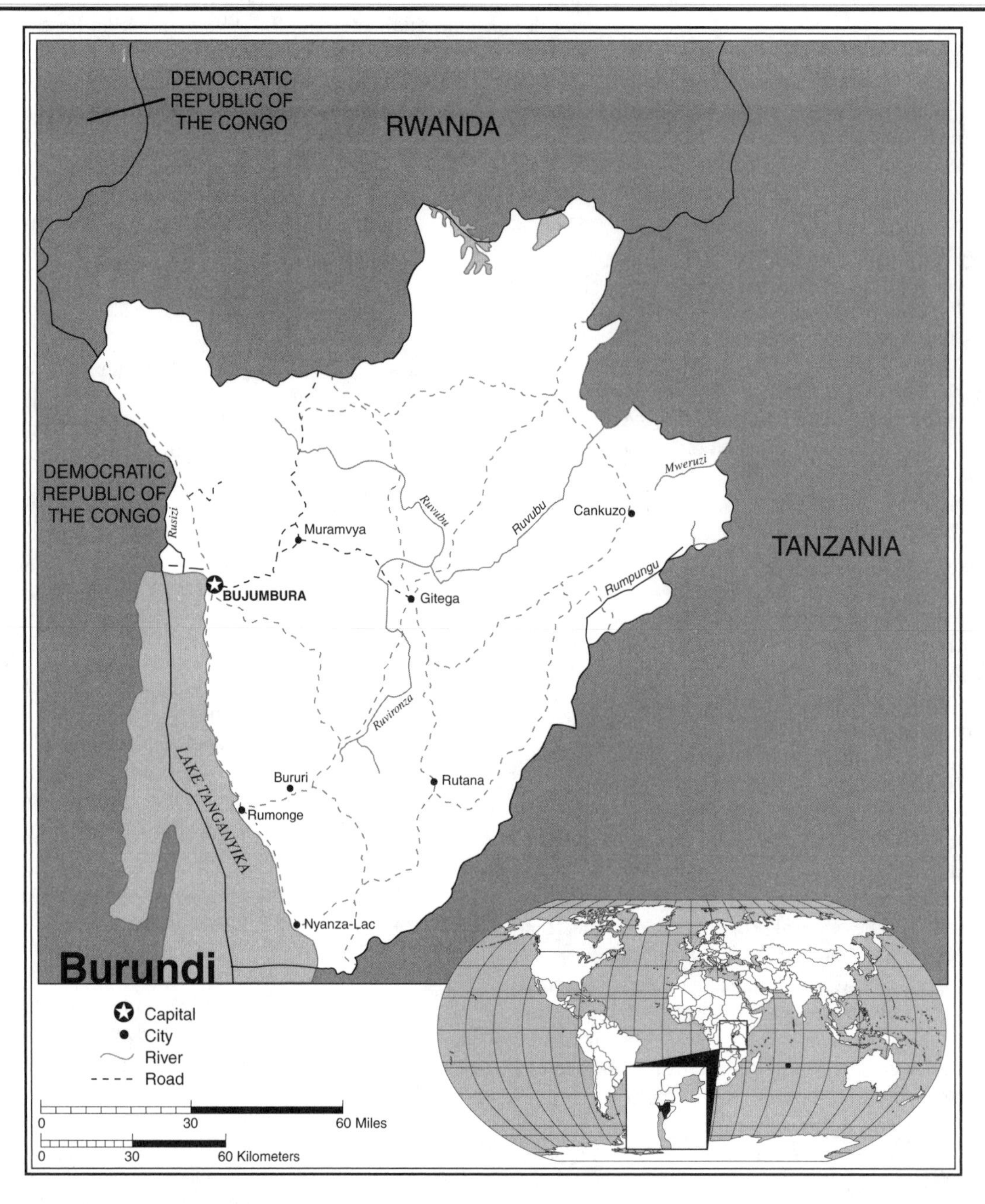

Burundi Statistics

GEOGRAPHY

Area in Square Miles (Kilometers): 10,759 (27,834) (about the size of Maryland)

Capital (Population): Bujumbura (346,000)

Environmental Concerns: soil erosion; deforestation; habitat loss

Geographical Features: hilly and mountainous, dropping to a plateau in the east; some plains; landlocked

Climate: tropical to temperate; temperature varies with altitude

PEOPLE

Population

Total: 6,373,000

Annual Growth Rate: 2.36%

Rural/Urban Population Ratio: 92/8

Major Languages: Kirundi; French; Swahili; others

Ethnic Makeup: 85% Hutu; 14% Tutsi; 1% Twa and others

Religions: 67% Christian; 23% indigenous beliefs; 10% Muslim

Health

Life Expectancy at Birth: 45 years (male); 47 years (female)

Infant Mortality: 69.7/1,000 live births

Physicians Available: 1/31,777 people

HIV/AIDS Rate in Adults: 11.32%

Education

Adult Literacy Rate: 35.3%
Compulsory (Ages): 7–13; free

COMMUNICATION

Telephones: 23,900 main lines
Televisions: 7/1,000 people
Internet Users: 14,000 (2003)

TRANSPORTATION

Highways in Miles (Kilometers): 8,688 (14,480)
Railroads in Miles (Kilometers): none
Usable Airfields: 7
Motor Vehicles in Use: 20,000

GOVERNMENT

Type: republic
Independence Date: July 1, 1962 (from UN trusteeship under Belgian administration)
Head of State/Government: President Domitien Ndayizeye is both head of state and head of government
Political Parties: Union for National Progress; Burundi Democratic Front; others
Suffrage: universal for adults

MILITARY

Military Expenditures (% of GDP): 5.3%
Current Disputes: severe interethnic conflict, transcending national borders

ECONOMY

Currency ($ U.S. equivalent): 1,072 francs = $1
Per Capita Income/GDP: $600/$3.7 billion
GDP Growth Rate: -1.3%
Inflation Rate: 10.7%
Population Below Poverty Line: 70%
Natural Resources: nickel; uranium; rare earth oxides; peat; cobalt; copper; platinum; vanadium; arable land; hydropower
Agriculture: coffee; cotton; tea; corn; sorghum; sweet potatoes; bananas; manioc; livestock
Industry: light consumer goods; assembly of imported components; public-works construction; food processing
Exports: $24 million (primary partners European Union, United States, Kenya)
Imports: $125 million (primary partners European Union, Tanzania, Zambia)

SUGGESTED WEB SITE

http://www.sas.upenn.edu/African_Studies/Country_Specific/Burundi.html

Burundi Country Report

Notwithstanding the tireless diplomatic efforts of South African mediators, Burundi has in recent years remained divided between the forces loyal to its transitional government and the predominantly ethnic-Hutu rebels of the Forces for the Defense of Democracy (FDD) and the National Liberation Forces (FNL). Talks among the groupings throughout 2002 failed to achieve national reconciliation.

DEVELOPMENT

Burundi's sources of wealth are limited. There is no active development of mineral resources, although nickel has been located and may be mined soon. There is little industry, and the coffee industry, which contributes 75% to 90% of export earnings, has declined.

Hutu rebels have been fighting with Burundi's government since 1993, when Tutsi paratroopers assassinated the small Central African country's first democratically elected president, who was a Hutu. Despite being in the minority, Tutsis have effectively controlled the nation of 6 million people for all but a few months since independence in 1961. The current transitional government took office in November 2001 to implement a power-sharing agreement, mediated in August 2000 by former South African president Nelson Mandela, between Hutu and Tutsi political parties. But the rebels maintain that true power remains in the hands of the Tutsi-dominated military.

Burundi's current divisions have deeper roots. In 1972 and again in 1988, tens of thousands of people, mostly Hutus, perished in genocidal attacks. In more recent years, the situation has been further complicated by the escalation of conflict between Tutsis and Hutus in the neighboring states of Rwanda and the Democratic Republic of the Congo (D.R.C.).

In the past the violence was initiated by members of the Tutsi governing elite seeking to maintain their privileged status through brutal military control. Today, the army's hold on the countryside is increasingly being challenged by an armed movement of the FDD and FNL. This movement is spreading counterterror in the name of the country's Hutu majority. What has remained the same over the years is the general indifference of the outside world to Burundi's horrific record of ethnic conflict.

In 1993, the country seemed poised to enter a bright new era when, in their first democratic elections, Burundians chose their first Hutu head of state, Melchior Ndadaye, and a Parliament dominated by the Hutu Front for Democracy in Burundi (FRODEBU) party. But within months Ndadaye was assassinated, setting the scene for subsequent Hutu–Tutsi violence, in which at least 200,000 people have been killed. In early 1994, Parliament elected another Hutu, Cyprien Ntaryamira, as president. However, he was killed when a plane he was traveling in was sabotaged in April—the same incident that killed the reformist president of neighboring Rwanda.

FREEDOM

Beset by on-going genocide, there is currently no genuine freedom in Burundi, for either its ethnic majority or minority populations. People continue to flee the country by the tens of thousands.

After talks among the main parties, another Hutu, Sylvestre Ntibantunganya, was appointed president in October 1994. But within months the mainly Tutsi Union for National Progress (UPRONA) party withdrew from the government and Parliament, sparking off a new wave of ethnic violence. In the process the capital city, Bujumbura, was largely emptied of its Hutu majority, while many ordinary Tutsis fled from much of the countryside. In July 1996, the army overthrew Ntibantunganya, bring back to power the Tutsi general Pierre Buyoyo, who had ruled the country from 1987 to 1993. Subsequent talks among the Burundian political parties, mediated first by former Tanzanian president Julius Nyerere and then by Mandela, failed to reach agreement on crucial issues. These included the role of the Burundian Army and the dismantling of "regroupment camps," which are said to hold more than 800,000 Hutu civilians.

A DIVIDED SOCIETY

Burundi's population is ethnosocially divided into three distinctive groups. At the bottom of the social hierarchy are the Twa, commonly stereotyped as "pygmies." Believed to be the earliest inhabitants of the country, today the Twa account for only about 1 percent of the population. The largest group, constituting 85 percent of the population, are the Hutus, most of whom subsist as farmers. The dominant group are the Tutsis, 14 percent of the population.

Among the Tutsis, who are subdivided into clans, status has long been associated with cattle-keeping. Leading Tutsis continue to form an aristocratic ruling class over the whole of Burundi society. Until 1966, the leader of Burundi's Tutsi aristocracy was the *Mwami,* or king.

The Burundi kingdom goes back at least as far as the sixteenth century. By the late 1800s, when the kingdom was incorporated into German East Africa, the Tutsis had subordinated the Hutus, who became clients of local Tutsi aristocrats, herding their cattle and rendering other services. The Germans and subsequently the Belgians, who assumed paramount authority over the kingdom after World War I, were content to rule through Burundi's established social hierarchy. But many Hutus as well as Tutsis were educated by Christian missionaries.

HEALTH/WELFARE

Much of the educational system has been in private hands, especially the Roman Catholic Church. Burundi lost many educated and trained people during the Hutu massacres in the 1970s and 1980s.

In the late 1950s, Prince Louis Rwagazore, a Tutsi, tried to accommodate Hutu as well as Tutsi aspirations by establishing the nationalist reform movement known as UPRONA. Rwagazore was assassinated before independence, but UPRONA led the country to independence in 1962, with King Mwambutsa IV retaining considerable power as head of state. The Tutsi elite remained dominant, but the UPRONA cabinets contained representation from the two major groups. This attempt to balance the interests of the Tutsis and Hutus broke down in 1965, when Hutu politicians within both UPRONA and the rival People's Party won 80 percent of the vote and the majority of the seats in both houses of the bicameral Legislature. In response, the king abolished the Legislature before it could convene. A group of Hutu army officers then attempted to overthrow the government. Mwambutsa fled the country, but Tutsi officers, led by Michel Micombero, crushed the revolt in a countercoup.

In the aftermath of the uprising, Micombero took power amid a campaign of reprisals in which, it is believed, some 5,000 Hutus were killed. He deposed Mwambutsa's son, Ntare V, from the kingship and set up a "Government of Public Safety," which set about purging Hutu members from the government and the army. The political struggle involved interclan competition among the Tutsis as well as the maintenance of their hegemony over the Hutus.

Under Micombero, Burundi continued to be afflicted with interethnic violence, occasional coup attempts, and pro-monarchist agitation. A major purge of influential Hutus was carried out in 1969. In 1972, Ntare V was lured to Uganda by Idi Amin, who turned him over to Micombero. Ntare was placed under arrest upon his arrival and was subsequently murdered by his guards.

ACHIEVEMENTS

Burundians were briefly united in July 1996 by the victory of their countryman Venuste Niyongabo in the men's 5,000-meter race at the Atlanta Summer Olympic Games. He dedicated his gold medal (the first for a Burundi citizen) to the hope of national reconciliation.

A declaration of martial law then set off another explosion of violence. In response to an alleged uprising involving the deaths of up to 2,000 Tutsis, government supporters began to massacre large numbers of Hutus. Educated Hutus were especially targeted in a two-month campaign of selective genocide, which is generally estimated to have claimed 200,000 victims (estimates range from 80,000 to 500,000 deaths for the entire period, with additional atrocities being reported through 1973). More than 100,000 Hutus fled to Uganda, Rwanda, Zaire (present-day D.R.C.), and Tanzania. Among the governments of the world, only Tanzania and Rwanda showed any deep concern for the course of events. China, France, and Libya used the crisis to significantly upgrade their military aid to the Burundi regime.

In 1974, Micombero formally transformed Burundi into a single-party state under UPRONA. Although Micombero was replaced two years later in a military coup by Colonel Jean-Baptiste Bagaza, power remained effectively in the hands of members of the Tutsi elite who controlled UPRONA, the civil service, and the army. In 1985, Bagaza widened existing state persecution of Seventh Day Adventists and Jehovah's Witnesses to include the Roman Catholic Church, to which a majority of Burundi's population belong, suspecting it of fostering seditious—that is, pro-Hutu—sympathies. (The overthrow of Bagaza by Pierre Buyoyo, in a 1987 military coup, led to a lifting of the anti-Catholic campaign.)

Timeline: PAST

1795
Mwami Ntare Rugaamba expands the boundaries of the Nkoma kingdom

1919
The area is mandated to Belgium by the League of Nations after the Germans lose World War I

1958–1961
Prince Louis Rwagazore leads a nationalist movement and founds UPRONA

1961
Rwagazore is assassinated; independence is achieved

1965–1966
A failed coup results in purges of Hutus in the government and army; Michel Micombero seizes power

1972
Government forces massacre 200,000 Hutus

1976
Jean-Baptiste Bagaza comes to power in a military coup

1987
Bagaza is overthrown in a military coup led by Pierre Buyoyo

1990s
Buyoyo loses in multiparty elections; Melchior Ndadye becomes Burundi's first Hutu president; Buyoyo regains power in a military coup

PRESENT

2000s
Unrest in neighboring states complicates the ethnic conflict in Burundi

Nelson Mandela mediates for national reconciliation

Domitien Ndayizeye elected president

Ethnic violence erupted again in 1988. Apparently some Tutsis were killed by Hutus in northern Burundi, in response to rumors of another massacre of Hutus. In retaliation, the army massacred between 5,000 and 25,000 Hutus. Another 60,000 Hutus took temporary refuge in Rwanda, while more than 100,000 were left homeless. In 1991, the revolutionary Party for the Liberation of the Hutu People, or Palipehutu, launched its own attacks on Tutsi soldiers and civilians, leading to further killing on all sides.

LAND ISSUES

Burundi is one of the poorest countries in the world, despite its rich volcanic soils and generous international development assistance (it has been one of the highest per capita aid recipients on the African continent). In addition to the dislocations caused by cycles of interethnic violence, the nation's development prospects are seriously compromised by geographic isolation and population pressure on the land. About 25 percent of Burundi's land is under cultivation—generally by individual farmers trying to subsist on plots of no more than three acres. Another 60 percent of the country is devoted to pasture for mostly Tutsi livestock. Hutu farmers continue to be tied by patron–client relationships to Tutsi overlords.

In the 1980s, the government tied its rural development efforts to an unpopular villagization scheme. This issue has complicated on-going attempts to reach some kind of accommodation between the Tutsi elite and Hutu masses. Having cautiously increased Hutu participation in his government while reserving ultimate power in the hands of the all-Tutsi Military Committee of National Salvation, Buyoyo agreed to the restoration of multiparty politics in 1991. A new Constitution was approved in March 1992; it allowed competition between approved, ethnically balanced, parties. In the resulting July 1993 elections, Buyoyo's UPRONA was defeated by the Front for Democracy in Burundi. FRODEBU's leader, Ndadaye, was sworn in as the head of a joint FRODEBU–UPRONA government. His subsequent assassination by Tutsi hard-liners in the military set off a new wave of interethnic killings. The firm stand against the coup by Buyoyo and the Tutsi/UPRONA prime minister, Sylvie Kinigi, helped to calm the situation, but attempts to make a fresh start collapsed in 1994 when a plane carrying Ntaryamira, Burundi's newly elected head of state, and his Rwandan counterpart was shot down over Rwanda. The latest coup followed UPRONA's withdrawal from the government following the massacre of more than 300 Tutsis by FDD, who by September 1996 were attempting to besiege the capital.

During a four-week period from late October to November 1996, the Tutsi-led Burundian military massacred at least 1,000 civilians. The government forces fought with Hutu rebels, as some 50,000 Hutus returned from camps that had been closed in Zaire. The Tutsi-dominated military set up more than a dozen "protection zones" for Hutu civilians while soldiers continued battling Hutu rebels. Strife continued as an estimated 200,000 Burundians were living in refugee camps in Tanzania.

BURUNDI TODAY

Burundi is a nation in search of its national identity. The country has been torn by ethnic violence since independence in 1961. Throughout the post independence era the dominant Tutsi minority and the Hutu majority have not been able to find a way to live together in peace. A ceasefire between the warring groups was signed in 2002 but the fighting continues.

Another South African power-sharing deal was signed in 2004, allocating government and national assembly posts to members of the Hutu majority and the Tutsi minority. During June 2004 the UN took over peacekeeping duties from African Union troops. After years of war and the installation of a new provisional president, Domitien Ndayizeye, most of the country is now beginning to reap the dividends of a peace process.

With only one rebel group remaining active in the countryside around the capital and a strong effort by the international community, there is hope that negotiations to resolve differences will succeed. Delayed presidential elections are due in April 2005, and will follow a planned referendum on a new constitution. President Ndayizeye, a senior figure in the largest Hutu party Frodebu, has faced the formidable challenge of maintaining good relations with Burundi's Tutsi-led government army while persuading Hutu rebels to stop fighting.

Comoros (Union of Comoros)

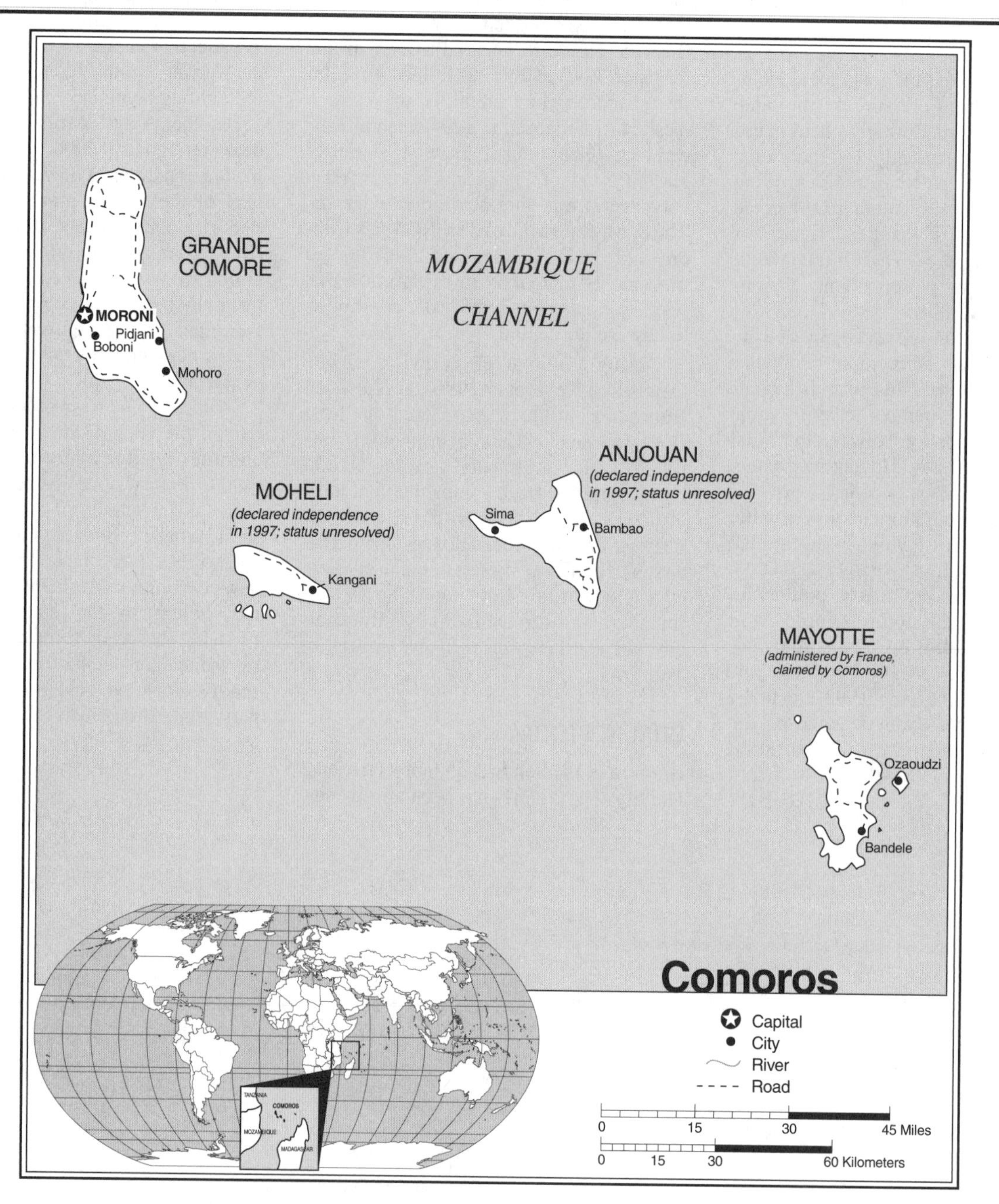

Comoros Statistics

GEOGRAPHY

Area in Square Miles (Kilometers): 838 (2,171) (about 12 times the size of Washington, D.C.)

Capital (Population): Moroni (49,000)

Environmental Concerns: soil degradation and erosion; deforestation

Geographical Features: volcanic islands; interiors vary from steep mountains to low hills

Climate: tropical marine

PEOPLE

Population

Total: 631,901

Annual Growth Rate: 2.99%

Rural/Urban Population Ratio: 69/31

Major Languages: Arabic; French; Comoran

Ethnic Makeup: Antalote; Cafre; Makoa; Oimatsaha; Sakalava

Religions: 98% Sunni Muslim; 2% Roman Catholic

Health

Life Expectancy at Birth: 58 years (male); 62 years (female)

Infant Mortality: 81.7/1,000 live births

Physicians Available: 1/6,600 people

HIV/AIDS Rate in Adults: 0.12%

Education

Adult Literacy Rate: 57.3%
Compulsory (Ages): 7–16

COMMUNICATION

Telephones: 13,200 main lines
Internet Users: 5,000 (2003)

TRANSPORTATION

Highways in Miles (Kilometers): 522 (870)
Railroads in Miles (Kilometers): none
Usable Airfields: 4

GOVERNMENT

Type: republic
Independence Date: July 6, 1975 (from France)
Head of State/Government: President Azali Assoumani
Political Parties: Rassemblement National pour le Development; Front National pour la Justice
Suffrage: universal at 18

MILITARY

Military Expenditures (% of GDP): 3%
Current Disputes: Comoros claims the French-administered island of Mayotte; Moheli and Anjouan seek independence

ECONOMY

Currency ($ U.S. Equivalent): 435 francs = $1
Per Capita Income/GDP: $700/$441 million
GDP Growth Rate: 2%
Inflation Rate: 3.5%
Unemployment Rate: 20%; extreme underemployment
Labor Force by Occupation: 80% agriculture
Population Below Poverty Line: 60%
Natural Resources: negligible
Agriculture: perfume essences; copra; coconuts; cloves; vanilla; bananas; cassava
Industry: tourism; perfume distillation
Exports: $35.3 million (primary partners France, United States, Singapore)
Imports: $44.9 million (primary partners France, South Africa, Kenya)

SUGGESTED WEB SITES

http://www.arabji.com/Comoros/index.htm
http://www.sas.upenn.edu/African_Studies/Country_Specific/Comoros.html
http://www.cia.gov/cia/publications/factbook/geos/cn.html

Comoros Country Report

A small archipelago consisting of three main islands—Grande Comore, Moheli, and Anjouan (a fourth island, Mayotte, has voluntarily remained under French rule)—in recent years Comoros has struggled to maintain its fragile unity. In 1997, separatists seized control of Anjouan and Moheli, subsequently declaring independence. But after years of failed mediation efforts by other African states, in December 2001 voters throughout Comoros were able to overwhelmingly agree on a new Constitution designed to reunite their country as a loose federation. This followed the seizure of power by a "military committee" on Anjouan that was committed to reunification. In April 2002, Azali Assoumani, who had previously seized power on Grande Comore, was sworn in as the president of the new "Union of Comoros," and Prime Minister Bolero was appointed Minister of External Defense and Territorial Security. But his authority was soon challenged by Mze Abdou Soule Elbak, who a month later was elected as the president of Grande Comore. A military standoff thereafter developed on Grande Comore between followers of Azali and Elbak, further threatening the islands' prospects of ever achieving political stability. President Azali has not yet appointed a Prime Minister.

The years since independence from France, in 1975, have not been kind to Comoros, which has been consistently listed by the United Nations as one of the world's least-developed countries. Lack of economic development has been compounded at times by natural disasters, eccentric and authoritarian leadership, political violence, and external interventions. The 1990 restoration of multiparty democracy, along with subsequent elections in 1992–1993, has so far failed to provide a basis for national consensus.

DEVELOPMENT

One of the major projects undertaken since independence has been the ongoing expansion of the port at Mutsamundu, to allow large ships to visit the islands. Vessels of up to 25,000 tons can now dock at the harbor. In recent years, there has been a significant expansion of tourism to Comoros.

Meanwhile, the entire archipelago remains impoverished. While many Comorans remain underemployed as subsistence farmers, more than half of the country's food is imported. As a result, many Comorans have questioned the wisdom of independence, but appeals by Anjouan and Moheli islanders for a return of French control have been rejected by Paris.

The Comoros archipelago was populated by a number of Indian Ocean peoples, who—by the time of the arrival of Europeans during the early 1500s—had combined to form the predominantly Muslim, Swahili-speaking society found on the islands today. In 1886, the French proclaimed a protectorate over the three main islands that currently constitute the Union of Comoros (France had ruled Mayotte since 1843). Throughout the colonial period, Comoros was especially valued by the French for strategic reasons. A local elite of large landholders prospered from the production of cash crops. Life for most Comorans, however, remained one of extreme poverty.

A month after independence, the first Comoran government, led by Ahmed Abdullah Abderemane, was overthrown by mercenaries, who installed Ali Soilih in power. He promised a socialist transformation of the nation and began to implement land reform, but he rapidly lost support both at home and abroad—under his leadership, gangs of undisciplined youths terrorized society, while the basic institutions and services of government all but disappeared. In 1977, the situation was made even worse by a major volcanic eruption, which left 20,000 people homeless, and by the arrival of 16,000 Comoran refugees following massacres in neighboring Madagascar.

FREEDOM

Freedom was abridged after independence under both Ahmed Abdullah and Ali Soilih. The government elected in 1990 ended human-rights abuses.

In 1978, another band of mercenaries—this time led by the notorious Bob Denard, whose previous exploits in Zaire (present-day Democratic Republic of the Congo or D.R.C.), Togo, and elsewhere had made his name infamous throughout Africa—

overthrew Soilih and restored Abdullah to power. Denard, however, remained the true power behind the throne.

HEALTH/WELFARE

Health statistics improved during the 1980s, but a recent World Health Organization survey estimated that 10% of Comoran children ages 3 to 6 years are seriously malnourished and another 37% are moderately malnourished.

The Denard–Abdullah government enjoyed close ties with influential right-wing elements in France and South Africa. Connections with Pretoria were manifested through the use of Comoros as a major conduit for South African supplies to the Renamo rebels in Mozambique. Economic ties with South Africa, especially in tourism and sanctions-busting, also grew. The government also established good relations with Saudi Arabia, Kuwait, and other conservative Arab governments while attracting significant additional aid from the international donor agencies.

In 1982, the country legally became a one-party state. Attempted coups in 1985 and 1987 aggravated political tensions. Many Comorans particularly resented the overbearing influence of Denard and his men. By November 1989, this group included President Abdullah himself. With the personal backing of President François Mitterand of France and President F. W. de Klerk of South Africa, Abdullah moved to replace Denard's mercenaries with a French-approved security unit. But before this move could be implemented, Abdullah was murdered following a meeting with Denard.

The head of the Supreme Court, Said Mohamed Djohar, was appointed interim president in the wake of the assassination. After a period of some confusion, during which popular protests against Denard swelled, Djohar quietly sought French intervention to oust the mercenaries. With both Paris and Pretoria united against him, Denard agreed to relinquish power, in exchange for safe passage to South Africa. The removal of Denard and temporary stationing of a French peacekeeping force was accompanied by the lifting of political restrictions in preparation for presidential elections. In 1990, a runoff resulted in a 55 percent electoral mandate for Djohar.

ACHIEVEMENTS

Comoros has long been the world's leading exporter of ylang-ylang, an essence used to make perfume. It is also the second-leading producer of vanilla and a major grower of cloves. Together, these cash crops account for more than 95% of export earnings. Unfortunately, the international prices of these crops have been low for the past 2 decades.

In September 1995, Denard's men returned to overthrow Djohar. But the mercenaries were soon forced to surrender to French forces, who installed Caambi el Yachourtu, rather than Djohar, as acting president. At the end of 1996, Mohamed Taki Abdulkarim replaced Yachourtu as president. In November 1998, Taki died suddenly and was replaced by Tadjiddine Ben Said Massounde as the head of a ruling military committee. Massounde's government was overthrown in a bloodless coup on April 30, 1999. Azali Assoumani was subsequently installed as president.

Timeline: PAST

1500s
Various groups settle in the islands, which become part of a Swahili trading network

1886
A French protectorate over the remaining Comoros islands is proclaimed

1914–1946
The islands are ruled as part of the French colony of Madagascar

1975
Independence is followed by a mercenary coup, which installs Ali Soilih

1978
Ali Soilih is overthrown by mercenaries; Ahmed Abdullah is restored

1980s
Abdullah proclaims a one-party state; real power remains in the hands of mercenary leader Bob Denard

1990s
The assassination of Abdullah leads to the removal of Denard and to multiparty elections

PRESENT

2000s
The country is renamed "Union of Comoros"

Despite the name change, Comoros's political unity has not been achieved

Djibouti (Republic of Djibouti)

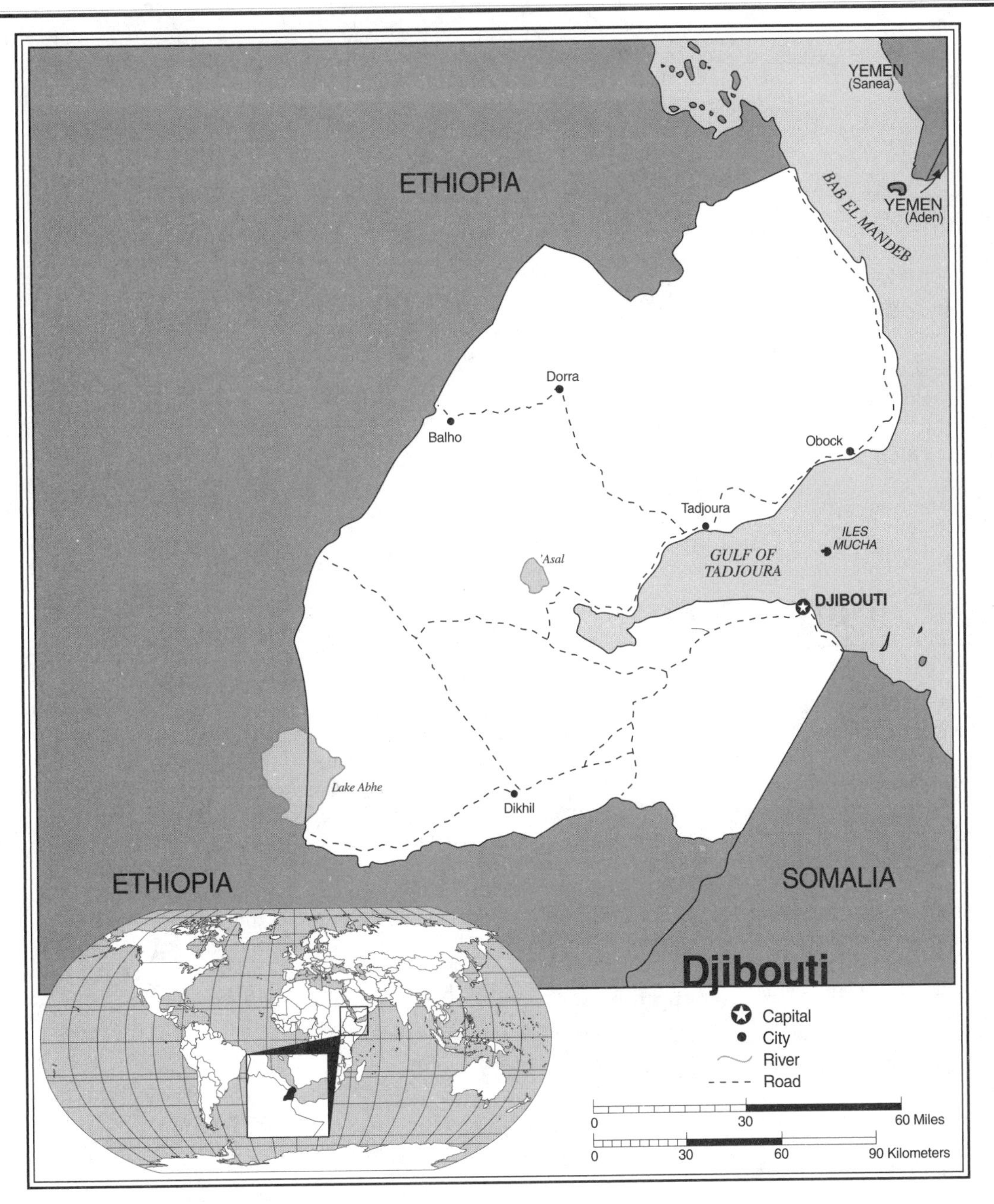

Djibouti Statistics

GEOGRAPHY

Area in Square Miles (Kilometers): 8,492 (22,000) (about the size of Massachusetts)

Capital (Population): Djibouti (542,000)

Environmental Concerns: insufficient potable water; desertification

Geographical Features: coastal plain and plateau, separated by central mountains

Climate: desert

PEOPLE

Population

Total: 466,900

Annual Growth Rate: 2.1%

Rural/Urban Population Ratio: 17/83

Major Languages: French; Arabic; Somali; Afar

Ethnic Makeup: 60% Issa/Somali; 35% Afar; 5% French, Arab, Ethiopian, Italian

Religions: 94% Muslim; 6% Christian

Health

Life Expectancy at Birth: 50 years (male); 53 years (female)
Infant Mortality: 99.7/1,000 live births
Physicians Available: 1/3,790 people
HIV/AIDS Rate in Adults: 11.75%

Education

Adult Literacy Rate: 46.2%

COMMUNICATION

Telephones: 10,000 main lines
Televisions: 43/1,000 people
Internet Users: 6,500 (2003)

TRANSPORTATION

Highways in Miles (Kilometers): 1,801 (2,906)
Railroads in Miles (Kilometers): 60 (97)
Usable Airfields: 12
Motor Vehicles in Use: 16,000

GOVERNMENT

Type: republic
Independence Date: June 27, 1977 (from France)
Head of State/Government: President Ismail Omar Guellah; Prime Minister Dileita Mohamed Dileita
Political Parties: People's Progress Assembly; Democratic Renewal Party; Democratic National Party; others
Suffrage: universal for adults

MILITARY

Military Expenditures (% of GDP): 4.4%
Current Disputes: ethnic conflict; border clashes with Eritrea

ECONOMY

Currency ($ U.S. Equivalent): 177 francs = $1
Per Capita Income/GDP: $1,400/$586 million
GDP Growth Rate: 3.5%
Inflation Rate: 2%
Unemployment Rate: 50%
Population Below Poverty Line: 50%
Natural Resources: geothermal areas
Agriculture: livestock; fruits; vegetables
Industry: port and maritime support; construction
Exports: $260 million (primary partners Somalia, Yemen, Ethiopia)
Imports: $440 million (primary partners France, Ethiopia, Italy)

SUGGESTED WEB SITES

`http://www.sas.upenn.edu/African_Studies/Country_Specific/Djibouti.html`
`http://www.republique-djibouti.com`
`http://www.cia.gov/cia/publications/factbook/geos/dj.html`

Djibouti Country Report

After a decade of civil unrest, Djibouti has settled down under the leadership of its second president, Ismail Omar Guellah. In April 1999, Guellah succeeded the aging Hassan Gouled Aptidon, who stepped down due to ill health. While Guellah's main opponent, Musa Ahmed Idriss, was arrested after claiming massive electoral fraud, the new president has since consolidated his authority by building on the process of national reconciliation that had begun under his predecessor. In February 2000, this resulted in a peace agreement between the government and armed rebel holdouts of the Front for the Restoration of Democratic Unity (FRUD), resulting in the return of the rebel leader Ahmad Dini. A subsequent coup attempt, allegedly orchestrated by the chief of police, was crushed.

DEVELOPMENT

Recent discoveries of natural-gas reserves in Djibouti could result in a surplus for export. A number of small-scale irrigation schemes have been established. There is also a growing, though still quite small, fishing industry.

Political conflict in Djibouti has mirrored the country's ethnic tensions between the Somali-speaking Issas and the Afar-speakers. An earlier, 1997, power-sharing agreement was reached between the long-ruling, Issa-dominated Popular Rally for Progress Party (RPP) and a more moderate faction of the Afar-dominated FRUD. Although the FRUD moderates went on to win all 65 seats in December 1997 legislative elections, more radical elements of FRUD continued their armed resistance to President Aptidon. While the conflict has now ended, suspicions between Afars and Issas continue to threaten Djibouti's fragile political unity.

FREEDOM

The government continues to harass and detain its critics. Prison conditions are harsh, with the sexual assault of female prisoners being commonplace.

Since achieving its independence from France, Djibouti has also had to strike a cautious balance between the competing interests of its larger neighbors, Ethiopia and Somalia. In the past, Somalia has claimed ownership of the territory, based on the numerical preponderance of Djibouti's Somali population, variously estimated at 50 to 70 percent. However, local Somalis as well as Afars also have strong ties to communities in Ethiopia. Furthermore, Djibouti's location at the crossroads of Africa and Eurasia has made it a focus of continuing strategic concern to nonregional powers, particularly France, which maintains a large military presence in the country.

In January 2002, German warships and 1,000 sailors arrived in Djibouti to patrol shipping lanes in the Red Sea area, in support of U.S. actions in Afghanistan. Although Djibouti says it won't be used as a base for attacks against another country in the region, some 900 U.S. troops also set up camp in support of the U.S.-led war on terror. The effort by Djibouti's government to fight the war on terror had one major political consequence as the government, in September 2002, passed a law allowing three other parties to compete in elections, thus opening the way for full multiparty politics.

In January 2003, the Union for Presidential Majority Coalition candidate Ismael Omar Gelleh won Djibouti's first free multiparty elections since independence in 1977. Ismael Omar Gelleh succeeded his uncle and Djibouti's first president, Hassan Gouled Aptidon, in April 1999, at the age of 52. He took office after being elected in a multiparty ballot, which was not contested by Mr. Aptidon. A former head of security, he worked for many years in his uncle's office. He is known to favor continuing Djibouti's traditionally strong ties with France, and has played an important role in trying to reconcile the different factions in neighbouring Somalia. In September 2003, one of President Gelleh's first actions after assuming office was to begin a drive to detain and expel illegal immigrants, thought to make up 15 percent of the population.

Modern Djibouti's colonial genesis is a product of mid-nineteenth-century European rivalry over control of the Red Sea. In 1862, France occupied the town of Obock, across the harbor from the city of Djibouti. This move was taken in anticipation of the 1869 opening of the Suez Canal, which transformed the Red Sea into the major shipping route between Asia, East Africa, and Europe. In 1888, Paris, having acquired Djibouti city and its hinterland, proclaimed its authority over French Somaliland, the modern territory of Djibouti.

HEALTH/WELFARE

Progress has been made in reducing infant mortality, but health services are strained in this very poor country. However, on the positive side, school enrollment has expanded by nearly one third since 1987.

The independence of France's other mainland African colonies by 1960, along with the formation in that year of the Somali Democratic Republic, led to local agitation for an end to French rule. To counter the effects of Somali nationalism, the French began to favor the Afar minority in local politics and employment. French president Charles de Gaulle's 1966 visit was accompanied by large, mainly Somali, pro-independence demonstrations. As a result, a referendum was held on the question of independence. Colonial control of voter registration assured a predominantly Afar electorate, who, fearful of Somali domination, opted for continued French rule. French Somaliland was then transformed into the self-governing "Territory of Afars and Issas." The name reflected a continuing colonial policy of divide-and-rule; members of the Issas clan constituted just over half of the area's Somali-speakers.

ACHIEVEMENTS

Besides feeding its own refugees, the government of Djibouti has played a major role in assisting international efforts to relieve the effects of recurrent famines in Ethiopia, Somalia, and Sudan.

By the 1970s, neither Ethiopia nor France was opposed to Djibouti's independence but, for their own strategic reasons, both countries backed the Afar community in its desire for assurances that the territory would not be incorporated into Somalia. An ethnic power-sharing arrangement was established that in effect acknowledged local Somali preponderance. The empowerment of local Somalis, in particular Issas, was accompanied by diminished pan-Somali sentiment. On June 27, 1977, the Republic of Djibouti became independent. French troops remained in the country, however, supposedly as a guarantee of its sovereignty. Internally, political power was divided by means of ethnically balanced cabinets.

War broke out between Ethiopia and Somalia a few months after Djibouti's independence. Djibouti remained neutral, but ethnic tensions mounted with the arrival of Somali refugees. In 1981, the Afar-dominated Djiboutian Popular Movement was outlawed. The Issa-dominated Popular Rally for Progress (RPP) then became the country's sole legal party.

Refugees have poured into Djibouti for years now, fleeing conflict and famine in Ethiopia, Somalia, and Sudan. The influx has swelled the country's population by about one third and has deepened Djibouti's dependence on external food aid. Massive unemployment among Djibouti's largely urban population remains a critical problem.

Timeline: PAST

1862
France buys the port of Obock

1888
France acquires the port of Djibouti

1917
The Addis Ababa-Djibouti Railroad is completed

1958
Djibouti votes to remain part of Overseas France

1977
Independence; the Ogaden War

1980s
The underground Union of Movements for Democracy is formed as an interethnic, antigovernment coalition

1990s
Civil war rends the country; Ismail Omar Guellah is elected to replace President Hassan Gouled Aptidon

PRESENT

2000s
Ethnic conflict continues

Eritrea (State of Eritrea)

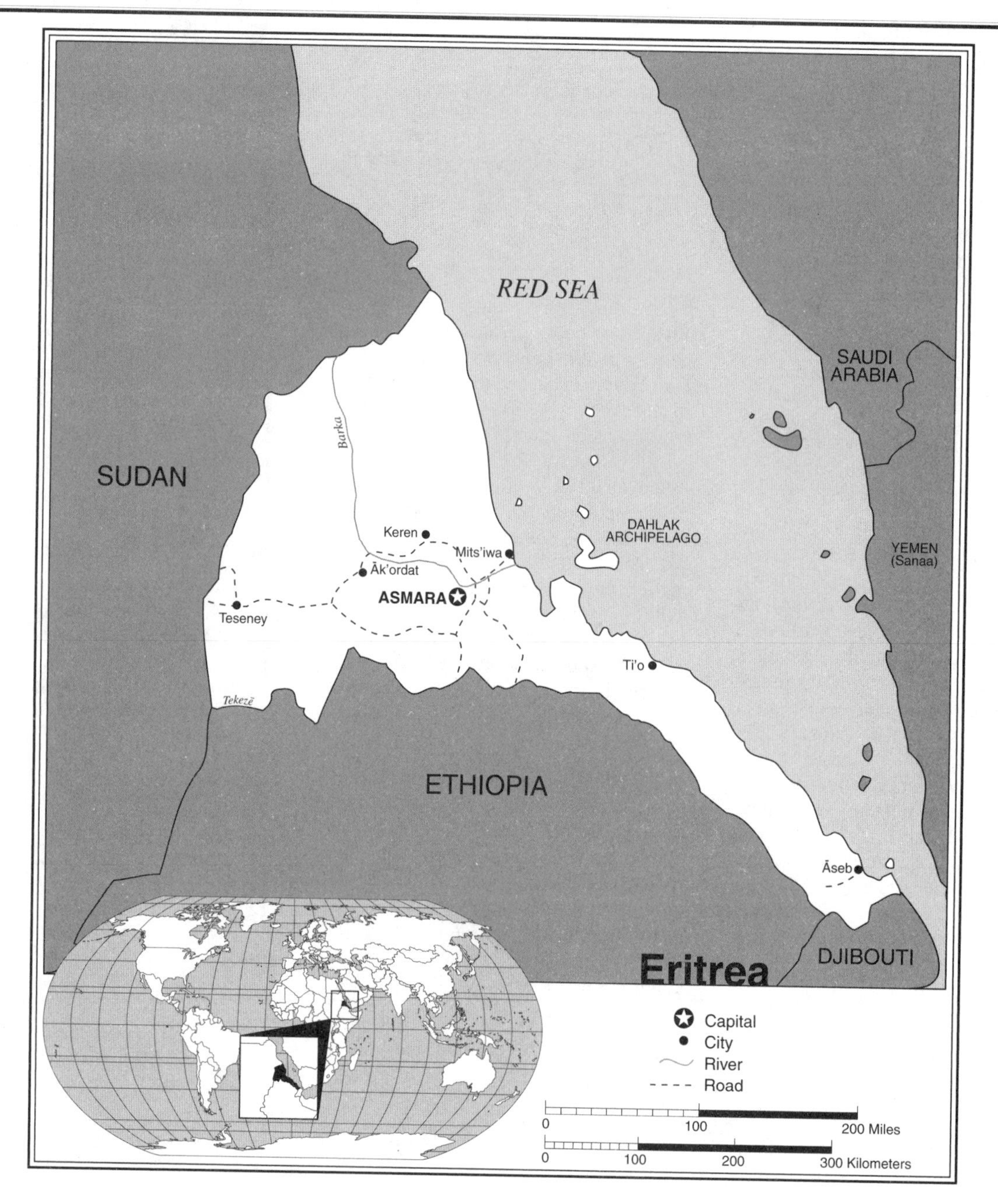

Eritrea Statistics

GEOGRAPHY

Area in Square Miles (Kilometers): 46,829 (121,320) (about the size of Pennsylvania)

Capital (Population): Asmara (503,000)

Environmental Concerns: deforestation; desertification; soil erosion; overgrazing

Geographical Features: north-south–trending highlands, descending on the east to a coastal desert plain, on the northwest to hilly terrain, and on the southwest to flat-to-rolling plains

Climate: hot, dry desert on the seacoast; cooler and wetter in the central highlands; semiarid in western hills and lowlands

PEOPLE

Population

Total: 4,466,000

Annual Growth Rate: 2.57%

Rural/Urban Population Ratio: 82/18

Major Languages: various, including Tigrinya, Tigre, Kunama, Arabic, Amharic, and Afar

Ethnic Makeup: 50% ethnic Tigrinya; 40% Tigre and Kunama; 4% Afar; 3% Saho; 3% others

Religions: Muslim; Coptic Christian; Roman Catholic; Protestant

Health

Life Expectancy at Birth: 54 years (male); 59 years (female)
Infant Mortality: 73.6/1,000 live births
Physicians Available: 1/36,000 people
HIV/AIDS Rate in Adults: 2.87%

Education

Adult Literacy Rate: 25%
Compulsory (Ages): 7–13; free

COMMUNICATION

Telephones: 38,100 main lines
Televisions: 6/1,000 people
Internet Users: 19,500 (2003)

TRANSPORTATION

Highways in Miles (Kilometers): 2,436 (3,930)
Railroads in Miles (Kilometers): 191 (307)
Usable Airfields: 21

GOVERNMENT

Type: transitional government
Independence Date: May 24, 1993 (from Ethiopia)
Head of State/Government: President Isaias Aferweki (or Afworki) is both head of state and head of government
Political Parties: People's Front for Democracy and Justice (only recognized party)
Suffrage: universal at 18

MILITARY

Military Expenditures (% of GDP): 19.8%
Current Disputes: border disputes with Yemen; uneasy cease-fire with Ethiopia

ECONOMY

Currency ($ U.S. Equivalent): n/a
Per Capita Income/GDP: $740/$3.2 billion
GDP Growth Rate: 2%
Inflation Rate: 12.3%
Labor Force by Occupation: 80% agriculture
Natural Resources: gold; potash; zinc; copper; salt; possibly petroleum and natural gas; fish
Agriculture: sorghum; lentils; vegetables; maize; cotton; tobacco; coffee; sisal; livestock; fish
Industry: food processing; beverages; clothing and textiles
Exports: $34.8 million (primary partners Sudan, Ethiopia, Japan)
Imports: $470 million (primary partners Italy, United Arab Emirates, Germany)

SUGGESTED WEB SITES

http://www.sas.upenn.edu/African_Studies/Country_Specific/Eritrea.html
http://www.eritrea.org
http://www.cia.gov/cia/publications/factbook/geos/er.html

Eritrea Country Report

In 1998, a border dispute between Eritrea and Ethiopia, around the town of Badme, erupted into open war. This formally ended with a cease-fire agreement in June 2000, but not before leaving thousands of soldiers dead on both sides. In December of that year, Eritrea and Ethiopia signed a further agreement establishing commissions to mark the border, exchange prisoners, return displaced people, and hear compensation claims. On February 6, 2001, Eritrea accepted the United Nations' plan for a temporary demilitarized zone along its border with Ethiopia. By the end of the month, Ethiopia had completed its troop withdrawal. A key provision of the peace agreement was met in April when Eritrea announced that its forces had pulled out of the border zone with Ethiopia. In May, Eritrea and Ethiopia agreed on a UN-proposed mediator to try to demarcate their disputed border. The completion of this task in 2002 has resulted in what will hopefully be a lasting peace between Eritrea and Ethiopia, after decades of conflict.

Eritrea became Africa's newest nation in May 1993, ending 41 years of union with Ethiopia. The origins of Eritrea's separation date back to September 1961, when a small group of armed men calling themselves the Eritrean Liberation Front (ELF) began a bitter independence struggle that would last for three decades.

Between 60,000 and 70,000 people perished as a result of that war, while another 700,000—then about one fifth of the total population—went into exile. What had been one of the continent's most sophisticated light-industrial infrastructures was largely reduced to ruins. Yet the war has also left a positive legacy, in the spirit of unity, self-reliance, and sacrifice that it engendered among Eritreans.

DEVELOPMENT

Since liberation, the government has concentrated its efforts on restoring agricultural and communications infrastructure. The railway and ports of Assab and Massawa are being rehabilitated. In 1991, 80% of the country was dependent on food aid, but subsequent good rains helped boost crop production.

There is no clear-cut reason why a nationalist sentiment should have emerged in Eritrea. Like most African countries, the boundaries of Eritrea are an artificial product of the late-nineteenth-century European scramble for colonies. Between 1869 and 1889, the territory fell under the rule of Italy. Italian influence survives today, especially in the overcrowded but elegant capital city of Asmara, which was developed as a showcase of neo-Roman imperialism. Italian rule came to an abrupt end in 1941, when British troops occupied the territory in World War II. The British withdrew only in 1952. In accordance with the wishes of the UN Security Council, the territory was then federated as an autonomous state within the "Empire of Ethiopia."

The federation did not come about through the wishes of the Eritreans. It was, rather, based on the dubious Ethiopian claim that Eritrea was an integral part of the Empire that had been alienated by the Italians. Among the Christians, there were historic cultural ties with their Ethiopian coreligionists, though the Tigrinya-speaking Copts of Eritrea were ethnically distinct from the Empire's then–politically dominant Amharic-speakers. The Muslim lowland areas had never been under any form of Ethiopian control. But, perhaps more important, developments under Italian rule had laid the basis for a sense that Eritrea had its own identity.

FREEDOM

The Eritrean government has pledged to uphold a bill of rights. While the government is dominated by the former EPLF, other parties and organizations participate in the 105-seat Provisional Council. Multiparty elections in 1997 confirmed former EPLF leader Isaias Aferweki as president.

In the face of growing dissatisfaction inside the territory, Ethiopia's emperor, Haile Selassie, ended Eritrea's autonomous status in 1962. Fighting intensified in the early 1970s, after a faction ultimately known as the Eritrean Popular Liberation

Front (EPLF) split from the ELF. The 1974 overthrow of Selassie briefly brought hopes of a peaceful settlement. But Ethiopia's new military rulers, known as the Dergue, committed themselves to securing the area by force. The ELF faded as the EPLF became increasingly effective in pinning down larger numbers of Ethiopian troops. In a major break with tradition, a large proportion of the EPLF's "Liberation Army," including many in command positions, was made up of women. In areas liberated by the EPLF, women were given the right to own land and choose their husbands, while the practice of female circumcision was discouraged.

HEALTH/WELFARE

A major challenge for the government has been the repatriation of hundreds of thousands of war refugees, mostly from neighboring Sudan. Rebuilding efforts were spearheaded by ex-combatants of the Liberation Army, who continued to work for virtually no pay. The EPLF established its own medical and educational services during the war.

Had it not been for the massive military support that the Dergue received from the Soviet Union and its allies, the conflict would have ended sooner. In the late 1980s, the EPLF began to work more closely with other groups inside Ethiopia proper that had taken up arms against their government. This resulted in an alliance between the EPLF and the Ethiopian People's Revolutionary Democratic Front (EPRDF), which was facilitated by the fact that leading members of both groups spoke Tigrinya. In May 1991, the Dergue collapsed, with the EPLF taking Asmara in the same month that EPRDF troops entered the Ethiopian capital of Addis Ababa. In July, the new EPRDF government agreed in principle to Eritrea's right to self-determination.

ACHIEVEMENTS

Eritrea's independence struggle and on-going national development efforts have been carried out against overwhelming odds, and with very little external support. During the war, self-reliance was manifested in the fact that most weapons and ammunition used by the EPLF were captured from Ethiopian forces.

In 1997, the EPLF, transformed as the People's Front for Democracy and Justice, claimed an overwhelming mandate in elections in which there was little effective opposition. A number of smaller parties, including remnants of the ELF, had joined the Front. Former EPLF leader Isaias Aferweki was confirmed as president of Eritrea.

The renewal of war in Ethiopia has had a devastating effect on Eritrea's economy, which had been making significant progress in the years following independence. The rehabilitation of the port of Massawa and other infrastructure had boosted trade. Light industries, mostly based in Asmara, had recovered. International investors have shown increased interest in the country's mineral wealth, especially offshore oil. In July 1997, the country introduced its own currency, the nakfa, which replaced the Ethiopian birr. Resulting exchange disputes between the two nations led to a souring of relations prior to the outbreak of the border war.

Timeline: PAST

1869
Italians occupy the Eritrean port of Assab

1889
Italians occupy all of Eritrea

1935–1936
Italians use Eritrea as a springboard for conquest of Ethiopia

1941–1952
Great Britain occupies Eritrea

1952
Eritrea is federated with Ethiopia

1961
The ELF begins the liberation struggle

1962
Federation ends; Eritrea is a province of Ethiopia

1990s
99.8% vote yes for Eritrea's independence; Isaias Aferweki becomes the newly independent nation's first president

PRESENT

2000s
Eritrea and Ethiopia try to forge a lasting peace

Ethiopia (Federal Democratic Republic of Ethiopia)

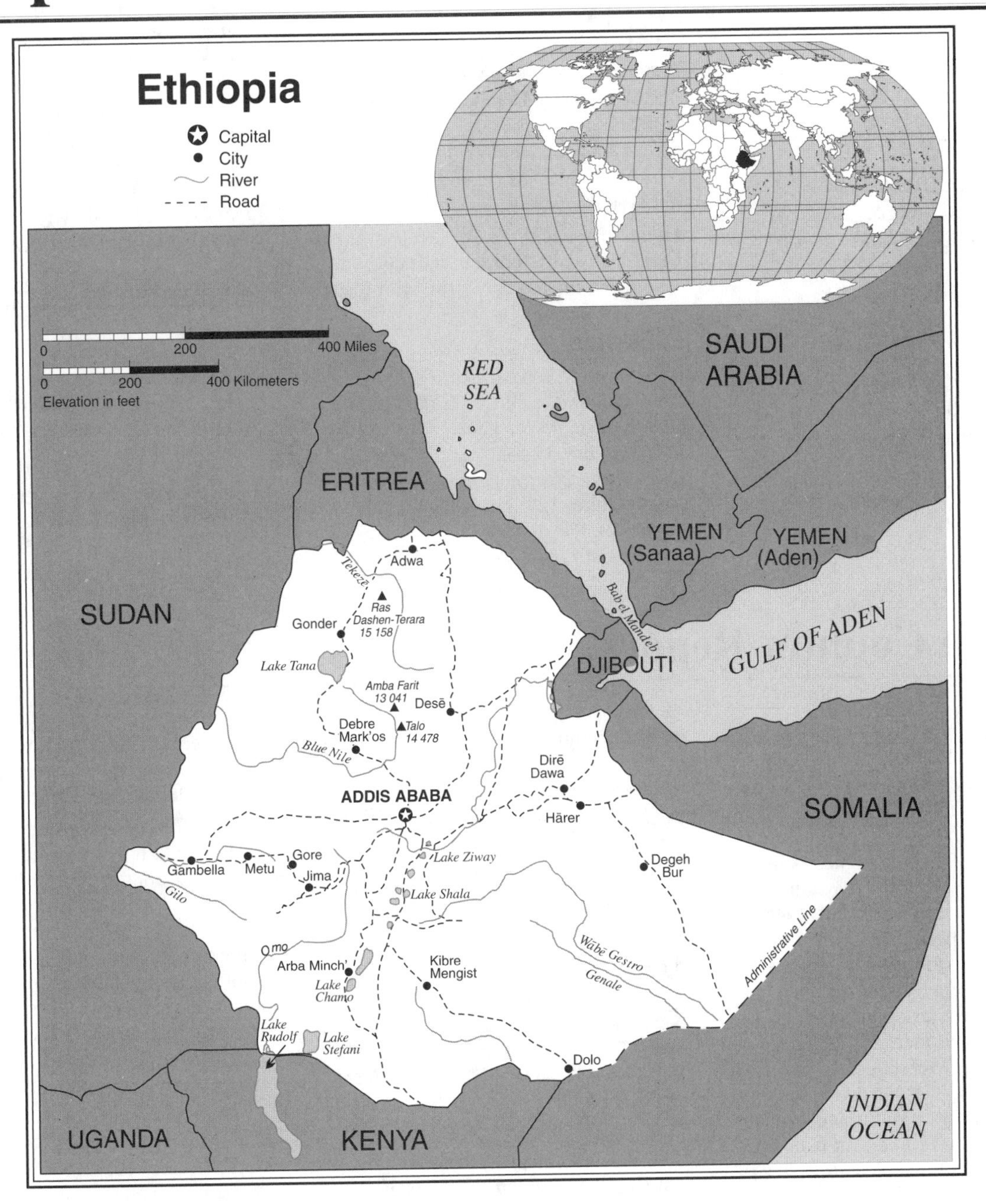

Ethiopia Statistics

GEOGRAPHY

Area in Square Miles (Kilometers): 435,071 (1,127,127) (about twice the size of Texas)

Capital (Population): Addis Ababa (2,753,000)

Environmental Concerns: deforestation; overgrazing; soil erosion; desertification

Geographical Features: a high plateau with a central mountain range divided by the Great Rift Valley; landlocked

Climate: tropical monsoon with wide topographic-induced variation

PEOPLE

Population

Total: 67,851,281

Annual Growth Rate: 1.89%

Rural/Urban Population Ratio: 83/17

Major Languages: Amharic; Tigrinya; Oromo; Somali; Arabic; Italian; English

Ethnic Makeup: 40% Oromo; 32% Amhara and Tigre; 9% Sidamo; 19% others

Religions: 45%–50% Muslims; 35%–40% Ethiopian Orthodox Christian; 12% animist; remainder others

Health

Life Expectancy at Birth: 43 years (male); 44 years (female)

Infant Mortality: 98.6/1,000 live births
Physicians Available: 1/36,600 people
HIV/AIDS Rate in Adults: 10.63%

Education

Adult Literacy Rate: 35.5%
Compulsory (Ages): 7–13; free

COMMUNICATION

Telephones: 435,000 main lines
Televisions: 4/1,000 people
Internet Users 75,000 (2003)

TRANSPORTATION

Highways in Miles (Kilometers): 17,016 (28,360)
Railroads in Miles (Kilometers): 425 (681)
Usable Airfields: 86
Motor Vehicles in Use: 66,000

GOVERNMENT

Type: federal republic
Independence Date: oldest independent country in Africa (at least 2,000 years)
Head of State/Government: President Girma Woldegiorgis; Prime Minister Meles Zenawi
Political Parties: Ethiopian People's Revolutionary Democratic Front; many others
Suffrage: universal at 18

MILITARY

Military Expenditures (% of GDP): 12.6%
Current Disputes: border conflicts with Somalia; uneasy cease-fire with Eritrea

ECONOMY

Currency ($ U.S. Equivalent): 8.72 birrs = $1
Per Capita Income/GDP: $700/$46 billion
GDP Growth Rate: -3.8%
Inflation Rate: 17.8%
Labor Force by Occupation: 80% agriculture; 12% government and services; 8% industry
Population Below Poverty Line: 64%
Natural Resources: gold; platinum, copper; potash; natural gas; hydropower
Agriculture: cereals; pulses; coffee; oilseed; sugarcane; vegetables; livestock
Industry: food processing; beverages; textiles; cement; building materials; hydropower
Exports: $442 million (primary partners Germany, Japan, Djibouti)
Imports: $1.54 billion (primary partners Saudi Arabia, United States, Japan)

SUGGESTED WEB SITES

http://www.ethiopians.com
http://www.ethiopianembassy.org
http://www.ethiopianspokes.net
http://www.ethiopiadaily.com
http://www.state.gov/www/background_notes/ethiopia_0398_bgn.html
http://www.cia.gov/cia/publications/factbook/geos/et.html

Ethiopia Country Report

The end of the year 2000 witnessed two major events in Ethiopia. In November, Haile Selassie, the former emperor of Ethiopia, was officially buried in Addis Ababa's Trinity Cathedral. And in December, Ethiopia and Eritrea signed a peace agreement in Algeria, formally ending two years of bloody armed conflict. The agreement, which followed a successful Ethiopian military offensive, established commissions to delineate the disputed border between the countries and provided for the exchange of prisoners and the return of displaced people. Subsequently the two countries officially accepted a new common border, drawn up by an independent commission in The Hague. But both sides made claims to the town of Badme.

DEVELOPMENT

There has been some progress in the country's industrial sector in recent years, after a sharp decline during the 1970s. Soviet-bloc investment resulted in the establishment of new enterprises in such areas as cement, textiles, and farm machinery.

In April 2001, thousands of demonstrators clashed with police in Addis Ababa, in protest against police brutality and in support of calls for political and academic freedom. A year earlier, President Meles Zenawi's Ethiopian People's Revolutionary Democratic Front (EPRDF) had won an easy victory in legislative elections against some 25 opposition parties. In almost half of the constituencies, EPRDF candidates ran unopposed.

FREEDOM

Despite its public commitment to freedom of speech and association, the EPRDF government has resorted to authoritarian measures against its critics. In 1995, Ethiopia had the highest number of jailed journalists in Africa. Basic freedoms are also compromised.

Following equally overwhelming and controversial election victories in 1994–1995, the EPRDF remains the country's dominant political group. The movement initially came to power in 1991 as an armed movement, following the successful overthrow of Ethiopia's Marxist-oriented military dictatorship, after years of struggle.

Since coming to power, the EPRDF has faced a wide spectrum of opponents. Some critics see its transformation of Ethiopia into a multiethnic federation of 14 self-governing regions as a threat to national unity. Others contend that its devolutionary structures are a sham designed to obscure its own determination to rule from the center as a virtual one-party state. International as well as domestic supporters, however, see Ethiopia's new Constitution as a bold experiment in institutionalizing a new model of multiethnic statehood.

The EPRDF had emerged in the 1980s as an umbrella movement fighting to liberate Ethiopia from the repressive misrule by the Provisional Military Administrative Council, popularly known as the *Dergue* (Amharic for "Committee"). The Dergue had come to power through a popular uprising against the country's former imperial order. It is still uncertain whether Ethiopia's second revolution in two decades will succeed where its first one failed.

Political instability has reduced Ethiopia from a developing breadbasket to a famine-ridden basket case. Interethnic conflict among an increasingly desperate population, many of whom have long had better access to arms and ammunition than to food and medicine, could lead to the state's disintegration. In the late 1990s, the problems facing the EPRDF government were intensified by the outbreak of the border war with Eritrea, which compromised the landlocked country's access to the sea.

AN IMPERIAL PAST

Ethiopia rivals Egypt as Africa's oldest country. For centuries, its kings claimed direct descent from the biblical King Solomon and the Queen of Sheba. Whether Ethiopia was the site of Sheba is uncertain,

(United Nations photo 122,841 by Muldoon)

From 1916 to 1974, Ethiopia was ruled by Haile Selassie, also known as Ras Tafari (from which today's term *Rastafarian* is derived). He is pictured above, on the left, shaking hands with the now infamous Idi Amin of Uganda.

as is the local claim that, prior to the birth of Christ, the country became the final resting place of the Ark of the Covenant holding the original Ten Commandments given to Moses (the Ark is said to survive in a local monastery).

Local history is better established from the time of the Axum Empire, which prospered from the first century. During the fourth century, the Axumite court adopted the Coptic Christian faith, which has remained central to the culture of Ethiopia's highland region. The Church still uses the Geez, the ancient Axumite tongue from which the modern Ethiopian languages of Amharic and Tigrinya are derived, in its services.

From the eighth century A.D., much of the area surrounding the highlands fell under Muslim control, all but cutting off the Copts from their European coreligionists. (Today, most Muslim Ethiopians live in the lowlands.) For many centuries, Ethiopia's history was characterized by struggles among the groups inhabiting these two regions and religions. Occasionally a powerful ruler would succeed in making himself truly "King of Kings" by uniting the Christian highlands and expanding into the lowlands. At other times, the mountains would be divided into weak polities that were vulnerable to the raids of both Muslim and non-Muslim lowlanders.

HEALTH/WELFARE

Ethiopia's progress in increasing literacy during the 1970s was undermined by the severe dislocations of the 1980s. By 1991, Ethiopia had some 500 government soldiers for every teacher.

MODERN HISTORY

Modern Ethiopian history began in the nineteenth century, when the highlands became politically reunited by a series of kings, culminating in Menilik II, who built up power by importing European armaments. Once the Coptic core of his kingdom was intact, Menilik began to spread his authority across the lowlands, thus uniting most of contemporary Ethiopia. In 1889 and 1896, Menilik also defeated invading Italian armies, thus preserving his empire's independence during the European partition of Africa.

From 1916 to 1974, Ethiopia was ruled by Ras Tafari (from which is derived the term *Rasta,* or *Rastafarian*), who, in 1930, was crowned Emperor Haile Selassie. The late Selassie remains a controversial figure. For many decades, he was seen both at home and abroad as a reformer who was modernizing his state. In 1936, after his kingdom had been occupied by Benito Mussolini, the leader of Italy, he made a memorable speech before the League of Nations, warning the world of the price it would inevitably pay for appeasing Fascist aggression. At the time, many African-Americans and Africans outside of Ethiopia saw Selassie as a great hero in the struggle of black peoples everywhere for dignity and empowerment. Selassie returned to his throne in 1941 and thereafter served as an elder statesman to the African nationalists of the 1950s and 1960s. However, by the latter decade, his own domestic authority was increasingly being questioned.

In his later years, Selassie could not, or would not, move against the forces that were undermining his empire. Despite its trappings of progress, the Ethiopian state remained quasi-feudal in character. Many of the best lands were controlled by the nobility and the Church, whose leading members lived privileged lives at the expense of the peasantry. Many educated people grew disenchanted with what they perceived as a

(United Nations photo 164612 by John Isaac)

Ethiopians experienced a brutal civil war from 1974 to 1991. The continuous fighting displaced millions of people. The problems of this forced migration were compounded by drought and starvation. The drought victims pictured above are gathered at one of the many relief camps.

reactionary monarchy and social order. Urban workers resented being paid low wages by often foreign owners. Within the junior ranks of the army and civil service, there was also great dissatisfaction with the way in which their superiors were able to siphon off state revenues for personal enrichment. But the empire's greatest weakness was its inability to accommodate the aspirations of the various ethnic, regional, and sectarian groupings living within its borders.

Ethiopia is a multiethnic state. Since the time of Menilik, the dominant group has been the Coptic Amhara-speakers, whose preeminence has been resented by their Tigrinya coreligionists as well as by predominantly non-Coptic groups such as the Afars, Gurages, Oromo, and Somalis. In recent years, movements fighting for ethnoregional autonomy have emerged among the Tigrinya of Tigray, the Oromo, and, to a lesser extent, the Afars, while many Somalis in Ethiopia's Ogaden region have long struggled for union with neighboring Somalia. Somali irredentism led to open warfare between the two principal Horn of Africa states in 1963–1964 and again in 1977–1978.

The former northern coastal province of Eritrea was a special case. From the late nineteenth century until World War II, it was an Italian colony. After the war, it was integrated into Selassie's empire. Thereafter, a local independence movement, largely united as the Eritrean People's Liberation Front (EPLF), waged a successful armed struggle, which led to Eritrea's full independence in 1993.

ACHIEVEMENTS

With a history spanning 2 millennia, the cultural achievements of Ethiopia are vast. Today, Addis Ababa is the site of the headquarters of the Organization of African Unity. Ethiopia's Kefe Province is the home of the coffee plant, from whence it takes its name.

REVOLUTION AND REPRESSION

In 1974, Haile Selassie was overthrown by the military, after months of mounting unrest throughout the country. A major factor triggering the coup was the government's inaction in 1972–1974, when famine swept across the northern provinces, claiming 200,000 lives. Some accused the Amhara government of using the famine as a way of weakening the predominantly Tigrinya areas of the empire. Others saw the tragedy simply as proof of the venal incompetence of Selassie's administration.

The overthrow of the old order was welcomed by most Ethiopians. Unfortunately, what began as a promising revolutionary transformation quickly degenerated into a repressive dictatorship, which pushed the nation into chronic instability and distress. By the end of 1974, after the first in a series of bloody purges within its ranks, the Dergue had embraced Marxism as its guiding philosophy. Revolutionary measures followed. Companies and lands were nationalized. Students were sent into the countryside to assist in land reforms and to teach literacy. Peasants and workers were organized into cooperative associations, called *kebeles.* Initial steps were also taken to end Amhara hegemony within the state.

Progressive aspects of the Ethiopian revolution were offset by the murderous nature of the regime. Power struggles within the Dergue, as well as its determination to eliminate all alternatives to its authority, contributed to periods of "red terror," during which thousands of supporters of the revolution as well as those associated with the old regime were killed. By 1977, the Dergue itself had been transformed from a collective decision-making body to a small clique loyal to Colonel

Mengistu Haile Mariam, who became a presidential dictator.

Mengistu sought for years to legitimize his rule through a commitment to Marxist-Leninism. He formally presided over a Commission for Organizing the Party of the Working People of Ethiopia, which, in 1984, announced the formation of a single-party state, led by the new Workers' Party. But real power remained in the hands of Mengistu's Dergue.

CIVIL WAR

From 1974 to 1991, Ethiopians suffered through civil war. In the face of oppressive central authority, ethnic-based resistance movements became increasingly effective in their struggles throughout much of the country. In the late 1970s, the Mengistu regime began to receive massive military aid from the Soviet bloc in its campaigns against the Eritreans and Somalis. Some 17,000 Cuban troops and thousands of other military personnel from the Warsaw Pact countries allowed the government temporarily to gain the upper hand in the fighting. The Ethiopian Army grew to more than 300,000 men under arms at any given time, the largest military force on the continent. Throughout the 1980s, military expenditures claimed more than half of the national budget.

Despite the massive domestic and international commitment on the side of the Mengistu regime, the rebels gradually gained the upper hand. Before 1991, almost all of northern Eritrea, except its besieged capital city of Asmara, had fallen to the EPLF, which had built up its own powerful arsenal, largely from captured government equipment. Local rebels had also liberated the province of Tigray and, as part of the EPRDF coalition, pushed south toward Ethiopia's capital city of Addis Ababa. In the south, independent Oromo and Somali rebels challenged government authority. There was also resistance to Mengistu from within the ranks of the national army. A major rebellion against his authority in 1989 was crushed, devastating military morale in the process. The regime was further undermined by the withdrawal of remaining Cuban and Soviet-bloc support.

Ethiopians have paid a terrible price for their nation's conflicts. Tens of thousands have been killed in combat, while many more have died from the side effects of war. In 1984–1985, the conscience of the world was moved by the images of mass starvation in the northern war zone. (At the time, however, the global media and concerned groups like Band Aid paid relatively little attention to the nonenvironmental factors that contributed to the crisis.) Up to 1 million lives were lost before adequate relief supplies reached the famine areas. Although drought and other environmental factors, such as soil erosion, contributed to the catastrophe, the fact that people continued to starve despite the availability of international relief can be attributed only to the use of food as a weapon of war.

There were other political constraints on local crop production. Having seized the lands of the old ruling class, the Mengistu regime, in accordance with its Marxist-Leninist precepts, invested most of its agricultural inputs in large state farms, whose productivity was abysmal. Peasant production also fell in nondrought areas, due to insecure tenure, poor producer prices, lack of credit, and an absence of consumer goods. Ethiopia's rural areas were further disrupted by the government's heavy-handed villagization and relocation schemes. In 1984–1985, thousands died when the government moved some 600,000 northerners to what were supposedly more fertile regions in the southwest. Many considered the scheme to be part of the central government's war effort against local communities resistant to its authority. By the same token, villagization has long been associated with counterinsurgency efforts; concentrated settlements allow occupying armies to exert greater control over potentially hostile populations.

UNCERTAIN PROSPECTS

The Dergue's demise has not as yet been accompanied by national reconciliation. Opposition to the EPRDF's attempt to transform Ethiopia into a multiethnic federation has been especially strong among Amharas, many of whom support the All-Amhara People's Organization. Others accuse the EPRDF—or, more especially, former Stalinists within the Tigrean People's Liberation Front (TPLF), which has been its predominant element—of trying to create its own monopoly of power.

In 1992, fighting broke out between the EPRDF and the forces of its former rebel partner, the Oromo Liberation Front (OLF), which claims to represent Ethiopia's largest ethnic group (Oromos constitute 40 percent of the population). The OLF was prominent among those who boycotted June 1992 local-government elections, which were further marred by allegations of vote-rigging and intimidation on behalf of the EPRDF. In December 1993, the OLF joined a number of other movements in a Council of Alternative Forces for Peace and Democracy (CAFPD). At its inaugural meeting, seven CAFPD delegates were detained for allegedly advocating the armed overthrow of the government. In April, another antigovernment coalition, the Ethiopian National Democratic party, was formed. Both movements called for an election boycott.

Another source of resistance to the EPRDF was the Ogadeni National Liberation Front (ONLF), which won strong support from Ogadeni Somalis in the June 1992 elections. In April, the Transitional Government removed ONLF's Hassan Jireh from power as the Ogaden region's elected administrator; in May, he was arrested. As a result, clashes occurred between the ONLF and EPRDF in the area, with the former boycotting the subsequent polls. Smaller uprisings and acts of terror, such as a January 1996 bombing of an Addis Ababa hotel, have posed further challenges for the government, which has also had to cope with drought. But while the extent of its electoral mandate is disputed, the EPRDF has demonstrated that it retains strong popular support, while its opposition is divided.

Timeline: PAST

1855
Emperor Tewodros begins the conquest and development of modern Ethiopia

1896
Ethiopia defeats Italian invaders at the Battle of Adowa

1936
Fascist Italy invades Ethiopia and rules until 1941

1961
The Eritrean liberation struggle begins

1972–1973
Famines in Tigray and Welo Provinces result in up to 200,000 deaths

1974
Emperor Haile Selassie is overthrown; the PMAC is established

1977
Diplomatic realignment and a new arms agreement with the Soviet Union

1980s
Massive famine, resulting from both drought and warfare

1990s
The Mengistu regime is overthrown by EPRDF rebels; Eritrea achieves independence of Ethiopia

PRESENT

2000s
Interethnic political tensions continue

Ethiopia and Eritrea work to forge a lasting peace

Kenya (Republic of Kenya)

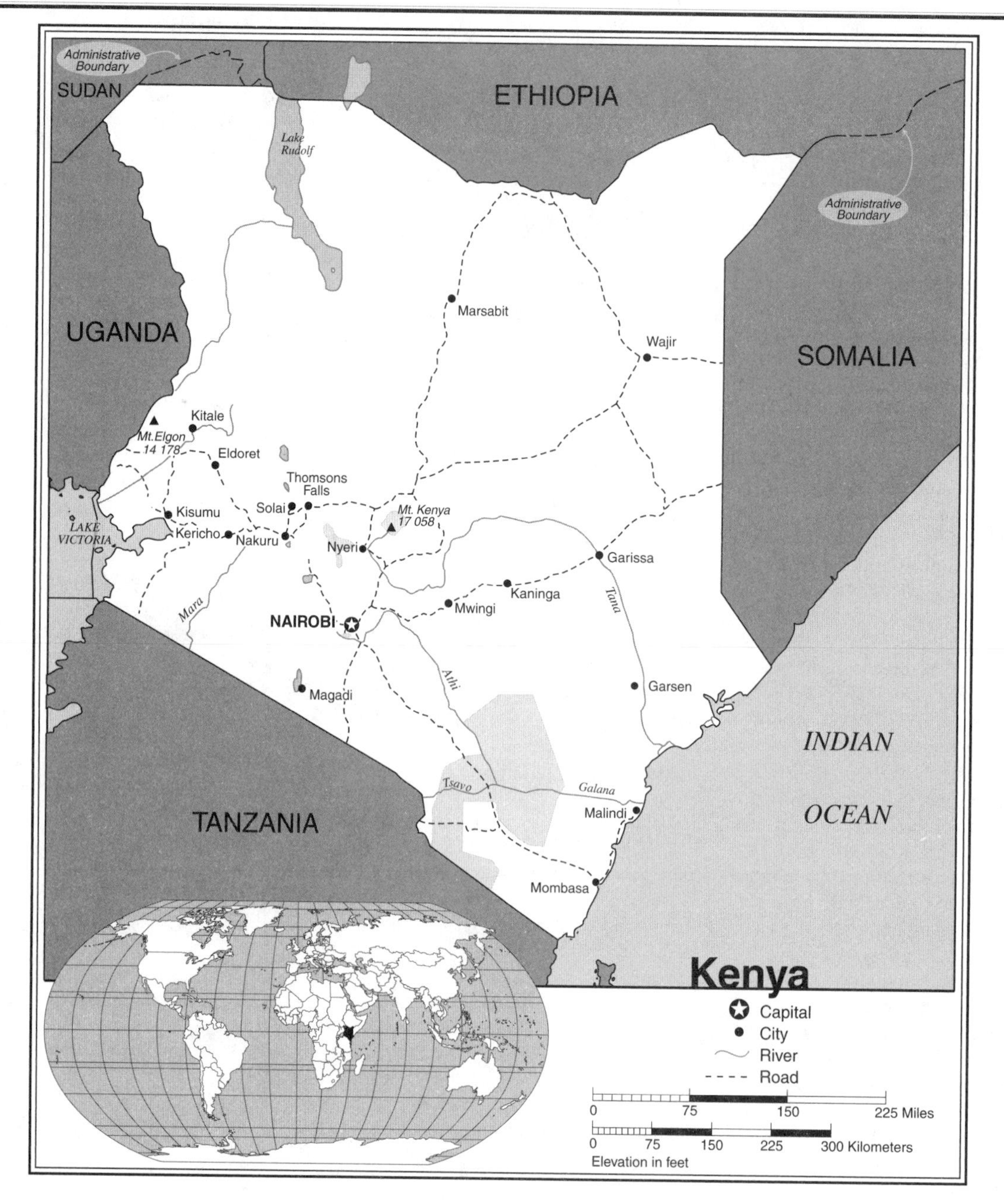

Kenya Statistics

GEOGRAPHY

Area in Square Miles (Kilometers): 224,900 (582,488) (twice the size of Nevada)

Capital (Population): Nairobi (2,343,000)

Environmental Concerns: water pollution; deforestation; soil erosion; poaching; water-hyacinth infestation in Lake Victoria

Geographical Features: low plains rise to central highlands bisected by the Great Rift Valley; fertile plateau in the west

Climate: tropical to arid

PEOPLE

Population

Total: 32,021,856

Annual Growth Rate: 1.15%

Rural/Urban Population Ratio: 68/32

Major Languages: English; Kiswahili; Maasai; others

Ethnic Makeup: 22% Kikuyu; 14% Luhya; 13% Luo; 12% Kalenjin; 11% Kamba; 28% others

Religions: 38% Protestant; 28% Catholic; 26% indigenous beliefs; 8% others

Health

Life Expectancy at Birth: 46 years (male); 48 years (female)
Infant Mortality: 67.2/1,000 live births
Physicians Available: 1/5,999 people
HIV/AIDS Rate in Adults: 13.5%

Education

Adult Literacy Rate: 78%
Compulsory (Ages): 6–14; free

COMMUNICATION

Telephones: 328,480 main lines
Televisions: 18/1,000 people
Internet Users: 400,000 (2002)

TRANSPORTATION

Highways in Miles (Kilometers): 38,198 (63,663)
Railroads in Miles (Kilometers): 1,654 (2,650)
Usable Airfields: 231
Motor Vehicles in Use: 357,000

GOVERNMENT

Type: republic
Independence Date: December 12, 1963 (from the United Kingdom)
Head of State/Government: President Mwai Kibaki is both head of state and head of government
Political Parties: Kenya African National Union; Forum for the Restoration of Democracy; Democratic Party of Kenya; others
Suffrage: universal at 18

MILITARY

Military Expenditures (% of GDP): 1.8%
Current Disputes: border conflict with Sudan; civil unrest and interethnic violence; tensions with Somalia

ECONOMY

Currency ($ U.S. Equivalent): 77.2 shillings = $1
Per Capita Income/GDP: $1,000/$31 billion
GDP Growth Rate: 1.5%
Inflation Rate: 9.8%
Unemployment Rate: 40%
Labor Force by Occupation: 75%–80% agriculture
Population Below Poverty Line: 50%
Natural Resources: gold; limestone; soda ash; salt barites; rubies; fluorspar; garnets; wildlife; hydropower
Agriculture: coffee; tea; corn; wheat; sugarcane; fruit; vegetables; livestock and dairy products
Industry: small-scale consumer goods; agricultural processing; oil refining; cement; tourism
Exports: $1.8 billion (primary partners Uganda, United Kingdom, Tanzania)
Imports: $3.1 billion (primary partners United Kingdom, United Arab Emirates, Japan)

SUGGESTED WEB SITES

http://www.kenyaweb.com
http://www.kenyaembassy.com
http://www.kentimes.com

Kenya Country Report

In December 2002, the 24-year rule of Kenya's second president, Daniel arap Moi, ended when Kenyans elected a new leader. Opposition-party candidate Mwai Kibaki won by a landslide. The defeat of the former ruling party under Moi will undoubtedly have a wide ranging set of implications for Kenya. The most important steps taken by the new government was the July 2004 proposal of a new constitution which is aimed at curtailing the powers of the president and creating a new post of Prime Minister.

Late in 2002, Moi had informed Kenyans that Uhuru Kenyatta, the son of Kenya's first president, Jomo Kenyatta, would serve as the ruling Kenyan African National Union (KANU) party presidential candidate. Uhuru Kenyatta had entered politics a year earlier, when Moi appointed him to both the Parliament and the cabinet.

DEVELOPMENT

Under its director Richard Leakey, the Kenyan Wildlife Service during the early 1990s cracked down on poachers while placing itself in the forefront of the global campaign to ban all ivory trading. As a result, local wildlife populations began to recover.

The designation of Kenyatta as Moi's successor caused a number of resignations by members of cabinet and other senior KANU leaders, who then formed a "National Rainbow Coalition." In October, the grouping combined with most of the other established opposition parties in rallying around former vice-president Kibaki as its presidential candidate.

The elections took place against a backdrop of economic adversity. With a major portion of its economy devoted to tourism, since the terrorist attacks in the United States on September 11, 2001, Kenya has been faced with the worst recession since independence. The election took place during a period of stress for the Kenyan tourism sector after the Novemeber 2002 bomb attack in which ten Kenyans and 3 Israelis were killed. A simultaneous rocket attack on an Israeli airliner failed. A statement, reportedly from Al-Qaeda, claimed responsibility. In addition, years of drought in northern and central Kenya have left more than 1 million people dependent on food supplies from the government.

The designation of Uhuru Kenyatta as his successor was but the latest in a series of surprise political moves by Moi. In June 2001, he reshuffled the cabinet, appointing opposition-party leader Raila Odinga as energy minister in the first coalition government in Kenya's history. Earlier, in 1999, President Moi had surprised many by appointing one of his best-known critics, the internationally renowned archaeologist and conservationist Richard Leakey, as head of Kenya's civil service, with an apparent mandate to restore the country's flagging economic fortunes by rooting out endemic corruption and lethargy in the public sector. While Leakey and his "Recovery Team" could claim some progress in shaking up the system, entrenched interests had largely frustrated its efforts.

FREEDOM

In 1997–1998, Kenya's already poor human-rights record deteriorated, with extra-judicial killings, beatings, and torture by the police. The National Youth Service, intended to provide young Kenyans with vocational training, was co-opted to block opposition political meetings. The worst atrocities, however, were linked to interethnic violence.

In August 1998, the world's attention had focused on the bombing of the U.S. Embassy building in Nairobi, Kenya's capital city. Scores of people in the vicinity of the blast were killed or injured. For Kenyans, it was an unprecedented act of international terrorism on their soil. It was also a further blow to the country's already troubled tourist industry, which had been hard hit by a continuing upward spiral of both criminal and political violence. The strength of the tourist industry—a critical foreign-exchange earner—coupled with years of growth in manufacturing and services had made Kenya, and especially Nairobi, the commercial center of East Africa. But today, after four decades of independence, most Kenyans remain the impoverished citizens of a state struggling to develop as a nation. And the recent restoration of multiparty politics has so far served only to intensify interethnic conflict.

In the precolonial past, Kenyan communities belonged to relatively small-scale, but economically interlinked, societies. Predominantly pastoral groups, such as the Maasai and Turkana, exchanged cattle for the crops of the Kalinjin, Kamba, Kikuyu, Luo, and others. Swahili city-states developed on the coast. In the 1800s, caravans of Arab as well as Swahili traders stimulated economic and political changes. However, the outsiders who had the greatest modern impact on the Kenyan interior were European settlers, who began to arrive in the first decade of the twentieth century. By the 1930s, much of the temperate hill country around Nairobi had become the "White Highlands." More than 6 million acres of land—Maasai pasture and Kikuyu and Kamba farms—were stolen by the settlers. African communities were often displaced to increasingly overcrowded reserves. Laborers, mostly Kikuyu migrants from the reserves, worked for the new European owners, sometimes on lands that they had once farmed for themselves.

HEALTH/WELFARE

Kenya's social infrastructure has been burdened by the influx of some 300,000 refugees from the neighboring states of Ethiopia, Somalia, and Sudan. Circumcision of girls under age 17 was banned in 2001. The lack of rainfall during the last four years has caused major crop failure and a major food crisis for consumers.

By the 1950s, African grievances had been heightened by increased European settlement and the growing removal of African "squatters" from their estates. There were also growing class and ideological differences among Africans, leading to tensions between educated Christians with middle-class aspirations and displaced members of the rural underclass. Many members of the latter group, in particular, began to mobilize themselves in largely Kikuyu oathing societies, which coalesced into the Mau Mau movement.

Armed resistance by Mau Mau guerrillas began in 1951, with isolated attacks on white settlers. In response, the British proclaimed a state of emergency, which lasted for 10 years. Without any outside support, the Mau Mau held out for years by making effective use of the highland forests as sanctuaries. Nonetheless, by 1955, the uprising had largely been crushed. Although the name Mau Mau became for many outsiders synonymous with antiwhite terrorism, only 32 European civilians actually lost their lives during the rebellion. In contrast, at least 13,000 Kikuyu were killed. Another 80,000 Africans were detained by the colonial authorities, and more than 1 million were resettled in controlled villages. While the Mau Mau were overwhelmed by often ruthless counterinsurgency measures, they achieved an important victory: The British realized that the preservation of Kenya as a white-settler–dominated colony was militarily and politically untenable.

In the aftermath of the emergency, the person who emerged as the charismatic leader of Kenya's nationalist movement was Jomo Kenyatta, who had been detained and accused by the British—without any evidence—of leading the resistance movement. At independence, in 1963, he became the president. He held the office until his death in 1978.

ACHIEVEMENTS

Each year, Kenya devotes about half of its government expenditures to education. Most Kenyan students can now expect 12 years of schooling. Tertiary education is also expanding.

To many, the situation in Kenya under Kenyatta looked promising. His government encouraged racial harmony, and the slogan Harambee (Swahili for "Pull together") became a call for people of all ethnic groups and races to work together for development. Land reforms provided plots to 1.5 million farmers. A policy of Africanization created business opportunities for local entrepreneurs, and industry grew. Although the Kenya African National Union was supposedly guided by a policy of "African Socialism," the nation was seen by many as a showcase of capitalist development.

POLITICAL DEVELOPMENT

Kenyatta's Kenya quickly became a de facto one-party state. In 1966, the country's first vice-president, Oginga Odinga, resigned to form an opposition party, the Kenyan People's Union (KPU). Three years later, however, the party was banned and its leaders, including Odinga, were imprisoned. Thereafter KANU became the focus of political competition, and voters were allowed to remove sitting members of Parliament, including cabinet ministers. But politics was marred by intimidation and violence, including the assassinations of prominent critics within government, most notably Economic Development Minister Tom Mboya, in 1969, and Foreign Affairs Minister J. M. Kariuki, in 1975. Constraints on freedom of association were justified in the interest of preventing ethnic conflict—much of the KPU support came from the Luo group. However, ethnicity has always been important in shaping struggles within KANU itself.

Under Daniel arap Moi, Kenyatta's successor, the political climate grew steadily more repressive. In 1982, his government was badly shaken by a failed coup attempt, in which about 250 people died and approximately 1,500 others were detained. The air force was disbanded and the university, whose students came out in support of the coup-makers, was temporarily closed.

In the aftermath of the coup, all parties other than KANU were formally outlawed. Moi followed this step by declaring, in 1986, that KANU was above the government, the Parliament, and the judiciary. Press restrictions, detentions, and blatant acts of intimidation became common. Those members of Parliament brave enough to be critical of Moi's imperial presidency were removed from KANU and thus Parliament. Political tensions were blamed on the local agents of an ever-growing list of outside forces, including Christian missionaries and Muslim fundamentalists, foreign academics and the news media, and Libyan and U.S. meddlers.

A number of underground opposition groups emerged during the mid-1980s, most notably the socialist-oriented Mwakenya movement, whose ranks included such prominent exiles as the writer Ngugi wa Thiong'o. In 1987, many of these groups came together to form the United Movement for Democracy, or UMOJA (Swahili for "unity"). But in the immediate aftermath of the 1989 KANU elections, which in many areas were blatantly rigged, Moi's grip on power appeared strong.

The early months of 1990, however, witnessed an upsurge in antigovernment unrest. In February, the murder of Foreign Minister Robert Ouko touched off rioting in Nairobi and his home city of Kisumu. Another riot occurred when squatters were forcibly evicted from the Nairobi shantytown of Muoroto. Growing calls for the restoration of multiparty democracy fueled a cycle of unrest and repression. The detention in July of two former cabinet ministers, Kenneth Matiba and Charles Rubia, for their part in the democracy agitation sparked nationwide rioting, which left at least 28 people dead and 1,000 arrested. Opposition movements, most notably the Forum for the Restoration of Democracy

(FORD), began to emerge in defiance of the government's ban on their activities.

Under mounting external pressure from Western donor countries as well as from his internal opponents, Moi finally agreed to the legalization of opposition parties, in December 1991. Unfortunately, this move failed to diffuse Kenya's increasingly violent political, social, and ethnic tensions.

Continued police harassment of the opposition triggered renewed rioting throughout the country. There was also a rise in interethnic clashes in both rural and urban areas, which many, even within KANU, attributed to government incitement. In the Rift Valley, armed members of Moi's own Kalinjin grouping attacked other groups for supposedly settling on their land. Hundreds were killed and thousands injured and displaced in the worst violence since the Mau Mau era. In the face of the government's cynical resort to divide-and-rule tactics, the fledgling opposition movement betrayed the hopes of many of its supporters by becoming hopelessly splintered. New groups, such as the Islamic Party of Kenya, openly appealed for support along ethnoreligious lines. More significantly, a leadership struggle in the main FORD grouping between Matiba and the veteran Odinga split the party into two, while a proposed alliance with Mwai Kibaki's Democratic Party failed to materialize. Although motivated as much by personal ambitions and lingering mistrust between KANU defectors and long-term KANU opponents, the FORD split soon took on an ethnic dimension, with many Kikuyu backing Matiba, while Luo remained solidly loyal to Odinga.

Taking skillful advantage of his opponents' disarray, Moi called elections in December 1992, which resulted in his plurality victory, with 36 percent of the vote. KANU was able to capture 95 of the 188 parliamentary seats that were up for grabs. The two FORD factions won 31 seats each, while the Democratic Party captured 23 seats. Notwithstanding voting irregularities, most independent observers blamed the divided opposition for sowing the seeds of its own defeat. The death of the widely respected Odinga in January 1994 coincided with renewed attempts to form a united opposition to KANU. In 1996, a group of opposition Members of Parliament was established.

Kenya's politics reflects class as well as ethnic divisions. The richest 10 percent of the population own an estimated 40 percent of the wealth, while the poorest 30 percent own only 10 percent. Past economic growth has failed to alleviate poverty. Kenya's relatively large middle class has grown resentful of increased repression and the evident corruption at the very top, but it is also fearful about perceived anarchy from below.

Although its rate of growth has declined during much of the past two decades, the Kenyan economy has expanded since independence. Nairobi is now the leading center of industrial and commercial activity in East Africa. While foreign capital has played an important role in industrial development, the largest share of investment has come from government and the local private sector.

A significant percentage of foreign-exchange earnings has come from agriculture. A wide variety of cash crops is exported, a diversity that has buffered the nation's economy to some degree from the uncertainties associated with single-commodity dependence. While large plantations—now often owned by wealthy Kenyans—have survived from the colonial era, much of the commercial production is carried out by small landholders.

A major challenge to Kenya's well-being has been its rapidly expanding population. Although there are some hopeful signs that women are beginning to plan for fewer children than in the past, the nation has been plagued with one of the highest population growth rates in the world, until very recently hovering around 3 percent per year (it is now estimated at 1.15 percent). More than half of all Kenyans are under age 15. Pressure on arable land is enormous. It will be difficult to create nonagricultural employment for the burgeoning rural-turned-urban workforce, even in the context of democratic stability and renewed economic growth.

THE ROAD AHEAD

Since the election there have been major steps taken to eliminate corruption in Kenya. Immediately following the election in January 2003, the government presented, and passed, a bill creating an anti-corruption commission. Subsequently, a Moi critic John Githongo was appointed as anti-graft czar. In appreciation for their efforts, in November 2003 the International Monetary Fund (IMF) resumed lending after a three-year gap, citing effective anti-corruption measures. One month later in an effort for peace and reconciliation, the government decided to grant former president Daniel arap Moi immunity from prosecution on corruption charges.

Timeline: PAST

1895
The British East African Protectorate is proclaimed

1900–1910
British colonists begin to settle in the Highlands area

1951
Mau Mau, a predominantly Kikuyu movement, resists colonial rule

1963
Kenya gains independence under the leadership of Jomo Kenyatta

1978
Daniel arap Moi becomes president upon the death of Kenyatta

1980s
A coup attempt by members of the Kenyan Air Force is crushed; political repression grows

1990s
Prodemocracy agitation leads to a return of multiparty politics; interethnic violence threatens democratic transition; President Daniel arap Moi is reelected; the U.S. Embassy building in Nairobi is bombed

PRESENT

2000s
Kenya seeks to root out public corruption

Kenya works to strengthen its economy and international reputation

Opposition candidate Mwai Kibaki wins the presidency, ending KANU rule

Wangari Maathi wins the Nobel Prize for her contribution towards the preservation of the Kenyan natural environment

Madagascar (Republic of Madagascar)

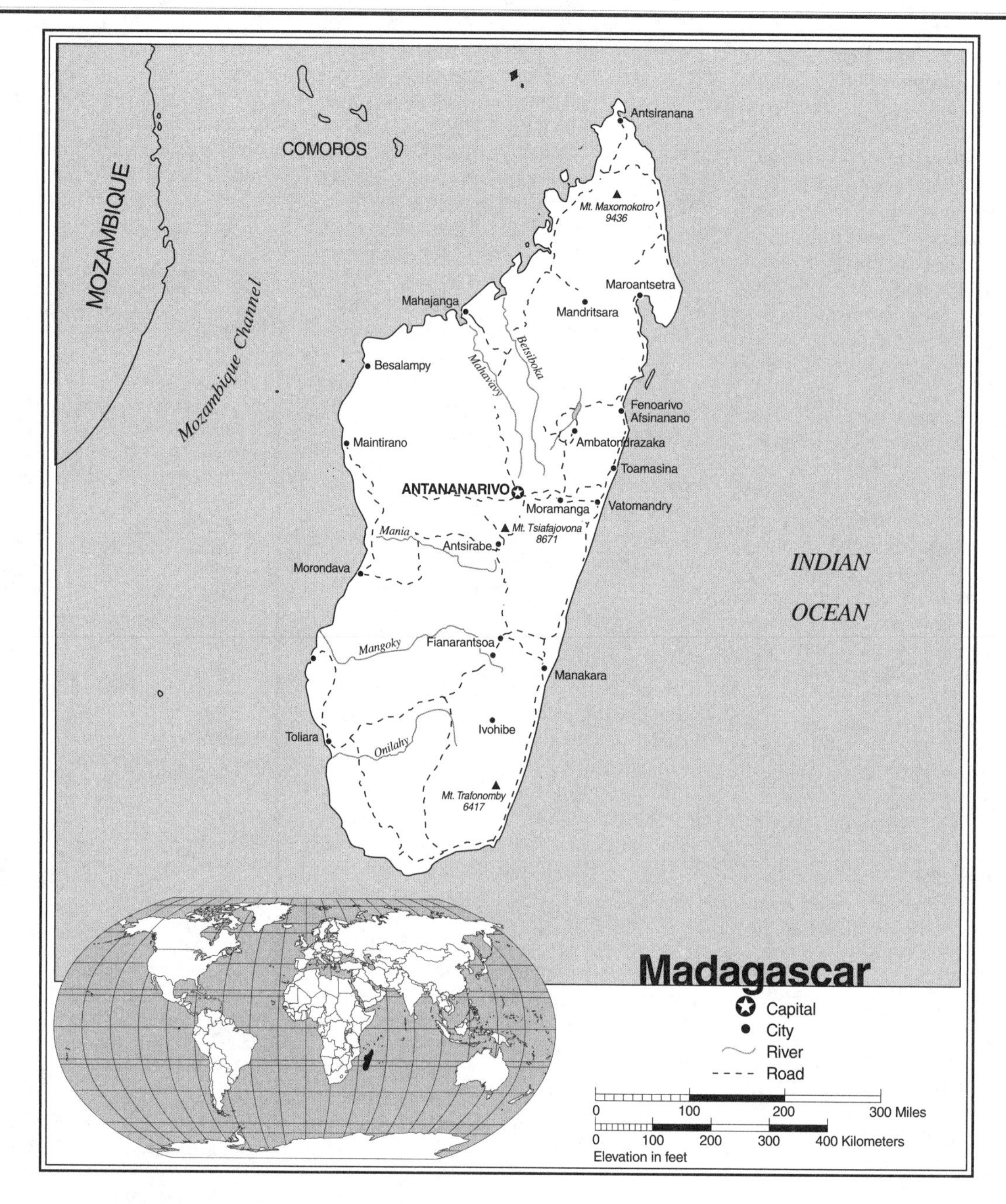

Madagascar Statistics

GEOGRAPHY

Area in Square Miles (Kilometers): 226,658 (587,041) (about twice the size of Arizona)

Capital (Population): Antananarivo (1,689,000)

Environmental Concerns: soil erosion resulting from deforestation and overgrazing; desertification; water contamination; endangered species

Geographical Features: narrow coastal plain; high plateau and mountains in the center; the world's fourth-largest island

Climate: tropical along the coast; temperate inland; arid in the south

PEOPLE

Population

Total: 17,501,871

Annual Growth Rate: 3.02%

Rural/Urban Population Ratio: 71/29

Major Languages: Malagasy; French

Ethnic Makeup: Malayo Indonesian; Cotiers; French; Indian; Creole; Comoran

Religions: 52% indigenous beliefs; 41% Christian; 7% Muslim

Health

Life Expectancy at Birth: 53 years (male); 58 years (female)

Infant Mortality: 81.9/1,000 live births
Physicians Available: 1/8,628 people
HIV/AIDS Rate in Adults: 0.15%

Education

Adult Literacy Rate: 46%
Compulsory (Ages): for 5 years between 6–13

COMMUNICATION

Telephones: 59,600 main lines
Daily Newspaper Circulation: 4/1,000 people
Televisions: 20/1,000 people
Internet Users: 70,500 (2003)

TRANSPORTATION

Highways in Miles (Kilometers): 29,900 (49,837)
Railroads in Miles (Kilometers): 530 (883)
Usable Airfields: 130
Motor Vehicles in Use: 76,000

GOVERNMENT

Type: republic
Independence Date: June 26, 1960 (from France)
Head of State/Government: President Marc Ravalomanana; Prime Minister Jacques Sylla
Political Parties: I Love Madagascar; Movement for the Progress of Madagascar; Renewal of the Social Democratic Party; others
Suffrage: universal at 18

MILITARY

Military Expenditures (% of GDP): 1.2%
Current Disputes: territorial disputes over islands administered by France; civil strife

ECONOMY

Currency ($ U.S. Equivalent): 6,161 francs = $1
Per Capita Income/GDP: $870/$14 billion
GDP Growth Rate: 6%
Inflation Rate: 8%
Population Below Poverty Line: 70%
Natural Resources: graphite; chromite; coal; bauxite; salt; quartz; tar sands; semiprecious stones; mica; fish; hydropower
Agriculture: coffee; vanilla; sugarcane; cloves; cocoa; rice; cassava (tapioca); beans; bananas; livestock products
Industry: meat processing; soap; breweries; tanneries; sugar; textiles; glassware; cement; automobile assembly; paper; petroleum; tourism
Exports: $680 million (primary partners France, United States, Germany)
Imports: $919 million (primary partners France, Hong Kong, China)

SUGGESTED WEB SITES

http://www.embassy.org
http://www.madagascarnews.com
http://www.cia.gov/cia/publications/factbook/geos/ma.html

Madagascar Country Report

Madagascar has been called the "smallest continent"; indeed, many geologists believe that it once formed the core of a larger landmass, whose other principal remnants are the Indian subcontinent and Australia. The world's fourth-largest island remains a world unto itself in other ways. Botanists and zoologists know it as the home of flora and fauna not found elsewhere. The island's culture is also distinctive. The Malagasy language, which with dialectical variations is spoken throughout the island, is related to the Malay tongues of distant Indonesia. But despite their geographic separation from the African mainland and their Asiatic roots, the Malagasy are very aware of their African identity.

DEVELOPMENT

In 1989, "Export Processing Zones" were established to attract foreign investment through tax and currency incentives. The government especially hopes to attract business from neighboring Mauritius, whose success with such zones has led to labor shortages and a shift toward more value-added production.

While the early history of Madagascar is the subject of much scholarly debate, it is clear that by the year A.D. 500, the island was being settled by Malay-speaking peoples, who may have migrated via the East African coast rather than directly from Indonesia. The cultural imprint of Southeast Asia is also evident in such aspects as local architecture, music, cosmology, and agricultural practices. African influences are equally apparent. During the precolonial period, the peoples of Madagascar were in communication with communities on the African mainland, and the waves of migration across the Mozambique channel contributed to the island's modern ethnic diversity.

During the early nineteenth century, most of Madagascar was united by the rulers of Merina. In seeking to build up their realm and preserve the island's independence, the Merina kings and queens welcomed European (mostly English) missionaries, who helped introduce new ideas and technologies. As a result, many Malagasy, including the royalty, adopted Christianity. The kingdom had established diplomatic relations with the United States and various European powers and was thus a recognized member of the international community. Foreign businesspeople were attracted to invest in the island's growing economy, while the rapid spread of schools and medical services, increasingly staffed by Malagasy, brought profound changes to the society.

The Merina court hoped that its "Christian civilization" and modernizing army would deter European aggression. But the French were determined to rule the island. The 1884–1885 Franco–Malagasy War ended in a stalemate, but a French invasion in 1895 led to the Merina kingdom's destruction. It was not an easy conquest. The Malagasy Army, with its artillery and modern fortifications, held out for many months; eventually, however, it was outgunned by the invaders. French sovereignty was proclaimed in 1896, but "pacification" campaigns continued for another decade.

FREEDOM

Respect for human rights has improved since 1993, and there has been little political violence since the 1996 election. There are isolated reports of police brutality against criminal suspects and detainees, as well as instances of arbitrary arrest and detention. Prison conditions are often life threatening, with women experiencing abuse, including rape. New judges are being appointed in an effort to relieve the overburdened judiciary.

French rule reduced what had been a prospering state into a colonial backwater. The pace of development slowed as the local economy was restructured to serve the interests of French settlers, whose numbers had swelled to 60,000 by the time of World War II. Probably the most important French contribution to Madagascar was the encouragement their misrule gave to the

(United Nations photo 140091 by L. Rajaonina)

Madagascar has a unique ethnic diversity, created by migrations from the African mainland and Southeast Asia. The varied ethnic makeup of the population can be seen in the faces of these schoolchildren.

growth of local nationalism. By the 1940s, a strong sense of Malagasy identity had been forged through common hatred of the colonialists.

HEALTH/WELFARE

Primary-school enrollment is now universal. Thirty-six percent of the appropriate age group attend secondary school, while 5% of those ages 20 to 24 are in tertiary institutions. Malaria remains a major health challenge. Madagascar's health and education facilities are underfunded.

The local overthrow of Vichy power by the British in 1943 created an opening for Malagasy nationalists to organize themselves into mass parties, the most prominent of which was the Malagasy Movement for Democratic Renewal (MRDM). In 1946, the MRDM elected two overseas deputies to the French National Assembly, on the basis of its call for immediate independence. France responded by instructing its administrators to "fight the MRDM by every means." Arrests led to resistance. In March 1947, a general insurrection began. Peasant rebels, using whatever weapons they could find, liberated large areas from French control. French troops countered by destroying crops and blockading rebel areas, in an effort to starve the insurrectionists into surrendering. Thousands of Malagasy were massacred. By the end of the year, the rebellion had been largely crushed, although a state of siege was maintained until 1956. No one knows precisely how many Malagasy lost their lives in the uprising, but contemporary estimates indicate about 90,000.

INDEPENDENCE AND REVOLUTION

Madagascar gained its independence in 1960. However, many viewed the new government, led by Philibert Tsiranana of the Social Democratic Party (PSD), as a vehicle for continuing French influence; memories of 1947 were still strong. Lack of economic and social reform led to a peasant uprising in 1971. This Maoist-inspired rebellion was suppressed, but the government was left weakened. In 1972, new unrest, this time spearheaded by students and workers in the towns, led to Tsiranana's overthrow by the military. After a period of confusion, radical forces consolidated power around Lieutenant Commander Didier Ratsiraka, who assumed the presidency in 1975.

Under Ratsiraka, a new Constitution was adopted that allowed for a controlled process of multiparty competition, in which all parties were grouped within the National Front. Within this framework, the largest party was Ratsiraka's Vanguard of the Malagasy Revolution (AREMA). Initially, all parties were expected to support the president's Charter of the Malagasy Revolution, which called for a Marxist-oriented socialist transformation. In accordance with the Charter, foreign-owned banks and financial institutions were nationalized. A series of state enterprises were also established to promote industrial development, but few proved viable.

Although 80 percent of the Malagasy were employed in agriculture, investment in rural areas and concerns was modest. The government attempted to work through *fokonolas* (indigenous village-management bodies). State farms and collectives were also established on land expropriated from French settlers. While these efforts led to some improvements, such as increased mechanization, state marketing monopolies and planning constraints contributed to shortfalls. Efforts to keep consumer prices low were blamed for a drop in rice production, the Malagasy staple, while cash-crop production, primarily coffee, vanilla, and cloves, suffered from falling world prices.

ACHIEVEMENTS

A recently established wildlife preserve will allow the unique animals of Madagascar to survive and develop. Sixty-six species of land animals are found nowhere else on earth, including the aye-aye, a nocturnal lemur that has bat ears, beaver teeth, and an elongated clawed finger, all of which serve the aye-aye in finding food.

Since 1980, Madagascar has experienced grave economic difficulties, which have given rise to political instability. Food shortages in towns have led to rioting, while frustrated peasants have abandoned their fields. Ratsiraka's government turned increasingly from socialism to a greater reliance on market economics. But the economy has remained impoverished.

In 1985, having abandoned attempts to make the National Front into a vehicle for a single-party state, Ratsiraka presided over a loosening of his once-authoritarian control. In February 1990, most remaining restrictions on multiparty politics were lifted. But the regime's opponents, including a revived PSD, became militant in their demands for a new constitution. After six months of crippling strikes and protests, Ratsiraka formally ceded many of his powers to a transitional government, headed by Albert Zafy, in November 1991. In February 1993, Zafy won the presidency by a large margin. But subsequent divisions with Parliament over his rejection of an International Monetary Fund austerity plan, accompanied by allegations of financial irregularities, led to his impeachment in August 1996. In elections held at the end of the year, Ratsiraka made a comeback, narrowly defeating Zafy. More than half of the population, however, stayed away from the polls.

Marc Ravalomanana shocked most outside observers when, under the banner of his newly organized I Love Madagascar (TIM) party, he claimed outright victory in presidential elections in December 2001. A bitter seven-month struggle for power ensued with the president of 27 years, Didier Ratsiraka.

Ravalomanana argued that, having won a majority of the votes in the first round of the elections, there need not be another round of voting. Ratsiraka, on the other hand, sought solace in the Constitution and demanded another round of voting. since many saw the elections as rigged in favor of Ratsiraka, it was widely perceived in Madagascar that Ravalomanana was the winner and that Ratsiraka was merely stalling for time and acting as an impediment to hold onto power. In April 2002, after a recount, the High Constitutional Court named Ravalomanana the winner of the December 2001 polls, but Ratsiraka ignored the verdict.

Once the demand for new elections was turned down, Ratsiraka mobilized the army and took over various areas of Madagascar. Without Madagascar Army support, Ravalomanana organized a motley crew of reserves and attacked the army-held positions. The result was chaotic. For seven months the country had two presidents and two capitals. Widespread demonstrations and worker strikes paralyzed trade, commerce, and industry, with political discourse becoming intransigent. The dispute caused immediate economic depression, as Ratsiraka's allies put up economic barricades throughout much of the country. Gaoline and medical supplies were in short supply, as foreign reserves held in U.S. banks were frozen.

In July 2002, shortly after the United States and France recognized Ravalomanana as the legitimate leader, Ratsiraka and various family members flew into exile in the Seychelles, and his forces on the island either surrendered or switched sides. Ravalomanana has promised to use his entrepreneurial flair to fight the poverty and unemployment that afflict many Madagascans. But he has inherited an economy that is suffering after months of economic disruption and political violence.

The African Union has so far refused to accept Ravalomanana's presidency and did not admit him to the initiation of the organization in September 2002. The AU had demanded that new elections be held to resolve the issue. The lack of AU recognition, however, has been offset by recognition from France, Britain, and the United States.

New elections were not held and as the president took office he began to search out and prosecute corruption in the country. During February 2003 the former head of the armed forces was arrested and charged with attempting a coup against President Ravalomanana. In August 2003, exiled former president Didier Ratsiraka was sentenced to 10 years hard labor for embezzling public funds. In December former Prime Minister Tantely Andrianarivo was also sentenced to 12 years hard labor for abuse of office. For his efforts at rooting out corruption, the World Bank and International Monetary Fund wrote off nearly half of Madagascar's debt—around two billion dollars.

Timeline: PAST

1828
Merina rulers gain sovereignty over other peoples of the island

1884–1885
Franco-Malagasy War

1904
The French complete the conquest of the island

1947–1948
A revolt is suppressed by the French, with great loss of life

1960
Independence from France; Philibert Tsiranana becomes the first president

1972
A coup leads to the fall of the First Malagasy Republic

1975
Didier Ratsiraka becomes president by military appointment

1980s
Economic problems intensify

1990
Elections in 1989–1990 strengthen multiparty democracy

PRESENT

2000s
Ratsiraka finally cedes power to Marc Ravalomanana

Mauritius (Republic of Mauritius)

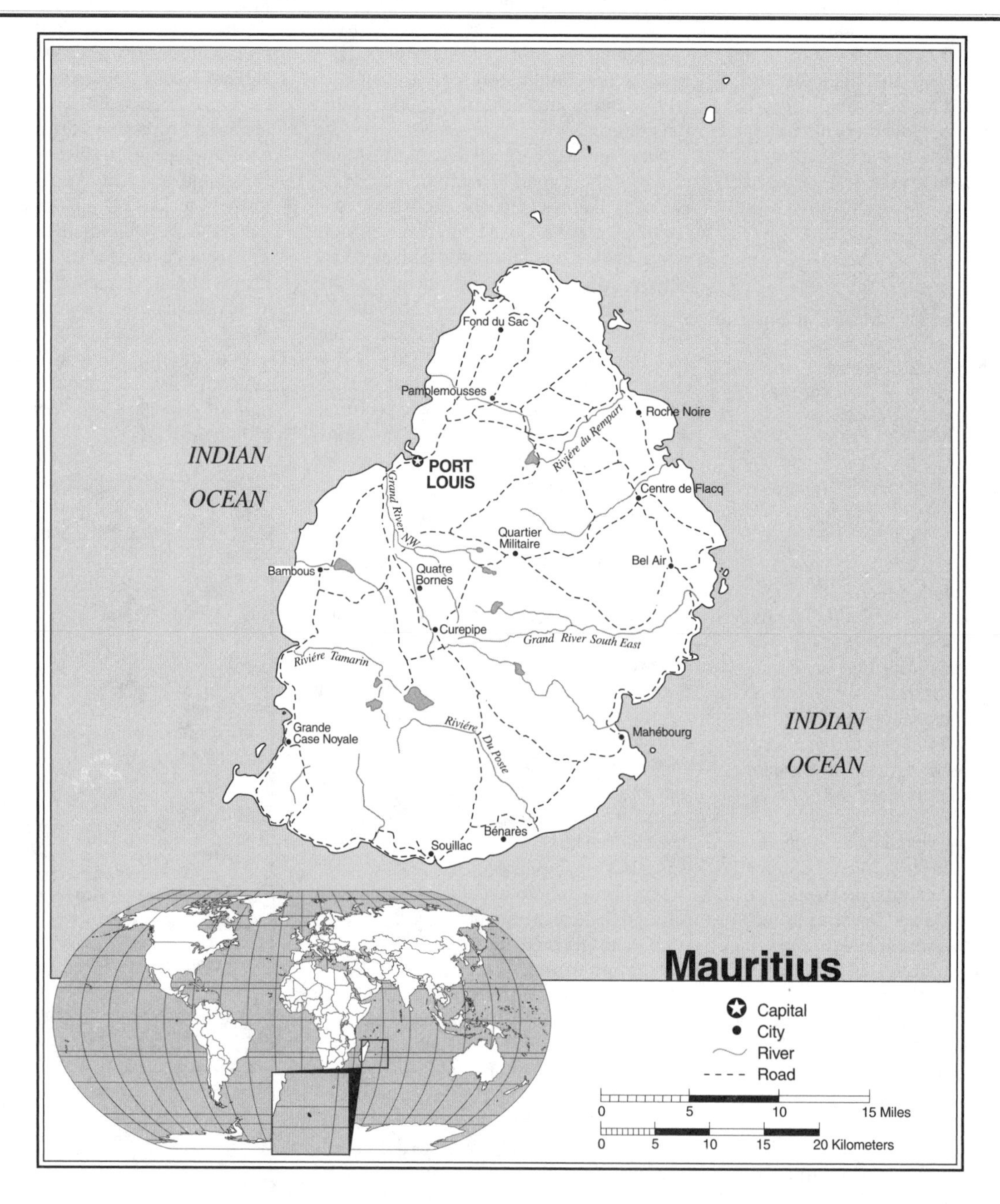

Mauritius Statistics

GEOGRAPHY

Area in Square Miles (Kilometers): 720 (1,865) (about 11 times the size of Washington, D.C.)
Capital (Population): Port Louis (176,000)
Environmental Concerns: water pollution; population pressures on land and water resources
Geographical Features: a small coastal plain rising to discontinuous mountains encircling a central plateau
Climate: tropical

PEOPLE

Population

Total: 1,200,000
Annual Growth Rate: 0.86%
Rural/Urban Population Ratio: 59/41
Major Languages: English; Creole; French; Hindi; Urdu; Hakka; Bojpoori
Ethnic Makeup: 68% Indo-Mauritian; 27% Creole; 3% Sino-Mauritian; 2% Franco-Mauritian
Religions: 52% Hindu; 28% Christian; 17% Muslim; 3% others

Health

Life Expectancy at Birth: 67 years (male); 75 years (female)
Infant Mortality: 16.6/1,000 live births
Physicians Available: 1/1,182 people
HIV/AIDS Rate in Adults: 0.08%

Education

Adult Literacy Rate: 83%
Compulsory (Ages): 5–12

COMMUNICATION

Telephones: 348,200 main lines
Daily Newspaper Circulation: 49/1,000 people
Televisions: 150/1,000 people
Internet Users: 150,000 (2003)

TRANSPORTATION

Highways in Miles (Kilometers): 1,126 (1,877)
Railroads in Miles (Kilometers): none
Usable Airfields: 5
Motor Vehicles in Use: 82,500

GOVERNMENT

Type: parliamentary democracy
Independence Date: March 12, 1968 (from the United Kingdom)
Head of State/Government: President Sir Anerood Jugnauth; Prime Minister Paul Berenger
Political Parties: Mauritian Labor Party; Mauritian Militant Movement; Militant Socialist Movement; Mauritian Militant Renaissance; Mauritian Social Democratic Party; Organization of the People of Rodrigues; Hizbullah; others
Suffrage: universal at 18

MILITARY

Military Expenditures (% of GDP): 0.2%
Current Disputes: territorial disputes with France and the United Kingdom

ECONOMY

Currency ($ U.S. equivalent): 27.4 rupees = $1
Per Capita Income/GDP: $11,400/$13.85 billion
GDP Growth Rate: 4.1%
Inflation Rate: 4.2%
Unemployment Rate: 9.8%
Labor Force by Occupation: 36% construction and industry; 24% services; 14% agriculture; 26% others
Population Below Poverty Line: 10%
Natural Resources: arable land; fish
Agriculture: sugarcane; tea; corn; potatoes; bananas; pulses; cattle; goats; fish
Industry: food processing; textiles; apparel; chemicals; metal products; transport equipment; nonelectrical machinery; tourism
Exports: $1.6 billion (primary partners United Kingdom, France, United States)
Imports: $2 billion (primary partners South Africa, France, India)

SUGGESTED WEB SITES

http://www.mauritius-info.com
http://www.ncb.intnet.ma/govt/
http://www.sas.upenn.edu/African_Studies/Country_Specific/Mauritius.html

Mauritius Country Report

Although it was not permanently settled until 1722, today Mauritius is home to 1.2 million people of South Asian, Euro-African, Chinese, and other origins. Out of this extraordinary human diversity has emerged a society that in recent decades has become a model of democratic stability and economic growth as well as ethnic, racial, and sectarian tolerance.

DEVELOPMENT

The success of the Mauritian EPZ along with the export-led growth of various Asian economies has encouraged a growing number of other African countries, such as Botswana, Cape Verde, and Madagascar, to launch their own export zones.

Mauritius was first settled by the French, some of whom achieved great wealth by setting up sugar plantations. From the beginning, the plantations prospered through their exploitation of slave labor imported from the African mainland. Over time, the European and African communities merged into a common Creole culture; that membership currently accounts for one quarter of the Mauritian population. A small number claim pure French descent. For decades, members of this latter group have formed an economic and social elite. More than half the sugar acreage remains the property of 21 large Franco–Mauritian plantations; the rest is divided among nearly 28,000 small landholdings. French cultural influence remains strong. Most of the newspapers on the island are published in French, which shares official-language status with English. Most Mauritians also speak a local, French-influenced, Creole language. Most Mauritian Creoles are Roman Catholics.

In 1810, Mauritius was occupied by the British; they ruled the island until 1968. (After years of debate, in 1992, the country cut its ties with Great Britain to become a republic.) When the British abolished slavery, in 1835, the plantation owners turned to large-scale use of indentured labor from what was then British India. Today nearly two thirds of the population are of South Asian descent and have maintained their home languages. Most are Hindu, but a substantial minority are Muslim. Other faiths, such as Buddhism, are also represented.

Although the majority of Mauritians gained the right to vote after World War II, the island has maintained an uninterrupted record of parliamentary rule since 1886. Ethnic divisions have long been important in shaping political allegiances. But ethnic constituency-building has not led, in recent years, to communal polarization. Other factors—such as class, ideology, and opportunism—have also been influential. All postindependence governments have been multiethnic coalitions.

FREEDOM

Political pluralism and human rights are respected on Mauritius, but problem areas remain. There are reports of police abuse of suspects and delayed access to defense counsel. Child labor exists. Legislation outlawing domestic violence has been passed recently. The nation has more than 30 political parties, of which about a half dozen are important at any given time. The Mauritian labor movement is one of the strongest in all of Africa.

While government in Mauritius has been characterized by shifting coalitions, with no single party winning a majority of seats, there has been relative stability in leadership at the top. Over the past half-century, Mauritius has had only three prime ministers. In September 2000, a coalition led by former prime minister Sir Aneerood Jugnauth won a landslide victory, ousting the rival coalition of Navin Ramgoolam, who had ousted Jugnauth five years earlier. Jugnauth had first come to power in 1982 by defeating Seewoosagar Ramgoolam, Navin's father.

HEALTH/WELFARE

Medical and most educational expenses are free. Food prices are heavily subsidized. Rising government deficits, however, threaten future social spending. Mauritius has a high life expectancy rate and a low infant mortality rate. Human-rights education has been introduced in secondary schools.

In February 2002 a major turn of political events occurred. Political power changed hands in Mauritius as Cassam Uteem resigned as president after refusing to sign the controversial anti-terrorism bill. The vice president, who also refused to sign the bill, resigned as well. Later in the year Karl Hoffman was elected president by National Assembly, but stepped down in September 2003. President Anerood Jugnauth took his place on October 7, 2003.

Although most major political parties have in the past espoused various shades of socialism, Mauritius's economic success in recent decades has created a strong consensus in favor of export-oriented market economics. Until the 1970s, the Mauritian economy was almost entirely dependent on sugar. While 45 percent of the island's total landmass continues to be planted with the crop, sugar now ranks below textiles and tourism in its contribution to export earnings and gross domestic product. The transformation of Mauritius from monocrop dependency into a fledging industrial state with a strong service sector has made it one of the major economic success stories of the developing world. Mauritian growth has been built on a foundation of export-oriented manufacturing. At the core of the Mauritian takeoff is its island-wide Export Processing Zone (EPZ), which has attracted industrial investment through a combination of low wages, tax breaks, and other financial incentives. Although most of the EPZ output has been in the field of cheap textiles, the economy has begun to diversify into more capital- and skill-intensive production. The year 2002 saw the Cyber Cities plan launched to create concentrations of hi-tech facilities and to boost the economy by adding thousands of jobs. In 1989, Mauritius also entered the international financial services market by launching Africa's first offshore banking center.

ACHIEVEMENTS

Perhaps Mauritius's most important modern achievement has been its successful efforts to reduce its birth rate. This has been brought about by government-backed family planning as well as by increased economic opportunities for women.

The success of the Mauritian economy is measured in relative terms. Mauritius is still considered a middle-income country. In reality, however, there are, as with most developing societies, great disparities in the distribution of wealth. Nonetheless, quality-of-life indicators confirm a rising standard of living for the population as a whole. While great progress has been made toward eliminating poverty and disease, concern has also grown about the environmental capacity of the small, crowded country to sustain its current rate of development. There is also a general recognition that Mauritian prosperity is—and will for the foreseeable future remain—extremely vulnerable to global-market forces. This was demonstrated after the September 11, 2001, terrorist attacks in the United States, when tourism took a precipitous dive, and previously with the declining level of sugar production caused by the cyclone in 1999. The much-acclaimed ethnic diversity of Mauritius also came under examination as there were racially inspired riots in 2000 between Creoles and Hindu communities after the death while in police custody of a Creole Rastafarian pop star, Rasta Karya.

Timeline: PAST

1600s
The Dutch claim, but abandon, Mauritius

1722
French settlers arrive, and slaves imported from the African mainland

1814
The Treaty of Paris formally cedes Mauritius to the British

1835
Slavery is abolished; South Asians arrive

1937
Rioting on sugar estates shakes the political control of the Franco-Mauritian elite

1948
An expanded franchise allows greater democracy

1968
Independence

1979
A cyclone destroys homes as well as much of the sugar crop

1982
Aneerood Jugnauth replaces Seewoosagar Ramgoolam as prime minister

1990s
Mauritius becomes a republic; Mauritius becomes the 12th member of the SADC

PRESENT

2000s
Jugnauth regains the prime ministership in a landslide electoral victory

Mauritius continues its focus on market economics

Anerood Jugnauth is made president after Karl Offmann steps down

Rwanda (Rwandese Republic)

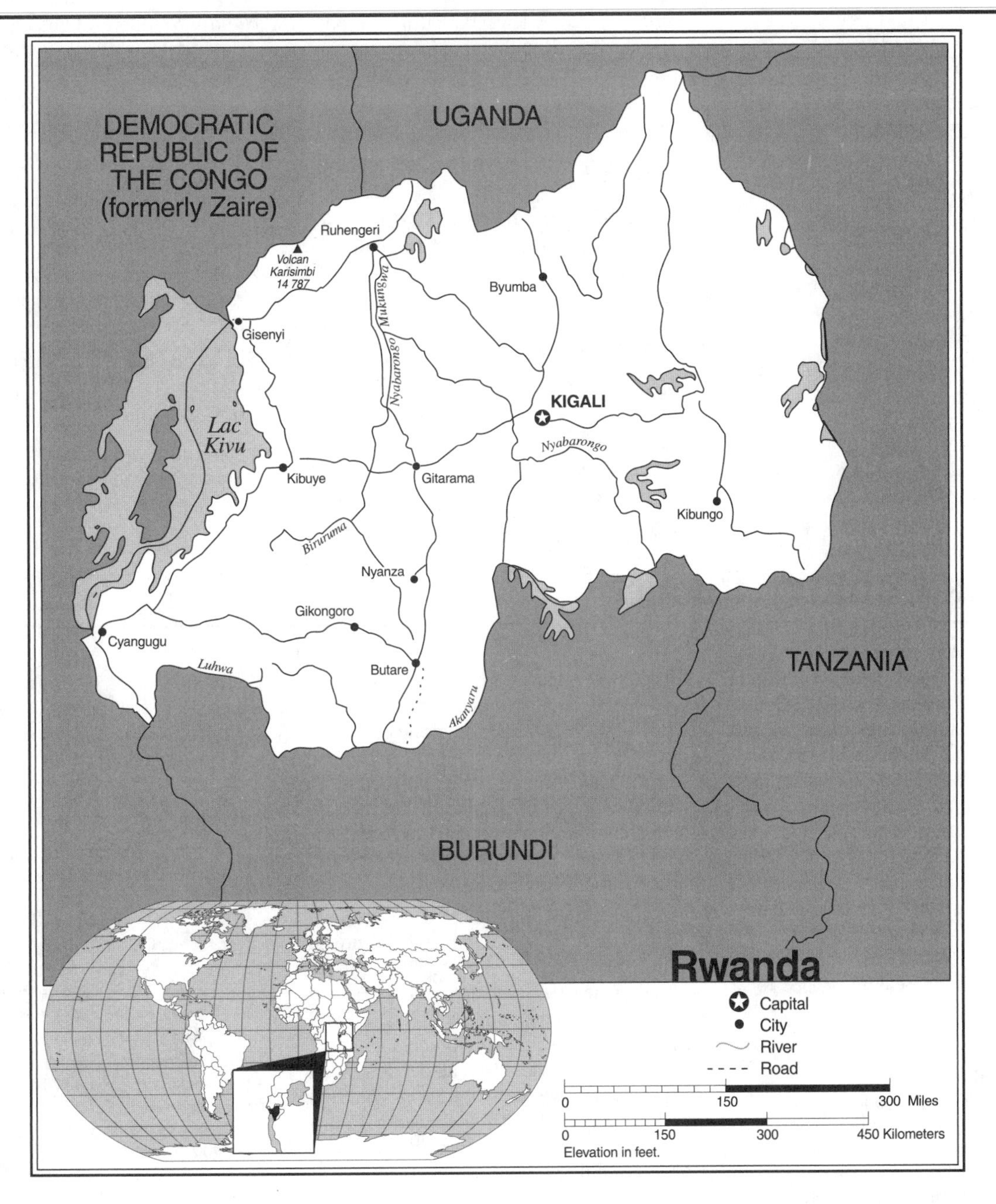

Rwanda Statistics

GEOGRAPHY

Area in Square Miles (Kilometers): 10,169 (26,338) (about the size of Maryland)

Capital (Population): Kigali (412,000)

Environmental Concerns: deforestation; overgrazing; soil exhaustion and erosion; poaching

Geographical Features: mostly grassy uplands and hills; landlocked

Climate: temperate

PEOPLE

Population

Total: 7,954,013

Annual Growth Rate: 1.82%

Rural/Urban Population Ratio: 94/6

Major Languages: Kinyarwanda; French; Kiswahili; English

Ethnic Makeup: 84% Hutu; 15% Tutsi; 1% Twa

Religions: 95% Christian; 4% Muslim; 2% none or other

Health

Life Expectancy at Birth: 38 years (male); 39 years (female)

Infant Mortality: 118/1,000 live births

Physicians Available: 1/50,000 people

HIV/AIDS Rate in Adults: 11.21%

Education

Adult Literacy Rate: 60.5%
Compulsory (Ages): 7–14

COMMUNICATION

Telephones: 23,200 main lines
Internet Users: 25,000 (2002)

TRANSPORTATION

Highways in Miles (Kilometers): 7,200 (12,000)
Railroads in Miles (Kilometers): none
Usable Airfields: 8
Motor Vehicles in Use: 28,000

GOVERNMENT

Type: republic
Independence Date: July 1, 1962 (from Belgian-administered UN trusteeship)
Head of State/Government: President Paul Kagame; Prime Minister Bernard Makuza
Political Parties: Rwanda Patriotic Front; Democratic Republican Movement; Liberal Party; Democratic Socialist Party; others
Suffrage: universal at 18

MILITARY

Military Expenditures (% of GDP): 3.1%
Current Disputes: continued internal ethnic violence; conflict in the Democratic Republic of the Congo

ECONOMY

Currency ($ U.S. equivalent): n/a
Per Capita Income/GDP: $1,300/$10 billion
GDP Growth Rate: 3.5%
Inflation Rate: 7.5%
Labor Force by Occupation: 90% agriculture
Population Below Poverty Line: 70%
Natural Resources: tungsten ore; tin; cassiterite; methane; hydropower; arable land
Agriculture: coffee; tea; pyrethrum; bananas; sorghum; beans; potatoes; livestock
Industry: agricultural-products processing; mining; light consumer goods; cement
Exports: $61 million (primary partners Europe, Pakistan, United States)
Imports: $248 million (primary partners Kenya, Europe, United States)

SUGGESTED WEB SITES

http://www.rwanda1.com/government/
http://www.rwandanews.org
http://www.cia.gov/cia/publications/factbook/geos/rw.html

Rwanda Country Report

In July 2002, the presidents of Rwanda and the Democratic Republic of the Congo (D.R.C.) signed a peace agreement to end their four-year war. With South Africa and the United Nations acting as guarantors, the deal committed Rwanda to the withdrawal of troops from eastern D.R.C., and the D.R.C. to helping disarm Rwandan Hutu refugees in its territory who were responsible for the 1994 genocide that targeted Rwanda's Tutsi minority. By October 2002, Rwanda had honored its commitment, bringing cautious hope for a lasting peace.

DEVELOPMENT

Hydroelectric stations meet much of the country's energy needs. Before the 1994 genocide, plans were being made to exploit methane-gas reserves under Lake Kivu.

Rwandans have been living a nightmare. The 1994 genocide ranks as one of the world's greatest tragedies. A small country, only about the size of Maryland, Rwanda had 8 million inhabitants early in 1994, making it continental Africa's most densely populated state. This population was divided into three groups: the Hutu majority (89 percent); the Tutsis (10 percent); and the Twa, commonly stereotyped as "Pygmies" (1 percent). By September 1994, civil war involving genocidal conflict between Hutus and Tutsis had nearly halved the country's resident population while radically altering its group demography. Up to 1 million people, including women and children, were killed and an additional 2 million displaced. The international community, especially France, the United States, and the United Nations, did little to stop the genocide.

Rebel forces of the ousted government that initiated the genocide have continued to threaten Rwanda, from the D.R.C. In response, since 1997 Rwanda has intervened against its opponents in the D.R.C. as well as in Rwanda itself. In the process, much of eastern D.R.C. has come under Rwandan occupation. Rwanda's ability to send and sustain some 50,000 well-armed troops into the D.R.C. would probably not have been possible without the tacit support of its donors. A UN report released in October 2002 confirmed the involvement of the Rwandan, as well as the Ugandan and Zimbabwean, militaries in the exploitation of the D.R.C.'s natural resources.

The dominant figure in Rwanda over the past decade has been Paul Kagame, who in April 2000 was inaugurated as president, following the resignation of Pasteur Bizimungu. The change at the top had little immediate effect. As vice-president and minister of defense, Kagame, through his largely Tutsi-controlled Rwanda Patriotic Front (RPF), had dominated the "Government of National Unity" since its installation in 1994. While the resignation of Bizimungu and other members of the mainly Hutu Republican Democratic Movement (RDM) raised eyebrows, the RDM itself opted to remain within the government.

FREEDOM

The Rwandan government and its opponents continue to be responsible for serious human-rights abuses, including massacres. More than 120,000 prisoners are in overcrowded jails, most accused of participating in the 1994 genocide. Genocide trials, which began at the end of 1996, have made little progress and are expected to take years to complete. Hutu death squads, composed of members of the defeated former Rwandan Armed Forces and Interahamwe genocide gangs, continue to target Tutsis and foreigners.

At the end of 2001, voting to elect members of traditional *gacaca* (courts) began. The courts, in which ordinary Rwandans judge their peers, aim to clear the backlog of 1994 genocide cases. Just as important, justice can be seen to be done by the people themselves, based on their traditional standards.

THE 1994 GENOCIDE

The genocide began within a half-hour of the April 6, 1994, death of the country's democratizing dictator, President Juvenal Habyarimana. Along with president Cyprien

Ntaryamira of neighboring Burundi, Habyarimana was killed when his plane was shot down. While the identity of the culprits remains a matter of speculation, Belgian troops reported that rockets were fired from Kanombe military base, which was then controlled by the country's Presidential Guard, known locally as the Akuza. In broadcasts over the independent Radio Libre Mille Collins, Hutu extremists then openly called for the destruction of the Tutsis—"The graves are only half full, who will help us fill them up?" Because the tirades were in idiomatic Kinyarwanda, the national language, they initially escaped the attention of most international journalists on the spot. Meanwhile, Akuza and regular army units set up roadblocks and began systematically to massacre Tutsi citizens in the capital, Kigali. Even greater numbers perished in the countryside, because their names appeared on death lists that had been prepared with the help of local Hutu chiefs.

By July, more than 500,000 people had been murdered. While most were Tutsis, Hutus who had supported moves toward ethnic reconciliation and democratization had also been targeted. During December 2003, three former media directors were found guilty of inciting Hutus to kill Tutsis during the 1994 genocide and received lengthy jail sentences. In addition to elements within the Hutu-dominated military, the killings were carried out by youth-wing militias of the ruling party, the National Revolutionary Movement for Development (MRND) and the Coalition for Defense of Freedom (CDR). Known respectively as the Interahamwe and Impuzamugambi, the ranks of these two all-Hutu militias had mushroomed in the aftermath of an August 1993 agreement designed to return the country to multiparty rule. The extent of the killings was apparent in neighboring Tanzania, where thousands of corpses were televised being carried downstream by the Kagara River. At a rate of 80 an hour, they entered Lake Victoria, more than 100 miles from the Rwandan border.

SYSTEMATIC PLANS FOR MASS MURDER

Preparations for the genocide had been going on for months. According to Amnesty International, Hutu "Zero Network" death squads had already murdered some 2,300 people in the months leading up to the crisis. Although this information was the subject of press reports, no action was taken by the 2,500 peacekeeping troops who had been stationed in the country since June 1993 as the United Nations Assistance Mission to Rwanda (UNAMIR). Once the crisis began, most of UNAMIR's personnel were hastily withdrawn. French and Belgian paratroopers arrived for a brief time to evacuate their nationals.

Rwanda's genocide did not end with the destruction of a third or more of the Tutsi minority. Enraged by the massacres of their brethren, the 14,000-man Tutsi-dominated Rwanda Patriotic Front, which had been waging an armed struggle against the Habyarimana regime since October, launched a massive offensive. The 35,000-man regular army, along with the militias, crumbled. In July 1994, the RPF took full control of Kigali and drove the remnants of the government and its army eastward into Zaire (since 1997, the D.R.C.). Two million panic-stricken Hutu civilians also fled across the border. By then, about 1 million Rwandans were already in exile. Another 2.5 million people were crowded into a "safe zone" created by the French military. As the French prepared to pull out, the fate of these refugees was uncertain.

HEALTH/WELFARE

The health system has collapsed. International relief agencies in Rwanda and neighboring countries are attempting to feed and care for the population. Diseases like cholera have spread rapidly, to deadly effect.

In depopulated Kigali, the RPF set up a "Provisional Government" with a Hutu president, Pasteur Bizimungu, and prime minister Faustin Twagiramungu. Its most powerful figure, however, was the RPF commander, Major General Paul Kagame, who became both the vice-president and minister of defense. In June 2004, former president Pasteur Bizimungu was sentenced to 15 years in jail for embezzlement, inciting violence and associating with criminals.

HUTUS AND TUTSIS

The roots of Hutu–Tutsi animosity in Rwanda (as well as Burundi) run deep. Yet it is not easy for an outsider to differentiate between the two groups. Their members both speak Kinyarwanda and look the same physically, notwithstanding the stereotype of the Tutsis being exceptionally tall; intermarriage between the two groups has taken place for centuries. By some accounts, the Tutsi arrived as northern Nilotic conquerors, perhaps in the fifteenth century. But others believe that the two groups have always been defined by class or caste rather than by ethnicity.

In the beginning, according to one epic Kinyarwanda poem, the godlike ruler Kigwa fashioned a test to choose his successor. He gave each of his sons a bowl of milk to guard during the night. His son Gatwa drank the milk. Yahutu slept and spilled the milk. Only Gatutsi guarded it well. The myth justifies the old Rwandan social order, in which the Twas were the outcasts, the Hutus servants, and the Tutsis aristocrats. Historically, Hutu serfs herded cattle and performed various other services for their Tutsi "protectors." At the top of the hierarchy was the Mwami, or king.

ACHIEVEMENTS

Abbé Alexis Kagame, a Rwandan Roman Catholic churchman and scholar, has written studies of traditional Rwandan poetry and has written poetry about many of the traditions and rituals. Some of his works have been composed in Kinyarwanda, an official language of Rwanda, and translated into French. He has gained an international reputation among scholars.

THE COLONIAL ERA: HUTU AND TUTSI ANIMOSITIES CONTINUE

Rwanda's feudal system survived into the colonial era. German and, later, Belgian administrators opted to rule through the existing order. But the social order was subtly destabilized by the new ideas emanating from the Catholic mission schools and by the colonialists' encouragement of the predominantly Hutu peasantry to grow cash crops, especially coffee. Discontent also grew due to the ever-increasing pressure of people and herds on already crowded lands.

In the late 1950s, under UN pressure, Belgium began to devolve political power to Rwandans. The death of the Mwami in 1959 sparked a bloody Hutu uprising against the Tutsi aristocracy. Tens of thousands, if not hundreds of thousands, were killed. Against this violent backdrop, pre-independence elections were held in 1961. These resulted in a victory for the first president, Gregoire Kayibanda's, Hutu Emancipation Movement (better known as Parmehutu). Thus, at independence, in 1962, Rwanda's traditionally Tutsi-dominated society was suddenly under a Hutu-dominated government.

In 1963 and 1964, the continued interethnic competition for power exploded into more violence, which resulted in the flight of hundreds of thousands of ethnic Tutsis to neighboring Burundi, Tanzania, and Uganda. Along with their descendants, this refugee population today numbers about 1 million. Successive Hutu-dominated governments have barred their return, questioning their citizenship and citing extreme land pressure as barriers to their re-absorption. But the implied hope that the refugees would integrate into their host societies has failed to materialize. The RPF was originally formed in Uganda by Tutsi exiles, many of whom were hardened veterans of that country's past conflicts. The repatriation of all Rwandan Tutsis has been a key RPF demand.

HABYARIMANA TAKES POWER

Major General Juvenal Habyarimana, a Hutu from the north, seized power in a military coup in 1973. Two years later, he institutionalized his still army-dominated regime as a one-party state under the MRND, in the name of overcoming ethnic divisions. Yet hostility between the Hutus and Tutsis remained. Inside the country, a system of ethnic quotas was introduced, which formally limited the remaining Tutsi minority to a maximum of 14 percent of the positions in schools and at the workplace. In reality, the Tutsis were often allocated less, while the MRND's critics maintained that the best opportunities were reserved for Hutus from Habyarimana's northern home area of Kisenyi.

POPULAR DISCONTENT

In the 1980s, many Hutus, as well as Tutsis, grew impatient with their government's corrupt authoritarianism. The post-1987 international collapse of coffee prices, Rwanda's major export-earner, led to an economic decline, further fueling popular discontent. Even before the armed challenge of the RPF, the MRND had agreed to give up its monopoly of power, though this pledge was compromised by continued repression. Prominent among the new parties that then emerged were the Democratic Republic Movement (MDR), the Social Democrats (PSD), and the Liberals (PL). The PL and PSD were able to attract both Hutu and Tutsi support. As a result, many of their Hutu as well as Tutsi members were killed in 1994. The MDR was associated with southern-regional Hutu resentment at the MRND's supposed northern bias.

A political breakthrough occurred in March 1992 with the formation of a "Transitional Coalition Government," headed by the MDR's Dismas Nsengiyaremye, which also included MRND, PSD, and PL ministers. Habyarimana remained as president. With French military assistance, including the participation of several hundred French "advisers," Habyarimana's interim government of national unity was able to halt the RPF's advance in 1992. A series of cease-fires was negotiated with the RPF, leading up to the promise of (but never-realized) UN–supervised elections. But from the beginning, progress toward national reconciliation was compromised by hard-line Hutus within the ruling military/MRND establishment and the extremist CDR. Ironically, these elements, who conspired to carry out the anti-Tutsi genocide in order to maintain control, were pushed out of the country by the RPF. These Hutu officials, soldiers, and militarymen, thought to be responsible for massacring hundreds of thousands of Tutsis and moderate Hutus during the 1994 Civil War, were exiled to camps in Zaire (D.R.C.) and Tanzania. Soon these militants returned and began a two-month wave of killings in western Rwanda in an apparent attempt to stop the Tutsis from testifying at genocide trials being conducted by the Rwandan government and the United Nations.

As Rwanda civil unrest intensified, refugees continued to flow into strife-torn Zaire. Estimates are that between 100,000 and 350,000 Hutu citizens are still in camps there. In March 1997, some 70,000 Rwandan Hutu refugees were gathered in Ubunda, a town 80 miles south of Kisangani on the Zaire River. But civil war in the D.R.C. has caused these people to be pushed back into Rwanda, where they are faced with the persistent and serious unrest in their home country. In October 2004 the government sent nearly 400 Rwandan troops as part of a peacekeeping mission in Sudan's Darfus region.

RWANDA TODAY

The years 2002–2004 were extremely important in Rwanda. During this period a new constitution, designed to prevent another genocide, was approved by the electorate. The document bans the incitement of ethnic hatred and calls for regular elections of political officials. In August 2003, in the first presidential elections since the 1994 genocide, Paul Kagame won a landslide victory. With his election as president, Kagame's role in Rwandan life became legitimized. October 2003 witnessed the first multi-party parliamentary elections, and President Kagame's Rwandan Patriotic Front won an absolute majority of the seats.

Although the election results seem to indicate support for President Kagame and his ruling party, they came with a tremendous amount of controversy. European Union (EU) observers noted that the 2003 elections had been marred by massive fraud and irregularities. During March 2004 an independent French report concluded that President Kagame ordered the 1994 attack on the president's plane, which in turn sparked the genocide. Although President Kagame immediately rejected the conclusion of the French report, its implications shall haunt the nation for many years to come.

Timeline: PAST

A.D. 1860–1895
Mwami Kigeri Rwabugiri expands and consolidates the kingdom

1916
Belgium rules Rwanda as a mandate of the League of Nations

1959
The Hutu rebellion

1962
Rwanda becomes independent; Gregoire Kayibana is president

1973
Juvenal Habyarimana seizes power

1975
The National Revolutionary Movement for Development is formed

1978
A new Constitution is approved in a nationwide referendum; Habyarimana is reelected president

1990s
Genocidal conflict results in a dramatic drop in the country's resident population; millions are killed or displaced; French relief workers withdraw from Rwanda; Tutsi massacre of Hutu at Kibeho refugee camp

PRESENT

2000s
Paul Kagame becomes president

Rwandans occupy much of eastern Democratic Republic of the Congo

Seychelles (Republic of Seychelles)

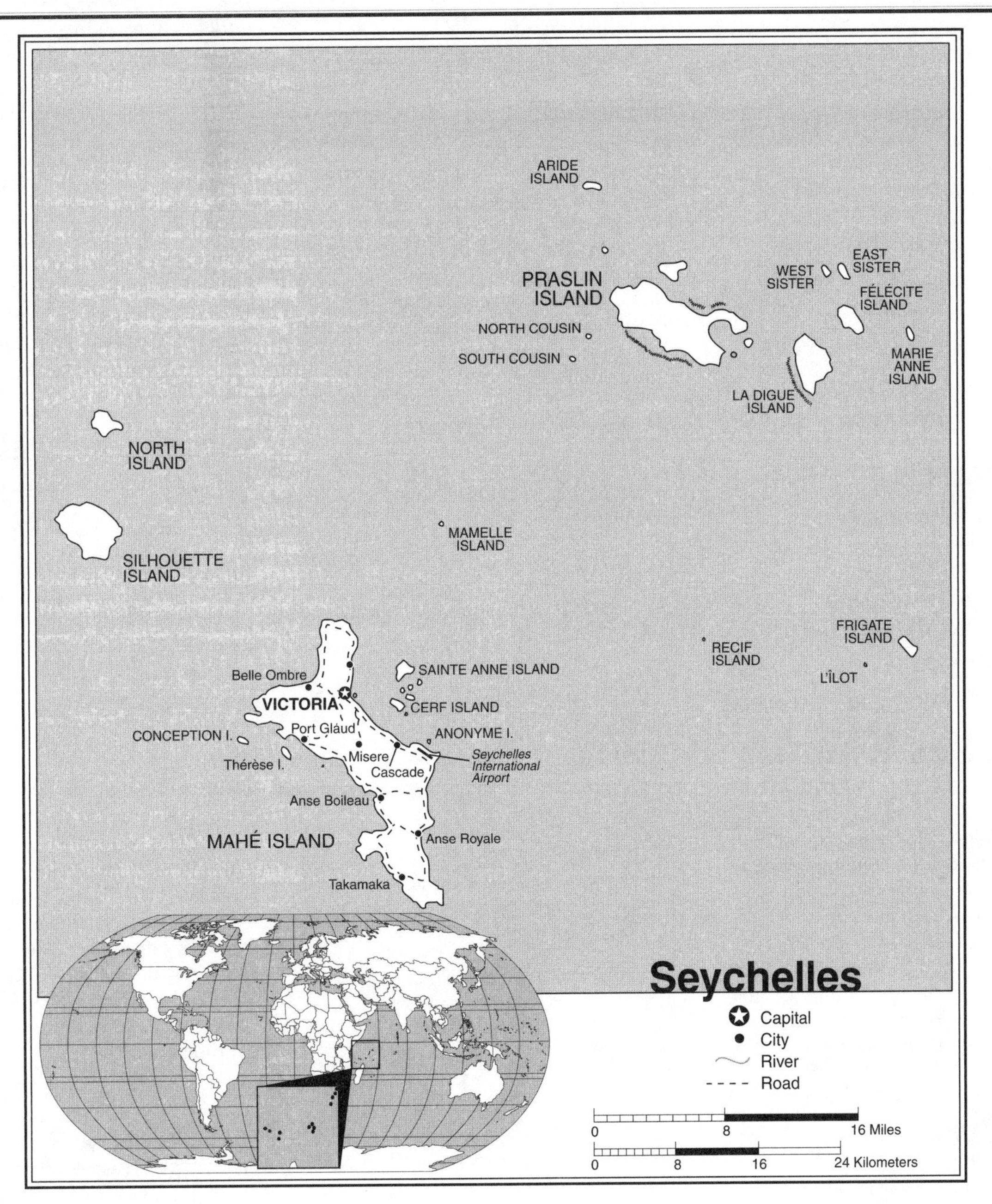

Seychelles Statistics

GEOGRAPHY

Area in Square Miles (Kilometers): 185 (455) (2.5 times the size of Washington, D.C.)

Capital (Population): Victoria (30,000)

Environmental Concerns: uncertain freshwater supply

Geographical Features: Mahé Group is granitic, narrow coastal strip; rocky, hilly; other islands are coral, flat, elevated reefs

Climate: tropical marine

PEOPLE

Population

Total: 80,100

Annual Growth Rate: 0.45%
Rural/Urban Population Ratio: 37/63
Major Languages: English; French; Creole
Ethnic Makeup: Seychellois (mixture of Asians, Africans, and Europeans)
Religions: 96% Christian; 4% others

Health

Life Expectancy at Birth: 65 years (male); 76 years (female)
Infant Mortality: 16.8/1,000 live births
Physicians Available: 1/906 people
HIV/AIDS Rate in Adults: na

Education

Adult Literacy Rate: 58%
Compulsory (Ages): 6–15; free

COMMUNICATION

Telephones: 21,700 main lines
Daily Newspaper Circulation: 41/1,000 people
Televisions: 173/1,000 people
Internet Users: 11,700 (2002)

TRANSPORTATION

Highways in Miles (Kilometers): 162 (270)
Railroads in Miles (Kilometers): none
Usable Airfields: 14
Motor Vehicles in Use: 8,600

GOVERNMENT

Type: republic
Independence Date: June 29, 1976 (from the United Kingdom)
Head of State/Government: President James Michel is both head of state and head of government
Political Parties: Seychelles People's Progressive Front; Seychelles National Party; Democratic Party
Suffrage: universal at 17

MILITARY

Military Expenditures (% of GDP): 1.8%
Current Disputes: claims Chagos Archipelago

ECONOMY

Currency ($ U.S. Equivalent): 5.66 rupees = $1
Per Capita Income/GDP: $7,800/$626 million
GDP Growth Rate: 1.5%
Inflation Rate: 3.3%
Labor Force by Occupation: 71% services; 19% industry; 10% agriculture
Natural Resources: fish; copra; cinnamon trees
Agriculture: vanilla; coconuts; sweet potatoes; cinnamon; cassava; bananas; chickens; fish
Industry: tourism; fishing; copra and vanilla processing; coconut oil; boat building; printing; furniture; beverages
Exports: $182.6 million (primary partners United Kingdom, Italy, France)
Imports: $360 million (primary partners Italy, South Africa, France)

SUGGESTED WEB SITES

http://www.seychelles-online.com/sc
http://www.sas.upenn.edu/African_Studies/Country_Specific/Seychelles.html
http://www.cia.gov/cia/publications/factbook/geos/se.html

Seychelles Country Report

Africa's smallest country in terms of both size and population, the Republic of Seychelles consists of a number of widely scattered archipelagos off the coast of East Africa. Over the last quarter-century, Seychellois have enjoyed enormous economic and social progress. According to the United Nations, today they enjoy the highest standard of living in Africa.

FREEDOM

Since the restoration of multiparty democracy, there has been greater political freedom in Seychelles. The opposition nonetheless continues to complain of police harassment and to protest about the government's control over the broadcast media.

DEVELOPMENT

Seychelles has declared an Exclusive Economic Zone of 200 miles around all of its islands in order to promote the local fishing industry. Most of the zone's catch is harvested by foreign boats, which are supposed to pay licensing fees to Seychelles.

But for many years, the country's politics was bitterly polarized between supporters of President James Mancham and his successor, Albert René. The holding of multiparty elections in 1992 and 1993, after 15 years of single-party rule under René, has been accompanied by a significant degree of reconciliation between the partisans of these two long-time rivals.

The roots of Seychelles' modern political economy go back to 1963, when Mancham's Democratic Party and René's People's United Party were established. The former originally favored private enterprise and the retention of the British imperial connection, while the latter advocated an independent socialist state. Electoral victories in 1970 and 1974 allowed Mancham to pursue his dream of turning Seychelles into a tourist paradise and a financial and trading center by aggressively seeking outside investment. Tourism began to flourish following the opening of an international airport on the main island of Mahe in 1971, fueling an economic boom. Between 1970 and 1980, per capita income rose from nearly $150 to $1,700 (today it is about $7,600).

In 1974, Mancham, in an about-face, joined René in advocating the islands' independence. The Democratic Party, despite its modest electoral and overwhelming parliamentary majority, set up a coalition government with the People's United Party. On June 29, 1976, Seychelles became independent, with Mancham as president and René as prime minister.

HEALTH/WELFARE

A national health program has been established; private practice has been abolished. Free-lunch programs have raised nutritional levels among the young. Education is also free, up to age 15.

On June 5, 1977, with Mancham out of the country, René's supporters, with Tanzanian assistance, staged a successful coup in which several people were killed. Thereafter René assumed the presidency and suspended the Constitution. A period of rule by decree gave way in 1979, without the benefit of referendum, to a new constitutional framework in which the

People's Progressive Front Seychelles (SPPF), successor to the People's United Party, was recognized as the nation's sole political voice. The first years of one-party government were characterized by continued economic growth, which allowed for an impressive expansion of social-welfare programs.

ACHIEVEMENTS

Seychelles has become a world leader in wildlife preservation. An important aspect of the nation's conservation efforts has been the designation of one island as an international wildlife refuge.

Political power since the coup has largely remained concentrated in the hands of René. The early years of his regime, however, were marked by unrest. In 1978, the first in a series of unsuccessful countercoups was followed, several months later, by violent protests against the government's attempts to impose a compulsory National Youth Service, which would have removed the nation's 16- and 17-year-olds from their families in order to foster their sociopolitical indoctrination in accordance with the René government's socialist ideals. Another major incident occurred in 1981, when a group of international mercenaries, who had the backing of authorities in Kenya and South Africa as well as exiled Seychellois, were forced to flee in a hijacked jet after an airport shootout with local security forces. Following this attempt, Tanzanian troops were sent to the islands. A year later, the Tanzanians were instrumental in crushing a mutiny of Seychellois soldiers.

Despite its success in creating a model welfare state, which undoubtedly strengthened its popular acceptance, for years René continued to govern in a repressive manner. Internal opposition was not tolerated by his government, and exiled activists were largely neutralized. About one fifth of the islands' population now live overseas (not all of these people left the country, however, for political reasons).

In 1991, René gave in to rising internal and external pressures for a return to multiparty democracy. In July 1992, his party won 58 percent of the vote for a commission to rewrite the Constitution. Mancham's Democrats received just over a third of the vote. But in November 1992, voters heeded Mancham's call, rejecting the revised constitution proposed by the pro-René commission. Faced with a possible deadlock, the two parties reached consensus on new proposals, which were ratified in a June 1993 referendum. Presidential and parliamentary elections held the following month confirmed majority support for René's party.

In elections in 1998 and 2001, the SPPF gained renewed mandates. These latter contests are perhaps more significant for the emergence of the United Opposition coalition, led by Wavel Ramkalawan, as the main opposition party in place of the aging Mancham's Democrats. Increasing support of the United Opposition coalition was apparent in September 2001, when President René won another term in office, with 54 percent of the votes, against 45 percent for Ramkalawan.

Timeline: PAST

1771
French settlement begins

1814
British rule is established

1830
The British end slavery

1903
Seychelles is detached from Mauritius by the British and made a Crown colony

1948
Legislative Council with qualified suffrage is introduced

1967
Universal suffrage

1976
Independence

1977
An Albert René coup against James Mancham

1980s
An Amnesty International report alleges government fabrication of drug-possession cases for political reasons

1990s
René agrees to a multiparty system; René and his party are approved in presidential and parliamentary elections

PRESENT

2000s
The tourism sector accounts for 30% of Seychellois employment and 70% of hard-currency earnings

The government seeks to diversify the economy

France Albert René steps down as president

Somalia

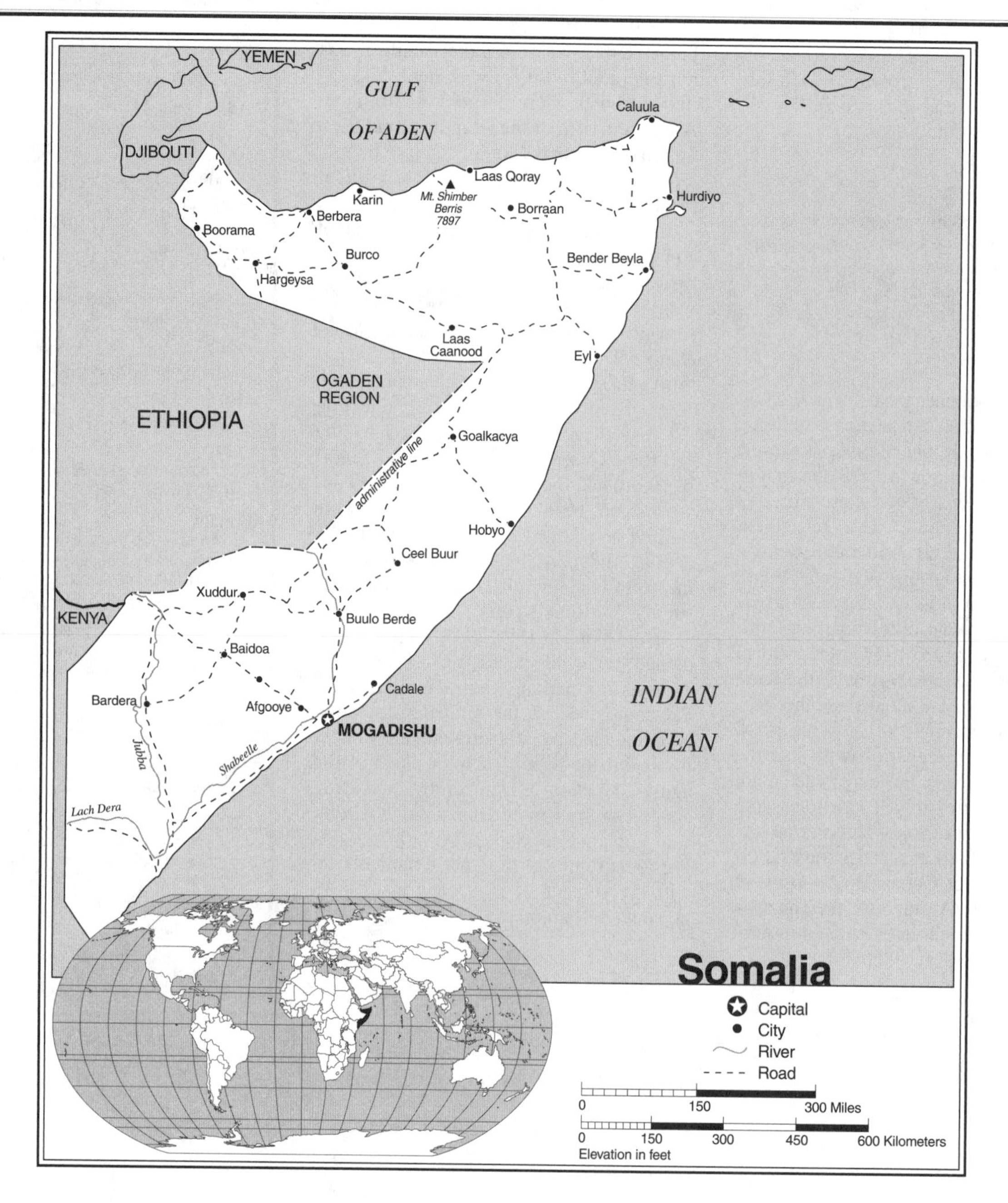

Somalia Statistics

GEOGRAPHY

Area in Square Miles (Kilometers): 246,331 (638,000) (about the size of Texas)

Capital (Population): Mogadishu (1,212,000)

Environmental Concerns: famine; contaminated water; deforestation; overgrazing; soil erosion; desertification

Geographical Features: principally desert; mostly flat to undulating plain, rising to hills in the north

Climate: arid to semiarid

PEOPLE

Population

Total: 8,304,601*

Annual Growth Rate: 3.46%

Rural/Urban Population Ratio: 73/27

Major Languages: Somali; Arabic; Italian; English

Ethnic Makeup: 85% Somali; Bantu; Arab

Religion: Sunni Muslim

Health

Life Expectancy at Birth: 45 years (male); 47 years (female)

Infant Mortality: 122/1,000 live births

Physicians Available: 1/19,071 people

Education

Adult Literacy Rate: 38%
Compulsory (Ages): 6–14; free

COMMUNICATION

Telephones: 100,000 main lines
Televisions: 18/1,000 people
Internet Users: 89,000 (2002)

TRANSPORTATION

Highways in Miles (Kilometers): 13,702 (22,100)
Railroads in Miles (Kilometers): none
Usable Airfields: 54
Motor Vehicles in Use: 20,000

GOVERNMENT

Type: "Transitional National Government"
Independence Date: July 1, 1960 (from a merger of British Somaliland and Italian Somaliland)
Head of State/Government: President Abdullahi Yusuf Ahmed; Prime Minister Ali Muhammad Ghedi
Political Parties: none
Suffrage: universal at 18

MILITARY

Military Expenditures (% of GDP): 0.9%
Current Disputes: civil war; border and territorial disputes with Ethiopia

ECONOMY

Currency ($ U.S. Equivalent): 11,000 shillings = $1
Per Capita Income/GDP: $550/$4.1 billion
Inflation Rate: NA
Labor Force by Occupation: 71% agriculture; 29% industry and services
Natural Resources: uranium; iron ore; tin; gypsum; bauxite; copper; salt
Agriculture: livestock; bananas; sugarcane; cotton; cereals; corn; sorghum; mangoes; fish
Industry: sugar refining; textiles; limited petroleum refining
Exports: $186 million (primary partners Saudi Arabia, United Arab Emirates, Yemen)
Imports: $314 million (primary partners Djibouti, Kenya, India)

SUGGESTED WEB SITES

http://www.unsomalia.org
http://www.somalianews.com
http://somalinet.com
http://somaliawatch.org
http://www.cia.gov/cia/publications/factbook/geos/so.html

*Note: Population statistics in Somalia are complicated by the large number of nomads and by refugee movements in response to famine and clan warfare.

Somalia Country Report

Wracked by violence, ruled by warlords, torn asunder not just by tribal but also petty clan disputes (some of which go back centuries), there seems to be nowhere to look for a spark of hope towards peace in Somalia. The list of transgressions committed by Somalia's leaders over the last fifteen years is immense. The country has thrown out their traditional allies, waged war with their regional neighbors, killed numerous international volunteers (who were supplying the starving residents with food and medical supplies), and fought viciously with one another in an unprecedented anarchy.

In what looked like a chance for peace in Somalia, there was a breakthrough in January 2004 at peace talks held in Kenya when the warlords and a few politicians signed a deal to set up a new parliament. Within four months however, renewed fighting broke out, killing 100 people as ethnic militias clashed in the southern town of Bula Hawo. The violence notwithstanding, a new transitional parliament was inaugurated in August 2004 at a ceremony in Kenya. In October, the body elected Abdullahi Yusuf as president of Somalia. The election took place in Kenya because the Somali capital was regarded as being too dangerous.

President Yusuf pledged to do his best to promote reconciliation and to set about rebuilding the country. He called on the international community to provide aid and peacekeepers. The former army officer and faction leader led a guerrilla movement in the 1970s aimed at ousting the Somali dictator Siad Barre. In the 1990s President Yusuf emerged as the pre-eminent leader of his native Puntland region; which he declared autonomous in 1998. President Yusuf's leadership style is said to be authoritarian. Clearly, in the eyes of the assembled representatives of Somalia, what the country needed in 2004 was another separatist dictator, who arose from the ranks of the warlords themselves, to resolve the pressing issues of the Somalian state.

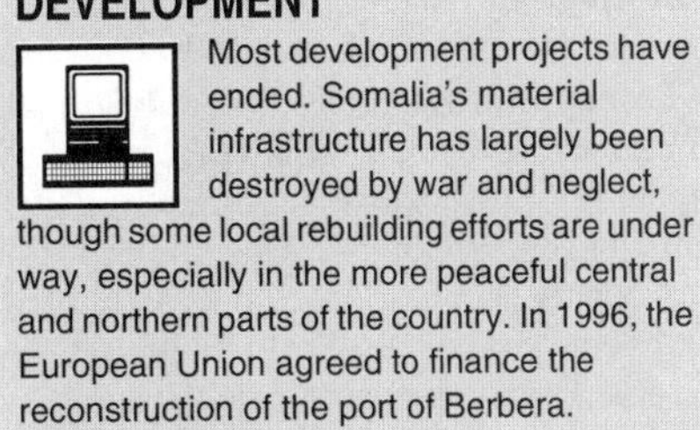

DEVELOPMENT

Most development projects have ended. Somalia's material infrastructure has largely been destroyed by war and neglect, though some local rebuilding efforts are under way, especially in the more peaceful central and northern parts of the country. In 1996, the European Union agreed to finance the reconstruction of the port of Berbera.

Somalia has in effect been without a central government since 1991, when after the overthrow of President Siad Barre, the country entered a period of chaos from which it has never recovered. The country has been dominated by "warlords" reponsible for small fiefdoms supported by heavily armed militias under their control. The resulting intermilitia fighting, added to the inability to deal with famine and disease, has led to the deaths of up to 1 million people. There appears to be little hope for an early resolution to the continuing conflicts in Somalia. The international donor community has long vanished, fundamentally giving up efforts to work out a peaceful solution to the problems of the nation. To compound matters, the northern portion of the country has broken away from the south and is now called Somaliland.

A "Transitional National Government" led by Abdiqassim Salah Hassan was put in place in August 2000. Hassan was elected by a "Transitional National Assembly," which had been formed at an internationally backed peace conference in neighboring Djibouti. After months of negotiations, the new government was accepted by a broad cross-section of Somali society. But a number of key groupings, in addition to still-powerful military leaders, remained outside of the accord. The northern "Puntland" and "Somaliland" governments were among those that boycotted the talks. The latter administration has declared itself an independent state, despite international opposition. In the breakaway Somaliland, Dahir Riyale Kahin was selected as president during May 2002 after the death of Mohamed Ibrahim Egal. He pledged to preserve the sovereignty of the state after narrowly winning the presidential election.

For much of the outside world, Somalia has become a symbol of failure of both international peacekeeping operations and the postcolonial African state. For the Somalis themselves, Somalia is an ideal that has ceased to exist—but may yet be re-created. Literally hundreds of thousands of Somalis starved to death in 1991–1992 before a massive U.S.–led United Nations intervention—officially known as UNITAF but labeled "Operation Restore Hope" by the Americans—assured the delivery of relief supplies. The 1994 withdrawal of most UN forces (the last token units left in March 1995), following UNITAF's failure to disarm local militias while supporting the creation of a "Transitional National Council," led to the termination of relief efforts in many areas, but widespread famine was averted in 1994–1995. Repeated attempts to reach a settlement between the various armed factions have continued to fail, despite the death in August 1996 of Somalia's most powerful warlord, General Mohammed Farah Aideed.

FREEDOM

Plagued by persistent hunger and internal violence, and with the continuing threat of governance by the anarchic greed of the warlords, the living have no true freedom in Somalia.

SOMALI SOCIETY

The roots of Somalia's suffering run deep. Somalis have lived with the threat of famine for centuries, as the climate is arid even in good years. Traditionally, most Somalis were nomadic pastoralists, but in recent years, this way of life has declined dramatically. Prior to the 1990s crisis, about half the population were still almost entirely reliant on livestock. Somali herds have sometimes been quite big: In the early 1980s, more than 1 million animals, mostly goats and sheep, were exported annually. Large numbers of cattle and camels have also been kept. But hundreds of thousands of animals were lost due to lack of rain during the mid-1980s; and since 1983, reports of rinderpest led to a sharp drop in exports, due to the closing of the once-lucrative Saudi Arabian market to East African animals.

A quarter of the Somali population have long combined livestock-keeping with agriculture. Cultivation is possible in the area between the Juba and Shebelle Rivers and in portions of the north. Although up to 15 percent of the country is potentially arable, only about 1 percent of the land has been put to plow at any given time. Bananas, cotton, and frankincense have been major cash crops, while maize and sorghum are subsistence crops. Like Somali pastoralists, farmers walk a thin line between abundance and scarcity, for locusts as well as drought are common visitors.

HEALTH/WELFARE

Somalia's small health service has almost completely disappeared, leaving the country reliant on a handful of international health teams. By 1986, education's share of the national budget had fallen to 2%. Somalia had 525 troops per teacher, the highest such ratio in Africa.

The delicate nature of Somali agriculture helps to explain recent urbanization. One out of every four Somalis lives in the large towns and cities. The principal urban center is Mogadishu, which, despite being divided by war, still houses well over a million people. Unfortunately, as Somalis have migrated in from the countryside, they have found little employment. Even before the recent collapse, the country's manufacturing and service sectors were small. By 1990, more than 100,000 Somalis had become migrant workers in the Arab/Persian Gulf states. (In 1990–1991, many were repatriated as a result of the regional conflict over Kuwait.)

Until recently, many outsiders assumed that Somalia possessed a greater degree of national coherence than most other African states. Somalis do share a common language and a sense of cultural identity. Islam is also a binding feature. However, competing clan and subclan allegiances have long played a divisive political role in the society. Membership in all the current armed factions is congruent with blood loyalties. Traditionally, the clans were governed by experienced, wise men. But the authority of these elders has now largely given way to the power of younger men with a surplus of guns and a surfeit of education and a lack of moral decency.

Past appeals to greater Somali nationalism have also been a source of conflict by encouraging irredentist sentiments against Somalia's neighbors. During the colonial era, contemporary Somalia was divided. For about 75 years, the northern region was governed by the British, while the southern portion was subject to Italian rule. These colonial legacies have complicated efforts at nation-building. Many northerners feel that their region has been neglected and would benefit from greater political autonomy or independence.

Somalia became independent on July 1, 1960, when the new national flag, a white, five-pointed star on a blue field, was raised in the former British and Italian territories. The star symbolized the five supposed branches of the Somali nation—that is, the now-united peoples of British and Italian Somalilands and the Somalis still living in French Somaliland (modern Djibouti), Ethiopia, and Kenya.

THE RISE AND FALL OF SIAD BARRE

Siad Barre came to power in 1969, through a coup promising radical change. As chairman of the military's Supreme Revolutionary Council, Barre combined Somali nationalism and Islam with a commitment to "scientific socialism." Some genuine efforts were made to restructure society through the development of new local councils and worker management committees. New civil and labor codes were written. The Somali Revolutionary Socialist Party was developed as the sole legal political party.

Initially, the new order seemed to be making progress. The Somali language was alphabetized in a modified form of Roman script, which allowed the government to launch mass-literacy campaigns. Various rural-development projects were also implemented. In particular, roads were built, which helped to break down isolation among regions.

The promise of Barre's early years in office gradually faded. Little was done to follow through the developments of the early 1970s, as Barre increasingly bypassed the participatory institutions that he had helped to create. His government became one of personal rule; he took on emergency powers, relieved members of the governing council of their duties, surrounded himself with members of his own Marehan branch of the Darod clan, and isolated himself from the public. Barre also isolated Somalia from the rest of Africa by pursuing irredentist policies in order to unite the other points of the Somali star under his rule. To accomplish this task, he began to encourage local guerrilla movements among the ethnic Somalis living in Kenya and Ethiopia.

ACHIEVEMENTS

Somalia has been described as a "nation of poets." Many scholars attribute the strength of the Somali poetic tradition not only to the nomadic way of life, which encourages oral arts, but to the role of poetry as a local social and political medium.

In 1977, Barre sent his forces into the Ogaden region to assist the local rebels of the Western Somali Liberation Front. The invaders achieved initial military success against the Ethiopians, whose forces had been weakened by revolutionary strife and battles with Eritrean rebels. However, the intervention of some 17,000 Cuban troops and other Soviet-bloc personnel on the side of the Ethiopians quickly turned the tide of battle. At the same time, the Somali incursion was condemned by all members of the Organization of African Unity.

The intervention of the Soviet bloc on the side of the Ethiopians was a bitter disappointment to Barre, who had enjoyed Soviet support for his military buildup. In exchange, he had allowed the Soviets to establish a secret base at the strategic northern port of Berbera. However, in 1977, the Soviets decided to shift their allegiances decisively to the then–newly established revolutionary government in Ethiopia. Barre in turn tried to attract U.S. support with offers of basing rights at Berbera, but the Carter administration was unwilling to jeopardize its interests in either Ethiopia or Kenya by backing Barre's irredentist adventure. American–Somali relations became closer during the Reagan administration, which signed a 10-year pact giving U.S. forces access to air and naval facilities at Berbera, for which the United States increased its aid to Somalia, including limited arms supplies.

In 1988, Barre met with Ethiopian leader Mengistu Mariam. Together, they pledged to respect their mutual border. This understanding came about in the context of growing internal resistance to both regimes. By 1990, numerous clan-based armed resistance movements were enjoying success against Barre.

Growing resistance was accompanied by massive atrocities on the part of government forces. Human-rights concerns were cited by the U.S. and other governments in ending their assistance to Somalia. In March 1990, Barre called for national dialogue and spoke of a possible end to one-party rule. But continuing atrocities, including the killing of more than 100 protesters at the national stadium, fueled further armed resistance.

In January 1991, Barre fled Mogadishu, which was seized by resistance forces of the United Somali Congress (USC). The USC set up an interim administration, but its authority was not recognized by other groups. By the end of the year, the USC itself had split into two warring factions. A 12-faction "Manifesto Group" recognized Ali Mahdi as the country's president. But Mahdi's authority was repudiated by the four-faction Somali National Alliance (SNA), led by Farah Aideed. Much of Mogadishu was destroyed in inconclusive fighting between the two groupings. Other militias, including forces still loyal to Barre (that is, the Somali National Front, or SNF), also continued to fight one another. In the north, the Somali National Movement (SNM) declared its zone's sovereign independence as "Somaliland."

Continued fighting coincided with drought. As failed crops and dying livestock resulted in countrywide famine, international relief efforts were unable to supply sufficient quantities of outside food to those most in need, due to the prevailing state of lawlessness. In mid-1992, the International Red Cross estimated that, of southern Somalia's 4.5 million people, 1.5 million were in danger of starvation. Another 500,000 or so had fled the country. More than 300,000 children under age five were reported to have perished.

As Somalia's suffering grew and became publicized in the Western media, many observers suggested the need for the United Nations to intervene. A small UN presence, known as UNISOM, was established in August 1992, but its attempts to police the delivery of relief supplies proved to be ineffectual. Conceived as a massive U.S.–led military operation, initially consisting of 30,000 troops (22,000 Americans), UNITAF averted catastrophe by assuring the delivery of food and medical supplies to Somalia's starving millions. Still, the foreign troops' mission was unclear.

In 1993, a bloody clash between Aideed's SNA militia and Pakistani troops in Mogadishu led to full-scale armed conflict. Efforts by UNITAF forces to capture Aideed and neutralize his men were unsuccessful. After a U.S. helicopter was shot down in October 1993, President Bill Clinton decided to end American involvement in UNITAF by March 1994. By then, much higher losses had been suffered by several other nations participating in the UNITAF–UNISOM coalition, causing them also to reassess their commitments. Outgunned and demoralized, the remaining UN forces (officially labeled UNISOM II) remained largely confined to their compounds until their withdrawal.

Timeline: PAST

1886–1887
The British take control of northern regions of present-day Somalia

1889
Italy establishes a protectorate in the eastern areas of present-day Somalia

1960
Somalia is formed through a merger of former British and Italian colonies under UN Trusteeship

1969
Siad Barre comes to power through an army coup; the Supreme Revolutionary Council is established

1977–1978
The Ogaden war in Ethiopia results in Somalia's defeat

1980s
SNM rebels escalate their campaign in the north; government forces respond with genocidal attacks on the local Issaq population

1990s
The fall of Barre leaves Somalia without an effective central government; U.S.–led UN intervention feeds millions while attempting to restore order
In the face of mounting losses, foreign troops pull out of Somalia

PRESENT

2000s
Chaos still reigns in Somalia, causing untold suffering

Abdiqassim Salah Hassan is named interim president

2004
275-member Transitional Federal Government replaces the Transitional National Government

Sudan (Republic of the Sudan)

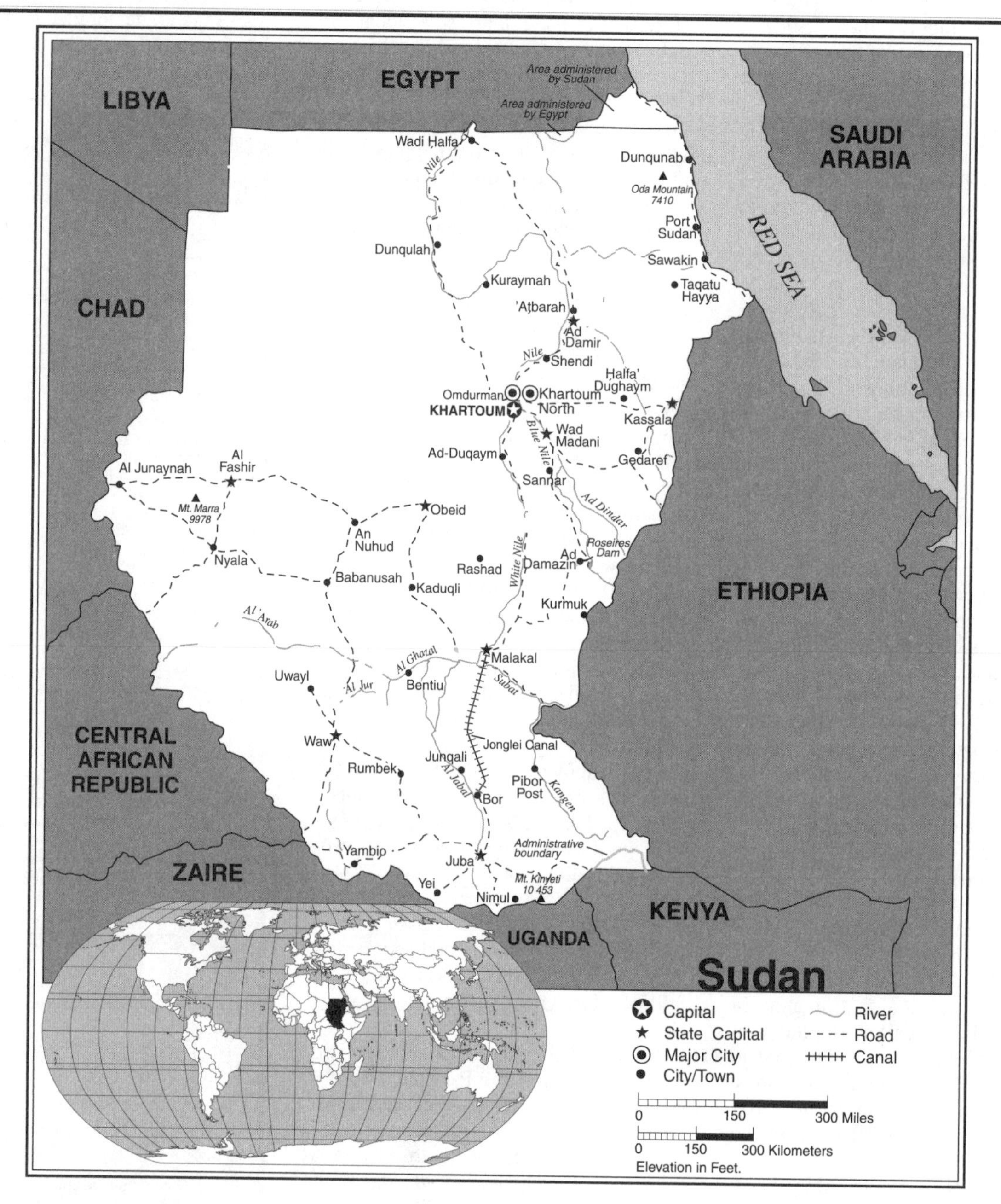

Sudan Statistics

GEOGRAPHY

Area in Square Miles (Kilometers): 967,247 (2,505,810) (about 1/4 the size of the United States)

Capital (Population): Khartoum (2,853,000)

Environmental Concerns: insufficient potable water; excessive hunting of wildlife; soil erosion; desertification

Geographical Features: generally flat, featureless plain; mountains in the east and west

Climate: arid desert to tropical

PEOPLE

Population

Total: 39,148,162

Annual Growth Rate: 2.73%

Rural/Urban Population Ratio: 65/35

Major Languages: Arabic; Sudanic languages; Nubian; English; others

Ethnic Makeup: 52% black; 39% Arab; 6% Beja; 3% others

Religions: 70% Sunni Muslim, especially in north; 25% indigenous beliefs; 5% Christian

Health

Life Expectancy at Birth: 56 years (male); 58 years (female)
Infant Mortality: 67/1,000 live births
Physicians Available: 1/11,300 people
HIV/AIDS Rate in Adults: 0.99%

Education

Adult Literacy Rate: 46%

COMMUNICATION

Telephones: 900,000 main lines
Daily Newspaper Circulation: 21/1,000 people
Televisions: 8.2/1,000 people
Internet Users: 300,000 (2003)

TRANSPORTATION

Highways in Miles (Kilometers): 7,198 (11,610)
Railroads in Miles (Kilometers): 3,425 (5,516)
Usable Airfields: 65
Motor Vehicles in Use: 75,000

GOVERNMENT

Type: transitional
Independence Date: January 1, 1956 (from Egypt and the United Kingdom)
Head of State/Government: President Umar Hasan Ahmad al-Bashir is both head of state and head of government
Political Parties: National Congress Party; Popular National Congress; Umma; Sudan People's Liberation Movement (Army); others
Suffrage: universal at 17

MILITARY

Military Expenditures (% of GDP): 2.5%
Current Disputes: civil war; border disputes and clashes with Egypt and Kenya

ECONOMY

Currency ($ U.S. Equivalent): 260 pounds = $1
Per Capita Income/GDP: $1,900/$70.95 billion
GDP Growth Rate: 5.9%
Inflation Rate: 8.8%
Unemployment Rate: 18.7%
Labor Force by Occupation: 80% agriculture; 13% government; 7% industry and commerce
Natural Resources: petroleum; iron ore; chromium ore; copper; zinc; tungsten; mica; silver; gold; hydropower
Agriculture: cotton; sesame; gum arabic; sorghum; millet; wheat; sheep; groundnuts
Industry: textiles; cement; cotton ginning; edible oils; soap; sugar; shoes; petroleum refining
Exports: $2.1 billion (primary partners Japan, China, Saudi Arabia)
Imports: $1.6 billion (primary partners China, Saudi Arabia, United Kingdom)

SUGGESTED WEB SITES

http://www.sudan.net
http://sudanhome.com
http://www.sudmer.com
http://www.sunanews.net
http://www.cia.gov/cia/publications/factbook/geos/su.html

Sudan Country Report

The name Sudan comes from the Arabic *bilad al-sudan*, or "land of the blacks." Today, Sudan is Africa's largest country. Apart from an 11-year period of peace, it has been torn since its indpendence in 1956 by civil war between the mainly Muslim north and the animist and Christian south. Sudan's tremendous size as well as its great ethnic and religious diversity have frustrated the efforts of successive postindependence governments to build a lasting sense of national unity.

DEVELOPMENT

Many ambitious development plans have been launched since independence, but progress has been limited by political instability. The periodic introduction and redefinition of "Islamic" financial procedures have complicated long-term planning.

The current president, Omar Bashir, was reelected in 2001 for another five years. The elections were boycotted by the main opposition parties. The Machakos Protocol of July 2002, which was signed by both the government and the two largest southern rebel groups, calls for a six-year interval period, after which there will be a referendum held on self-determination for the south. However, the Muslim-led Sudanese government has continued to attack the southern rebels; through the age-old tactic of divide-and-conquer, it has been able to make periodic inroads into the rebels' African strongholds. Ethnic groups are pitted against one another. Meanwhile, there has been evidence of the widespread enslavement of blacks in the south. In December 2001, for example, more than 14,550 slaves, mainly blacks, were freed following campaigning by human-rights activists.

Listed by the U.S. government as a major supporter of terrorism, until 1997 Sudan's Islamic fundamentalist government provided refuge for Osama bin Laden. Since then, the government has been keen to overcome its image as a pariah state. The challenges facing Sudan have also been complicated by the discovery of major oil fields in the south. The government has sought to establish safe enclaves for the exploitation of the oil fields, at the cost of relocating people who were living in the area. The oil fields may make Sudan rich, but they remain a primary source of alienation, as funds generated by the government from oil revenues have been used to purchase weapons against the southern rebels.

The future looks like continued civil war until Sudan ceases to enslave its people, grants the southerners self-determination, and ceases trying to impose an Islamic state on its religiously mixed population.

FREEDOM

The current regime rules through massive repression. In 1992, Africa Watch accused it of practicing genocide against the Nuba people. Elsewhere, tales of massacres, forced relocations, enslavement, torture, and starvation are commonplace. The insurgent groups have also been responsible for numerous atrocities.

HISTORY

Sudan, like its northern neighbor Egypt, is a gift of the Nile. The river and its various branches snake across the country, providing water to most of the 80 percent of Sudanese who survive by farming. From ancient times, the Upper Nile region of northern Sudan has been the site of a series of civilizations whose histories are closely intertwined with those of Egypt. There has been constant human interaction between

(United Nations photo 157661 by Milton Grant)

Millions of Sudanese have been displaced by war and drought. The effects on the population have been devastating, and even the best efforts of the international community have met with only limited success.

the two zones. Some groups, such as the Nubians, expanded northward into the Egyptian lower Nile.

The last ruler to unite the Nile Valley politically was the nineteenth-century Turko–Egyptian ruler Muhammad Ali. After absorbing northern Sudan, by then predominantly Arabized Muslim, into his Egyptian state, Ali gradually expanded his authority to the south and west over non-Arabic and, in many cases, non-Muslim groups. This process, which was largely motivated by a desire for slave labor, united for the first time the diverse regions that today make up Sudan. In the 1880s, much of Sudan fell under the theocratic rule of the Mahdists, a local anti-Egyptian Islamic movement. The Mahdists were defeated by an Anglo–Egyptian force in 1898. Thereafter, the British dominated Sudan until its independence, in 1956.

Sudanese society has remained divided ever since. There has been strong pan-Arab sentiment in the north, but 60 percent of Sudanese, concentrated in the south and west, are non-Arab. About a third of Sudanese, especially in the south, are also non-Muslim. Despite this fact, many, but by no means all, Sudanese Muslims have favored the creation of an Islamic state. Ideological divisions among various socialist- and nonsocialist-oriented factions have also been important. Sudan has long had a strong Communist Party (whether legal or not), drawing on the support of organized labor, and an influential middle class.

The division between northern and southern Sudan has been especially deep. A mutiny by southern soldiers prior to independence escalated into a 17-year rebellion by southerners against what they perceived to be the hegemony of Muslim Arabs. Some 500,000 southerners perished before the Anya Nya rebels and the government reached a compromise settlement, recognizing southern autonomy in 1972.

HEALTH/WELFARE

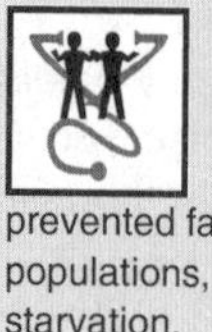

Civil strife and declining government expenditures have resulted in rising rates of infant mortality. Warfare has also prevented famine relief from reaching needy populations, resulting in instances of mass starvation.

In northern Sudan, the first 14 years of independence saw the rule of seven different civilian coalitions and six years of military rule. Despite this chronic instability, a tradition of liberal tolerance among political factions was generally maintained. Government became increasingly authoritarian during the administration of Jaafar Nimeiri, who came to power in a 1969 military coup.

Nimeiri quickly moved to consolidate his power by eliminating challenges to his government from the Islamic right and the Communist left. His greatest success was ending the Anya Nya revolt, but his subsequent tampering with the provisions of the peace agreement led to renewed resistance. In 1983, Nimeiri decided to impose Islamic law throughout Sudanese society. This led to the growth of the Sudanese People's Liberation Army (SPLA), under the leadership of John Garang, which quickly seized control of much of the southern Sudanese countryside. Opposition to Nimeiri had also been growing in the north, as more people became alienated by the regime's increasingly heavy-handed ways and inability to manage the declining economy. Finally, in 1985, he was toppled in a coup.

The holding of multiparty elections in 1986 seemed to presage a restoration of Sudan's tradition of pluralism. With the SPLA preventing voting in much of the south, the two largest parties were the northern-based Umma and Democratic Union (DUP). The National Islamic Front was the third-largest vote-getter, with eight other parties plus a number of independents gaining parliamentary seats. The major challenge facing the new coalition government, led by Umma, was reconciliation with the SPLA. Because the SPLA, unlike the earlier Anya Nya, was committed to national unity, the task did not appear insurmountable. However, arguments within the government over meeting key SPLA demands, such as the repeal of Islamic law, caused the war to drag on. A hard-line faction within Umma and the NIF sought to resist a return to secularism. In March 1989, a new government, made up of Umma and the DUP, committed itself to accommodating the SPLA. However, a month later, on the day the cabinet was to ratify an agreement with the rebels, there was a coup by pro-NIF officers.

ACHIEVEMENTS

Although his music is banned in his own country, Mohammed Wardi is probably Sudan's most popular musician. Now living in exile, he has been imprisoned and tortured for his songs against injustice, which also appeal to a large international audience, especially in North Africa and the Middle East.

Besides leading to a breakdown in all efforts to end the SPLA rebellion, the NIF/military regime has been responsible for establishing the most intolerant, repressive government in Sudan's modern history. Extra-judicial executions have become commonplace. Instances of pillaging and enslavement of non-Muslim communities by government-linked militias have increased. NIF-affiliated security groups have become a law unto themselves, striking out at their perceived enemies and intimidating Muslims and non-Muslims alike to conform to their fundamentalist norms. Islamic norms are also being invoked to justify a radical campaign to undermine the status of women.

In 1990, most of the now-banned political parties (including Umma, the DUP, and the Communists) aligned themselves with the SPLA as the National Democratic Alliance. But opposition by the northern-based parties proved ineffectual, leading to the formation of a new, Eritrean-based armed movement—the Sudan Alliance of Forces, headed by Abdul Azizi Khalid.

Beginning in 1991, the SPLA was weakened by a series of splits. Two factions—Kerubino Kuanyin Bol's SPLA–Bahr al-Ghazal group and Riek Macher's Southern Sudan Independence Army (SSIA)—accepted a government peace plan in April 1996. But the plan was rejected by John Garang's SPLA (Torit faction), which remains the most powerful southern group. After a number of years of being on the defensive, Garang's forces began making significant advances in 1996, partially as a result of increased support from neighboring countries that have come to look upon the Khartoum regime as a regional threat. (In June 1995, the regime was implicated in an attempt to assassinate Egyptian president Hosni Mubarak in Ethiopia, which resulted in the imposition of UN antiterrorism sanctions. Border clashes have since occurred with Eritrea, Kenya, and Uganda as well as Egypt.)

ECONOMIC PROSPECTS

Although it has great potential, political conflict has left Sudan one of the poorest nations in the world. Persistent warfare and lack of financing are blocking needed infrastructural improvements. Sudan's unwillingness to pay its foreign debt has led to calls for its expulsion from the International Monetary Fund.

Nearly 7 million Sudanese (out of a total population then of 23 million) had been displaced by 1988—more than 4 million by warfare, with drought and desertification contributing to the remainder. Sudan has been a major recipient of international emergency food aid for years, but warfare, corruption, and genocidal indifference have often blocked help from reaching the needy. In 1994, the United Nations estimated that 700,000 southern Sudanese faced the prospect of starvation.

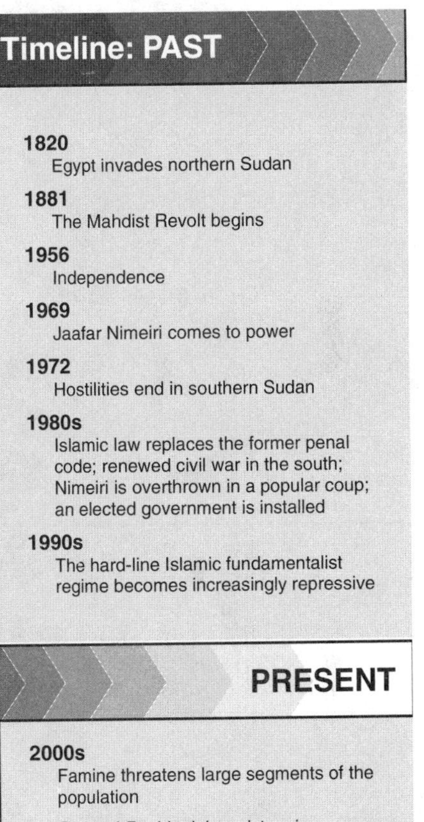

Timeline: PAST

1820
Egypt invades northern Sudan

1881
The Mahdist Revolt begins

1956
Independence

1969
Jaafar Nimeiri comes to power

1972
Hostilities end in southern Sudan

1980s
Islamic law replaces the former penal code; renewed civil war in the south; Nimeiri is overthrown in a popular coup; an elected government is installed

1990s
The hard-line Islamic fundamentalist regime becomes increasingly repressive

PRESENT

2000s
Famine threatens large segments of the population

Omar al-Bashir claims victory in boycotted elections

Tanzania (United Republic of Tanzania)

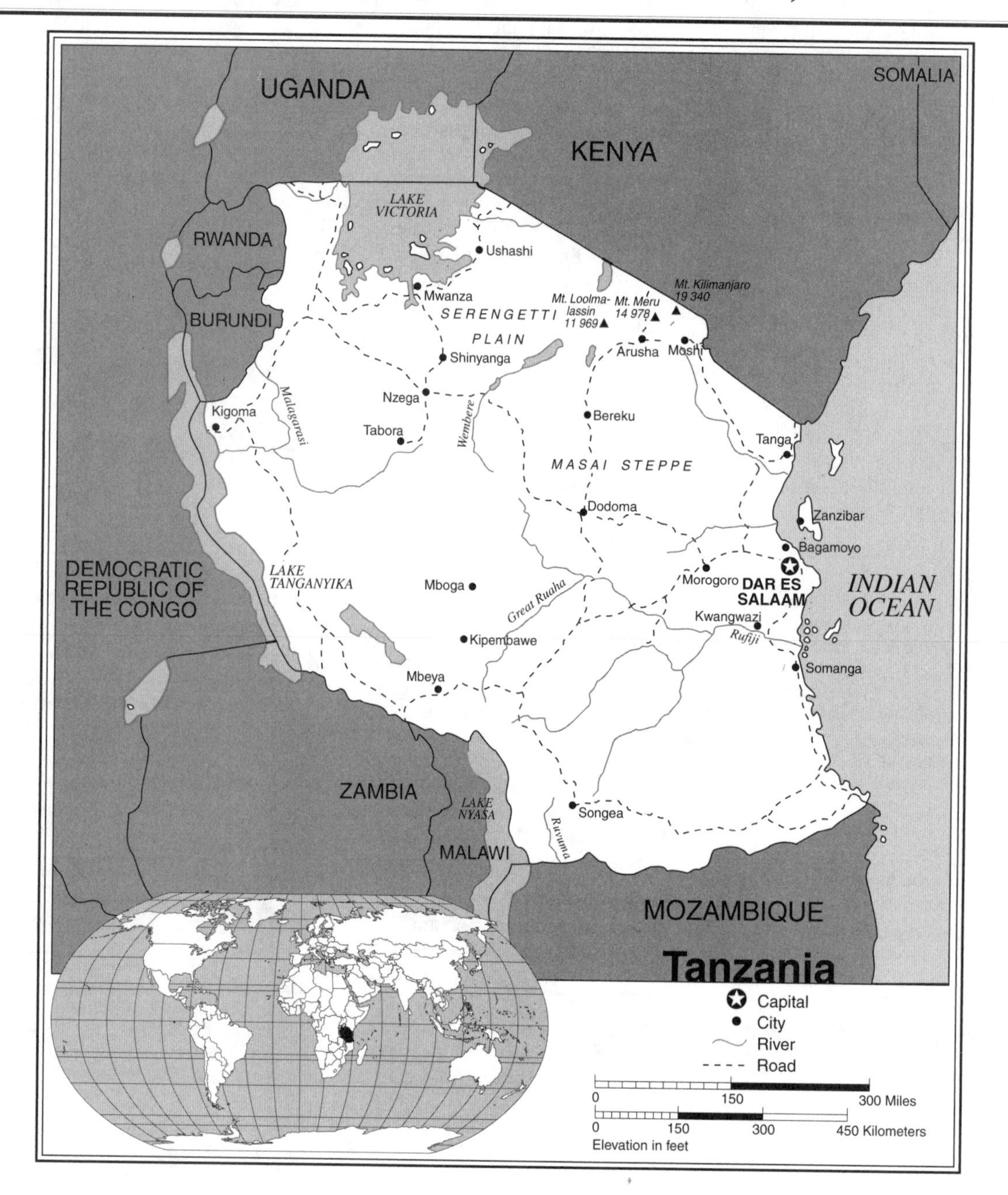

Tanzania Statistics

GEOGRAPHY

Area in Square Miles (Kilometers): 363,950 (939,652) (about twice the size of California)

Capital (Population): Dar es Salaam (2,347,000); Dodoma (to be new capital) (180,000)

Environmental Concerns: soil degradation; deforestation; desertification; destruction of coral reefs and marine environment

Geographical Features: plains along the coast; central plateau; highlands in the north and south

Climate: tropical to temperate

PEOPLE

Population

Total: 36,588,225

Annual Growth Rate: 1.95%

Rural/Urban Population Ratio: 68/32

Major Languages: Kiswahili; Chagga; Gogo; Ha; Haya; Luo; Maasai; English; others

Ethnic Makeup: 99% African; 1% others

Religions: indigenous beliefs; Muslim; Christian; Hindu

Health

Life Expectancy at Birth: 51 years (male); 53 years (female)
Infant Mortality: 77.8/1,000 live births
Physicians Available: 1/20,511 people
HIV/AIDS Rate in Adults: 8.09%

Education

Adult Literacy Rate: 68%
Compulsory (Ages): 7–14; free

COMMUNICATION

Telephones: 149,000 main lines
Televisions: 2.8/1,000 people
Internet Users: 250,000 (2003)

TRANSPORTATION

Highways in Miles (Kilometers): 52,800 (85,161)
Railroads in Miles (Kilometers): 2,141 (3,453)
Usable Airfields: 125
Motor Vehicles in Use: 134,000

GOVERNMENT

Type: republic
Independence Date: December 9, 1961 (from United Nations trusteeship)
Head of State/Government: President Benjamin William Mkapa is both head of state and head of government
Political Parties: Revolutionary Party; National Convention for Construction and Reform; Civic United Front; Union for Multiparty Democracy; Democratic Party; United Democratic Party; others
Suffrage: universal at 18

MILITARY

Military Expenditures (% of GDP): 0.2%
Current Disputes: boundary disputes with Malawi; civil strife

ECONOMY

Currency ($ U.S. Equivalent): 966 shillings = $1
Per Capita Income/GDP: $610/$22.1 billion
GDP Growth Rate: 5%
Inflation Rate: 4.4%
Labor Force by Occupation: 80% agriculture; 20% services and industry
Population Below Poverty Line: 51%
Natural Resources: hydropower; tin; phosphates; iron ore; coal; diamonds; gemstones; gold; natural gas; nickel
Agriculture: coffee; sisal; tea; cotton; pyrethrum; cashews; tobacco; cloves; wheat; fruits; vegetables; livestock
Industry: agricultural processing; mining; oil refining; shoes; cement; textiles; wood products; fertilizer; salt
Exports: $827 million (primary partners United Kingdom, India, Germany)
Imports: $1.55 billion (primary partners South Africa, Japan, United Kingdom)

SUGGESTED WEB SITES

http://www.tanzania.go.tz
http://www.tanzanianews.com
http://www.cia.gov/cia/publications/factbook/geos/tz.html
http://www.tanzania_online.gov.uk

Tanzania Country Report

After winning a second term of office in 2000, taking nearly 72 percent of the vote, President Benjamin Mkapa and the ruling Chama Cha Mapinduzi (CCM) party were confronted with a series of challenges. The government banned opposition rallies, which were demanding new elections. Police staged a raid in January 2001 on the offices of the Civic United Front (CUF), the main opposition party in Zanzibar, and killed two in the process. Widespread protests in Zanzibar ensued, during which at least 31 people were killed. One hundred were arrested, and CUF chairman Ibrahim Lupimba was charged with unlawful assembly and disturbing the peace.

The government sent in troop reinforcements, but the solution to the unrest ultimately was to prove to be a political one. CCM and the CUF agreed in March 2001 to the formation of a joint committee to restore calm to Zanzibar, and also to encourage the return of around 2,000 refugees who had fled to Kenya.

Still, tensions remained high throughout Tanzania, and the opposition parties picked up a great deal of support. In April, opposition parties staged the first major political demonstrations against the government in more than 20 years. Some 50,000 supporters of the opposition marched in Dar es Salaam.

In November 2001 the presidents of Tanzania, Uganda, and Kenya launched a regional parliament and court of justice in Arusha. They will legislate on matters of common interest such as trade and immigration rules.

DEVELOPMENT

The Bulyanhulu gold mine near the northern town of Mwanza opens, making Tanzania Africa's third-largest producer of gold.

In October 1999, more than 3 million Tanzanians joined leaders from around the world in filing through a temporary mausoleum housing the body of the country's late first president, Julius Nyerere. It was an overwhelming tribute to the man who was known at home and abroad as *Mwalimu*—Swahili for "teacher." Nyerere voluntarily relinquished power in 1985, but the legacy of his nation-building efforts can be found throughout the country. After independence, until recently Tanzania enjoyed internal unity and an expansion of social services. The Kiswahili language helped bind the nation together. But economic growth has remained modest; Tanzania continues to be one of the poorest nations in the world.

After a period of harsh German rule followed by paternalistic British trusteeship, the Tanzanian mainland gained its independence, as Tanganyika, in 1961. In 1964, it merged with the small island state of Zanzibar, which had been a British protectorate, to form the "United Republic of Tanzania." Political activity in Tanzania was restricted to the Chama Cha Mapinduzi party, which joined the former Tanganyika African National Union with its Zanzibar partner, the Afro-Shirazi Party.

FREEDOM

Civil rights in Tanzania, especially on the island of Zanzibar, remain circumscribed, with police often harassing supporters of the political opposition. Arbitrary arrest and torture remain commonplace on the island, but on the mainland there has been a steady opening up of society since 1995, though there still exist restrictions on freedoms of speech and association.

In February 1992, the CCM agreed to compete with other "national parties"—provided they did not "divide the people along tribal, religious or racial lines." Multiparty elections were held in October–December 1995, with the CCM claiming victory over a divided opposition in a poll

(UN photo 154919 by Ray Witlin)

The Tanzanian economy is primarily agriculture-based. However, rainfall for most of the country is sporadic. This, coupled with wide swings in world-market demand for its cash crops, has led to economic pressures. These men in the village of Lumeji are receiving seed grains needed to develop Tanzania's food production.

characterized by massive irregularities. The disputed official results were CCM, 60 percent of the vote and 187 members of Parliament (MPs), including 128 seats where the results were still being legally contested months later; the National Committee for Constitutional Reform (NCCR), 25 percent of the vote and 15 MPs; and the Civic United Front, 24 MPs but only a small percentage of the vote, concentrated in Zanzibar and Pemba Islands. The new CCM leader, Ben Mkapa, replaced as president Ali Hassan Mwinyi, who was forced to step down after having served two terms. The dominant personality in the CCM, however, remained former president Nyerere.

By 1967, the CCM's predecessors had already eliminated legal opposition, when they proclaimed their commitment to the Arusha Declaration, a blueprint for "African Socialism." At the time, Tanzania was one of the least-developed countries in the world. It has remained so. Beyond this fact, there is much controversy over the degree to which the goals of the declaration have been achieved. To some critics, the Arusha experiment has been responsible for reducing a potentially well-off country to ruin. Supporters often counter that it has led to a stable society in which major strides have been made toward greater democracy, equality, and human development. Both sides exaggerate.

HEALTH/WELFARE

The Tanzanian Development Plan calls for the government to give priority to health and education in its expenditures. This reflects a recognition that early progress in these areas has been undermined to some extent in recent years. Malnutrition remains a critical problem.

Like many African states, Tanzania has a primarily agrarian economy that is constrained by a less than optimal environment. Although some 80 percent of the population are employed in agriculture, only 8 percent of the land is under cultivation. Rainfall for most of the country is low and erratic, and soil erosion and deforestation are critical problems in many areas. But geography and environmental problems are only one facet of Tanzania's low agricultural productivity. There has also been instability in world-market demand for the nation's principal cash crops: coffee, cotton, cloves, sisal, and tobacco. The cost of imported fuel, fertilizers, and other inputs has risen simultaneously.

Government policies have also been responsible for underdevelopment. Perhaps the greatest policy disaster was the program of villagization. Tanzania hoped to relocate its rural and unemployed urban populations into *ujaama* (Swahili for "familyhood") villages, which were to become the basis for agrarian progress. In the early 1970s, coercive measures were adopted to force the pace of resettlement. Agricultural production is estimated to have fallen as much as 50 percent during the initial period of ujaama dislocation, transforming the nation from a grain exporter to a grain importer.

Another policy constraint was the exceedingly low official produce prices paid by the government to farmers. Many peasants withdrew from the official market, while others turned to black-market sales. Since 1985, the official market has been liberalized, and prices have risen. This has been accompanied by a modest rise in production, yet the lack of consumer goods in rural areas is widely seen as a disincentive to greater development.

All sectors of the Tanzanian economy have suffered from deteriorating infrastructure. Here again there are both external and internal causes. Balanced against rising imported-energy and equipment costs have been inefficiencies caused by poor planning, barriers to capital investment, and a relative neglect of communications and transport. Even when crops or goods are available, they often cannot reach their destination. Tanzania's few bituminized roads have long been in a chronic state of disrepair, and there have been frequent shutdowns of its railways. In particular, much of the southern third of the country is isolated from access to even inferior transport services.

ACHIEVEMENTS

The government has had enormous success in its program of promoting the use of Kiswahili (Swahili) as the national language throughout society. Mass literacy in Kiswahili has facilitated the rise of a national culture, helping to make Tanzania one of the more cohesive African nations.

Manufacturing declined from 10 to 4 percent of gross domestic product in the 1980s, with most sectors operating at less than half of their capacity. Inefficiencies also grew in the nation's mining sector. Diamonds, gold, iron, coal, and other minerals are exploited, but production has been generally falling and now accounts for less than 1 percent of GDP. Lack of capital investment has led to a deterioration of existing operations and an inability to open up new deposits.

As with agriculture, the Tanzanian government has in recent years increasingly abandoned socialism in favor of market economics, in its efforts to rehabilitate and expand the industrial and service sectors of the economy. A number of state enterprises are being privatized, and better opportunities are being offered to outside investors. Tourism is now being actively promoted, after decades of neglect. As the country liberalized its economy it has begun to play a more active role in regional economic affairs. During March 2004 the presidents of Tanzania, Uganda, and Kenya signed a protocol in Arusha over proposed customs unions, intended to boost trade.

Tanzania has made real progress in extending health, education, and other social services to its population since independence, though the statistical evidence is inadequate and official claims exaggerated. Some 1,700 health centers and dispensaries have been built since 1961, but they have long been plagued by shortages of medicines, equipment, and even basic supplies such as bandages and syringes. Although the country has a national health service, patients often end up paying for material costs.

Much of the progress that has been made in human services is a function of outside donations. Despite the Arusha Declaration's emphasis on self-reliance, Tanzania has for decades been either at or near the top of the list of African countries in per capita receipt of international aid.

Even before the recent opening to multipartyism, Tanzania's politics was in a state of transition. Political life was dominated from the 1950s by Julius Nyerere, who was the driving personality behind the Arusha experiment. However, in 1985, he gave up the presidency in favor of Ali Hassan Mwinyi, and, in 1990, Nyerere resigned as chairman of the CCM, without having to give up his leading influence in the party.

The move to multiparty politics is complicated by the omnipresent CCM. The party has sought to control all organized social activity outside of religion. A network of community and workplace cells has assured that all Tanzanians have at least one party official responsible for monitoring their affairs.

In 1993, a dozen new opposition parties were registered, though others, notably the Democratic Party of Reverend Christopher Mtikila, remain banned. Opposition disunity contributed to subsequent CCM election victories. In addition, the CCM still enjoys a near media monopoly, occasionally invoking the National Security Act to harass independent journalists. Overt political repression has been most notable on Zanzibar and Pemba Islands.

Timeline: PAST

1820
The sultan of Oman transfers the capital to Zanzibar as Arab commercial activity increases

1885
Germany declares a protectorate over the area

1905–1906
The Maji Maji rebellion unites many ethnic groups against German rule

1919
Tanganyika becomes a League of Nations mandate under the United Kingdom

1961
Tanganyika becomes independent; Julius Nyerere is the leader

1964
Tanzania is formed of Tanganyika and Zanzibar

1967
The Arusha Declaration establishes a Tanzanian socialist program for development

1985
Nyerere retires; Ali Hassan Mwinyi succeeds as president

1990s
The CCM wins disputed multiparty elections

PRESENT

2000s
Tanzania continues to look for ways to diversify its economy

Tensions erupt between the government and the opposition

President Ben Mkapa is criticized for ordering a $21 million presidential plane

Uganda (Republic of Uganda)

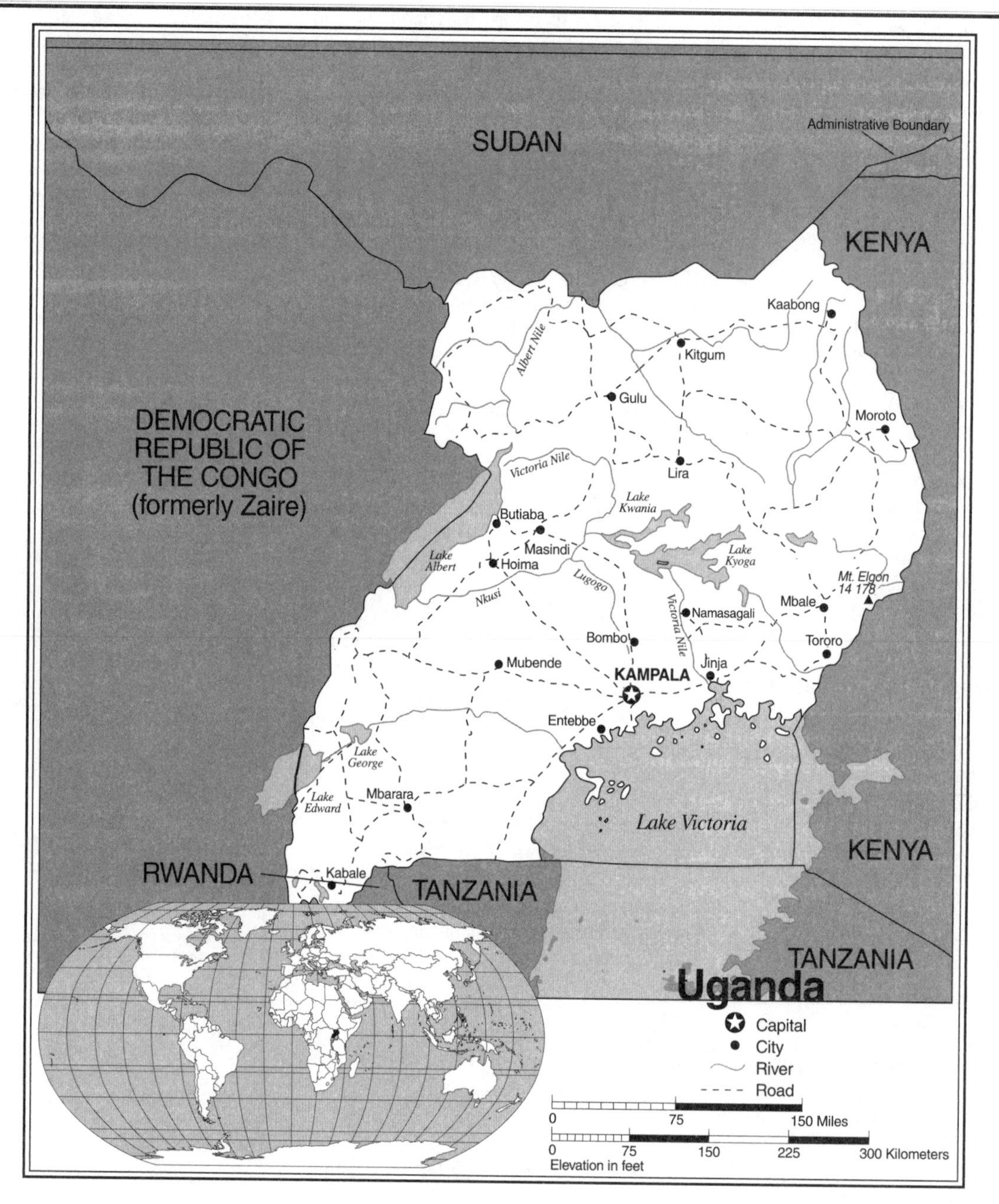

Uganda Statistics

GEOGRAPHY

Area in Square Miles (Kilometers): 91,076 (235,885) (about the size of Oregon)

Capital (Population): Kampala (1,274,000)

Environmental Concerns: draining of wetlands; deforestation; overgrazing; soil erosion; widespread poaching

Geographical Features: mostly plateau, with a rim of mountains

Climate: generally tropical, but semiarid in the northeast

PEOPLE

Population

Total: 26,404,543

Annual Growth Rate: 2.94%

Rural/Urban Population Ratio: 87/13

Major Languages: English; Swahili; Bantu languages; Nilotic languages

Ethnic Makeup: Bantu; Nilotic; Nilo-Hamitic; Sudanic

Religions: 66% Christian; 18% indigenous beliefs; 16% Muslim

Health

Life Expectancy at Birth: 39 years (male); 40 years (female)

Infant Mortality: 89.3/1,000 live births

Physicians Available: 1/20,700 people

HIV/AIDS Rate in Adults: 6.1%

Education

Adult Literacy Rate: 62%

COMMUNICATION

Telephones: 81,000 main lines
Televisions: 27/1,000 people
Internet Users: 125,000 (2003)

TRANSPORTATION

Highways in Miles (Kilometers): 16,200 (27,000)
Railroads in Miles (Kilometers): 745 (1,241)
Usable Airfields: 27
Motor Vehicles in Use: 51,000

GOVERNMENT

Type: republic
Independence Date: October 9, 1962 (from the United Kingdom)
Head of State/Government: President Yoweri Kaguta Museveni is both head of state and head of government
Political Parties: National Resistance Movement (only organization allowed to operate unfettered)
Suffrage: universal at 18

MILITARY

Military Expenditures (% of GDP): 2.1%
Current Disputes: continuing ethnic strife in the region

ECONOMY

Currency ($ U.S. Equivalent): 1,963 Uganda shillings = $1
Per Capita Income/GDP: $1,400/$36 billion
GDP Growth Rate: 4.4%
Inflation Rate: 7.9%
Labor Force by Occupation: 82% agriculture; 13% services; 5% industry
Population Below Poverty Line: 35%
Natural Resources: copper; cobalt; salt; limestone; hydropower; arable land
Agriculture: coffee; tea; cotton; tobacco; cassava; potatoes; corn; millet; pulses; livestock
Industry: sugar; brewing; tobacco; textiles; cement
Exports: $367 million (primary partner Europe)
Imports: $1.26 billion (primary partners Kenya, United States, India)

SUGGESTED WEB SITES

http://www.ugandaweb.com
http://www.government.go.ug
http://www.monitor.co.ug
http://www.mbendi.co.za/cyugcy.htm
http://www.uganda.co.ug/
http://www.cia.gov/cia/publications/factbook/geos/ug.html

Uganda Country Report

In 2000, Ugandans voted to reject multiparty politics in favor of continuing President Yoweri Museveni's "no-party" system. This action was followed in 2001 by another victory for Museveni in the presidential elections, in which he defeated his rival, Kizza Besigye, by 69 percent to 28 percent. Museveni has remained in power since 1985 through his tight control over the military and governing institutions. But his position has been challenged by his involvement in regional conflicts as well as by domestic rebels, most notably the Lords Resistance Army (LRA), which continues to terrorize northern Uganda. This led to a series of signifiant diplomatic initiatives in 2002.

DEVELOPMENT

In the past few years, Uganda's economy has been growing at an average annual rate of about 5%, boosted by increased investment. Foreign economic assistance nonetheless accounts for approximately 29% of government spending.

In March 2002, Sudan and Uganda signed an agreement aimed at containing the LRA, which was active along their common border. The deal allowed Uganda to carry out limited operations in Sudan. Although the LRA thereafter came under sustained attack in what the Ugandan military dubbed "Operation Iron Fist," the movement not only survived but also was able to regain control over areas in northern Uganda. In July, the Ugandan government broke off the offensive and agreed to begin negotiations with the LRA to end the war.

Finding common ground with the LRA may not be easy. A self-proclaimed "prophet" named Joseph Kony, who says he wants to run Uganda in conformity with the biblical Ten Commandments, leads the movement. But, in practice, the LRA has been responsible for the abduction of thousands of boys and girls. The boys are indoctrinated to kill and rape, while the girls are consigned to serve as laborers and targets of sexual gratification.

Over the past five years, Uganda's army has also been heavily involved in fighting in the Democratic Republic of the Congo (D.R.C., the former Zaire). In March 2001, Uganda classified Rwanda—its former ally in the D.R.C.'s civil war—as a hostile nation after several months of clashes between the two countries' armed forces over key areas of eastern D.R.C., which they had collectively occupied since 1998. Both countries, along with Zimbabwe, have been accused of pillaging the D.R.C.'s natural wealth. Tensions were eased between the two states in February 2002, when Museveni met the Rwandan president, Paul Kagame, as part of an on-going, British-backed effort to defuse tensions.

In August 2002, Museveni signed a peace accord with D.R.C. president Joseph Kabila, which had been brokered by the Angolan president Jose Eduardo dos Santos. This resulted in the rapid withdrawal of Ugandan forces from Eastern D.R.C. Uganda pulled out the last of its troops from eastern D.R.C. in May 2004 as tens of thousands of D.R. Congo civilians seek asylum in Uganda.

FREEDOM

The human-rights situation in Uganda remains poor, with government security forces linked to torture, extra-judicial executions, and other atrocities. Freedom of speech and association are curtailed. Insurgent groups are also associated with atrocities; the Lord's Resistance Army continues to kill, torture, maim, and abduct large numbers of civilians, enslaving numerous children.

Uganda's foreign and domestic conflicts pose a potential threat to the very real progress that the country has made since the coming to power of Museveni's National Resistance Movement (NRM). After years of repressive rule accompanied by massive interethnic violence, Uganda is still struggling for peace and reconciliation. A land rich in natural and human resources, Uganda suffered dreadfully during the despotic regimes of Milton Obote (1962–1971, 1980–1985) and Idi Amin (1971–1979). Under these two dictators, hundreds of thousands of Ugandans were murdered by the state. Former dictator Idi Amin died in a hospital in Jedaah, Saudi Arabia in August 2003. Amin was not mourned in Uganda and had no state funeral.

HEALTH/WELFARE

Millions of Ugandans live below the poverty line. Uganda's traditionally strong school system was damaged but not completely destroyed under Amin and Obote. In 1986, some 70% of primary-school children attended classes. The killing and exiling of teachers have resulted in a serious drop of standards at all levels of the education system, but progress is under way. The adult literacy rate has risen to 62%.

The country had reached a state of general social and political collapse by 1986, when the NRM seized power. The new government soon made considerable progress in restoring a sense of normalcy in most of the country, except for the north. In May 1996, Museveni officially received 74 percent of the vote in a contested presidential poll. Despite charges of fraud by his closest rival, Paul Ssemogerere, most independent observers accepted the poll as an endorsement of Museveni's leadership, including his view that politics should remain organized on a nonparty basis. There has since, however, emerged growing international criticism of his intolerance of genuine political pluralism.

HISTORIC GEOGRAPHY

The breakdown of Uganda is an extreme example of the disruptive role of ethnic and sectarian competition, which was fostered by policies of both its colonial and postcolonial governments. Uganda consists of two major zones: the plains of the northeast and the southern highlands. It has been said that you can drop anything into the rich volcanic soils of the well-watered south and it will grow. Until the 1960s, the area was divided into four kingdoms—Buganda, Bunyoro, Ankole, and Toro—populated by peoples using related Bantu languages. The histories of these four states stretch back hundreds of years. European visitors of the nineteenth century were impressed by their sophisticated social orders, which the Europeans equated with the feudal monarchies of medieval Europe.

When the British took over, they integrated the ruling class of the southern highlands into a system of "indirect rule." By then, missionaries had already succeeded in converting many southerners to Christianity; indeed, civil war among Protestants, Catholics, and Muslims within Buganda had been the British pretext for establishing their overrule.

The Acholi, Langi, Karamojang, Teso, Madi, and Kakwa peoples, who are predominant in the northeast, lack the political heritage of hierarchical state-building found in the south. These groups are also linguistically separate, speaking either Nilotic or Nilo-Hamitic languages. The British united the two regions as the Uganda Protectorate during the 1890s (the word *Uganda,* which is a corruption of "Buganda," has since become the accepted name for the larger entity). But the zones developed separately under colonial rule.

Cash-crop farming, especially of cotton, by local peasants spurred an economic boom in the south. The Bugandan ruling class benefited in particular. Increasing levels of education and wealth led to the European stereotype of the "progressive" Bugandans as the "Japanese of Africa." A growing class of Asian entrepreneurs also played an important role in the local economy, although its prosperity, as well as that of the Bugandan elite, suffered from subordination to resident-British interests.

The south's growing economy stood in sharp contrast to the relative neglect of the northeast. Forced to earn money to pay taxes, many northeasterners became migrant workers in the south. They were also recruited, almost exclusively, to serve in the colonial security forces.

ACHIEVEMENTS

The Ugandan government was one of the first countries in Africa (and the world) to acknowledge the seriousness of the HIV/AIDS epidemic within its borders. It has instituted public-information campaigns and welcomed outside support.

As independence approached, many Bugandans feared that other groups would compromise their interests. Under the leadership of their king, Mutesa II, they sought to uphold their separate status. Other groups feared that Bugandan wealth and educational levels could lead to their dominance. A compromise federal structure was agreed to for the new state. At independence, the southern kingdoms retained their autonomous status within the "United Kingdom of Uganda." The first government was made up of Mutesa's royalist party and the United People's Congress (UPC), a largely non-Bugandan coalition, led by Milton Obote, a Langi. Mutesa was elected president and Obote prime minister.

Timeline: PAST

1500s
Establishment of the oldest Ugandan kingdom, Bunyoro, followed by the formation of Buganda and other kingdoms

1893
A British protectorate over Uganda is proclaimed

1962
Uganda becomes independent

1966
Milton Obote introduces a new unitary Constitution and forces Bugandan compliance

1971
Idi Amin seizes power

1978—1979
Amin invades Tanzania; Tanzania invades Uganda and overturns Amin's government

1980s
The rise and fall of the second Obote regime; the NRM takes power under Yoweri Museveni

1990s
Recovery produces slow gains; unrest continues in the northeast

PRESENT

2000s
Uganda addresses the HIV/AIDS pandemic

Museveni retains power

THE REIGN OF TERROR

In 1966, the delicate balance of ethnic interests was upset when Obote used the army—still dominated by fellow northeasterners—to overthrow Mutesa and the Constitution. In the name of abolishing "tribalism," Obote established a one-party state and ruled in an increasingly dictatorial fashion. However, in 1971, he was overthrown by his army chief, Idi Amin. Amin began his regime with widespread public support but alienated himself by favoring fellow Muslims and Kakwa. He expelled the 40,000-member Asian community and distributed their property to his cronies. The Langi, suspected of being pro-Obote, were also early targets of his persecution, but his attacks soon spread to other members of Uganda's Christian community, at the time about 80 percent of the total population. Educated people in particular were purged. The number of

Ugandans murdered by Amin's death squads is unknown; the most commonly cited figure is 300,000, but estimates range from 50,000 to 1 million. Many others went into exile. Throughout the world, Amin's name became synonymous with despotic rule.

A Ugandan military incursion into Tanzania led to war between the two countries in 1979. Many Ugandans joined with the Tanzanians in defeating Amin's army and its Libyan allies.

Unfortunately, the overthrow of Amin, who fled into exile, did not lead to better times. In 1980, Obote was returned to power, through a fraudulent vote count. His second administration was characterized by a continuation of the violence of the Amin years. Obote's security forces massacred an estimated 300,000 people, mostly southerners, while an equal number fled the country. Much of the killing occurred in the Bugandan area known as the Luwero triangle, which was completely depopulated; its fields are still full of skeletons today. As the killings escalated, so did the resistance of Museveni's NRM guerrillas, who had taken to the bush in the aftermath of the failed election. In 1985, a split between Ancholi and Langi officers led to Obote's overthrow and yet another pattern of interethnic recrimination. Finally, in 1986, the NRM gained the upper hand.

Thereafter a new political order began to emerge based on Museveni's vision of a "no-party government." His position was strengthened in March 1994, when elections to a Constituent Assembly resulted in his supporters' capturing more than two thirds of the seats. In another controversial initiative, Museveni allowed the restoration of traditional offices, including Bugandan kingship.

THE STRUGGLE CONTINUES

Museveni's National Resistance Movement administration has faced enormous challenges in trying to bring about national reconstruction. The task has been complicated by continued warfare in the northeast by armed factions representing elements of the former regimes, independent Karamojong communities, and followers of prophetic religious movements. In 1987, an uprising of the Holy Spirit rebels of Alice Lakwena was crushed, at the cost of some 15,000 lives.

The restoration of peace to most of the country has promoted economic growth. Western-backed economic reforms produced an annual growth rate of 13 percent between 1990 and 1998. The rate of inflation also improved, falling from 200 to 7 percent in the same period. On the regional level, Museveni has championed the formation of a new "East African Community" (EAC), which is intended to lay the groundwork for economic and ultimately monetary integration with Tanzania and Kenya.

While rebuilding their shattered country, Ugandans have had to cope with an especially severe outbreak of HIV/AIDS. Thousands have died of the disease in the last decade; it is believed that literally hundreds of thousands of Ugandans are HIV-positive. The government's bold acknowledgment and proactive efforts to address the crisis, however, have been credited with helping to contain the pandemic.

North Africa

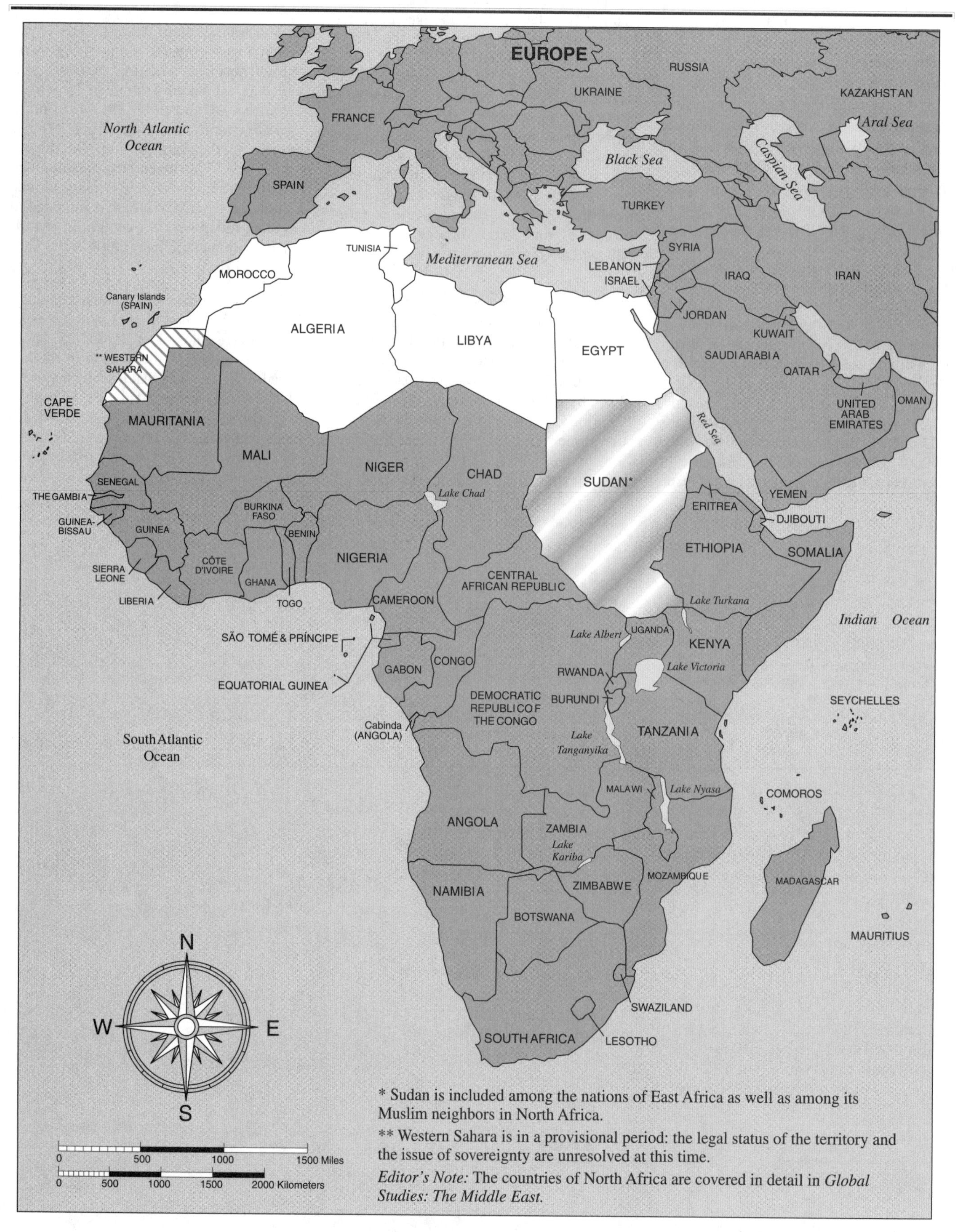

* Sudan is included among the nations of East Africa as well as among its Muslim neighbors in North Africa.

** Western Sahara is in a provisional period: the legal status of the territory and the issue of sovereignty are unresolved at this time.

Editor's Note: The countries of North Africa are covered in detail in *Global Studies: The Middle East.*

North Africa: The Crossroads of the Continent

Located at the geographical and cultural crossroads between Europe, Asia, and the rest of Africa, North Africa has served since ancient times as a link between the civilizations of sub-Saharan Africa and the rest of the world. Traders historically carried the continent's products northward, either across the Sahara Desert or up the Nile River and Red Sea, to the great port cities of the Mediterranean coast. Goods also flowed southward. In addition, the trade networks carried ideas: Islam, for example, spread from coastal North Africa across much of the rest of the continent to become the religion of at least one third of all Africans.

While there are millions of people now living in the nations of North Africa who would most willingly claim that they are not Africans at all but Arabs, it important to remember that the designation Arab is cultural/ethnic in origin. Africa is a continent not a particular cultural group. In this sense the people of North Africa are Africans, no matter where their particular ethnic group might originally have come from. North Africans occupy a geographically integral part of Africa. Due to the color of their skins they may look like they are foreigners to the continent, but they have been living in Africa for thousands of years and they don't seem to be ready to leave their African homes for awhile. As much as the Egyptians Tunisians and Algerians would like to become integrated into the European Union its wise to note that the Europeans are quick to tell these fine North Africans that they are not Europeans, they have never been Europeans, and the North Africans will never be Europeans. The Arabs are merely another tribe of people living in Africa. The pyramids are African, built by Africans in Africa, to worship African gods and spiritual entities that reflect a particular cosmological view of the world. The Arabs who now rule Egypt (and would like to rule Sudan) are in this sense guardians of the African temples.

The whole idea of race being dominant over geography is a historical and counterproductive to our understanding of the people of Africa, the land in which they reside, and the struggles they face in their everyday lives, past, present and future. Moreover ethnic/racial and cultural discrimination serves little purpose in any contemporary analysis of the continent. The former colonial masters did not create divisions of the African people, they merely used them to separate one African group from another and rule them all. Claiming one particular ethnic group and the way in which they live as cause for superiority over another, is no better than what the colonial regimes did for generations. It takes us nowhere but to war and conflict. To claim that an Arab is somehow superior to an African, is just as bad as saying that Hutu are superior to the Tutsi, and the Colored are superior to the Tswana in South Africa. The consequences of such definitions are fraught with danger and might well lead to attempts at genocide. Until such time as the people of North Africa leave North Africa and go somewhere else they are going to remain Africans.

North Africa's role as the continent's principal window to the Western world gradually declined after the year A.D. 1500, as the trans-Atlantic trade increased. (The history of East Africa's participation in Indian Ocean trade goes back much further.) However, the countries of North Africa have continued to play an important role in the greater continent's development.

The countries of North Africa—Morocco, Algeria, Tunisia, Libya, and Egypt—and their millions of people differ from one another, but they share a predominant, overarching Arab-Islamic culture that both distinguishes them from the rest of Africa and unites them with the Arabic-speaking nations of the Middle East. To begin to understand the societies of North Africa and their role in the rest of the continent, it is helpful to examine the area's geography. The region's diverse environment has long encouraged its inhabitants to engage in a broad variety of economic activities: pastoralism, agriculture, trading, crafts, and, later, industry.

GEOGRAPHY AND POPULATION

Except for Tunisia, which is relatively small, the countries of North Africa are sprawling nations. Algeria, Libya, and Egypt are among the biggest countries on the African continent, and Morocco is not far behind. Their size can be misleading, for much of their territories is encompassed by the largely barren Sahara Desert. The approximate populations of the five states today range from Egypt's 71 million people to Libya's 6 million; Morocco has 31 million, Algeria 32 million, and Tunisia almost 10 million citizens. All these populations are increasing at a rapid rate; indeed, well over half of the region's citizens are under age 21.

Due to its scarcity, water is the region's most precious resource, so most people live either in valleys near the Mediterranean coast or along the Nile. The latter courses through the desert for thousands of miles, creating a narrow green ribbon that is the home of the 95 percent of Egypt's population who live within 12 miles of its banks. More than 90 percent of the people of Algeria, Libya, Morocco, and Tunisia live within 200 miles of either the Mediterranean or, in the case of Morocco, the Atlantic coast.

Besides determining where people live, the temperate, if often too dry, climate of North Africa has always influenced local economies and lifestyles. There is intensive agriculture along the coasts and rivers. Algeria, Morocco, and Tunisia are well known for their tree and vine crops, notably citrus fruits, olives, and wine grapes. The intensively irrigated Nile Valley has been a leading source of high-quality cotton as well as locally consumed foodstuffs since the time of the American Civil War, which temporarily removed U.S.–produced fiber from the world market. In the oases that dot the Sahara Desert, date palms are grown for their sweet fruits, which are almost a regional staple. Throughout the steppelands between the fertile coasts and the desert, pastoralists follow flocks of sheep and goats or herds of cattle and camels in constant search of pasture. Although now few in number, it was these nomads who in the past developed the trans-Saharan trade. As paved roads and air-

(Photo by Wayne Edge)

Camels are still used in North Africa to move through the desert.

ports have replaced their caravan routes, long-distance nomadism has declined. But the traditions it bred, including a love of independence, remain an important part of North Africa's cultural heritage.

Urban culture has flourished in North Africa since the ancient times of the Egyptian pharaohs and the mercantilist rulers of Carthage. Supported by trade and local industries, the region's medieval cities, such as Cairo, Fez, and Kairouan, were the administrative centers of great Islamic empires, whose civilizations shined during Europe's dark ages. In the modern era, the urban areas are bustling industrial centers, ports, and political capitals.

Geography—or, more precisely, geology—has helped to fuel economic growth in recent decades. Although agriculture continues to provide employment in Algeria and Libya for as much as a third of the labor force, discoveries of oil and natural gas in the 1950s dramatically altered these two nations' economic structures. Between 1960 and 1980, Libya's annual per capita income jumped from $50 to almost $10,000, transforming it from among the poorest to among the richest countries in the world. Algeria has also greatly benefited from the exploitation of hydrocarbons, although less dramatically than Libya. Egypt and Tunisia have developed much smaller oil industries, which nonetheless provide for their domestic energy needs and generate much-needed foreign exchange. The decline in oil prices during the 1980s, however, reduced revenues, increased unemployment, and contributed to social and political unrest, especially in Algeria. While it has no oil, Morocco profits from its possession of much of the world's phosphate production, which is concentrated in the Moroccan-occupied territory of Western Sahara.

CULTURAL AND POLITICAL HERITAGES

The vast majority of the inhabitants of North Africa are Arabic-speaking Muslims. Islam and Arabic both became established in the region between the seventh and eleventh centuries A.D. Thus, by the time of the Crusades in the eastern Mediterranean, the societies of North Africa were thoroughly incorporated into the Muslim world, even though the area had earlier been the home of many Christian scholars. Except for Egypt, where about 5 percent of the population remain loyal to the Coptic Church, there is virtually no Christianity among modern North Africans. Until recently, important Jewish communities existed in all the region's countries, but their numbers have dwindled as a result of mass emigration to Israel.

With Islam came Arabic, the language of the Koran—the holy book of Islam. Today, Egypt and Libya are almost exclusively Arabic-speaking. In Algeria, Morocco, and Tunisia, Arabic coexists with various local minority languages, which are collectively known as Berber (from which the term *Barbary*, as in Barbary Coast, was derived). As many as a third of the Moroccans speak a form of Berber as their first language. Centuries of interaction between the Arabs and Berbers as well as their common adherence to Islam have promoted a sense of cultural unity between the two communities, although ethnic disputes have developed in Algeria and Morocco over demands that Berber be included in local school curriculums. As was the case almost everywhere else on the continent, the linguistic situation in North Africa was further complicated by the introduction of European languages during the colonial era. Today, French is particularly important as a language of technology and administration in Algeria, Morocco, and Tunisia.

By the seventeenth century, all the countries of North Africa, except Morocco, were autonomous provinces of the Ottoman Empire, which was based in present-day Turkey and also incorporated most of the Middle East. Morocco was an independent state; indeed, it was one of the earliest to recognize the independence of the United States from Britain. From 1830, the European powers gradually encroached upon the Ottoman Empire's North African realm. Thus, like most of their sub-Saharan counterparts, all the states of North Africa fell under European imperial control. Algeria's conquest by the French began in 1830 but took decades to accomplish, due to fierce local resistance. France also seized Tunisia in 1881 and, along with Spain, partitioned Morocco in 1912. Britain occupied Egypt in 1882, and Italy invaded Libya in 1911, although anti-Italian resistance continued until World War II, when the area was liberated by Allied troops.

The differing natures of their European occupations have influenced the political and social characters of each North African state. Algeria, which was directly incorporated into France as a province for 120 years, did not win its independence until 1962, after a protracted war of liberation. Morocco, by contrast, was accorded independence in 1956, after only 44 years of Franco–Spanish administration, during which the local monarchy continued to reign. Tunisia's 75 years of French rule also ended in 1956, as a strong nationalist party took the reins of power. Egypt, although formally independent of Great Britain, did not win genuine self-determination until 1952, when a group of nationalist army officers came to power by overthrowing the British-supported monarchy. Libya became a temporary ward of the United Nations after Italy was deprived of its colonial empire during World War II. The nation was granted independence by the United Nations in 1951, under a monarch whose religious followers had led much of the anti-Italian resistance.

Editor's Note: The countries of North Africa are covered in detail in *Global Studies: The Middle East.*

Algeria (Peoples' Democratic Republic of Algeria)

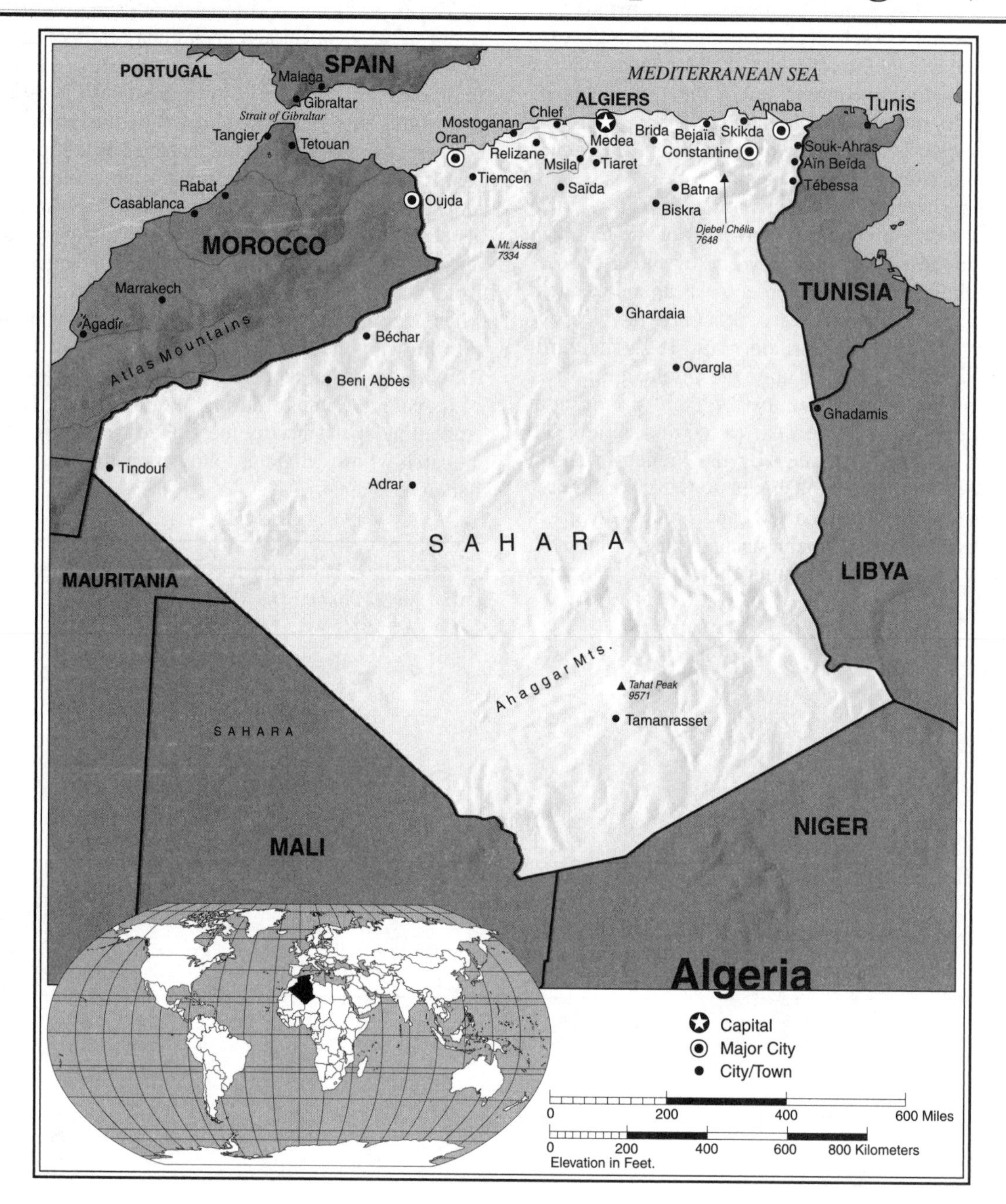

Algeria Statistics

GEOGRAPHY

Area in Square Miles (Kilometers): 919,352 (2,381,740) (about 3 1/2 times the size of Texas)

Capital (Population): Algiers (3,705,000)

Environmental Concerns: soil erosion; desertification; water pollution; inadequate potable water

Geographical Features: mostly high plateau and desert; some mountains; narrow, discontinuous coastal plain

Climate: arid to semiarid; mild winters and hot summers on coastal plain; less rain and cold winters on high plateau; considerable temperature variation in desert

PEOPLE

Population

Total: 32,277,942

Annual Growth Rate: 1.68%

Rural/Urban Population Ratio: 44/56

Major Languages: Arabic; Berber dialects; Ahaggar (Tuareg); French

Ethnic Makeup: 99% Arab-Berber; less than 1% European
Religions: 99% Sunni Muslim (Islam is the state religion); 1% Shia Muslim, Christian, and Jewish

Health

Life Expectancy at Birth: 68.97 years (male); 71.67 years (female)
Infant Mortality Rate (Ratio): 39/1,000 live births
Physicians Available (Ratio): 1/1,066 people

Education

Adult Literacy Rate: 61.6%
Compulsory (Ages): 6–15

COMMUNICATION

Telephones: 2,300,000 (1998); 500,000 new lines being connected in 2003, plus 33,500 celular phones
Daily Newspaper Circulation: 52 per 1,000 people
Televisions: 71 per 1,000 people
Internet Users: 2 (2000)

TRANSPORTATION

Highways in Miles (Kilometers): 63,605 (102,424)
Railroads in Miles (Kilometers): 2,963 (4,772)
Usable Airfields: 136
Motor Vehicles in Use: 920,000

GOVERNMENT

Type: republic
Independence Date: July 5, 1962 (from France)
Head of State/Government: President Abdelaziz Bouteflika; Prime Minister Ali Benfis
Political Parties: National Liberation Front (FLN), majority party; National Democratic Rally (RND), National Reform Movement, chief minority parties; others include Movement for a Peaceful society; Islamic Salvation Front (FIS) outlawed since April 1992
Suffrage: universal at 18

MILITARY

Military Expenditures (% of GDP): 4.1%
Current Disputes: disputed southeastern border with Libya; Algeria supports Polisario Front which seeks to establish an independent Western Sahara, currently occupied by Morocco

ECONOMY

Currency ($ U.S. Equivalent): 78 Algerian dinars = $1
Per Capita Income/GDP: $5,600/$177 billion
GDP Growth Rate: 3.8%
Inflation Rate: 3%
Unemployment Rate: 34%
Labor Force: 9,400,000
Natural Resources: petroleum; natural gas; iron ore; phosphates; uranium; lead; zinc
Agriculture: wheat; barley, oats; grapes; olives; citrus fruits; sheep; cattle
Industry: petroleum; natural gas; light industries; mining; electrical; petrochemicals; food processing
Exports: $19.6 billion (primary partners Italy, United States, France)
Imports: $10.6 billion (primary partners France, United States, Italy)

Algeria Country Report

The modern state of Algeria occupies the central part of North Africa, a geographically distinctive and separate region of Africa that includes Morocco, Tunisia, and Libya. The name of the country comes from the Arabic word *al-Jaza'ir,* "the islands," because of the rocky islets along this part of the Mediterranean coast. The name of the capital, Algiers, has the same origin.

The official name of the state is the Democratic and Popular Republic of Algeria. It is the second-largest nation in Africa (after Sudan). The overall population density is low, but the population is concentrated in the northern third of the country. The vast stretches of the Algerian Sahara are largely unpopulated. The country had an extremely high birth rate prior to 1988, but government-sponsored family-planning programs have significantly reduced the rate.

GEOGRAPHY

Algeria's geography is a formidable obstacle to broad economic and social development. About 80 percent of the land is uncultivable desert, and only 12 percent is arable without irrigation. Most of the population live in a narrow coastal plain and in a fertile, hilly inland region called the Tell (Arabic for "hillock"). The four Saharan provinces have only 3 percent of the population but comprise more than half the land area.

The mineral resources that made possible Algeria's transformation in two decades from a land devastated by civil war to one of the developing world's success stories are all located in the Sahara. Economic growth, however, has been uneven, generally affecting the rural and lower-class urban populations unfavorably. The large-scale exodus of rural families into the cities, with consequent neglect of agriculture, has resulted in a vast increase in urban slums. Economic disparities were a major cause of riots in 1988, that led to political reforms and the dismantling of the socialist system responsible for Algerian development since independence.

Algeria is unique among newly independent Middle Eastern countries in that it gained its independence through a civil war. For more than 130 years (1830–1962), it was occupied by France and became a French department (similar to a U.S. state). With free movement from mainland France to Algeria and vice versa, the country was settled by large numbers of Europeans, who became the politically dominant group in the population although they were a minority. The modern Algerian nation is the product of the interaction of native Muslim Algerians with the European settlers, who also considered Algeria home.

The original inhabitants of the entire North African region were Berbers, a people of unknown origin grouped into various tribes. Berbers make up about 30 percent of the total population. The majority live in the eastern Kabylia region and the Aures (Chaouia), with a small, compact group in the five cities of the Mzab, in the Algerian Sahara. The Tuareg, a nomadic Berber people spread across southern Algeria, Mali, and Niger, are the only ones with a written script, called Tifinagh. In the past, they were literally "lords of the desert," patrolling the caravan routes on their swift camels and collecting tolls for safe passage as guides for caravaneers. Today, Tuareg are more likely to be found pumping gas in cities or doing low-wage work in the oil fields than patrolling the desert.

The Arabs, who brought Islam to North Africa in the seventh century A.D., converted the Algerian Berbers after a fierce resistance. The Arabs brought their language as a unifying feature, and religion linked the Algerians with the larger Islamic world. Today, most follow Sunni Islam, but a significant minority, about 100,000, are Shia Muslims. They refer to themselves as *Ibadis*, from their observance of an ancient Shia rite, and live in five "holy cities" clustered in a remote Saharan valley where centuries ago they took refuge from Sunni rulers of northern Algeria. One of many pressures on the government today is that of an organized Kabyle movement, which seeks greater autonomy for the region and an emphasis on Berber language in schools, along with the revitalization of Kabyle culture.

HISTORY

The Corsair Regency

The foundations of the modern Algerian state were laid in the sixteenth century, with the establishment of the Regency of Algiers, an outlying province of the Ottoman Empire. Algiers in particular, due to its natural harbor, was developed for use by the Ottomans as a naval base for wars against European fleets in the Mediterranean. The Algerian coast was the farthest westward extent of Ottoman power. Consequently, Algiers and Oran, the two major ports, were exposed to constant threats of attack by Spanish and other European fleets. They could not easily be supported, or governed directly, by the Ottomans. The regency, from its beginnings, was a state geared for war.

Outside of Algiers and its hinterland, authority was delegated to local chiefs and religious leaders, who were responsible for tax collection and remittances to the dey's treasury. The chiefs were kept in line with generous subsidies. It was a system well adapted to the fragmented society of Algeria and one that enabled a small military group to rule a large territory at relatively little cost.[1]

The French Conquest

In 1827, the dey of Algiers, enraged at the French government's refusal to pay an old debt incurred during Napoleon's wars, struck the French consul on the shoulder with a fly-whisk in the course of an interview. The king of France, Charles X, demanded an apology for the "insult" to his representative. None was forthcoming, so the French blockaded the port of Algiers in retaliation. But the dey continued to keep silent. In 1830, a French army landed on the coast west of the city, marched overland, and entered it with almost no resistance. The dey surrendered and went into exile.[2]

The French, who had been looking for an excuse to expand their interests in North Africa, now were not sure what to do with Algiers. The overthrow of the despotic Charles X in favor of a constitutional monarchy in France confused the situation even further. But the Algerians considered the French worse than the Turks, who were at least fellow Muslims. In the 1830s, they rallied behind their first national leader, Emir Abd al-Qadir.

Abd al-Qadir was the son of a prominent religious leader and, more important, was a descendant of the Prophet Muhammad. Abd al-Qadir had unusual qualities of leadership, military skill, and physical courage. From 1830 to 1847, he carried on guerrilla warfare against a French army of more than 100,000 men with such success that at one point the French signed a formal treaty recognizing him as head of an Algerian nation in the interior. Abd al-Qadir described his strategy in a prophetic letter to the king of France:

> France will march forward, and we shall retire. But France will find it necessary to retire, and we shall return. We shall weary and harry you, and our climate will do the rest.[3]

In order to defeat Abd al-Qadir, the French commander used "total war" tactics, burning villages, destroying crops, killing livestock, and levying fines on peoples who continued to support the emir. These measures, called "pacification" by France, finally succeeded. In 1847, Abd al-Qadir surrendered to French authorities. He was imprisoned for several years, in violation of a solemn commitment, and was then released by Emperor Napoleon III. He spent the rest of his life in exile.

DEVELOPMENT

Algeria ranks 5th in the world in natural gas reserves and second in gas exports. The hydrocarbons sector generates 60 percent of revenues, 30 percent of GDP. In 2001 the country signed an association trety with the European Union and has applied for membership in the World Trade Organization.

Although he did not succeed in his quest, Abd al-Qadir is venerated as the first Algerian nationalist, able by his leadership and Islamic prestige to unite warring groups in a struggle for independence from foreign control. Abd al-Qadir's green and white flag was raised again by the Algerian nationalists during the second war of independence (1954–1962), and it is the flag of the republic today.

Algérie Française

After the defeat of Abd al-Qadir, the French gradually brought all of present-day Algerian territory under their control. The Kabyles, living in the rugged mountain region east of Algiers, were the last to submit. The Kabyles had submitted in 1857, but they rebelled in 1871 after a series of decrees by the French government had made all Algerian Muslims subjects but not citizens, giving them a status inferior to French and other European settlers.

The Kabyle rebellion had terrible results, not only for the Kabyles but for all Algerian Muslims. More than a million acres of Muslim lands were confiscated by French authorities and sold to European settlers. A special code of laws was enacted to treat Algerian Muslims differently from Europeans, with severe fines and sentences for such "infractions" as insulting a European or wearing shoes in public. (It was assumed that a Muslim caught wearing shoes had stolen them.)

In 1871, Algeria legally became a French department. But in terms of exploitation of natives by settlers, it may as well have remained a colony. One author notes that "the desire to make a settlement colony out of an already populated area led to a policy of driving the indigenous people out of the best arable lands."[4] Land confiscation was only part of the exploitation of Algeria by the *colons* (French settlers). They developed a modern Algerian agriculture integrated into the French economy, providing France with much of its wine, citrus, olives, and vegetables. Colons owned 30 percent of the arable land and 90 percent of the best farmland. Special taxes were imposed on the Algerian Muslims; the colons were exempted from paying most taxes.

Jules Cambon, governor general of Algeria in the 1890s, once described the country as having "only a dust of people left her." What he meant was that the ruthless treatment of the Algerians by the French during the pacification had deprived them of their natural leaders. A group of leaders developed slowly in Algeria, but it was made up largely of *evolués*—persons who had received French educations, spoke French better than Arabic, and accepted French citizenship as the price of status.[5]

Other Algerians, several hundred thousand of them, served in the French Army in the two world wars. Many of them became aware of the political rights that they were supposed to have but did not. Still others, religious leaders and teachers, were influenced by the Arab nationalist movement for independence from foreign control in Egypt and other parts of the Middle East.

On May 8, 1945, the date of the Allied victory over Nazi Germany, a parade of Muslims celebrating the event but also demanding equality led to violence in the city

(UN/UNDP Photo 155705/Ruth Massey)

Agriculture is important in raising the living standards of Algeria. These farmers are harveting forage peas, which will be used for animal feed.

of Sétif. Several colons were killed; in retaliation, army troops and groups of colon vigilantes swept through Muslim neighborhoods, burning houses and slaughtering thousands of Muslims. From then on, Muslim leaders believed that independence through armed struggle was the only choice left to them.

The War for Independence

November 1 is an important holiday in France. It is called Toussaint (All Saints' Day). On that day, French people remember and honor all the many saints in the pantheon of French Catholicism. It is a day devoted to reflection and staying at home.

In the years after the Sétif massacre, there had been scattered outbreaks of violence in Algeria, some of them created by the so-called Secret Organization (OS), which had developed an extensive network of cells in preparation for armed insurrection. In 1952, French police accidentally uncovered the network and jailed most of its leaders. One of them, a former French Army sergeant named Ahmed Ben Bella, subsequently escaped and went to Cairo, Egypt.

Eventually there were 400,000 French troops in Algeria, as opposed to just 6,000 guerrillas. But the French consistently refused to consider the situation in Algeria a war. They called it a "police action." Others called it the "war without a name."[6] Despite their great numerical superiority, they were unable to defeat the FLN.

The war was settled not by military action but by political negotiations. The French people and government, worn down by the effects of World War II and their involvement in Indochina, grew sick of the slaughter, the plastic bombs exploding in public places (in France as well as Algeria), and the brutality of the army in dealing with guerrilla prisoners. A French newspaper editor expressed the general feeling: "Algeria is ruining the spring. This land of sun and earth has never been so near us. It invades our hearts and torments our minds."[7]

THE AGONY OF INDEPENDENCE

With the collapse of the OAS campaign against the FLN as well as its own government, the way was clear for Algeria to become an independent nation for the first time in its history. This became a reality on July 5, 1962, with the signing of a treaty with France.

The first leader to emerge from intraparty struggle to lead the nation was Ahmed Ben Bella, who had spent the war in exile in Egypt but had great prestige as the political brains behind the FLN. Ben Bella laid the groundwork for an Algerian political system centered on the FLN as a single legal political party, and in September 1963, he was elected president. Ben Bella introduced a system of *autogestion* (workers' self-management), by which tenant farmers took over the management of farms abandoned by their colon owners and restored them to production as cooperatives. Autogestion became the basis for Algerian socialism—the foundation of development for decades.

Ben Bella did little else for Algeria, and he alienated most of his former associates with his ambitions for personal power. In June 1965, he was overthrown in a military coup headed by the defense minister, Colonel Houari Boumedienne.

Boumedienne declared that the coup was a "corrective revolution, intended to reestablish authentic socialism and put an end to internal divisions and personal rule."[8] The government was reorganized under a Council of the Revolution, all military men, headed by Boumedienne, who subsequently became president of the republic. After a long period of preparation and gradual assumption of power by the reclusive and taciturn Boumedienne, a National Charter (Constitution) was approved by voters in 1976. The Charter defined Algeria as a socialist state with Islam as the state religion, basic citizens' rights guaranteed, and leadership by the FLN as the only legal political party. A National Popular Assembly (the first elected in 1977) was responsible for legislation.

In theory, the Algerian president had no more constitutional powers than the U.S. president. However, in practice, Boumedienne was the ruler of the state, being president, prime minister, and commander of the armed forces rolled into one. In November 1978, he became ill from a rare blood disease; he died in December.

(UN photo/KataBader)

The rapid growth in the population of Algeria, coupled with urban migration, has created a serious housing shortage, as this crowded apartment building in Algiers testifies.

The FLN closed ranks and named Colonel Chadli Bendjedid to succeed Boumedienne as president for a five-year term. In 1984, Bendjedid was reelected. But the process of ordered socialist development was abruptly and forcibly interrupted in October 1988. A new generation of Algerians, who had come of age long after the war for independence, took to the streets, protesting high prices, lack of jobs, inept leadership, a bloated bureaucracy, and other grievances.

The riots accelerated the process of Algeria's "second revolution" toward political pluralism and dismantling of the single-party socialist system. President Bendjedid initially declared a state of emergency; and for the first time since independence, the army was called in to restore order. Some 500 people were killed in the rioting, most of them jobless youths. Another constitutional change, also effective in 1989, made the cabinet and prime minister responsible to the National Assembly.

The president retained his popularity during the upheaval and was reelected for a third term, winning 81 percent of the votes. A number of new parties were formed in 1989 to contest future Assembly elections. They represented a variety of political and social positions.

For its part, the government sought to revitalize the FLN as a genuine mass party on the order of the Tunisian Destour, while insisting that it would not duplicate its neighbor country's *democratie de façade* but would instead embark on real political reforms.

FOREIGN POLICY

During the first decade of independence, Algeria's foreign policy was strongly nationalistic and anti-Western. Having won their independence from one colonial power, the Algerians were vocally hostile toward the United States and its allies, calling them enemies of popular liberation. Algeria supported revolutionary movements all over the world, providing funds, arms, and training. The Palestine Liberation Organization, rebels against Portuguese colonial rule in Mozambique, Muslim guerrillas fighting the Christian Ethiopian government in Eritrea—all benefited from active Algerian support.

The government broke diplomatic relations with the United States in 1967, due to American support for Israel, and did not restore them for a decade. In the mid-1970s, Algeria moderated its anti-Western stance in favor of nonalignment and good relations with both East and West. Relations improved thereafter to such a point that Algerian mediators were instrumental in resolving the 1979–1980 American hostage crisis in Iran, since Iran regarded Algeria as a suitable mediator—Islamic yet nonaligned. However, Algeria's subsequent alignment with Iraq (in sympathy for Iraq as a fellow-Arab state) during the Iran–Iraq War caused a break in diplomatic relations with the Islamic Republic. They were not restored until 2000.

Until recently, Algeria's relations with Morocco were marked by suspicion, hostility, and periodic conflict. The two countries clashed briefly in 1963 over ownership of iron mines near Tindouf, on the border. Algeria also supported the Western Saharan nationalist movement fighting for independence for the former Spanish colony against Moroccan occupation. After Morocco annexed the territory, Algeria provided bases, sanctuary, funds, and weapons to the Polisario Front, the military wing of the movement. The Bendjedid government recognized the self-declared Sahrawi Arab Democratic Republic in 1980 and sponsored SADR membership in the Organization for African Unity.

However, Algeria's own economic problems, along with moves to open the political system to multiparty activity, sharply reduced Algerian support for the Polisario in the 1980s. Polisario offices in Algiers were closed, and relations with Morocco improved after meetings between Benjedid and Hassan II, in which the latter accepted "in principle" a UN–sponsored referendum on the disputed territory.

The success of Algerian mediators in resolving international disputes has been duplicated in recent years in conflicts involving its other neighbors. In 1987, they succeeded in influencing Libyan leader Muammar al-Qadhafi to provide compensation for Tunisian workers expelled from Libya. A 1989 peace treaty between Libya and Chad also resulted from Algerian mediation.

THE ECONOMY

Algeria's oil and gas resources were developed by the French. Commercial production and exports began in 1958 and continued through the war for independence; they were not affected, since the Sahara was governed under a separate military administration. The oil fields were turned over to Algeria after independence but continued to be managed by French technicians until 1970, when the industry was nationalized.

Today, the hydrocarbons sector provides the bulk of government revenues and 90 percent of exports. New oil discoveries in 1996 and 2001 are expected to increase oil production, currently 852,000 barrels per day. Algeria provides 29 percent of the liquefied natural gas (LNG) imported by European countries, much of it through undersea pipelines to Italy and Spain.

FREEDOM

Algeria's constituion, issued in 1976 and amended 3 times, defines the country as a multi-party republic. However, it specified that no political association may be formed that is based on religious, linguistic, race, gender or regional differences. The gradual restoration of parliamentary democracy was expedited with the 2002 election for a National Popular Assembly and election by regional and municipal assemblies of two-thirds of the 144 members of the upper house (Council of Nations). And in what was described by outside observers as the cleanest, most open and free election in history in the Arab world, Bouteflika was re-elected president in April 2004, by an 83 percent majority, defeating five fivals including the prime minister and an avowed Islamist candidate. For the first time since their suppression of the 1992 elections military leaders declared their strict neutrality, in effect leaving politics to the politicians and reflecting the will of the people.

After a number of years of negative economic growth, the government initiated an austerity program in 1992. Imports of luxury products were prohibited and several new taxes introduced. The program was approved by the International Monetary Fund, Algeria's main source of external financing. In 1995, the IMF loaned $1.8 billion to cover government borrowing up to 60 percent under the approved austerity program to make the required "structural adjustment." In August of that year, the Paris Club—the international consortium that manages most of Algeria's foreign indebtedness—rescheduled $7 billion of the country's foreign debts due in 1996–1997, including interest payments, to ease the strain on the economy.

The agricultural sector employs 47 percent of the labor force and accounts for 12 percent of gross domestic product. But inasmuch as Algeria must import 70 percent of its food, better agricultural production is essential to overall economic development. Overall agricultural production growth has averaged 5 percent annually since 1990. The autogestion system introduced as a stop-gap measure after independence and enshrined later in FLN economic practice, when it seemed to work, was totally abandoned. In 1988, some 3,500 state farms were converted to collective farms, with individuals holding title to lands.

The key features of Bouteflika's economic reform program, one designed to attract foreign investment, include banking reforms, reduction of the huge government bureaucracy, favorable terms for foreign companies and privatization of state-owned enterprises. The telecommunications industry was privatized in 2000 and the government-owned cement and steel industries in 2002.

Privatization of state-owned enterprises is a key feature of the government's plan to attract foreign investment. The telecommunications sector was privatized in August 2000, and some 200 other public enterprises were in process of transfer to private ownership.

THE FUNDAMENTALIST CHALLENGE

Despite the growing appeal of Islamic fundamentalism in numerous Arab countries in recent years, Algeria until very recently seemed an unlikely site for the rise of a strong fundamentalist movement. The country's long association with France, its lack of historic Islamic identity as a nation, and several decades of single-party socialism militated against such a development. But the failure of successive Algerian governments to resolve severe economic problems, plus the lack of representative political institutions nurtured within the ruling FLN, brought about the rise of fundamentalism as a political. Fundamentalists took an active part in the 1988 riots; and with the establishment of a multiparty system, they organized a political party, the Islamic Salvation Front (FIS). It soon claimed 3 million adherents among the then 25 million Algerians.

HEALTH/WELFARE

The 1984 Family Law improved women's rights in marriage, education and work opportunities. But professional women and, more recenty, rural women and their chidren have become special targets of Islamic violence. Some 400 professional women were murdered in 1995 and more than 400 were killed in a one-day rampage in January 1998.

FIS candidates won 55 percent of urban mayoral and council seats in the 1989 local and municipal elections. The FLN conversely managed to hold on to power largely in the rural areas. Fears that FIS success might draw army intervention and spark another round of revolutionary violence led the government to postpone for six months the scheduled June 1991 elections for an enlarged 430-member National People's Assembly. An interim government, under the technocrat prime minister Sid Ahmed Ghozali, was formed to oversee the transition process.

In accordance with President Bendjedid's commitment to multiparty democracy, the first stage of Assembly elections took place on December 26, 1991, with FIS candidates winning 188 out of 231 contested seats. But before the second stage could take place, the army stepped in. FIS leaders were arrested, and the elections were postponed indefinitely. President Bendjedid resigned on January 17, 1992, well ahead of the expiration (in 1993) of his third five-year term. He said that he did so as a sacrifice in the interest of restoring stability to the nation and preserving democracy. Mohammed Boudiaf, one of the nine historic chiefs of the Revolution, returned from years of exile in Morocco to become head of the Higher Council of State, set up by military leaders after the abortive elections and resignation of President Bendjedid. FIS headquarters was closed and the party declared illegal by a court in Algiers. Local councils and provincial assemblies formed by the FIS after the elections were dissolved and replaced by "executive delegations" appointed by the Higher Council.

Subsequently, Boudiaf named a 60-member Consultative Council to work with the various political factions to reach a consensus on reforms. However, the refusal of such leaders as former president Ben Bella and Socialist Forces Front (FFS) leader Hocine Ait Ahmed to participate limited its effectiveness. Boudiaf was also suspected of using it to build a personal power base. On June 29, 1992, he was assassinated, reportedly by a member of his own presidential guard.

With Boudiaf gone, Algeria's generals turned to their own ranks for new leadership. In 1994, General Liamine Zeroual, the real strongman of the regime, was named head of state by the Higher Council. Zeroual pledged that elections for president would be held in November 1995 as a first step toward the restoration of parliamentary government. The top FIS leaders, Abbas Madani and Ali Belhaj, who had been given 12-year jail sentences for " endangering state security" were released but had their sentences commuted to house arrest, on the assumption that in return for dialogue, they would call a halt to the spiraling violence.

However, the dialogue proved inconclusive, and Zeroual declared that the presidential elections would be held on schedule. Earlier, leaders of the FIS, FFS, FLN, and several smaller parties had met in Rome, Italy, under the sponsorship of Sant-Egidio, a Catholic service agency, and announced a "National Contract." It called for the restoration of FIS political rights in return for an end to violence, multiparty democracy, and exclusion of the military from government. The Algerian "personality" was defined in the Contract as Islamic, Arab, and Berber.

Military leaders rejected the National Contract out of hand, due to the FIS's participation. However, the November 1995 presidential election was held as scheduled, albeit under massive army protection—soldiers were stationed within 65 feet of every polling place. Zeroual won handily, as expected, garnering 61 percent of the votes. But the fact that the election was held at all, despite a boycott call by several party leaders and threats of violence from the Armed Islamic Group (GIA), was impressive.[9]

THE KILLING FIELDS

The conflict between the armed wing of the FIS, the Armed Islamic Group (GIA), and the military regime reached a level of violence in the period after the 1995 "election" that left no room for compromise. The GIA targeted not only the army and police but also writers, journalists, government officials and other public figures, professional women, even doctors and dentists. Ironically, one of its victims was the head of the Algerian League for Human Rights, which had protested the detention without trial of some 9,000 FIS members in roofless prisons deep in the Sahara, under appallingly harsh conditions.

The GIA widened its circle of violence in the rest of the decade and on into the new century. In addition to Algerians it carried out attacks on foreigners, killing tourists as well as long-term foreign residents, notably Trappist monks. Rural villages were a favorite target, since they had neither police nor army protection. Entire village populations were massacred in a manner eerily reminiscent of the war for independence. The army and security forces did their share of killings, often arresting people and holding them indefinitely without charges. In 2003 the international organization Human Rights Watch reported that in addition to 120,000 deaths, some 7,000 persons had simply disappeared, never to be seen again by their families.

ACHIEVEMENTS

A new pipeline from the vast Hassi Berkine oil filed, the African continent's largest, went into production in 1998, with exports of 300,000 barres per day. Increased foreign investment due to expansion of the hydrocarbons sector, privatization of state-owned enterprises, and favorable terms for foreign companies increased GDP growth to 5% in 2001.

As the violence continued, the GIA also attacked foreigners, killing among others seven Trappist monks and the bishop of Oran. Rural villages were a favorite target since they lacked police or army protection. Men, women, and children in these villages were massacred under conditions of appalling brutality. A UN Human Rights subcommittee visited the country in 2000 and reported that the GIA and government troops were almost equally responsible for the casualties. By 2001, it was estimated that more than 120,000 people had been killed.

The violence tapered off in 2001, due in part to newly elected president Bouteflikas's amnesty plan, withdrawal of popular support for the GIA, and loss of its main base in the Algiers Casbah. Unfortunately, few GIA fighters responded to the amnesty offer, as it would have required them to surrender their weapons. Violence broke out again in the coastal provinces. The amnesty offer expired in 2001, and with 5,000 guerrillas at large, the prospects for peace in Algeria seemed more remote than ever. In the first half of the year, there were more casualties (2,500) than in all of 1999.

Timeline: PAST

1518–1520
Establishment of the Regency of Algiers

1827–1830
The French conquest, triggered by the "fly-whisk incident"

1847
The defeat of Abd al-Qadir by French forces

1871
Algeria becomes an overseas department of France

1936
The Blum-Viollette Plan, for Muslim rights, is annulled by colon opposition

1943
Ferhat Abbas issues the Manifesto of the Algerian Peope

1954–1962
Civil war, ending with Algerian independence

1965
Ben Bella is overthrown by Boumedienne

1976
The National Charter commits Algeria to revolutionary socialist deveopment

1978
President Boumedienne dies

1980s
Land reform is resumed with the breakup of 200 large farms into smaller units; Arabization campaign

1990s
President Bendjedid steps down; the Islamic Salvation Front becomes a force and eventually is banned; the economy undergoes an austerity program; civil war

PRESENT

2000s
Efforts to restore the mutliparty system

Continued civil conflict

WHAT PRICE DIALOGUE?

Algerians have been described as having one of two personality types: those "tolerant in matters of religion and way of life, multicultural in languages and traditions, open to the diversities of location at the great hub of the Mediterranean world" and those "secretive, violent, enemies of Islamic secularism, irreverence (toward Islam) and modernism." The establishment of the first group in control of the nation resulted largely from Zeroual's efforts. After his election in 1995, he ordered a referendum on key revisions to the Constitution. These included a ban on religious, linguistic, regional, and gender-based political parties; a limit of two five-year terms for presidents; and commitments to Islam as the state religion and Arabic as its official language. Another revision established a bicameral legislature with an appointed upper house (the Council of the Nation) and a lower house, the National Assembly, popularly elected under a system of proportional representation. The revisions were approved in November by 85.8 percent of eligible voters.

Zeroual next set June 5, 1997, for elections to the 380-member Assembly, the first national election since the abortive 1992 one. Despite a meager 65.5 percent voter turnout due to fears of violence, Zeroual's newly formed party, the National Democratic Rally, won 115 seats. Along with the FLN's 64 seats, the results gave the regime a slim majority.

Two "moderate" Islamist parties (so called because they rejected violence) also participated in the elections. The Movement for a Peaceful Society ran second to the government party, with 69 seats; and An-Nahdah won 34 seats, giving at least a semblance of opposition in the Assembly. The first local and municipal elections in Algeria's modern history were held in 1997 as well, continuing the trend as government-backed candidates won the majority of offices.

In 1999, Zeroual resigned and scheduled open presidential elections for April. Seven candidates filed; they included former foreign minister Abdelaziz Bouteflika, who had lived in Switzerland for many years. Subsequently all the other candidates withdrew, citing irregularities in the election process. But the election went off as scheduled, and Bouteflika was declared the winner, with 74 percent of the vote. The names of the other candidates remained on the ballot (the two "moderate" Islamist candidates received 17 percent of the vote).

Despite the apparently endless violence, the government continues to move slowly toward restoration of representative government. The first local elections in the country's history were held in 1997. Elections for a 389-seat National Popular Assembly took place in May 2002, with candidates elected by popular vote. The FLN won the majority of seats, 199, followed by the RND with 48, the National Reform Movement with 43 and the Movement for a Peaceful Society with 38. Although the FIS continued to be excluded from political participation, its leaders, Madani and Belhaj, were released from house arrest in July 2003. They are still prohibited from political activity.

Bouteflika's first term has been marked by improved security. The death tol! from violence averages 100 monthly, compared with 1,200 per month in the mid-90s. Some 85 percent of GIA militants accepted the government's 1999 amnesty offer. However, riots and demonstrations in the Kabylia region, with its predominantly Berber population, injected an ethnic component into national politics.

Since then, and to his credit, President Bouteflika has made a great effort to restore the multiparty system that had prevailed before the 1991 military takover. In January 2001, elections were held for the Senate, the upper house of the Legislature. Its 144 seats are two-thirds elected and one-third appointed by the president. Suffrage was limited to the 15,000 members of communal and provincial popular assemblies elected in 1997. The National Rally for Democracy (RND), which holds a majority of seats in the lower house, won 78 Senate seats, to 31 for its main opposition, the Movement for a Peaceful Society (MSP).

NOTES

1. Raphael Danziger, *Abd al-Qadir and the Algerians* (New York: Holmes and Meier, 1977), notes that Turkish intrigue kept the tribes in a state of near-constant tribal warfare, thereby preventing them from forming dangerous coalitions, p. 24.
2. The usual explanation for the quick collapse of the regency after 300 years is that its forces were prepared for naval warfare but not for attack by land. *Ibid.*, pp. 36–38.
3. Quoted in Harold D. Nelson, *Algeria, A Country Study* (Washington, D.C.: American University, Foreign Area Studies, 1979), p. 31.
4. Marnia Lazreg, *The Emergence of Classes in Algeria* (Boulder, CO: Westview Press, 1976), p. 53.
5. For Algerian Muslims to become French citizens meant giving up their religion, for all practical purposes, since Islam recognizes only Islamic law and to be a French citizen means accepting French laws. Fewer than 3,000 Algerians became French citizens during the period of French rule. Nelson, *op. cit.*, pp. 34–35.
6. John E. Talbott, *The War Without a Name: France in Algeria, 1954–1962* (New York: Alfred A. Knopf, 1980).
7. Georges Suffert, in *Esprit*, 25 (1957), p. 819.
8. Nelson, *op. cit.*, p. 68.

Egypt (Arab Republic of Egypt)

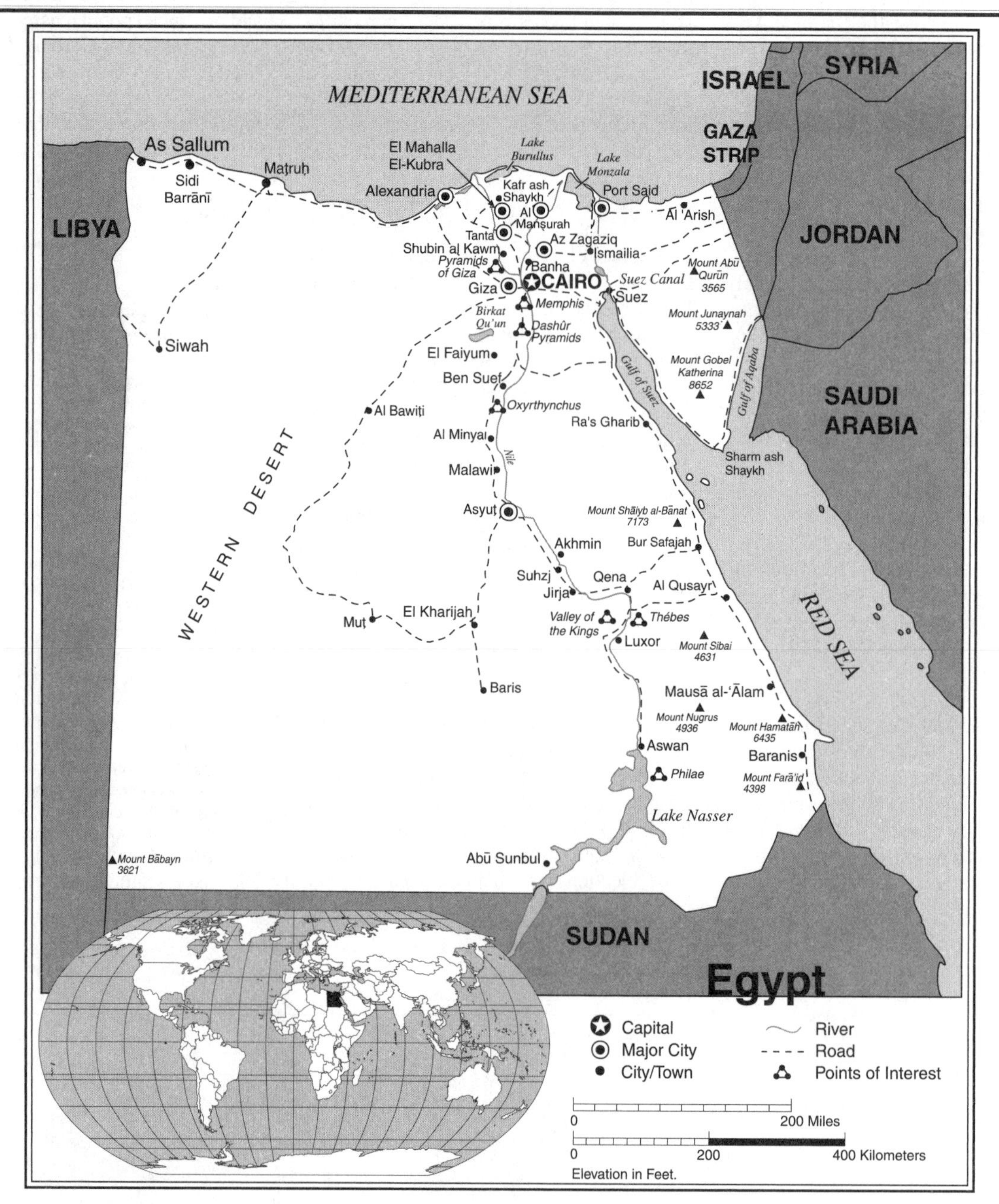

Egypt Statistics

GEOGRAPHY

Area in Square Miles (Kilometers): 386,258 (1,001,258) (about 3 times the size of New Mexico)

Capital (Population): Cairo (6,800,000)

Environmental Concerns: loss of agricultural land; increasing soil salinization; desertification; oil pollution threatening coral reefs and marine habitats; other water pollution; rapid population growth

Geographical Features: a vast desert plateau interrupted by the Nile Valley and Delta

Climate: desert; dry, hot summers; moderate winters

PEOPLE

Population

Total: 70,712,345

Annual Growth Rate: 1.66%

Rural/Urban Population Ratio: 55/45

Major Languages: Arabic; English

Ethnic Makeup: 99% Eastern Hamitic (Egyptian, Bedouin, Arab, Nubian); 1% others
Religions: 94% Muslim (mostly Sunni); 6% Coptic Christian and others

Health

Life Expectancy at Birth: 62 years (male); 66 years (female)
Infant Mortality Rate (Ratio): 58.6/1,000 live births
Physicians Available (Ratio): 1/472 people

Education

Adult Literacy Rate: 51.4%
Compulsory (Ages): for 5 years between 6 and 13

COMMUNICATION

Telephones: 3,972,000 main lines
Daily Newspaper Circulation: 43 per 1,000 people
Televisions: 110 per 1,000 people
Internet Users: 600,000

TRANSPORTATION

Highways in Miles (Kilometers): 39,744 (64,000)
Railroads in Miles (Kilometers): 2,973 (4,955)
Usable Airfields: 92
Motor Vehicles in Use: 1,703,000

GOVERNMENT

Type: republic
Independence Date: July 23, 1952, for the republic; February 28, 1922, marking the end of British rule
Head of State/Government: President Mohammed Hosni Mubarak; Prime Minister Atef Obeid
Political Parties: National Democratic Party (NDP), majority party; others are New Wafd; Tagammu (National Progressive Unionist Group); Nasserist Arab Democratic Party; Socialist Liberal Party. NDP holds 88 percent majority in Peoples Assembly
Suffrage: universal and compulsory at 18

MILITARY

Military Expenditures (% of GDP): 4.1%
Current Disputes: territorial dispute with Sudan over the Hala'ib Triangle

ECONOMY

Currency ($ U.S. Equivalent): 5.99 Egyptian pounds = $1
Per Capita Income/GDP: $3,700/$258 billion
GDP Growth Rate: 2.5%
Inflation Rate: 2.3%
Unemployment Rate: 12%
Labor Force: 20,600,000
Natural Resources: petroleum; natural gas; iron ore; phosphates; manganese; limestone; gypsum; talc; asbestos; lead; zinc
Agriculture: cotton; sugarcane; rice; corn; wheat; beans; fruits; vegetables; livestock; fish
Industry: textiles; food processing; tourism; chemicals; petroleum; construction; cement; metals
Exports: $7.3 billion (primary partners European Union, Middle East, Afro-Asian countries)
Imports: $16.4 billion (primary partners European Union, United States, Afro-Asian countries)

SUGGESTED WEBSITE

http://lcweb2.loc.gov/frd/cs/egtoc.htm

Egypt Country Report

The Arab Republic of Egypt is located at the extreme northeastern corner of Africa, with part of its territory—the Sinai Peninsula—serving as a land bridge to Southwest Asia. The country's total land area is approximately 386,000 square miles. However, 96 percent of this is uninhabitable desert. Except for a few scattered oases, the only settled and cultivable area is a narrow strip along the Nile River. The vast majority of Egypt's population is concentrated in this strip, resulting in high population density. Migration from rural areas to cities has intensified urban density; Cairo's population is currently 6.8 million, with an estimated 1.7 million more in the metropolitan area. It is a city that is literally "bursting at the seams."

Egypt today identifies itself as an Arab nation and is a founding member of the League of Arab States (which has its headquarters in Cairo). But its "Arab" identity is relatively new. It was first defined by the late president Gamal Abdel Nasser, who as a schoolboy became aware of his "Arabness" in response to British imperialism and particularly Britain's establishment of a national home for Jews in Arab Palestine. But Egypt's incredibly long history as a distinct society has given its people a separate Egyptian identity and a sense of superiority over other peoples, notably desert people such as the Arabs of old.[1] Also, its development under British tutelage gave the country a headstart over other Arab countries or societies. Despite its people's overall low level of adult literacy, Egypt has more highly skilled professionals than do other Arab countries.

HISTORY

Although Egypt is a modern nation in terms of independence from foreign control, it has a distinct national identity and a rich culture that date back thousands of years. The modern Egyptians take great pride in their brilliant past; this sense of the past gives them patience and a certain fatalism that enable them to withstand misfortunes that would crush most peoples. The Egyptian peasants, the *fellahin,* are as stoic and enduring as the water buffaloes they use to do their plowing. Since the time of the pharaohs, Egypt has been invaded many times, and it was under foreign control for most of its history. When Nasser, the first president of the new Egyptian republic, came to power in 1954, he said that he was the first native Egyptian to rule the country in nearly 3,000 years.

DEVELOPMENT

Egypt's GDP growth rate, which held steady at 4–5 percent in the 90s, has been hard-hit by the 9/11 terrorist attacks in the U.S. and the U.S. invasion of Iraq in March 2003. the tourism industry, which provides normally 12 percent of revenues, has been especially affected. In March the government banned imports of all but essential goods for 3 months to conserve dwindling foreign exchange.

It is often said that Egypt is the "gift of the Nile." The mighty river, flowing north to the Mediterranean with an enormous annual spate that deposited rich silt along its banks, attracted nomadic peoples to settle there as early as 6000 B.C. They developed a productive agriculture based on the river's seasonal floods. They lived in plastered mud huts in small, compact villages.

(UN photo/John Isaac)

These pyramids at Giza are among the most famous mementos of Egypt's past.

Their villages were not too different from those one sees today in parts of the Nile Delta.

Each village had its "headman," the head of some family more prosperous or industrious (or both) than the others. The arrival of other nomadic desert peoples gradually brought about the evolution of an organized system of government. Since the Egyptian villagers did not have nearby mountains or wild forests to retreat into, they were easily governable.

The institution of kingship was well established in Egypt by 2000 B.C., and in the time of Ramses II (1300–1233 B.C.), Egyptian monarchs extended their power over a large part of the Middle East. All Egyptian rulers were called pharaohs, although there was no hereditary system of descent and many different dynasties ruled during the country's first 2,000 years of existence. The pharaohs had their capital at Thebes, but they built other important cities on the banks of the Nile. Recent research by Egyptologists indicate that the ancient Egyptians had an amazingly accurate knowledge of astronomy. The Pyramids of Giza, for example, were built so as to be aligned with true north. Lacking modern instruments, their builders apparently used two stars, Thaban and Draconis, in the Big Dipper, for their alignment, with a point equidistant from them to mark their approximation of true north. Only centuries later was the North Star identified as such. The pyramids and Sphinx were built with Egyptian labor and without more than rudimentary machinery. They underline the engineering expertise of this ancient people. The world's oldest irrigation canal, its base paved with limestone blocks, was unearthed near Giza in 1996; and in 1999, a 2,000-year-old cemetery, discovered by chance in Bahariya Oasis, was found to contain rows of mummified men, women, and children along with wall murals showing funeral ceremonies, all in a remarkable state of preservation. The recent discovery of the bones of the second-largest dinosaur ever identified, in the same area, indicates that 90 million years ago, long before the pharaohs, Bahariya was a swampy tropical region similar to the Florida Everglades.

Another important discovery, in November 1999, was that of inscriptions on the walls of *Wadi Hoi* ("Valley of Terror") that may well be the world's oldest written language, predating the cuneiform letters developed by the Sumerians in Mesopotamia.

In the first century B.C., Egypt became part of the Roman Empire. The city of Alexandria, founded by Alexander the Great, became a center of Greek and Roman learning and culture. Later, it became a center of Christianity. The Egyptian Coptic Church was one of the earliest organized churches. The Copts, direct descendants of the early Egyptians, are the principal minority group in Egypt today. (The name Copt comes from *aigyptos,* Greek for "Egyptian.") The Copts welcomed the Arab invaders who brought Islam to Egypt, preferring them to their oppressive Byzantine Christian rulers. Muslim rulers over the centuries usually protected the Copts as "Peoples of the Book," leaving authority over them to their religious leaders, in return for allegiance and payment of a small tax. But in recent years, the rise of Islamic fundamentalism has made life more difficult for Egypt's Christians. As a minority group, they are caught between the fundamentalists and government forces seeking to destroy them.

Egypt also had, until very recently, a small but long-established Jewish community that held a similar position under various Muslim rulers. Most of the Jews emigrated to Israel after 1948.

THE INFLUENCE OF ISLAM

Islam was the major formative influence in the development of modern Egyptian society. Islamic armies from Arabia invaded Egypt in the seventh century A.D. Large numbers of nomadic Arabs followed, settling the Nile Valley until, over time, they became the majority in the population. Egypt was under the rule of the caliphs ("successors" of the Prophet Muhammad) until the tenth century, when a Shia group broke away and formed a separate government. The leaders of this group also called themselves caliphs. To show their independence, they founded a new capital in the desert south of Alexandria. The name they chose for their new capital was prophetic: *al-Qahira*—"City of War"—the modern city of Cairo.

In the sixteenth century, Egypt became a province of the Ottoman Empire. It was then under the rule of the Mamluks, originally slaves or prisoners of war who were converted to Islam. Many Mamluk leaders had

(Photo by Wayne Edge)

Islam is the dominant religion of North Africa and no matter how remote one can always find a shrine.

been freed and then acquired their own slaves. They formed a military aristocracy, constantly fighting with one another for land and power. The Ottomans found it simpler to leave Egypt under Mamluk control, merely requiring periodic tribute and taxes.

EGYPT ENTERS THE MODERN WORLD

At the end of the eighteenth century, rivalry between Britain and France for control of trade in the Mediterranean and the sea routes to India involved Egypt. The French general Napoleon Bonaparte led an expedition to Egypt in 1798. However, the British, in cooperation with Ottoman forces, drove the French from Egypt. A confused struggle for power followed. The victor was Muhammad Ali, an Albanian officer in the Ottoman garrison at Cairo. In 1805, the Ottoman sultan appointed him governor of Egypt.

Muhammad Ali set up an organized, efficient tax-collection system. He suppressed the Mamluks and confiscated all the lands that they had seized from Egyptian peasants over the years, lifting a heavy tax burden from peasant backs. He took personal charge of all Egypt's exports. Cotton, a new crop, became the major Egyptian export and became known the world over for its high quality. Dams and irrigation canals were dug to improve cultivation and expand arable land.

Muhammad Ali's successors were named *khedives* ("viceroys"), in that they ruled Egypt in theory on behalf of their superior, the sultan. In practice, they acted as independent rulers. Under the khedives, Egypt was again drawn into European power politics, with unfortunate results. Khedive Ismail, the most ambitious of Muhammad Ali's descendants, was determined to make Egypt the equal of Western European nations. His major project was the Suez Canal, built by a European company and opened in 1869.

However, the expense of this and other grandiose projects bankrupted the country. Ismail was forced to sell Egypt's shares in the Suez Canal Company—to the British government!—and his successors were forced to accept British control over Egyptian finances. In 1882, a revolt of army officers threatened to overthrow the khedive. The British intervened and established a de facto protectorate, keeping the khedive in office in order to avoid conflict with the Ottomans.

EGYPTIAN NATIONALISM

The British protectorate lasted from 1882 to 1956. An Egyptian nationalist movement gradually developed in the early 1900s, inspired by the teachings of religious leaders and Western-educated officials in the khedives' government. They advocated a revival of Islam and its strengthening to enable Egypt and other Islamic lands to resist European control.

At the end of World War I, Egyptian nationalist leaders organized the *Wafd* (Arabic for "delegation"). In 1918, the Wafd presented demands to the British for the complete independence of Egypt. The British rejected the demands, saying that Egypt was not ready for self-government. The Wafd then turned to violence, organizing boycotts, strikes, and terrorist attacks on British soldiers and on Egyptians accused of cooperating with the British.

Under pressure, the British finally abolished the protectorate in 1922. But they retained control over Egyptian foreign policy, defense, and communications as well as the protection of minorities and foreign residents and of Sudan, which had been part of Egypt since the 1880s. Thus, Egypt's "independence" was a hollow shell.

THE EGYPTIAN REVOLUTION

During the years of the monarchy, the Egyptian Army gradually developed a corps of professional officers, most of them from lower- or middle-class Egyptian backgrounds. They were strongly patriotic and resented what they perceived to be British cultural snobbery as well as Britain's continual influence over Egyptian affairs.

The training school for these young officers was the Egyptian Military Academy, founded in 1936. Among them was Gamal Abdel Nasser, the eldest son of a village postal clerk. Nasser and his fellow officers were already active in anti-British demonstrations by the time they entered the academy. During World War II, the British, fearing a German takeover of Egypt, rein-

(UN photo)

In 1952, the Free Officers organization persuaded Egypt's King Farouk to abdicate. The monarchy was formally abolished in 1954, when Gamal Abdel Nasser (pictured above) became Egypt's president, prime minister, and head of the Revolutionary Command Council.

stated the protectorate. Egypt became the main British military base in the Middle East. This action galvanized the officers into forming a revolutionary movement. Nasser said at the time that it roused in him the seeds of revolt. "It made [us] realize that there is a dignity to be retrieved and defended."[2]

When Jewish leaders in Palestine organized Israel in May 1948, Egypt, along with other nearby Arab countries, sent troops to destroy the new state. Nasser and several of his fellow officers were sent to the front. The Egyptian Army was defeated; Nasser himself was trapped with his unit, was wounded, and was rescued only by an armistice. Even more shocking to the young officers was the evident corruption and weakness of their own government. The weapons that they received were inferior and often defective, battle orders were inaccurate, and their superiors proved to be incompetent in strategy and tactics.

Nasser and his fellow officers attributed their defeat not to their own weaknesses but to their government's failures. When they returned to Egypt, they were determined to overthrow the monarchy. They formed a secret organization, the Free Officers. It was not the only organization dedicated to the overthrow of the monarchy, but it was the best disciplined and had the general support of the army.

On July 23, 1952, the Free Officers launched their revolution. It came six months after "Black Saturday," the burning of Cairo by mobs protesting the continued presence of British troops in Egypt. The Free Officers persuaded King Farouk to abdicate, and they declared Egypt a republic. A nine-member Revolutionary Command Council (RCC) was established to govern the country.

EGYPT UNDER NASSER

In his self-analytical book *The Philosophy of the Revolution,* Nasser wrote, " I always imagine that in this region in which we live there is a role wandering aimlessly about in search of an actor to play it."[3] Nasser saw himself as playing that role. Previously, he had operated behind the scenes, but always as the leader to whom the other Free Officers looked up. By 1954, Nasser had emerged as Egypt's leader. When the monarchy was formally abolished in 1954, he became president, prime minister, and head of the RCC. Cynics said that Nasser came along when Egypt was ready for another king; the Egyptians could not function without one!

Nasser came to power determined to restore dignity and status to Egypt, to eliminate foreign control, and to make his country the leader of a united Arab world. It was an ambitious set of goals, and Nasser was only partly successful in attaining them. But in his struggles to achieve these goals, he brought considerable status to Egypt. The country became a leader of the "Third World" of Africa and Asia, developing nations newly freed from foreign control.

Nasser was successful in removing the last vestiges of British rule from Egypt. British troops were withdrawn from the Suez Canal Zone, and Nasser nationalized the canal in 1956, taking over the management from the private foreign company that had operated it since 1869. That action made the British furious, since the British government had a majority interest in the company. The British worked out a secret plan with the French and the Israelis, neither of whom liked Nasser, to invade Egypt and overthrow him. British and French paratroopers seized the canal in October 1956, but the United States and the Soviet Union, in an unusual display of cooperation, forced them to withdraw. It was the first of several occasions when Nasser turned military defeat into political victory. It was also one of the few times when Nasser and the United States were on the same side of an issue.

Between 1956 and 1967, Nasser developed a close alliance with the Soviet Union—at least, it seemed that way to the United States. Nasser's pet economic project was the building of a dam at Aswan, on the upper Nile, to regulate the annual flow of river water and thus enable Egypt to reclaim new land and develop its agriculture. He applied for aid from the United States through the World Bank to finance the project, but he was turned down, largely due to his publicly expressed hostility toward Israel. Again Nasser turned defeat into a victory of sorts. The Soviet Union agreed to finance the dam, which was completed in 1971, and subsequently to equip and train the Egyptian Army. Thousands of Soviet advisers poured into Egypt, and it seemed to U.S. and Israeli leaders that Egypt had become a dependency of the Soviet Union.

The lowest point in Nasser's career came in June 1967. Israel invaded Egypt and defeated his Soviet-trained army, along with those of Jordan and Syria, and occupied the Sinai Peninsula in a lightning six-day war. The Israelis were restrained from marching on Cairo only by a United Nations cease-fire. Nasser took personal responsibility for the defeat, calling it *al-Nakba* ("The Catastrophe"). He announced his resignation, but the Egyptian people refused to accept it. The public outcry was so great that he agreed to continue in office. One observer wrote, "The irony was that Nasser had led the country to defeat, but Egypt without Nasser was unthinkable."[4]

Nasser had little success in his efforts to unify the Arab world. One attempt, for example, was a union of Egypt and Syria, which lasted barely three years (1958–1961). Egyptian forces were sent to support a new republican government in Yemen after the overthrow of that country's autocratic ruler, but they became bogged down in a civil war there and had to be withdrawn. Other efforts to unify the Arab world also failed. Arab leaders respected Nasser but were unwilling to play second fiddle to him in an organized Arab state. In 1967, after the Arab defeat, Nasser lashed out bitterly at the other Arab leaders. He said, "You issue statements, but we have to fight. If you want to liberate [Palestine] then get in line in front of us."[5]

Inside Egypt, the results of Nasser's 18-year rule were also mixed. Although he talked about developing representative government, Nasser distrusted political parties and remembered the destructive rivalries under the monarchy that had kept Egypt divided and weak. The Wafd and all other political parties were declared illegal. Nasser set up his own political organization to replace them, called the Arab Socialist Union (ASU). It was a mass party, but it had no real power. Nasser and a few close associates ran the government and controlled the ASU. The associates took their orders directly from Nasser; they called him *El-Rais*—"The Boss."

Nasser died in 1970. Ironically, his death came on the heels of a major policy success: the arranging of a truce between the Palestine Liberation Organization and the government of Jordan. Despite his health problems, Nasser had seemed indestructible, and his death came as a shock. Millions of Egyptians followed his funeral cortege through the streets of Cairo, weeping and wailing over the loss of their beloved Rais.

ANWAR AL-SADAT

Nasser was succeeded by his vice-president, Anwar al-Sadat, in accordance with constitutional procedure. Sadat had been one of the original Free Officers and had worked with Nasser since their early days at the Military Academy. In the Nasser years, Sadat had come to be regarded as a lightweight, always ready to do whatever The Boss wanted.

Many Egyptians did not even know what Sadat looked like. A popular story was told of an Egyptian peasant in from the country to visit his cousin, a taxi driver. As they drove around Cairo, they passed a large poster of Nasser and Sadat shaking hands. "I know our beloved leader, but who is the man with him?" asked the peasant. "I think he owns that café across the street," replied his cousin.

When Sadat became president, however, it did not take long for the Egyptian people to learn what he looked like. Sadat introduced a "revolution of rectification," which he said was needed to correct the errors of his predecessor.[6] These included too much dependence on the Soviet Union, too much government interference in the economy, and failure to develop an effective Arab policy against Israel. He was a master of timing, taking bold action at unexpected times to advance Egypt's international and regional prestige. Thus, in 1972 he abruptly ordered the 15,000 Soviet advisers in Egypt to leave the country, despite the fact that they were training his army and supplying all his military equipment. His purpose was to reduce Egypt's dependence on one foreign power, and as he had calculated, the United States now came to his aid.

A year later, in October 1973, Egyptian forces crossed the Suez Canal in a surprise attack and broke through Israeli defense lines in occupied Sinai. The attack was coordinated with Syrian forces invading Israel from the east, through the Golan Heights. The Israelis were driven back with heavy casualties on both fronts, and although they eventually regrouped and won back most of the lost ground, Sadat felt he had won a moral and psychological victory. After the war, Egyptians believed that they had held their own with the Israelis and had demonstrated Arab ability to handle the sophisticated weaponry of modern warfare. On the 25th anniversary of the 1973 October War, Egypt held its first military parade in 17 years, and 250 young couples were married in a mass public wedding ceremony at the Pyramids to remind the new generation—a third of the population are under age 15—of Egypt's great "victory."

(UN photo/Muldoon)

Nasser died in 1970 and was succeeded by Vice-President Anwar al-Sadat. Sadat, initially virtually unknown by the Egyptian people, took many bold steps in cementing his role as leader of Egypt.

Anwar al-Sadat's most spectacular action took place in 1977. It seemed to him that the Arab–Israeli conflict was at a stalemate. Neither side would budge from its position, and the Egyptian people were angry at having so little to show for the 1973 success. In November, he addressed a hushed meeting of the People's Assembly and said, "Israel will be astonished when it hears me saying ... that I am ready to go to their own house, to the Knesset itself, to talk to them."[7] And he did so, becoming for a second time the "Hero of the Crossing,"[8] but this time to the very citadel of Egypt's enemy.

Sadat's successes in foreign policy, culminating in the 1979 peace treaty with Israel, gave him great prestige internationally. Receipt of the Nobel Peace Prize, jointly with Israeli prime minister Menachem Begin, confirmed his status as a peacemaker. His pipe-smoking affability and sartorial elegance endeared him to U.S. policymakers.

The view that more and more Egyptians held of their world-famous leader was less flattering. Religious leaders and conservative Muslims objected to Sadat's luxurious style of living. The poor resented having to pay more for basic necessities. The educated classes were angry about Sadat's claim that the political system had become more open and democratic when, in fact, it had not. The Arab Socialist Union was abolished and several new political parties were allowed to organize. But the ASU's top leaders merely formed their own party, the National Democratic Party, headed by Sadat. For all practical purposes, Egypt under Sadat was even more of a single-party state under an authoritarian leader than it had been in Nasser's time.

Sadat's economic policies also worked to his disadvantage. In 1974, he announced a new program for postwar recovery, *Infitah* ("Opening"). It would be an open-door policy, bringing an end to Nasser's state-run socialist system. Foreign investors would be encouraged to invest in Egypt, and foreign experts would bring their technological knowledge to help develop industries. Infitah, properly applied, would bring an economic miracle to Egypt.

Rather than spur economic growth, however, Infitah made fortunes for just a few, leaving the great majority of Egyptians no better off than before. Chief among those who profited were members of the Sadat family. Corruption among the small ruling class, many of its members newly rich contractors, aroused anger on the part of the Egyptian people. In 1977, the economy was in such bad shape that the government increased bread prices. Riots broke out, and Sadat was forced to cancel the increase.

On October 6, 1981, President Sadat and government leaders were reviewing an armed-forces parade in Cairo to mark the eighth anniversary of the Crossing. Suddenly, a volley of shots rang out from one of the trucks in the parade. Sadat fell, mortally wounded. The assassins, most of them young military men, were immediately arrested. They belonged to *Al Takfir Wal Hijra* ("Repentance and Flight from Sin"), a secret group that advocated the reestablishment of a pure Islamic society in Egypt—by violence, if necessary. Their leader declared that the killing of Sadat was an essential first step in this process.

Islamic fundamentalism developed rapidly in the Middle East after the 1979 Iranian Revolution. The success of that revolution was a spur to Egyptian fundamentalists. They accused Sadat of favoring Western capitalism through his Infitah policy, of making peace with the "enemy of Islam" (Israel), and of not being a good Muslim. At their trial, Sadat's assassins said that they had acted to rid Egypt of an unjust ruler, a proper action under the laws of Islam.

HEALTH/WELFARE

Egypt's women won a significant victory in 1999 when the Court of Cassation upheld a government law banning female circumcision, a time-honored practice in many African societies, including Egypt. Women also won the right to file for divorce under the new 2001 Family Law, and in 2003 the first female judge was appointed in the court system.

Sadat may have contributed to his early death (he was 63) by a series of actions taken earlier in the year. About 1,600 people were arrested in September 1981 in a massive crackdown on religious unrest. They included not only religious leaders but also journalists, lawyers, intellectuals, provincial governors, and leaders of the country's small but growing opposition parties. Many of them were not connected with any fundamentalist Islamic organization. It seemed to most Egyptians that Sadat had overreacted, and at that point, he lost the support of the nation. In contrast to Nasser's funeral, few tears were shed at Sadat's. His funeral was attended mostly by foreign dignitaries. One of them said that Sadat had been buried without the people and without the army.

MUBARAK IN POWER

Vice-President Hosni Mubarak, former Air Force commander and designer of Egypt's 1973 success against Israel, succeeded Sadat without incident. Mubarak dealt firmly with Islamic fundamentalism at the beginning of his regime. He was given emergency powers and approved death sentences for five of Sadat's assassins in 1982. But he moved cautiously in other areas of national life, in an effort to disassociate himself from some of Sadat's more unpopular policies. The economic policy of Infitah, which had led to widespread graft and corruption, was abandoned; stiff sentences were handed out to a number of entrepreneurs and capitalists, including Sadat's brother-in-law and several associates of the late president.

Mubarak also began rebuilding bridges with other Arab states that had been damaged after the peace treaty with Israel. Egypt was readmitted to membership in the Islamic Conference, the Islamic Development Bank, the Arab League, and other Arab regional organizations. In 1990, the Arab League headquarters was moved from Tunis back to Cairo, its original location. Egypt backed Iraq with arms and advisers in its war with Iran, but Mubarak broke with Saddam Hussain after the invasion of Kuwait, accusing the Iraqi leader of perfidy. Some 35,000 Egyptian troops served with the UN–U.S. coalition during the Gulf War; and as a result of these efforts, the country resumed its accustomed role as the focal point of Arab politics.

Despite the peace treaty, relations with Israel continued to be difficult. One bone of contention was removed in 1989 with the return of the Israeli-held enclave of Taba, in the Sinai Peninsula, to Egyptian control. It had been operated as an Israeli beach resort.

The return of Taba strengthened the government's claim that the 10-year-old peace treaty had been valuable overall in advancing Egypt's interests. The sequence of agreements between the Palestine Liberation Organization and Israel for a sovereign Palestinian entity, along with Israel's improved relations with its other Arab neighbors, contributed to a substantial thaw in the Egyptian "cold peace" with its former enemy. In March 1995, a delegation from Israel's Knesset arrived in Cairo, the first such parliamentary group to visit Egypt since the peace treaty.

But relations worsened after the election in 1996 of Benjamin Netanyahu as head of a new Israeli government. Egypt had strongly supported the Oslo accords for a Palestinian state, and it had set up a free zone for transit of Palestinian products in 1995. The Egyptian view that Netanyahu was not adhering to the accords led to a "war of words" between the two countries. Israeli tourists were discouraged from visiting Egypt or received hostile treatment when visiting Egyptian monuments, and almost no Egyptians opted for visits to Israel. The newspaper *Al-Ahram* even stopped carrying cartoons by a popular Israeli-American cartoonist because he had served in the Israeli Army. The two governments cooperated briefly in the return of a small Bedouin tribe, the Azazma, to its Egyptian home area in the Sinai. The tribe had fled into Israel following a dispute with another tribe that turned into open conflict.

Ehud Barak's election as Israeli prime minister and his overtures for peace with the Palestinians were well received in Egypt. However, his resounding defeat in the 2000 elections by Ariel Sharon reinstituted the deep freeze in relations. The Israeli–Palestinian conflict has generated a great increase in anti-Israeli sentiments among the Egyptians. In part to counter these sentiments, but also to provide Israelis with a better understanding of the Arab world, a new government television station began broadcasting in Hebrew in December 2001.

FREEDOM

The Islamic fundamentalist challenge to Egypt's secular government has caused the erosion of many rights and freedoms enshrined in the country's Constitution. A state of emergency first issued in 1981 is still in effect; it was renewed in 2001 for a 3-year period. In June 2003 the Peoples' Assembly approved establishment of a National Council for Human Rights that would monitor violations or misuse of government authority.

Internal Politics

Although Mubarak's unostentatious lifestyle and firm leadership encouraged confidence among the Egyptian regime, the system that he inherited from his predecessors remained largely impervious to change. The first free multiparty national elections held since the 1952 Revolution took place in 1984—although they were not entirely free, because a law requiring political parties to win at least 8 percent of the popular vote limited party participation. Mubarak was reelected easily for a full six-year term (he was the only candidate), and his ruling National Democratic Party won 73 percent of seats in the Assembly. The New Wafd Party was the only party able to meet the 8 percent requirement.

New elections for the Assembly in 1987 indicated how far Egypt's embryonic democracy had progressed under Mubarak. This time, four opposition parties aside from his own party presented candidates. Although the National Democratic Party's plurality was still a hefty 69.6 percent, 17 percent of the electorate voted for opposition candidates. The New Wafd increased its percentage of the popular vote to 10.9 percent, and a number of Muslim Brotherhood members were elected as independents. The National Progressive Unionist Group, the most leftist of the parties, failed to win a seat.

Mubarak was elected to a fourth six-year term in September 1999, making him Egypt's longest-serving head of state in the

(Photo by Wayne Edge)

As the tourism sector increases in importance new facilities are being built in North Africa to accommodate the expected visitors.

country's independent history. His victory margin was 94 percent, two points less than in 1993, when as per usual he was the only candidate. Some 79 percent of Egypt's 24 million registered voters cast their ballots.

The January 2001 elections provided more evidence of public disaffection with Mubarak and the ruling party. The elections were held in stages beginning in October 2000. In the first stage, NDP candidates won only 38 of 118 seats. In the second and third stages, they won 179 of the Assembly's 454 seats. Some 218 seats were won by self-declared independents, and the rest by various minority parties. The independents subsequently aligned with NDP members to give the ruling party an 85 percent majority in the Legislature. However, barely 50 percent of eligible voters cast their ballots, emphasizing public disillusionment with the government.

AT WAR WITH FUNDAMENTALISM

Egypt's seemingly intractable social problems—high unemployment, an inadequate job market flooded annually by new additions to the labor force, chronic budgetary deficits, and a bloated and inefficient bureaucracy, to name a few—have played into the hands of Islamic fundamentalists, those who would build a new Egyptian state based on the laws of Islam. Although they form part of a larger fundamentalist movement in the Islamic world, one that would replace existing secular regimes with regimes that adhere completely to spiritual law and custom (*Shari'a*), Egypt's fundamentalists do not harbor expansionist goals. Their goal is to replace the Mubarak regime with a more purely "Islamic" one, faithful to the laws and principles of the religion and dominated by religious leaders.

Egypt's fundamentalists are broadly grouped under the organizational name al-Gamaa al-Islamiya, with the more militant ones forming subgroups such as the Vanguard of Islam and Islamic Jihad, itself an outgrowth of al-Takfir wal-Hijra, which had been responsible for the assassination of Anwar Sadat. Ironically, Sadat had formed Al-Gamaa to counter leftist political groups. However, it differs from its parent organization, the Muslim Brotherhood, in advocating the overthrow of the government by violence in order to establish a regime ruled under Islamic law. During Mubarak's first term, he kept a tight lid on violence. But in the 1990s, the increasing strength of the Islamists and their popularity with the large number of educated but unemployed youth led to an increase in violence and destablized the nation.

Violence was initially aimed at government security forces, but starting in 1992, the fundamentalists' strategy shifted to vulnerable targets such as foreign tourists and the Coptic Christian minority. A number of Copts were killed and many Copt business owners were forced to pay "protection money" to al-Gamaa in order to continue in operation. Subsequently the Copts' situation improved somewhat, as stringent security measures were put in place to contain Islamic fundamentalist violence. Gun battles in 1999 between Muslim and Copt villagers in southern Egypt resulted in 200 Christian deaths and the arrests of a number of Muslims as well as Copts. Some 96 Muslims were charged with violence before a state-security court, but only four were convicted, and to short jail terms. However, Muslim–Coptic relations remained unstable. Early in 2002, two Coptic weekly newspapers, *Al Nabaa* and *Akbar Nabaa*, were shut down after the Superior Press Council, a quasi-government body, filed a lawsuit charging them with "offending Egyptians and undermining national unity."

Islamic Jihad, the major fundamentalist organization and the one responsible for Sadat's assassination, subsequently shifted its locale and objectives in order to evade the repression of the Mubarak government. Many of its members joined the fighters in Afghanistan who were resisting the Soviet occupation of that country. After the Soviet withdrawal in 1989, some 300 of them remained, forming the core of the Taliban force that eventually won control of 90 percent of Afghanistan. In that capacity, they became associated with Osama bin Laden and his al-Qaeda international terror network. Two of their leaders, Dr. Ayman al-Zawahiri (a surgeon) and Muhammad Atif, are believed to have planned the September 11, 2001, terrorist bombings in the United States. However, Islamic Jihad's chief aim is the overthrow of the Mubarak government and its replacement by an Islamic one. Its hostility to the United States stems from American support for that government and for the U.S. alliance with Israel against the Palestinians.

In targeting tourism in their campaign to overthrow the regime, fundamentalists have attacked tourist buses. Four tourists were killed in the lobby of a plush Cairo hotel in 1993. In November 1997, 64 tourists were gunned down in a grisly massacre at the Temple of Hatshepsut near Luxor, in the Valley of the Kings, one of Egypt's prize tourist attractions. Aftershocks from the terrorist attacks on the United States have decimated the tourist industry, which is Egypt's largest source of income ($4.3 billion in 2000, with 5.4 million visitors in that year). Egyptair, the national airline, lost $56 million in October 2001 alone; and cancellations of package tours, foreign-airline bookings, and hotel reservations led to a 45 percent drop in tourist revenues.

One important reason for the rise in fundamentalist violence stems from the government's ineptness in meeting social crises. After the disastrous earthquake of October 1992, Islamic fundamentalist groups were first to provide aid to the victims, distributing $1,000 to each family made homeless, while the cumbersome, multilayered government bureaucracy took weeks to respond to the crisis. Similarly, al-Gamaa established a network of Islamic schools, hospitals, clinics, day-care centers, and even small industries in poor districts such as Cairo's Imbaba quarter.

The Mubarak government's response to rising violence has been one of extreme repression. The death penalty may be imposed for "antistate terrorism." The state of emergency that was established after Anwar Sadat's assassination in 1981 has been renewed regularly, most recently in 2001 for a three-year extension, over the vehement protests of opposition deputies in the Assembly. Some 770 members of the Vanguard of Islam were tried and convicted of subversion in 1993. The crackdown left Egypt almost free from violence for several years. But in 1996, al-Gamaa and two other hitherto unknown Islamic militant groups, Assiut Terrorist Organization and Kotbion (named for a Muslim Brotherhood leader executed in 1966 for an attempt to kill President Nasser), resumed terrorist activities. Eighteen Greek tourists were murdered in April, and the State Security Court sentenced five Assiut members to death for killing police and civilians in a murderous rampage. At their trial they chanted "God make a staircase of our skulls to Your glory," waving Korans in their cage, in an eerie replay of the trials of Sadat's assassins.

An unfortunate result of government repression of the militants is that Egypt, traditionally an open, tolerant, and largely nonviolent society, has taken on many of the features of a totalitarian state. Human rights are routinely suspended, the prime offenders being officers of the dreaded State Security Investigation (SSI). Indefinite detention without charges is a common practice, and torture is used extensively to extract "confessions" from suspects or their relatives. All of al-Gamaa's leaders are either in prison, in exile, or dead; and with 20,000 suspected Islamists also jailed, the government could claim with some justification that it had broken the back of the 1990s insurgency. Its confidence was enhanced in March 1999 when al-Gamaa said that it would no longer engage in violence. Two previous cease-fire offers had been spurned, but this newest offer resulted in the release of several hundred Islamists to "test its validity."

Due to the extremism of methods employed by both sides, the conflict between the regime and the fundamentalists has begun to polarize Egyptian society. As a prominent judge noted, "Islam has turned from a religion to an ideology. It has become a threat to Egypt, to civilization and to humanity."[9] The fundamentalists, in struggling to overthrow the regime and replace it with a more legitimately Islamic one (in their view) have at times attacked intellectuals, journalists, writers and others who do not openly advocate similar views or even oppose them. The novelist Farag Foda, who strongly criticized Egypt's "creeping Islamization" in his works, was killed outside of his Cairo home in the early 1990s, and Haguib Mahfouz, the Arab world's only Nobel laureate in literature, was critically woulded in 1994 by a Gamaa gunman.

In 1995, the regime imposed further restrictions on Egypt's normally freewheeling press and journalistic bodies. A law would impose fines of up to $3,000 and five-year jail sentences for articles "harmful to the state." The long arm of the law reached into the educational establishment as well. A university professor and noted Koranic scholar was charged with apostasy by clerics at Al-Azhar University, on the grounds that he had argued in his writings that the Koran should be interpreted in its historical/linguistic context alongside its identification as the Word of God. The charge came under the Islamic principle of *hisba,* "accountability." He was found guilty and ordered to divorce his wife, since a Muslim woman may not be married to an apostate. A 1996 law prohibited the use of hisba in the courts, but the damage had been done; the professor and his wife were forced into exile to preserve their marriage.

ACHIEVEMENTS

Alexandria, founded in 332 B.C. by Alexander the Great, was one of the world's great cities in antiquity, with its Library, its Pharos (Lighthouse), its palaces, and other monuments. Most of them were destroyed by fire or sank into the sea long ago, as the city fell into neglect. Then, in 1995, underwater archaelogists discovered the ruins of the Pharos; its location had not been known previously. Other discoveries followed—the palace of Cleopatra, the remains of Napoleon's fleet (sunk by the British in the Battle of the Nile), Roman and Greek trading vessels filled with amphorae, etc. The restoration of the Library was completed in 2000, with half of its 11 floors under the Mediterranean; visitors in the main reading room are surrounded by water cascading down its windows. After centuries of decay, Alexandria is again a magnet for tourists.

Another distinguished professor ran afoul of the government's Al-Azhar–imposed limits on free speech, as Saad Eddine Ibrahim, director of the American University at Cairo's Ibn Khaldoun Center for Democracy, was arrested in July 2000. He was charged with "defaming" Egypt abroad. The charge resulted from a documentary film produced at the Center which was critical of parliamentary electoral process for encouraging fraud. Due in part to widespread criticism in Europe and particularly the United States—Ibrahim holds dual Egyptian-American citizenship—he was released. In May 2001 he was rearrested, tried and convicted by a special court, and given a 7-year jail sentence. After strong international criticism and a warning by U.S. president Bush that his administration would oppose any increase in the $2 billion annual aid program to Egypt, the Court of Cassation, the country's highest court, threw out the conviction and ordered him released.[10] However, further international criticism followed, including a statement by President George Bush that his administration would oppose any increase in the $2 billion annual aid program to Egypt. The Court of Cassation, Egypt's highest court, subsequently threw out the conviction and ordered Ibrahim released·

Recently the head of the Group for Developing Democracy, a civil rights watchdog agency in Cairo, noted that "Egypt doesn't want real democracy. The state wants us puppets in its big show of paper democracy." And in July 2003 the director of the Cairo office of the UN Development Program told a reporter: "People seem to be accepting an immoral tradeoff between human rights and security."[11]

Despite its huge majority in the Assembly and its ruthless pursuit of Islamic militants, the Mubarak regime thus far has failed to deal effectively with the political, economic, and social inequities and lack of freedoms that continue to hamper Egypt's development. Observers have commented on Mubarak's mindset about Islamic groups, arguing that he makes no distinction between militants and moderates. As a result, Islamists now control the trade and student unions, schools, even the judiciary, forcing the general public to choose between them and a repressive regime.

A STRUGGLING ECONOMY

Egypt's economy rests upon a narrow and unstable base, due to rapid demographic growth and limited arable land and because political factors have adversely influenced national development. The country has a relatively high level of education and, as a result, is a net exporter of skilled labor to other Arab countries. But the overproduction of university graduates has produced a bloated and inefficient bureaucracy, as the government is required to provide a position for every graduate who cannot find other employment.

Agriculture is the most important sector of the economy, accounting for about one third of national income. The major crops are long-staple cotton and sugarcane. Egyptian agriculture since time immemorial has been based on irrigation from the Nile River. In recent years, greater control of irrigation water through the Aswan High Dam, expansion of land devoted to cotton production, and improved planting methods have begun to show positive results.

A new High Dam at Aswan, completed in 1971 upstream from the original one built in 1906, resulted from a political decision by the Nasser government to seek foreign financing for its program of expansion of cultivable land and generation of electricity for industrialization. When Western lending institutions refused to finance the dam, also for political reasons, Nasser turned to the Soviet Union for help. By 1974, just three years after its completion, revenues had exceeded construction costs. The dam made possible the electrification of all of Egypt's villages as well as a fishing industry at Lake Nasser, its reservoir. It proved valuable in providing the agricultural sector with irrigation water during the prolonged 1980–1988 drought, although at sharply reduced levels. However, the increased costs of land reclamation and loss of the sardine fishing grounds along the Mediterranean coast have made the dam a mixed blessing for Egypt.

Egypt was self-sufficient in foodstuffs as recently as the 1970s but now must import 60 percent of its food. Such factors as rapid population growth, rural-to-urban migration with consequent loss of agricultural labor, and Sadat's open-door policy for imports combined to produce this negative food balance. Subsidies for basic commodities, which cost the government nearly $2 billion a year, are an important cause of inflation, since they keep the budget continuously in deficit. Fearing a recurrence of the 1977 Bread Riots, the government kept prices in check. However, inflation, which had dropped to 8 percent in 1995 due to International Monetary Fund stabilization policies required for loans, rose to 37 percent in 1999 as the new free-market policy produced a tidal wave of imports. As a result, the foreign trade deficit increased drastically.

Egypt has important oil and natural-gas deposits, and new discoveries continue to strengthen this sector of the economy. Oil reserves increased to 3.3 billion barrels in 2001, due to new fields being brought on stream in the Western Desert. Proven natural-gas reserves are 51 trillion cubic feet, sufficient to meet domestic needs for 30 years at current rates of consumption.

A 2001 agreement with Jordan would guarantee Jordan's purchase of Egyptian natural-gas supplies, contingent on completion of the pipeline under the Red Sea from Al-Arish to Aqaba. But an earlier agreement with Israel, Egypt's closest and potentially most lucrative gas market, has been put on hold due to the renewed Palestinian–Israeli conflict. Under the agreement, Egypt would have provided $300 million a year in gas, meeting 15 percent of Israel's electric-power needs.

Egypt also derives revenues from Suez Canal tolls and user fees, from tourism, and from remittances from Egyptian workers abroad, mostly working in Saudi Arabia

Timeline: PAST

2500–671 B.C.
Period of the pharaohs

671–30 B.C.
The Persian conquest, followed by Macedonians and rule by Ptolemies

30 B.C.
Egypt becomes a Roman province

A.D. 641
Invading Arabs bring Islam

969
The founding of Cairo

1517–1800
Egypt becomes an Ottoman province

1798–1831
Napoleon's invasion, followed by the rise to power of Muhammad Ali

1869
The Suez Canal opens to traffic

1882
The United Kingdom establishes a protectorate

1952
The Free Officers overthrow the monarchy and establish Egypt as a republic

1956
Nationalization of the Suez Canal

1958–1961
Union with Syria into the United Arab Republic

1967
The Six-Day War with Israel ends in Israel's occupation of the Gaza Strip and the Sinai Peninsula

1970
Gamal Abdel Nasser dies; Anwar Sadat succeeds as head of Egypt

1979
A peace treaty is signed at Camp David between Egypt and Israel

1980s
Sadat is assassinated; he is succeeded by Hosni Mubarak; a crackdown on Islamic fundamentalists

1990s
The government employs totalitarian tactics in its battle with fundamentalists

PRESENT

2000s
Deep social and economic problems persist

and other oil-producing Gulf states. The flow of remittances from the approximately 4 million expatriate workers was reduced and then all but cut off with the Iraqi invasion of Kuwait. Egyptians fled from both countries in panic, arriving home as penniless refugees. With unemployment already at 20 percent and housing in short supply, the government faced an enormous assimilation problem apart from its loss of revenue. The United States helped by agreeing to write off $4.5 billion in Egyptian military debts. By 1995, the expatriate crisis caused by returning workers had eased somewhat, with 1 million Egyptian workers employed in Saudi Arabia and smaller numbers in other Arab countries.

One encouraging sign of brighter days ahead is the expansion of local manufacturing industries, in line with government efforts to reduce dependence upon imported goods. A 10-year tax exemption plus remission of customs duties on imported machinery have encouraged a number of new business ventures, notably in the clothing industry.

Unfortunately one of the few enterprises affecting Egypt's poor directly was literally "dumped" in January 2003 when the government stopped renewing licenses to the Zabbaleen, a 60,000-member Coptic community that traditionally collects a third of Cairo's 10,000 daily tons of garbage and trash. Future collections of all garbage and trash are to be made by foreign companies under contract. The new system would have certain advantages of the Zabbaleen system, mainly in terms of sanitation. But the economic impact on them will be severe. Over the years the Zabbaleen have used profits from trash collection to fund neighborhood improvements, schools and jobs for a great number of women.

In 1987, Mubarak gained some foreign help for Egypt's cash-strapped economy when agreement was reached with the International Monetary Fund for a standby credit of $325 million over 18 months to allow the country to meet its balance-of-payments deficit. The Club of Paris, a group of public and private banks from various industrialized countries, then rescheduled $12 billion in Egyptian external debts over a 10-year period.

Expanded foreign aid and changes in government agricultural policy required by the World Bank for new loans helped spur economic recovery in the 1990s, especially in agriculture. Production records were set in 1996 in wheat, corn, and rice, meeting 50 percent of domestic needs. The cotton harvest for that year was 350,000 tons, with 50,000 tons exported. However, a new agricultural law passed in 1992 but not implemented until 1997 ended land rents, allowing landlords to set their own leases and in effect reclaim their properties taken over by the government during the Nasser era. The purpose is to provide an incentive for tenant farmers to grow export crops such as cotton and rice. But as a result, Egypt's 900,000 tenant farmers have faced the loss of lands held on long-term leases for several generations.

However, these economic successes must be balanced against Egypt's chronic social problems and the lack of an effective representative political system. The head of the Muslim Brotherhood made the astute observation in a 1993 speech that "the threat is not in the extremist movement. It is in the absence of democratic institutions." Until such institutions are firmly in place, with access to education, full employment, broad political participation, civil rights, and the benefits of growth spread evenly across all levels of society, unrest and efforts to Islamize the government by force are likely to continue.

By 2000, the government's harsh repression had seriously weakened the fundamentalist movement, albeit at a heavy cost. Some 1,200 police officers and militants had been killed during the 1990s, and 16,000 persons remained jailed without charges on suspicion of membership in Islamic Jihad or other organizations. However, public disaffection continues to grow and to involve inceasing numbers of non-fundamentalists. A bomb attack targeting Israeli tourists on the Sinai penisula killed more than 30 people in October 2004. In November 2004 the funeral of Palestinian leader Yasser Arafat was held in Cairo. Although it was a state funeral with all honors bestowed upon Arafat, it was held at the airport in Cairo under extraordinary security. The Egyptian public was not allowed to attend.

Egypt's own difficulties with fundamentalists caused some reluctance on its part when support for the U.S.–led international coalition against terrorism formed after the September 11, 2001, bombings of the World Trade Center in New York City and the Pentagon near Washington, D.C. The reluctance stemmed in part from public anger over continued American support for Israel against the Palestinians and the suffering of Iraq's fellow Arabs under the 11-year sanctions imposed on that country. Although the Egyptian government earns billions of U.S. dollars in various forms of aid and millions more from U.S. tourism, the people of Egypt remain extremely ambivalent, if not openly hostile, towards U.S. government policy in the Middle East. While Egypt has proven to be a reliable U.S. ally, the feeling of millions of Egyptians is one of anger and distrust of U.S. government policy. As the war in Iraq escalates, much of the goodwill the U.S. had in Egypt is now either being squandered or being paid for dearly in U.S. dollars.

In March 2004 the government reached agreement with Israel to set up a number of Qualifying Industrial Zones (Q.I.Z.) in an effort to boost its flaggging economy. Egyptian manufacturers, notable of textiles, will be able to export goods duty-free to the U.S. provided that 35 percent of goods exported were locally produced and a percentage reserved for Israeli products. Egypt's total exports to the U.S. of $3.3 billion inlcuded $336 million in textiles and clothing. The Q.I.Z.s will add significantly to this total.

NOTES

1. Leila Ahmed, in *A Border Passage* (New York: Farrar, Strauss & Giroux, 1999), deals at length with Egyptian vs. Arab identity from the perspective of growing up in British-controlled Egypt.
2. An English observer said, "In arms and firing they are nearly as perfect as European troops." Afaf L. Marsot, *Egypt in the Reign of Muhammad Ali* (Cambridge, England: Cambridge University Press, 1984), p. 132.
3. Quoted in P. J. Vatikiotis, *Nasser and His Generation* (New York: St. Martin's Press, 1978), p. 35.
4. Gamal Abdel Nasser, *The Philosophy of the Revolution* (Cairo: Ministry of National Guidance, 1954), p. 52.
5. Derek Hopwood, *Egypt: Politics and Society 1945–1981* (London: George Allen and Unwin, 1982), p. 77.
6. Quoted in Vatikiotis, *op. cit.*, p. 245.
7. Hopwood, *op. cit.*, p. 106.
8. David Hirst and Irene Beeson, *Sadat* (London: Faber and Faber, 1981), p. 255.
9. "Banners slung across the broad thoroughfares of central Cairo acclaimed The Hero of the Crossing (of the October 1973 War)." *Ibid.*, pp. 17–18.
10. Said Ashmawy, quoted in "In God He Trusts," *Jerusalem Post Magazine* (July 7, 1995).
11. Reported in *The New York Times*, July 2003.

Libya (Socialist People's Libyan Arab Jamahiriyya)

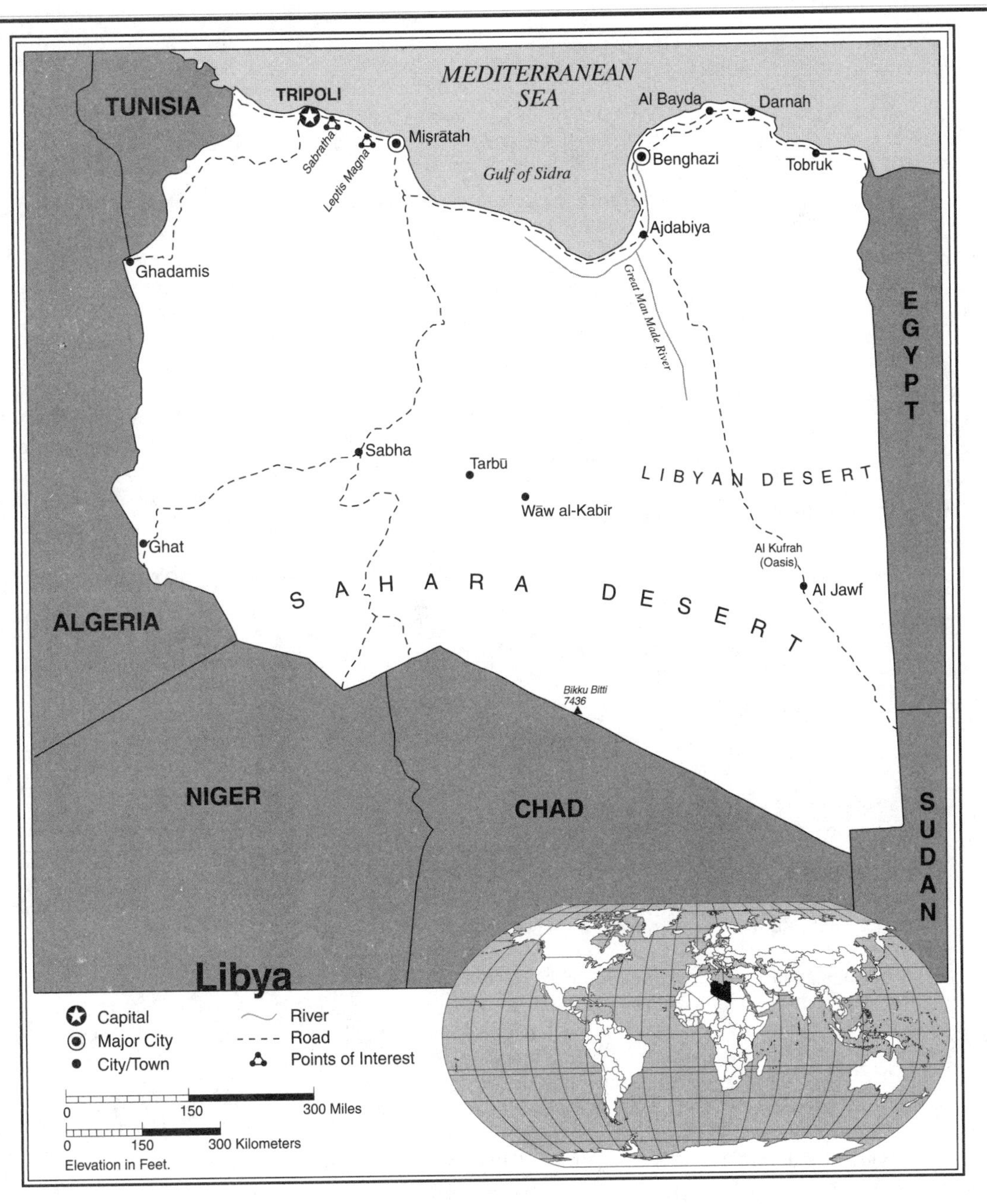

Libya Statistics

GEOGRAPHY

Area in Square Miles (Kilometers):

679,147 (1,759,450) (about the size of Alaska)

Capital (Population): Tripoli (1,681,000)

Environmental Concerns: desertification; very limited freshwater resources

Geographical Features: mostly barren, flat to undulating plains, plateaus, depressions

Climate: Mediterranean along the coast; dry, extreme desert in the interior

PEOPLE

Population

Total: 5,368,585 (includes 662,669 non-nationals of whom 500,000 are migrants from sub-Saharan Africa)

Annual Growth Rate: 2.4%

Rural/Urban Population Ratio: 14/86
Major Languages: Arabic; English; Italian
Ethnic Makeup: 97% Berber and Arab; 3% others
Religions: 97% Sunni Muslim; 3% others

Health

Life Expectancy at Birth: 74 years (male); 78 years (female)
Infant Mortality Rate: 27.9/1,000 live births
Physicians Available (Ratio): 1/948 people

Education

Adult Literacy Rate: 76.2%
Compulsory (Ages): 6–15

COMMUNICATION

Telephones: 380,000 main lines
Daily Newspaper Circulation: 15 per 1,000 people
Televisions: 105 per 1,000 people
Internet Users: 1 (2000)

TRANSPORTATION

Highways in Miles (Kilometers): 15,180 (24,484)
Railroads in Miles (Kilometers): none
Usable Airfields: 136
Motor Vehicles in Use: 904,000

GOVERNMENT

Type: officially a *Jamahiriyya* ("state belonging to the people") with government authority exercised by a General Peoples' Congress

Independence Date: December 24, 1951 (from Italy)

Head of State/Government: Revolutionary Leader Mahammad Au Minyar al-Qadhafi holds no official title but serves as de facto head of state; Mubarak al-Shamekh, Secretary of the GPC, is the equivalent to Prime Minister

Political Parties: none

Suffrage: universal and compulsory at 18

MILITARY

Military Expenditures (% of GDP): 3.9%

Current Disputes: Libya claims about 19,400 square kilometers of land in northern Niger and part of southeastern algeria; Both disputes currently dormant

ECONOMY

Currency ($ U.S. Equivalent): 1.2 dinars = $1. the official foreign trade rate was devalued in 2002 to 21.30 Libyan dinars = $1
Per Capita Income/GDP: $7,600/$40 billion
GDP Growth Rate: 3%
Inflation Rate: 18.5%
Unemployment Rate: 30%
Labor Force: 1,500,000
Natural Resources: petroleum; natural gas; gypsum
Agriculture: wheat; barley; olives; dates; citrus fruits; vegetables; peanuts; beef; eggs
Industry: petroleum; food processing; textiles; handicrafts; cement
Exports: $13.9 billion (primary partners Italy, Germany, Spain)
Imports: $7.6 billion (primary partners Italy, Germany, Tunisia)

SUGGESTED WEB SITES

`http://lcweb2.loc.gov/frd/cs/lytoc.html`
`http://home.earthlink.net/~dribrahim/`

Libya Country Report

The Socialist People's Libyan Arab Jamahiriya (Republic), commonly known as Libya, is the fourth largest of the Arab countries. Since it became a republic in 1969, it has played a role in regional and international affairs more appropriate to the size of its huge territory than to its small population.

Libya consists of three geographical regions: Tripolitania, Cyrenaica, and the Fezzan. Most of the population live in Tripolitania, the northwestern part of the country, where Tripoli, the capital and major port, is located. Cyrenaica, in the east along the Egyptian border, has a narrow coastline backed by a high plateau (2,400-feet elevation) called the Jabal al-Akhdar ("Green Mountain"). It contains Libya's other principal city, Benghazi. The two regions are separated by the Sirte, an extension of the Sahara Desert that reaches almost to the Mediterranean Sea. Most of Libya's oil fields are in the Sirte.

The Fezzan occupies the central part of the country. It is entirely desert, except for a string of widely scattered oases. Its borders are with Chad, Algeria, Niger, and Sudan. The border with Chad, established during French colonial rule in sub-Saharan Africa, was once disputed by Libya. The matter was settled through international mediation, with the border formally demarcated in 1994. Libya also claims areas in northern Niger and southeastern Algeria left over from the colonial period, when they formed part of the French West African empire. In the Libyan view, these areas should have been transferred to its control under the peace treaty that established the Libyan state and relinquishment of French control.

HISTORY

Until modern times, Libya did not have a separate identity, either national or territorial. It always formed a part of some other territorial unit and in most cases was controlled by outsiders. However, control was usually limited to the coastal areas. The Berbers of the interior were little affected by the passing of conquerors and the rise and fall of civilizations.

Libya's culture and social structure have been influenced more by the Islamic Arabs than by any other invaders. The Arabs brought Islam to Libya in the early seventh century. Arab groups settled in the region and intermarried with the Berber population to such an extent that the Libyans became one of the most thoroughly Arabized peoples in the Islamic world.

DEVELOPMENT

Although continued U.S. sanctions prohibit American firms from operating in Libya, improved relations with other countries have begun to generate diversification of the Libyan economy. An agreement with Ireland to import 50,000 live Irish cattle was concluded in March 2001, and Italy's export credit agency wrote off $230 million in Libyan debts to encourage investment by Italian firms. New oil discoveries in the Murzuq field have aided economic recovery.

Coastal Libya, around Tripoli, was an outlying province of the Ottoman Empire for several centuries. Like its urban neighbors Tunis and Algiers, Tripoli had a fleet of corsairs that made life dangerous for European merchant ships in the Mediterranean. When the United States became a

(Photo by Wayne Edge)

Throughout North Africa building materials are garnered from locally available resources like stone, wood, sand, and grass.

Mediterranean trading nation, the corsairs of Tripoli included American ships among their targets. The USS *Philadelphia* was sent to Tripoli to "teach the corsairs a lesson" in 1804, but it got stuck on a sandbar and was captured. Navy Lieutenant Stephen Decatur led a commando raid into Tripoli harbor and blew up the ship, inspiring the words to what would become the official U.S. Marine hymn: "From the halls of Montezuma to the shores of Tripoli...."

The Sanusiya Movement

At various stages in Islam's long history, new groups or movements have appeared committed to purifying or reforming Islamic society and taking it back to its original form of a simple community of believers led by just rulers. Several of these movements, such as the Wahhabis of Saudi Arabia, were important in the founding of modern Islamic states. The movement called the Sanusiya was formed in the nineteenth century. In later years, it became an important factor in the formation of modern Libya.

The founder, the Grand Sanusi, was a religious teacher from Algeria. He left Algeria after the French conquest and settled in northern Cyrenaica. The Grand Sanusi's teachings attracted many followers. He also attracted the attention of the Ottoman authorities, who distrusted his advocacy of a strong united Islamic world in which Ottomans and Arabs would be partners. In 1895, to escape from the Ottomans, the Grand Sanusi's son and successor moved Sanusiya headquarters to Kufra, a remote oasis in the Sahara.

The Sanusiya began as a peaceful movement interested only in bringing new converts to Islam and founding a network of *zawiyas* ("lodges") for contemplation and monastic life throughout the desert. But when European countries began to seize territories in North and West Africa, the Sanusi became warrior-monks and fought the invaders.

Italy Conquers Libya

The Italian conquest of Libya began in 1911. The Italians needed colonies, not only for prestige but also for the resettlement of poor and landless peasants from Italy's crowded southern provinces. The Italians expected an easy victory against a weak Ottoman garrison; Libya would become the "Fourth Shore" of a new Roman Empire from shore to shore along the Mediterranean. But the Italians found Libya a tougher land to subdue than they had expected. Italian forces were pinned to Tripoli and a few other points on the coast by the Ottoman garrison and the fierce Sanusi warrior-monks.

The Italians were given a second chance after World War I. The Ottoman Empire had been defeated, and Libya was ripe for the plucking. The new Italian government of swaggering dictator Benito Mussolini sent an army to occupy Tripolitania. When the Italians moved on Cyrenaica, the Grand Sa-

(UN photo/Rice)

After the 1969 Revolution, the government strove to develop many aspects of the country. These local chiefs are meeting to plan community development.

nusi crossed the Egyptian border into exile under British protection. The Italians found Cyrenaica much more difficult to control than Tripolitania. It is ideal guerrilla country, from the caves of Jabal al-Akhdar to the stony plains and dry, hidden *wadis* (river beds) of the south. It took nine years (1923–1932) for Italy to overcome all of Libya, despite Italy's vast superiority in troops and weapons. Sanusi guerrilla bands harried the Italians, cutting supply lines, ambushing patrols, and attacking convoys. Their leader, Shaykh Omar Mukhtar, became Libya's first national hero.

The Italians finally overcame the Sanusi by the use of methods that do not shock us today but seemed unbelievably brutal at the time. Cyrenaica was made into a huge concentration camp, with a barbed-wire fence along the Egyptian border. Nomadic peoples were herded into these camps, guarded by soldiers to prevent them from aiding the Sanusi. Sanusi prisoners were pushed out of airplanes, wells were plugged to deny water to the people, and flocks were slaughtered. In 1931, Omar Mukhtar was captured, court-martialed, and hanged in public. The resistance ended with his death.

The Italians did not have long to cultivate their Fourth Shore. During the 1930s, they poured millions of lire into the colony. A paved highway from the Egyptian to the Tunisian border along the coast was completed in 1937; in World War II, it became a handy invasion route for the British. A system of state-subsidized farms was set up for immigrant Italian peasants. Each was given free transportation, a house, seed, fertilizers, a mule, and a pair of shoes as inducements to come to Libya. By 1940, the Italian population had reached 110,000, and about 495,000 acres of land had been converted into productive farms, vineyards, and olive groves.[1]

Independent Libya

Libya was a major battleground during World War II, as British, German, and Italian armies rolled back and forth across the desert. The British defeated the Germans and occupied northern Libya, while a French army occupied the Fezzan. The United States later built an important air base, Wheelus Field, near Tripoli. Thus the three major Allied powers all had an interest in Libya's future. But they could not agree on what to do with occupied Libya.

Italy wanted Libya back. France wished to keep the Fezzan as a buffer for its African colonies, while Britain preferred self-government for Cyrenaica under the Grand Sanusi, who had become staunchly pro-British during his exile in Egypt. The Soviet Union favored a Soviet trusteeship over Libya, which would provide the Soviet Union with a convenient outlet in the Mediterranean. The United States waffled but finally settled on independence, which would at least keep the Soviet tentacles from enveloping Libya.

Due to lack of agreement, the Libyan "problem" was referred to the United Nations General Assembly. Popular demonstrations of support for independence in Libya impressed a number of the newer UN members; in 1951, the General Assembly approved a resolution for an independent Libyan state, a kingdom under the Grand Sanusi, Idris.

THE KINGDOM OF LIBYA

Libya has been governed under two political systems since independence: a constitutional monarchy (1951–1969); and a Socialist republic (1969–), which has no constitution because all power "belongs" to the people. Monarchy and republic have had almost equal time in power. But Libya's sensational economic growth and aggressive foreign policy under the republic need to be understood in relation to the solid, if unspectacular, accomplishments of the regime that preceded it.

At independence, Libya was an artificial union of the three provinces. The Libyan people had little sense of national identity or unity. Loyalty was to one's family, clan, village, and, in a general sense, to the higher authority represented by a tribal confederation. The only other loyalty linking Libyans was the Islamic religion. The tides of war and conquest that had washed over them for centuries had had little effect on their strong, traditional attachment to Islam.[2]

Political differences also divided the three provinces. Tripolitanians talked openly of abolishing the monarchy. Cyrenaica was the home and power base of King Idris; the king's principal supporters were the Sanusiya and certain important families. The distances and poor communication links between the provinces contributed to the impression that they should be separate countries. Leaders could not even agree on the choice between Tripoli and Benghazi for the capital. For his part, the king distrusted both cities as being corrupt and overly influenced by foreigners. He had his administrative capital at Baida, in the Jabal al-Akhdar.

The greatest problem facing Libya at independence was economics. Per capita income in 1951 was about $30 per year; in 1960, it was about $100 per year. Approximately 5 percent of the land was marginally usable for agriculture, and only 1 percent could be cultivated on a permanent basis. Most economists considered Libya to be a hopeless case, almost totally dependent on foreign aid for survival. (It is interesting to note that the Italians were seemingly able to force more out of the soil, but one must remember that the Italian government poured a great deal of money into the country to develop the plantations, and credit must also be given to the extremely hard-working Italian farmer.)

Despite its meager resources and lack of political experience, Libya was valuable to the United States and Britain in the 1950s and 1960s because of its strategic location. The United States negotiated a long-term lease on Wheelus Field in 1954, as a vital link in the chain of U.S. bases built around the southern perimeter of the Soviet Union due to the Cold War. In return, U.S. aid of $42 million sweetened the pot, and Wheelus became the single largest employer of Libyan labor. The British had two air bases and maintained a garrison in Tobruk.

Political development in the kingdom was minimal. King Idris knew little about parliamentary democracy, and he distrusted political parties. The 1951 Constitution provided for an elected Legislature, but a dispute between the king and the Tripolitanian National Congress, one of several Tripolitanian parties, led to the outlawing of all political parties. Elections were held every four years, but only property-owning adult males could vote (women were granted the vote in 1963). The same legislators were reelected regularly. In the absence of political activity, the king was the glue that held Libya together.

THE 1969 REVOLUTION

At dawn on September 1, 1969, a group of young, unknown army officers abruptly carried out a military coup in Libya. King Idris, who had gone to Turkey for medical treatment, was deposed, and a "Libyan Arab Republic" was proclaimed by the officers. These men, whose names were not known to the outside world until weeks after the coup, were led by Captain Muammar Muhammad al-Qadhafi. He went on Benghazi radio to announce to a startled Libyan population: "People of Libya ... your armed forces have undertaken the overthrow of the reactionary and corrupt regime.... From now on Libya is a free, sovereign republic, ascending with God's help to exalted heights."[3]

Qadhafi's new regime made a sharp change in policy from that of its predecessor. Wheelus Field and the British air bases were evacuated and returned to Libyan control. Libya took an active part in Arab affairs and supported Arab unity, to the extent of working to undermine other Arab leaders whom Qadhafi considered undemocratic or unfriendly to his regime.[4]

REGIONAL POLICY

To date, Qadhafi's efforts to unite Libya with other Arab states have not been successful. A 1984 agreement for a federal union with Morocco, which provided for separate sovereignty but a federated Assembly and unified foreign policies, was abrogated unilaterally by the late King Hassan II, after Qadhafi had charged him with "Arab treason" for meeting with Israeli leader Shimon Peres. Undeterred, Qadhafi tried again in 1987 with neighboring Algeria, receiving a medal from President Chadli Bendjedid but no other encouragement.

Although distrustful of the mercurial Libyan leader, other North African heads of state have continued to work with him on the basis that it is safer to have Qadhafi inside the circle than isolated outside. Tunisia restored diplomatic relations in 1987, and Qadhafi agreed to compensate the Tunisian government for lost wages of Tunisian workers expelled from Libya during the 1985 economic recession. Qadhafi also accepted International Court of Justice arbitration over Libya's dispute with Tunisia over oil rights in the Gulf of Gabes. In 1989, Libya joined with other North African states in the Arab Maghrib Union, which was formed to coordinate their respective economies. However, the AMU has yet to become a viable organization due to political differences among its members, in particular the Western Sahara dispute between Algeria and Morocco.

With little to show for his efforts to unite the Arab countries, Qadhafi turned his attention to sub-Saharan Africa. He had abolished the Secretariat for Arab Unity as a government ministry in 1997, and subsequently black African workers were invited to come and work in Libya. By 2000, nearly a million had arrived, most of them from Nigeria, Chad, and Ghana. Economic problems in sub-Saharan Africa caused thousands more to use Libya as an escape route for Europe, many of them also fleeing from civil war in Côte d'Ivoire and Sierra Leone. The flood of migrants generated tension between them and Libyan natives; the latter viewed the migrants as agents of social misbehavior ranging from prostitution to drug usage and AIDS. In August 2000, the Libyan government deported several thousand African workers. They were hauled to the Niger border in trucks and dumped across the border there. Qadhafi had announced earlier that a "United States of Africa" would come into existence in March 2001 under Libyan sponsorship. But for once the Libyan people did not agree with him; "We are native Arabs, not Africans," they told their leader.

(Gamma-Liaison/Christian Vioujard)

Muammar al-Qadhafi led a group of army officers in the military coup of 1969 that deposed King Idris. In later years, Qadhafi gained world-wide notoriety for his apparent sanction of terrorism.

SOCIAL REVOLUTION

Qadhafi's desert upbringing and Islamic education gave him a strong, puritanical moral code. In addition to closing foreign bases and expropriating properties of Italians and Jews, he moved forcefully against symbols of foreign influence. The Italian cathedral in Tripoli became a mosque, street signs were converted to Arabic, nightclubs were closed, and the production and sale of alcohol were prohibited.

But Qadhafi's revolution went far beyond changing names. In a three-volume work entitled *The Green Book,* he described his vision of the appropriate political system for Libya. Political parties would not be allowed, nor would constitutions, legislatures, even an organized court system. All of these institutions, according to Qadhafi, eventually become corrupt and unrepresentative. Instead, "people's committees" would run the government, business, industry, and even the universities. Libyan embassies abroad were renamed "people's bureaus" and were run by junior officers. (The takeover of the London bureau in 1984 led to counterdemonstrations by Libyan students and the killing of a British police officer by gunfire from inside the bureau. The Libyan bureau in Washington, D.C., was closed by the U.S. Federal Bureau of Investigation and the staff deported on charges of espionage and terrorism against Libyans in the United States.) The country was renamed the Socialist People's Libyan Arab Jamahiriya, and titles of government officials were eliminated. Qadhafi became "Leader of the Revolution," and each government department was headed by the secretary of a particular people's committee.

Qadhafi then developed a so-called Third International Theory, based on the belief that neither capitalism nor communism could solve the world's problems. What was needed, he said, was a "middle way" that would harness the driving forces of human history—religion and nationalism—to interact with each other to revitalize humankind. Islam would be the source of that middle way, because "it provides for the realization of justice and equity, it does not allow the rich to exploit the poor."[5]

THE ECONOMY

Modern Libya's economy is based almost entirely on oil exports. Concessions were granted to various foreign companies to explore for oil in 1955, and the first oil strikes were made in 1957. Within a decade, Libya had become the world's fourth-largest exporter of crude oil. During the 1960s, pipelines were built from the oil fields to new export terminals on the Mediterranean coast. The lightness and low sulfur content of Libyan crude oil make it highly desirable to industrialized countries, and, with the exception of the United States, differences in political viewpoint have had little effect on Libyan oil sales abroad.

After the 1969 Revolution, Libya became a leader in the drive by oil-producing countries to gain control over their petroleum industries. The process began in 1971, when the new Libyan government took over the interests of British Petroleum in Libya. The Libyan method of nationalization was to proceed against individual companies rather than to take on the "oil giants" all at once. It took more than a decade before the last company, Exxon, capitulated. However, the companies' $2 billion in assets were left in limbo in 1986, when the administration of U.S. president Ronald Reagan imposed a ban on all trade with Libya to protest Libya's involvement in international terrorism. President George Bush extended the ban for an additional year in 1990, although he expressed satisfaction with reduced Libyan support for terroristic activities, one example being the expulsion from Tripoli of the Palestine Liberation Front, a radical opponent of Yassir Arafat's Palestine Liberation Organization.

Recent discoveries have increased Libya's oil reserves 30 percent, to 29.5 billion barrels, and recoverable natural-gas reserves to 1.6 billion cubic meters. With oil production reaching a record 1.4 million barrels per day, Libya has been able to build a strong petrochemical industry. The Marsa Brega petrochemical complex is one of the world's largest producers of urea, although a major contract with India was canceled in 1996 due to UN sanctions on trade with Libya.

Until recently, industrial-development successes based on oil revenues enabled Libyans to enjoy an ever-improving standard of living, and funding priorities were shifted from industry to agricultural development in the budget. But a combination of factors—mismanagement, lack of a cadre of skilled Libyan workers, absenteeism, low motivation of the workforce, and a significant drop in revenues (from $22 billion in 1980 to $7 billion in 1988)—cast doubts on the effectiveness of Qadhafi's *Green Book* socialistic economic policies.

FREEDOM

The General People's Congress has the responsibility for passing laws and appointing a government. In 1994, the GPC approved legislation making Islamic law applicable in the country. They concerned retribution and blood money; rules governing wills, crimes of theft and violence, protection of society from things banned in the Koran, marriage, and divorce; and a ban on alcohol use.

In 1988, the leader began closing the book. As production incentives, controls on both imports and exports were eliminated, and profit sharing for employees of small businesses was encouraged. In 1990, the General People's Congress (GPC), Libya's equivalent of a parliament, began a restructuring of government, adding new secretariats (ministries) to help expand economic development and diversity the economy.

In January 2000, Qadhafi marched into a GPC meeting waving a copy of the annual budget. He tore up the copy and ordered most of the secretariats abolished. Their powers would be transferred to "provincial cells" outside of Tripoli. Only five government functions—finance, defense, foreign affairs, information, and African unity—would remain under central-government control. In October of that year, Qadhafi ordered further cuts, continuing his direct management of national affairs. For the first time he named a prime minister, Mubarak al-Shamekh, to head the stripped-down government. The secretariat for information was abolished, and the heads of the justice and finance secretariats summarily dismissed. The head of the National Oil Company (NOC), Libya's longest-serving government official, was transferred to a new post; Qadhafi had criticized the NOC for mismanagement of the oil industry and lack of vision.

Libya also started developing its considerable uranium resources. A 1985 agreement with the Soviet Union provided the components for an 880-megawatt nuclear-power station in the Sirte region. Libya has enough uranium to meet its foreseeable domestic needs. The German-built chemical-weapons plant at Rabta, described by Libyans as a pharmaceutical complex but confirmed as to its real function by visiting scientists, was destroyed in a mysterious fire in the 1980s. A Russian-built nuclear reactor at Tajoora, 30 miles from Tripoli, suffered a similar fate, not from fire but due to faulty ventilation and high levels of radiation. But the Lib-yans have pressed on. An underground complex at Mount Tarhuna, south of Tripoli, was completed in 1998 and closed subsequently to international inspection. Libya claims that it is part of the Great Man-Made River (GMR) project and thus not subject to such inspections. The country also refuses to sign the 1993 UN convention outlawing chemical weapons.

In addition to its heavy dependence on oil revenues, another obstacle to economic development in Libya is derived from an unbalanced labor force. One author observed, "Foreigners do all the work. Moroccans clean houses, Sudanese grow vegetables, Egyptians fix cars and drive trucks. Iraqis run the power stations and American and European technicians keep the equipment and systems humming. All the Libyans do is show up for makework government jobs."[6] Difficult climatic conditions and little arable land severely limit agricultural production; the country must import 75 percent of its food.

AN UNCERTAIN FUTURE

The revolutionary regime has been more successful than the monarchy was in making the wealth from oil revenues available to ordinary Libyans. Per capita income, which was $2,170 the year after the revolution, had risen to $10,900 by 1980. (With the drop in world oil prices and the residual effect of sanctions, it fell to $8,900 by 2001.)

This influx of wealth changed the lives of the people in a very short period of time. Seminomadic tribes such as the Qadadfas of the Sirte (Qadhafi's kin) have been provided with permanent homes, for example. Extensive social-welfare programs, such as free medical care, free education, and low-cost housing, have greatly enhanced the lives of many Libyans. However, this wealth has yet to be spread evenly across society. The economic downturn of the 1990s produced a thriving black market, along with price gouging and corruption in the public sector. In 1996, Libya organized "purification committees," mostly staffed by young army officers, to monitor and report instances of black-market and other illegal activities.

HEALTH/WELFARE

In addition to 1 million sub-Saharan African workers, Libya has made use of skilled workers as well as unskilled ones from many other Arab countries. Palestinian workers were expelled after the 1993 Oslo Agreement with Israel, which Qadhafi opposed vehemently. A GPC regulation issued in 2001 fixed the total number of skilled foreign workers at 40,000.

Until recently, opposition to Qadhafi was confined almost entirely to exiles abroad, centered on former associates living in Cairo, Egypt, who had broken with the Libyan leader for reasons either personal or related to economic mismanagement. But economic downturns and dissatisfaction with the leader's wildly unsuccessful foreign-policy ventures increased popular discontent at home. In 1983, Qadhafi had introduced two domestic policies that also generated widespread resentment: He called for the drafting of women into the armed services, and he recommended that all children be educated at home until age 10. The 200 basic "people's congresses," set up in 1977 to recommend policy to the national General People's Congress (which in theory is responsible for national policy), objected strongly to both proposals. Qadhafi then created 2,000 more people's congresses, presumably to dilute the opposition, but withdrew the proposals. In effect, suggested one observer, *The Green Book* theory had begun to work, and Qadhafi didn't like it.

Qadhafi's principal support base rests on the armed forces and the "revolutionary committees," formed of youths whose responsibility is to guard against infractions of *The Green Book* rules. "Brother Colonel" also relies upon a small group of collaborators from the early days of the Revolution, and his own relatives and members of the Qadadfa form part of the inner power structure. This structure is highly informal, and it may explain why Qadhafi is able to disappear from public view from time to time, as he did after the United States conducted an air raid on Tripoli in 1986, and emerge having lost none of his popularity and charismatic appeal.

In recent years, disaffection within the army has led to a number of attempts to overthrow Qadhafi. The most serious coup attempt took place in 1984, when army units allied with the opposition Islamic Front for the Salvation of Libya, based in Cairo and headed by several of Qadhafi's former associates, attacked the central barracks in Tripoli where he usually resides. The attackers were defeated in a bloody gun battle. A previously unknown opposition group based in Geneva, Switzerland, claimed in 1996 that its agents had poisoned the camel's milk that Qadhafi drinks while eating dates on his desert journeys, but proof of this claim is lacking.

However, the Libyan leader's elusiveness and penchant for secrecy make assessments of his continued leadership risky. According to the Tripoli rumor mill, someone attempts to assassinate Qadhafi every couple of months. But as yet no organized internal opposition has emerged, and the mercurial Libyan leader remains not only highly visible but also popular with his people.

INTERNAL CHANGES

Qadhafi has a talent for the unexpected that has made him an effective survivor. In 1988, he ordered the release of all political prisoners and personally drove a bulldozer through the main gate of Tripoli's prison to inaugurate "Freedom Day." Exiled opponents of the regime were invited to return under a promise of amnesty, and a number did so.

In June of that year, the GPC approved a "Charter of Human Rights" as an addendum to *The Green Book.* The charter outlaws the death penalty, bans mistreatment of prisoners, and guarantees every accused person the right to a fair trial. It also permits formation of labor unions, confirms the right to education and suitable employment for all Libyan citizens, and places Libya on record as prohibiting production of nuclear and chemical weapons. In March 1995, the country's last prison was destroyed and its inmates freed in application of the charter's guarantees of civil liberty.

THE WAR WITH CHAD

Libyan forces occupied the Aouzou Strip in northern Chad in 1973, claiming it as an integral part of the Libyan state. Occupation gave Libya access also to the reportedly rich uranium resources of the region. In subsequent years, Qadhafi played upon political rivalries in Chad to extend the occupation into a de facto one of annexation of most of its poverty-stricken neighbor.

But in late 1986 and early 1987, Chadian leaders patched up their differences and turned on the Libyans. In a series of spectacular raids on entrenched Libyan forces, the highly mobile Chadians, traveling mostly in Toyota trucks, routed the Libyans and drove them out of northern Chad. Chadian forces then moved into the Aouzou Strip and even attacked nearby air bases inside Libya. The defeats, with casualties of some 3,000 Libyans and loss of huge quantities of Soviet-supplied military equipment, exposed the weaknesses of the overequipped, undertrained, and poorly motivated Libyan Army.

ACHIEVEMENTS

The Great Man-Made River (GMR) called by Qadhafi the world's eighth wonder, went into its full operational phase in 2003, when subsurface water from deep in the Sahara Desert began flowing to Libya's cities through a network of 13-foot-diameter pipes from pumping stations. Excess water from the $27 billion project is to be stored in the Kufra basin, which has a capacity of 5,000 cubic metres. Pipeline blowouts continue to hamper completion. Once these problems are solved, water flow of 200 million cubic feet per day. Without the support of the GMR, Libya would remain essentially uninhabitable.

In 1989, after admitting his mistake, Qadhafi signed a cease-fire with then–Chadian leader Hissène Habré and agreed to submit the dispute over ownership of Aouzou to the International Court of Justice (ICJ). The ICJ affirmed Chadian sovereignty in 1994 on the basis of a 1955 agreement arranged by France as the occupying power there. Libyan forces withdrew from Aouzou in May, and since then the two countries have enjoyed a peaceful relationship. In 1998, the border was opened completely, in line with Qadhafi's policy of "strengthening neighborly relations."

FOREIGN POLICY

Libya's relations with the United States have remained hostile since the 1969 Revolution, which not only overthrew King Idris but also resulted in the closing of the important Wheelus Field air base. Despite

Qadhafi's efforts in more recent years to portray himself and Libya as respectable members of the world of nations, the country remains on the U.S. Department of State's list as one of the main sponsors of international terrorism. In 1986, U.S. war planes bombed Tripoli and Benghazi in retaliation for the bombing of a disco in Berlin, Germany, which killed two U.S. servicemen and injured 238 others. The retaliatory U.S. air attack on Libya killed 55 Libyan civilians, including Qadhafi's adopted daughter. After numerous delays and conflicting evidence about Libya's role in the Berlin bombing, a trial began in 1998 for four persons implicated in the attack. Only one, a diplomat in the embassy in East Berlin (now closed), was a Libyan national. The trial ended in 2001 with the conviction of the four; they were given 12- to 14-year sentences.

Libya resumed its old role of "pariah state" in 1992 by refusing to extradite two officers of its intelligence service suspected of complicity in the 1988 bombing of a Pan American jumbo jet over Lockerbie, Scotland. The United States, France, and Britain had demanded the officers' extradition and introduced a resolution to that effect in the UN Security Council; in the event of noncompliance on Libya's part, sanctions would be imposed on the country. *Resolution 748* passed by a 10-to-zero vote, with five abstentions. A concurrent ruling by the ICJ ordered Libya to turn over the suspects or explain in writing why it was not obligated to do so.

Qadhafi, however, refused to comply with *Resolution 748.* He argued that the suspects should be tried (if at all) in a neutral country, since they could not be given a fair trial either in Britain or the United States.

The Security Council responded by imposing partial sanctions on Libya. Despite the partial embargo, Libya's leader continued to reject compliance with the resolution. As a result, the Security Council in 1993 passed *Resolution 883,* imposing much stiffer sanctions on the country. The new sanctions banned all shipments of spare parts and equipment sales and froze Libyan foreign bank deposits. International flights to Libya were prohibited. The only area of the economy not affected was that of oil exports, since Britain and other Western European countries are dependent on low-sulfur Libyan crude for their economies.

Despite the sanctions and Libya's isolation, Qadhafi continued to refuse to surrender the two Lockerbie suspects. He maintained that they were innocent and could not receive a fair trial except in a neutral country under international law.

The tug-of-war between the United Nations and its recalcitrant member went on for six years. In March 1998, the United Nations, set a 60-day deadline for compliance. Subsequently Qadhafi reversed his stance on the Lockerbie suspects. While he insisted that the Libyan government was not involved, he agreed to turn over the suspects to be tried in a neutral court under Scottish law. The two were then flown to the Netherlands, where they were tried in a court set up in an abandoned Dutch air base, Camp Zeist. The trial was marked by intricate legal maneuverings and some questionable evidence. In 2000, one of the suspects was acquitted. The other, former Libyan intelligence agent Abdel Basset al-Megrahi, was found guilty and sentenced to life imprisonment. In August 2003 there was a major step toward resolution of the Lockerbie issue when Libya formally accepted the responsibility for the bombing in a letter to the UN Security Council. The acceptance would expedite a $2.7 billion settlement with the families of the 270 victims, with each family to receive between $5 and $10 million. Libya agreed to deposit the funds in an international bank. In turn, the UN would lift sanctions on the country imposed in 1991. However the sanctions imposed separately by the U.S. would remain in effect until certain "threatening elements" in Libyan behavior, such as intervention in sub-Saharan African conflicts, pursuit of weapons of mass destruction and poor human rights record, are removed.

After the UN Security Council voted to lift sanctions against Libya in September 2003, Libya made a series of efforts to repair the damage done to its image from the past two decades. In December 2003 Libya announced that it would abandon its programs to develop weapons of mass destruction, and in January of 2004 Libya agreed to compensate families of victims of the 1989 bombing of the French passenger aircraft over Sahara desert. In August 2004 Libya agreed to pay 35 million to compensate victims of the bombing of a Berlin nightclub in 1986. In recognition of these efforts, and to the surprise of many observers, British Prime Minister Tony Blair visited Libya in March 2004 and met with Libyan leader Muammar Gaddafi. It was the first visit since 1943.

PROSPECTS

The tide of fundamentalism sweeping across the Islamic world and challenging secular regimes has largely spared Libya thus far, although there were occasional clashes between fundamentalists and police in the 1980s, and in 1992, some 500 fundamentalists were jailed briefly. However, the bloody civil uprisings against the regimes in neighboring Algeria and Egypt caused Qadhafi in 1994 to reemphasize Libya's Islamic nature. New laws passed by the General People's Congress would apply Islamic law (Shari'a) and punishments in such areas as marriage and divorce, wills and inheritance, crimes of theft and violence (where the Islamic punishment is cutting off a hand), and for apostasy. Libya's tribal-based society and Qadhafi's own interpretation of Islamic law to support women's rights and to deal with other social issues continue to serve as obstacles to Islamic fundamentalism.

On September 7, 1999, the Libyan leader celebrated his 30th year in power with a parade of thousands of footsoldiers, along with long-range missiles and tanks, through the streets of Tripoli. Libyan jets, many of them piloted by women, flew overhead.

Timeline: PAST

1835
Tripoli becomes an Ottoman province with the Sanusiya controlling the interior

1932
Libya becomes an Italian colony, Italy's "Fourth Shore"

1951
An independent kingdom is set up by the UN under King Idris

1969
The Revolution overthrows Idris; the Libyan Arab Republic is established

1973–1976
Qadhafi decrees a cultural and social revolution with government by people's committees

1980s
A campaign to eliminate Libyan opponents abroad; the United States imposes economic sanctions in response to suspected Libya-terrorist ties; U.S. planes attack targets in Tripoli and Benghazi; Libyan troops are driven from Chad, including the Aouzou Strip

1990s
Libya's relations with its neighbors improve; the UN votes to impose sanctions on Libya for terrorist acts; Qadhafi comes to an agreement with the UN regarding the trial of the PanAm/Lockerbie bombing suspects

PRESENT

2000s
Qadhafi makes changes to governmental structure

At 60-plus the charismatic Libyan leader shows no sign of relinquishing power and seems in excellent health. In the absence of a formal succession process (Qadhafi has no official title), speculation centers on his oldest son, Muhammad Sayf al-Islam. However, a younger son, El-Saadi, represented Libya on an official visit to Japan in 2001.

The lifting of UN sanctions on the country resulted from Qadhafi's acceptance of international jurisdiction in the Lockerbie case. As a result, relations with Europe have been normalized. However, the United States continues to insist that the country is a sponsor of global terrorism. In July 2001, the U.S. Senate approved a five-year extension of the Iran–Libya Sanctions Act (ILSA). The act bars U.S. companies from doing business in Libya, imposing fines for those investing more than $20 million in Libyan development projects. But in December 2003 Qadhafi again confounded his critics by agreeing to discontinue Libya's nuclear weapons development program and open its facilities to international inspection. The country also signed the Nuclear Non-Proliferation Treaty. In March 2004 the Libyan leader ordered 3,300 chemical bombs destroyed and agreed to halt further production.

Although he remains hostile to the U.S. and equally to Israel, Qadhafi has cultivated an image of respectability in recent years and has vigorously promoted African unity. During the July 2003 meeting of the Organization for African Unity, he described HIV/AIDS as a "peaceful virus". Along with malaria and sleeping sickness, he said it was God's way of keeping white colonizers out of Africa. Salvos such as this one have earned him considerable support from other African nations. In 2003, with African backing, Libya was elected to the chairmanship of the UN Commission on Human Rights.

NOTES

1. "[I]rrigation, colonization and hard work have wrought marvels. Everywhere you see plantations forced out of the sandy, wretched soil." A. H. Broderick, *North Africa* (London: Oxford University Press, 1943), p. 27.
2. Religious leaders issued a *fatwa* ("binding legal decision") stating that a vote against independence would be a vote against religion. Omar el Fathaly, et al., *Political Development and Bureaucracy in Libya* (Lexington, KY: Lexington Books, 1977).
3. See *Middle East Journal,* vol. 24, no. 2 (Spring 1970), Documents Section.
4. John Wright, *Libya: A Modern History* (Baltimore, MD: Johns Hopkins University Press, 1982), pp. 124–126. Qadhafi's idol was former Egyptian president Nasser, a leader in the movement for unity and freedom among the Arabs. While he was at school in Sebha, in the Fezzan, he listened to Radio Cairo's Voice of the Arabs and was later expelled from school as a militant organizer of demonstrations.
5. *The London Times* (June 6, 1973).
6. Khidr Hamza, with Jeff Stein, *Saddam's Bombmaker* (New York: Scribner's, 2000), p. 289. The author was head of the Iraqi nuclear-weapons program before defecting to Libya and eventually the United States.
7. Donald G. McNeil Jr., in *The New York Times* (February 1, 2001).

Morocco (Kingdom of Morocco)

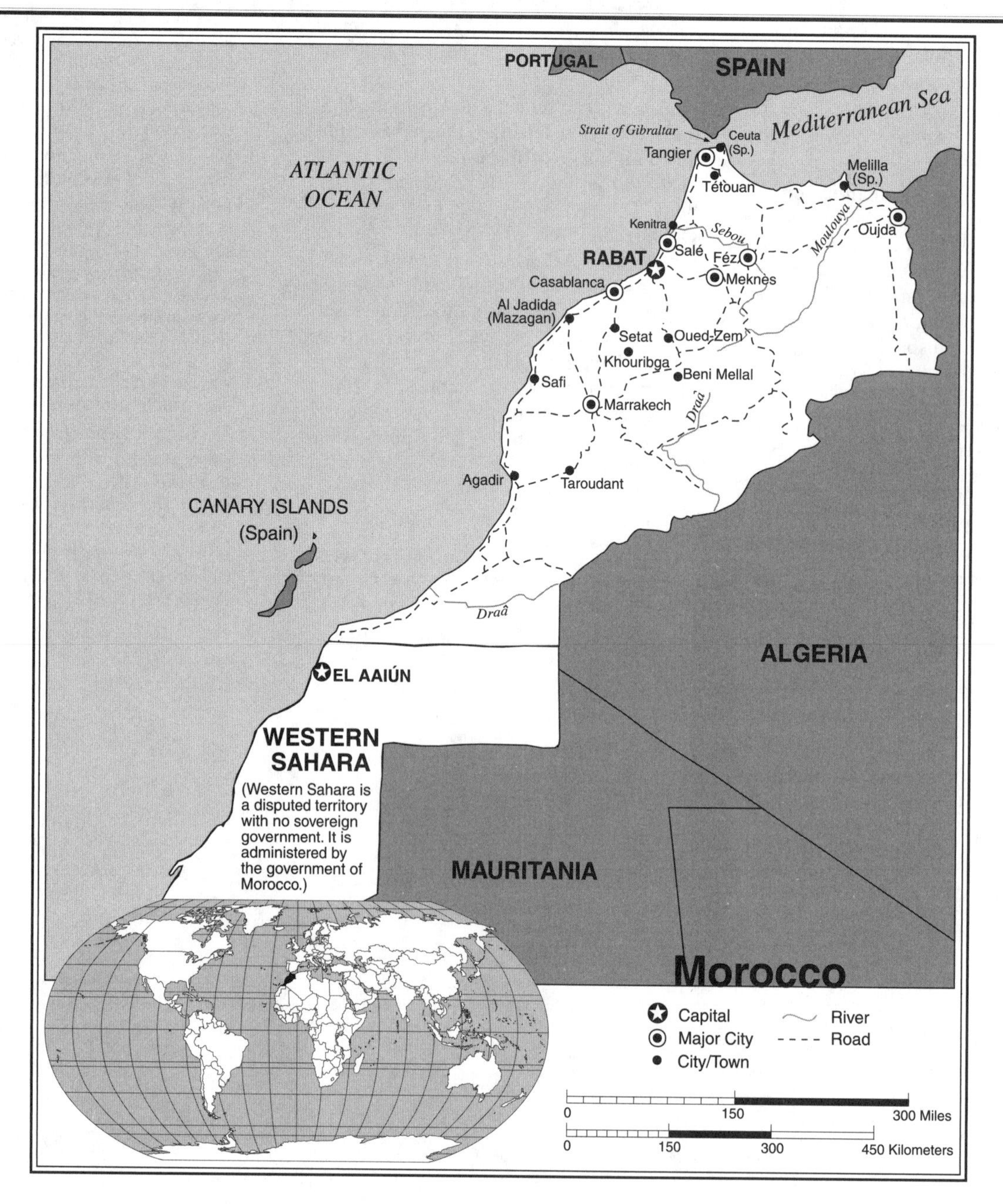

Morocco Statistics

GEOGRAPHY

Area in Square Miles (Kilometers): 274,400 (710,850) including the Western Sahara (102,675 square miles-266,000 sq.km.) about the size of California

Capital (Population): Rabat (1,293,000)

Environmental Concerns: land degradation; desertification; soil erosion; overgrazing; contamination of water supplies; oil pollution of coastal waters

Geographical Features: the northern coast and interior are mountainous, with large areas of bordering plateaux, intermontane valleys, and rich coastal plains; south, southeast and entire Western Sahara is desert

Climate: varies from Mediterranean to desert

PEOPLE

Population

Total: 31,167,783

Annual Growth Rate: 1.68%

Rural/Urban Population Ratio: 47/53

Major Languages: Arabic; Tama-zight; various Berber dialects; French

Ethnic Makeup: 64% Arab; 35% Berber; 1% non-Morroccan and Jewish Religions: 99% Sunni Muslim; 1% Christian and Jewish

Health

Life Expectancy at Birth: 67 years (male); 72 years (female)

Infant Mortality Rate: 47/1,000 live births

Physicians Available (Ratio): 1/2,923 people

Education

Adult Literacy Rate: 43.7%

Compulsory (Ages): 7–13

COMMUNICATION

Telephones: 1,515,000 main lines

Daily Newspaper Circulation: 13 per 1,000 people

Televisions: 93 per 1,000 people

Internet Service Providers: 8 (2000)

TRANSPORTATION

Highways in Miles (Kilometers): 37,649 (60,626)

Railroads in Miles (Kilometers): 1,184 (1,907)

Usable Airfields: 69

Motor Vehicles in Use: 1,278,000

GOVERNMENT

Type: constitutional monarchy

Independence Date: March 2, 1956 (from France)

Head of State/Government: King Muhammad VI; Prime Minister Driss Jettou

Political Parties: National Rally of Independents; Popular Movement; National Democratic Party; Constitutional Union; Socialist Union of Popular Forces; Istiqlal; Kutla Bloc; Party of Progress and Socialism; others

Suffrage: universal at 21

MILITARY

Military Expenditures (% of GDP): 4%

Current Disputes: final resolution on the status of Western Sahara remains to be worked out; from time to time Morocco demands the retrocession of Ceuta and Melilla, cities located physically within its territory but considered extensions of mainland Spain (plazas de soberaniá by the Spanish government)

ECONOMY

Currency ($ U.S. Equivalent): 9.37 dirhams = $1

Per Capita Income/GDP: $3,500/$105 billion

GDP Growth Rate: 8%

Inflation Rate: 2%

Unemployment Rate: 23%

Labor Force: 11,000,000

Natural Resources: phosphates, iron ore; manganese; lead; zinc; fish; salt

Agriculture: barley; wheat; citrus fruits; wine; vegetables; olives; livestock

Industry: phosphate mining and processing; food processing; leather goods; textiles; construction; tourism

Exports: $7.6 billion (primary partners France, Spain, United Kingdom)

Imports: $12.2 billion (primary partners France, Spain, Italy)

Morocco Country Report

The Kingdom of Morocco is the westernmost country in North Africa. Morocco's population is the second largest (after Egypt) of the Arab states. The country's territory includes the Western Sahara (a claim made under dispute), formerly two Spanish colonies, Rio de Oro and Saguia al-Hamra. Morocco annexed part in 1976 and the balance in 1978, after Mauritania's withdrawal from its share, as decided in an agreement with Spain. Since then, Morocco has incorporated the Western Sahara into the kingdom as its newest province.

Two other territories physically within Morocco remain outside Moroccan control. They are the cities of Ceuta and Melilla, both located on rocky peninsulas that jut out into the Mediterranean Sea. They have been held by Spain since the fifteenth century. (Spain also owns several small islands off the coast in Moroccan territorial waters in the Mediterranean.) Spain's support for Morocco's adminssion to the European Union (EU) as an associate memeber has eased tensions between them over the enclaves. An additional economic advantage to Morocco is that each day some 40,000 Moroccans cross legally into them for work.

In 1986 a Spanish law excluding Moroccan Muslim residents from Spanish citizenship led to protests among them. The Moroccan government did not pursue the protests, and in 1988 the question of citizenship became moot when the Spanish Cortes (Parliament) passed a law formally incorporating Ceuta and Melilla into Spain as overseas territories.

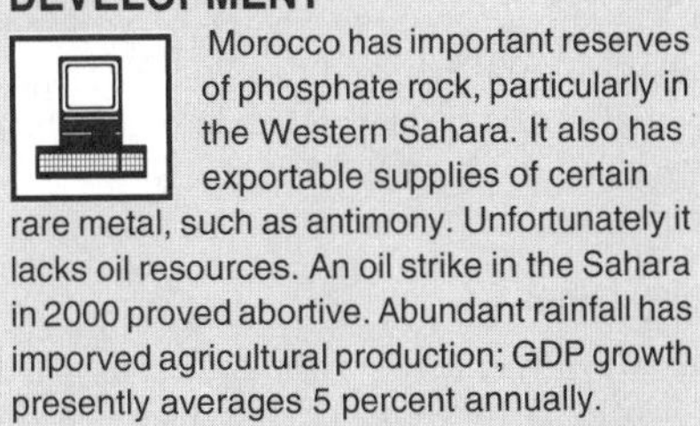

DEVELOPMENT

Morocco has important reserves of phosphate rock, particularly in the Western Sahara. It also has exportable supplies of certain rare metal, such as antimony. Unfortunately it lacks oil resources. An oil strike in the Sahara in 2000 proved abortive. Abundant rainfall has imporved agricultural production; GDP growth presently averages 5 percent annually.

In recent years Ceuta, in particular, has become a "migration station" for thousands of Moroccans and sub-Saharan Africans seeking jobs and a better life in Europe. Most of them are undocumented and use Ceuta as a base while working to cross the sea and reach safety in Spain, Italy or elsewhere in Europe. The influx puts a heavy strain on the city's facilities and services. Its land border is now encircled with towering barbed wire fences to slow down immigration. But as a Ceuta leader noted, "Before, the migration flow was exclusively from Morocco. Now emigrants from all of the world pour through."[1]

Moroccan geography explains the country's dual population structure. About 35 percent of the population are Berbers, descendants of the original North Africans. The Berbers were, until recently, grouped into tribes, often taking the name of a common ancestor, such as the Ait ("Sons of") 'Atta of southern Morocco.[2] Invading Arabs converted them to Islam in the eighth century but made few other changes in Berber life. Unlike the Berbers, the majority of the Arabs who settled in Morocco were, and are, town-dwellers. The Berbers, more than the Arabs, derived unity and support from their extended families rather than from state control, whether real or putative.

HISTORY

Morocco has a rich cultural history, with many of its ancient monuments more or less intact. It has been governed by some

(Hamilton Wright/government of Morocco)

Morocco has a rich history. The Karawiyyin Mosque at Fez was founded in the ninth century A.D. and is the largest mosque in North Africa. It is also the seat of one of Africa's oldest universities.

form of monarchy for over a thousand years, although royal authority was frequently limited or contested by rivals. The current ruling dynasty, the Alawis, assumed power in the 1600s. One reason for their long rule is the fact that they descend from the Prophet Muhammad. Thus, Moroccans have had a real sense of Islamic traditions and history through their rulers.

The first identifiable Moroccan "state" was established by a descendant of Muhammad named Idris, in the late eighth century. Idris had taken refuge in the far west of the Islamic world to escape civil war in the east. Because of his piety, learning, and descent from Muhammad, he was accepted by a number of Berber groups as their spiritual and political leader. His son and successor, Idris II, founded the first Moroccan capital, Fez. Father and son established the principle whereby descent from the Prophet was an important qualification for political power as well as social status in Morocco.

The Idrisids ruled over only a small portion of the current Moroccan territory, and, after the death of Idris II, their "nation" lapsed into decentralized family rule. In any case, the Berbers had no real idea of nationhood; each separate Berber group thought of itself as a nation. But in the eleventh and twelfth centuries, two Berber confederations developed that brought imperial grandeur to Morocco. These were the Almoravids and the Almohads. Under their rule, North Africa developed a political structure separate from that of the eastern Islamic world, one strongly influenced by Berber values.

The Almoravids began as camel-riding nomads from the Western Sahara who were inspired by a religious teacher to carry out a reform movement to revive the true faith of Islam. (The word *Almoravid* comes from the Arabic *al-Murabitun,* "men of the ribat," rather like the crusading religious orders of Christianity in the Middle Ages.) Fired by religious zeal, the Almoravids conquered all of Morocco and parts of western Algeria.

A second "imperial" dynasty, the Almohads, succeeded the Almoravids but improved on their performance. They were the first, and probably the last, to unite all of North Africa and Islamic Spain under one government. Almohad monuments, such as the Qutubiya tower, the best-known landmark of Marrakesh, and the Tower of Hassan in Rabat, still stand as reminders of their power and the high level of the Almohads' architectural achievements.

Mulay Ismail

The Alawis came to power and established their rule partly by force, but also as a result of their descent from the Prophet Muhammad. This link enabled them to win the support of both Arab and Berber populations. The real founder of the dynasty was Mulay Ismail, one of the longest-reigning and most powerful monarchs in Morocco's history.

Mulay Ismail unified the Moroccan nation. The great majority of the Berber groups accepted him as their sovereign. The sultan built watchtowers and posted permanent garrisons in Berber territories to make sure they continued to do so. He brought public security to Morocco also; it was said that in his time, a Jew or an unveiled woman could travel safely anywhere in the land, which was not the case in most parts of North Africa, the Middle East, and Europe.

In 1904, France, Britain, Spain, and Germany signed secret agreements partitioning the country. The French would be given the largest part of the country, while Spain would receive the northern third as a protectorate plus some territory in the Western Sahara. In return, the French and Spanish agreed to respect Britain's claim to Egypt and Germany's claim to East African territory.

FREEDOM

The "freest and fairest" elections in Moroccan history established a new 325-seat Chamber of Representatives equally balanced among the leading political parties. Some 35 women candidates were elected. King Muhammad VI's "National Action Plan" raises the legal marriageable age for women to 18 and gives other rights to them. In 2003 the king proposed revisions to the 1957 Mudawanna (Family Law) which were approved by the Chamber. They now have the right to file for divorce, share equally in family property, and travel without prior consent from male family members.

The French protectorate over Morocco covered barely 45 years (1912–1956). But in that brief period, the French introduced significant changes into Moroccan life. For the first time, southern Morocco was brought entirely under central government control, although the "pacification" of the Berbers was not complete until 1934.

Morocco's Independence Struggle

The movement for independence in Morocco developed slowly. The only symbol of national unity was the sultan, Muhammad ibn Yusuf. But he seemed ineffectual to most young Moroccans, particularly those educated in French schools, who began to question the right of France to rule a people against their will.

The hopes of these young Moroccans got a boost during World War II. The Western Allies, Great Britain and the United States, had gone on record in favor of the right of subject peoples to self-determination after the war. When U.S. president Franklin D. Roosevelt and British prime minister Winston Churchill came to Casablanca for an important wartime conference, the sultan was convinced to meet them privately and get a commitment for Morocco's independence. The leaders promised their support.

In 1953, the Glawi and his fellow qaids decided, along with the French, that the time was right to depose the sultan. The qaids demanded that he abdicate; they said that his presence was contributing to Moroccan instability. When he refused, he was bundled into a French plane and sent into exile. An elderly uncle was named to replace him.

The sultan's departure had the opposite effect from what was intended. In exile, he became a symbol for Moroccan resistance to the protectorate. Violence broke out, French settlers were murdered, and a Moroccan Army of Liberation began battling French troops in rural regions. Although the French could probably have contained the rebellion in Morocco, they were under great pressure in neighboring Algeria and Tunisia, where resistance movements were also under way. In 1955, the French abruptly capitulated. Sultan Muhammad ibn Yusuf returned to his palace in Rabat in triumph, and the elderly uncle retired to potter about his garden in Tangier.

INDEPENDENCE

Morocco became independent on March 2, 1956. (The Spanish protectorate ended in April, and Tangier came under Moroccan control in October, although it kept its free-port status and special banking and currency privileges for several more years.) It began its existence as a sovereign state with a number of assets—a popular ruler, an established government, and a well-developed system of roads, schools, hospitals, and industries inherited from the protectorate. Against these assets were the liabilities of age-old Arab–Berber and inter-Berber conflicts, little experience with political parties or democratic institutions, and an economy dominated by Europeans.

The sultan's goal was to establish a constitutional monarchy. His first action was to give himself a new title, King Muhammad V, symbolizing the end of the old autocratic rule of his predecessors.

Muhammad V did not live long enough to reach his goal. He died unexpectedly in 1961 and was succeeded by his eldest son, Crown Prince Hassan. Hassan II ruled until his death in 1999. While he fulfilled his father's promise immediately with a Constitution, in most other ways Hassan II set his own stamp on Morocco.

The Constitution provided for an elected Legislature and a multiparty political system. In addition to the Istiqlal, a number of other parties were organized, including one representing the monarchy. But the results of the French failure to develop a satisfactory party system soon became apparent. Berber–Arab friction, urban–rural distrust, city rivalries, and inter-Berber hostility all intensified. Elections failed to produce a clear majority for any party, not even the king's.

HEALTH/WELFARE

In October 2000, the International Labor Organization (ILO) ranked Morocco as the 3rd-highest country in the world, after China and India, in the exploitation of child labor. Moroccan children as young as 5, all girls, are employed in the carpet industry, working up to 10 hours per day weaving the carpets that are at present Morocco's major source of foreign currency. In recent years, more and more Moroccan minors have been leaving their families and migrating illegally to Europe through the Spanish port city of Ceuta. Some 3,500 did so in 2002.

In 1971, during a diplomatic reception, cadets from the main military academy invaded the royal palace near Rabat. A number of foreign diplomats were killed and the king held prisoner briefly before loyal troops could crush the rebellion. The next year, a plot by air-force pilots to shoot down the king's plane was narrowly averted. The two escapes helped confirm in Hassan's mind his invincibility under the protection of Allah.

But they also prompted him to reinstate the parliamentary system. A new Constitution issued in 1972 defined Morocco "as a democratic and social constitutional monarchy in which Islam is the established religion."[3] However, the king retained the constitutional powers that, along with those derived from his spiritual role as "Commander of the Faithful" and lineal descendant of Muhammad, undergirded his authority.

INTERNAL POLITICS

Morocco's de facto annexation of the Western Sahara has important implications for future national development due to the territory's size, underpopulation, and mineral resources, particularly shale oil and phosphates. But the annexation has been equally important to national pride and political unity. The "Green March" of 350,000 unarmed Moroccans into Spanish territory in 1975 to dramatize Morocco's claim was organized by the king and supported by all segments of the population and the opposition parties. In 1977, opposition leaders agreed to serve under the king in a "government of national union." The first elections in 12 years were held for a new Legislature; several new parties took part.

The 1984 elections continued the national unity process. The promonarchist Constitutional Union (UC) party won a majority of seats in the Chamber of Representatives (Parliament). A new party, the National Rally of Independents (RNI), formed by members with no party affiliations, emerged as the chief rival to the UC.

New elections were scheduled for 1989 but were postponed three times; the king said that extra time was needed for the economic-stabilization program to show results and generate public confidence. The elections finally took place in two stages in 1993: the first for election of party candidates, and the second for trade-union and professional-association candidates. The final tally showed 195 seats for center-right (royalist) candidates, to 120 for the Democratic-bloc opposition. As a result, coalition government became necessary. The two leading opposition parties, however—the Socialist Union of Popular Forces (USFP) and the Istiqlal—refused to participate, claiming election irregularities. Opposition from members of these parties plus the Kutla Bloc, a new party formed from the merger of several minor parties, blocked legislative action until 1994. At that point, the entire opposition bloc walked out of the Legislature and announced a boycott of the government.

King Hassan resolved the crisis by appointing then–USFP leader Abdellatif Filali as the new prime minister, thus bringing the opposition into the government. The king continued with this method

(Hamilton Wright/government of Morocco)

Tangier was once a free city and port. Just across the Strait of Gibraltar from Spain, it now is Morocco's northern metropolis. Modernization and expansion of port facilities to accommodate large cruise ships and tankers got under way in 1999.

of political reconciliation by appointing the new head of the USFP, Abderrahmane Youssoufi, to the position after the latter's return from political exile in 1998.

A referendum in 1996 approved several amendments to the constitution. One in particular replaced the unicameral legislature by a bicameral one. The Chamber of Representatives (lower house) is to be elected directly, for 5-year terms. The chamber of Counselors (upper house) is to be two-thirds elected and one-third appointed. In September 2002 elections were held for the 325-seat lower house. Some 26 parties, a dozen of them brand new, presented candidates. Described by the government as the first free and fair election in national history, they resulted in much higher turnout than the 58 percent of the vote-buying, tainted 1997 election.

The election results underlined the broad spectrum of Moroccan politics. The Socialist Union of Popular Forces (USFP), headed by then Prime Minister Youssoufi, won 50 seats, the venerable Istiqlal Party 48, the National Rally of Independents 41, the National Popular Movement 27. The Party of Justice and Development, which had replaced the banned Islamist Justice and Development, won 42 seats; its predecessor had held only 18 in the outgoing Chamber. Also noteworthy was the election of 35 women; a quota of 30 had been reserved for them.

FOREIGN RELATIONS

During his long reign, King Hassan II served effectively in mediating the long-running Arab–Israeli conflict. He took an active part in the negotiations for the 1979 Egyptian–Israeli peace treaty and for the treaty between Israel and Jordan in 1994. For these services he came to be viewed by the United States and by European powers as an impartial mediator. However, his absolute rule and suppression of human rights at home caused difficulties with Europe. The European Union suspended $145 million in aid in 1992; it was restored only after Hassan had released long-time political prisoners and pardoned 150 alleged Islamic militants. In 1995, Morocco became the second African country, after Tunisia, to be granted associate status in the EU. In February 2003, a Casablanca court sentenced three Saudi members of al-Qaeda to 10 years in prison after they were accused of plotting to attack U.S. and British warships in the Straits of Gibraltar. In May 2003, 41 people were killed and many more injured in a series of suicide bomb attacks in the business capital Casablanca. As a reward for their staunch support of U.S. policy and for their assistance, a free trade agreement with the U.S. came into effect in July 2004. The deal followed Washington's designation of Morocco as a major non-Nato ally.

Thus far, Morocco's only venture in "imperial politics" has been in the Western Sahara. This California-size desert territory, formerly a Spanish protectorate and then a colony after 1912, was never a part of the modern Moroccan state. Its only connection is historical—it was the headquarters and starting point for the Almoravid dynasty, camel-riding nomads who ruled western North Africa and southern Spain in the eleventh century. But the presence of so much empty land, along with millions of tons of phosphate rock and potential oil fields, encouraged the king to "play international politics" in order to secure the territory. The 1975 Green March has been followed up by large-scale settlement of Moroccans there in the past two decades. Like the American West in the nineteenth century, it was Morocco's "last frontier." Moroccans were encouraged to move there, with government pledges of free land, tools, seeds, and equipment for farmers, as well as housing.

Since the 1976 partition, ownership of the Western Sahara has been challenged by the Polisario, an independence movement backed by Algeria. Acting under the aegis of its responsibility for decolonization and self-government of colonized peoples, the United Nations established a peacekeeping force for the Western Sahara (MINURSO) in 1991. A UN resolution thereafter called for a referendum that would give the population a choice between independence and full integration with Morocco. Voter registration would precede the referendum, in order to determine eligibility of voters.

A decade later, the referendum seems less and less likely to be held. King Hassan II unilaterally named the territory Morocco's newest province, and by 2001 Moroccan settlers formed a majority in the population of 244,593. In December 2001, French president Jacques Chirac made an official visit to Morocco and saluted the country for the development of its "southern provinces." Earlier, the United Nations had appointed former U.S. secretary of state James Baker as mediator between Morocco and the Polisario. After several failed attempts at mediation, Baker submitted a plan for postponement of the referendum until 2006. In the interim, the Sahrawis would elect an autonomous governing body, with its powers limited to local and provincial affairs. The voting list would include all residents. The Security Council approved the Baker plan. But in view of the extensive Moroccanization of the territory, its self-government under Sahrawi leadership remained highly unlikely. King Muhammad made his first visit there in 2002, receiving a thunderous welcome from the Moroccan settler population. Emphasizing its integration into the kingdom as its newest province, he approved offshore oil-exploration concessions and released 56 Sahrawi political prisoners as a "sign of affection for the sons of the Sahara."

THE ECONOMY

Morocco has many of certain resources but too little of other, critical ones. It has two thirds of the world's known reserves of phosphate rock and is the top exporter of phosphates. The major thrust in industrial development is in phosphate-related industries. Access to deposits was one reason for Morocco's annexation of the Western Sahara, although to date there has been little extraction there due to the political conflict. The downturn in demand and falling prices in the global phosphates market brought on a debt crisis in the late 1980s. Increased phosphate demand globally and improved crop production following the end of several drought years have strengthened the economy. Privatization of the government-owned tobacco monopoly, the first industry to be so affected, generated a budgetary surplus in 2002.

The country also has important but undeveloped iron-ore deposits and a small but significant production of rare metals such as mercury, antimony, nickel, and lead. In the past, a major obstacle to development was the lack of oil resources. Prospects for oil improved in 2000 when the U.S. oil company Skidmore Energy was thought to have struck oil near Talsinnt, in the eastern Sahara. But the find, which the king had declared to be God's gift to Morocco, turned out to be mud. In 2001, the French oil company TotalFinaElf and Kerr-McGee of Texas were granted parallel concessions of 44,000 square miles offshore in Western Saharan waters near Dakhla.

ACHIEVEMENTS

A Moroccan runner, Abdelkader Mouaziz, won the 31st New York City Marathon, two and one-half minutes ahead of his nearest rival. Another Moroccan, Youssef el Aynaoui, had become one of the world's premier tennis players and competed well in major tournaments before his retirement.

Although recurring droughts have hampered improvement of the agricultural sector, it still accounts for 20 percent of gross domestic product and employs 50 percent of the labor force. Production varies widely from year to year, due to fluctuating rainfall. Abundant rains in 2001 resulted in bumper crops and a 25 percent increase in agricultural output, with 8 percent growth in GDP for that year.

The fisheries sector is equally important to the economy, with 2,175 miles of coastline and half a million square miles of territorial waters to draw from. Fisheries account for 16 percent of exports; annual production is approximately 1 million tons. The agreement with the European Union for associate status has been very beneficial to the industry. Morocco received $500 million in 1999–2001 from European countries in return for fishing rights for their vessels in Moroccan territorial waters.

But the economic outlook and social prospects remain bleak for most people. Although the birth rate has been sharply reduced, job prospects are limited for the large number of young Moroccans entering the labor force each year. The "suicide bombers," who attacked a Jewish community center, a hotel, foreign consulates and other structures in Casablanca in May 2003, killing some 41 persons, were said to belong to the radical Islamist organiation al-Sirat al-Mustakim (Righteous Path), believed to be linked with al-Qaeda. However, the fact that they came mostly from the impoverished Thomasville slum area of the city suggests that they acted not out of a desire to overthrow the monarchy, but out of frustration with the problems that face Morocco's youth today, namely unemployment, poverty and lack of opportunities in the workplace. As one observer noted of those arrested (only 12 were suicide bombers), little distinguished them from the group of young men idling in the streets or hawking designer sunglasses at intersections around town.

Timeline: PAST

788–790
The foundations of Moroccan nation are established by Idris I and II, with the capital at Fez

1062–1147
The Almoravid and Almohad dynasties, Morocco's "imperial period"

1672
The current ruling dynasty, the Alawi, establishes its authority under Mulay Ismail

1912
Morocco is occupied and placed under French and Spanish protectorates

1956
Independence under King Muhammad V

1961
The accession of King Hassan II

1975
The Green March into the Western Sahara dramatizes Morocco's claim to the area

1980s
Bread riots; agreement with Libya for a federal union; the king unilaterally abrogates the 1984 treaty of union with Libya

1990s
Elections establish parliamentary government; King Hassan dies and is succeeded by King Muhammad VI

PRESENT

2000s
King Muhammad VI works to improve human rights
The economic picture brightens

PROSPECTS

King Hassan II died in July 1999. The monarch had ruled his country for 38 years—the second-longest reign in the Middle East. Like King Hussein of Jordan, Hassan became identified with his country to such a degree that "Hassan was Morocco, and Morocco was Hassan." But unlike Jordan's ruler, Hassan combined religious with secular authority. Among his many titles was that of "Commander of the Faithful," and the affection felt for him by most Moroccans, particularly women and youth, was amply visible during his state funeral. His frequent reminders to the nation in speeches and broadcasts that "I am the person entrusted by God to lead you" clearly identified him in the public mind not only as their religious leader but also as head of the family.

The king's eldest son, Crown Prince Muhammad, succeeded him without incident as Muhammad VI. Morocco's new ruler began his reign with public commitments to reform human-rights protections and an effort to atone for some aspects of Hassan's autocratic rule. The king's declared commitment to human rights and political reform have been undercut to a large extent by the repressive structure inherited from his father. This structure, comprising the security services, army leaders, and a coterie of senior ministers, is a major obstacle to civil change. One of his first acts was the dismissal of longtime Interior Minister Driss Basri, considered the power behind the throne during Hassan II's reign. He was felt by most Moroccans to be in charge of everything from garbage collection to the torture of political prisoners, and even important foreign-policy decisions. But Basri's elaborate security apparatus was left largely intact.

Muhammad VI also publicly admitted the existence of the Tazmamat "death camp" and other camps in the Sahara, where rebel army officers and political prisoners were held, often for years and without trial or access to their families. (The family of General Oufkir, leader of the 1972 attempted coup who was later executed, were among those held, but they managed to escape.)[8] The new king also committed $3.8 million in compensation to the families of those who had been imprisoned.

The rise of Islamic fundamentalism as a political force is an obstacle to Muhammad VI's vision of Morocco. Hassan II kept fundamentalists on a tight rein. He suppressed the main Islamist movement, Adil wa Ihsan ("Justice and Charity"), and sent its leader to a mental institution. Hassan also used his religious credentials to pose as a fundamentalist leader of his people. Muhammad VI seems more willing to integrate the fundamentalists into the political structure. Recently he issued a "National Action Plan" guaranteeing rights for women, approving a free press, and promising other civil rights long absent from Moroccan society.

In March 2000, half a million people took to the streets to declare their support for the action plan. Independent, nongovernment newspapers appeared on the newstands. The king's popularity soared as he traveled about the country to learn about problems at first hand, and he was hailed as the "king of the poor" by the people.

NOTES

1. Sue Miller, "Migration Station", *Christian Science Monitor*, June 26, 2003.
2. See David M. Hart, *Dadda 'Atta and His Forty Grandsons* (Cambridge, England: Menas Press, 1981), pp. 8–11. Dadda 'Atta was a historical figure, a minor saint or marabout.
3. See Malika Oufkir, with Michele Fitoussi, *Stolen Lives: Twenty Years in a Desert Jail* (New York: Hyperion Books, 1999). Another prisoner, Ahmed Marzouki, recently published his memoir of life there. Entitled *Cell 10*, it has sold widely in Morocco.

Tunisia (Republic of Tunisia)

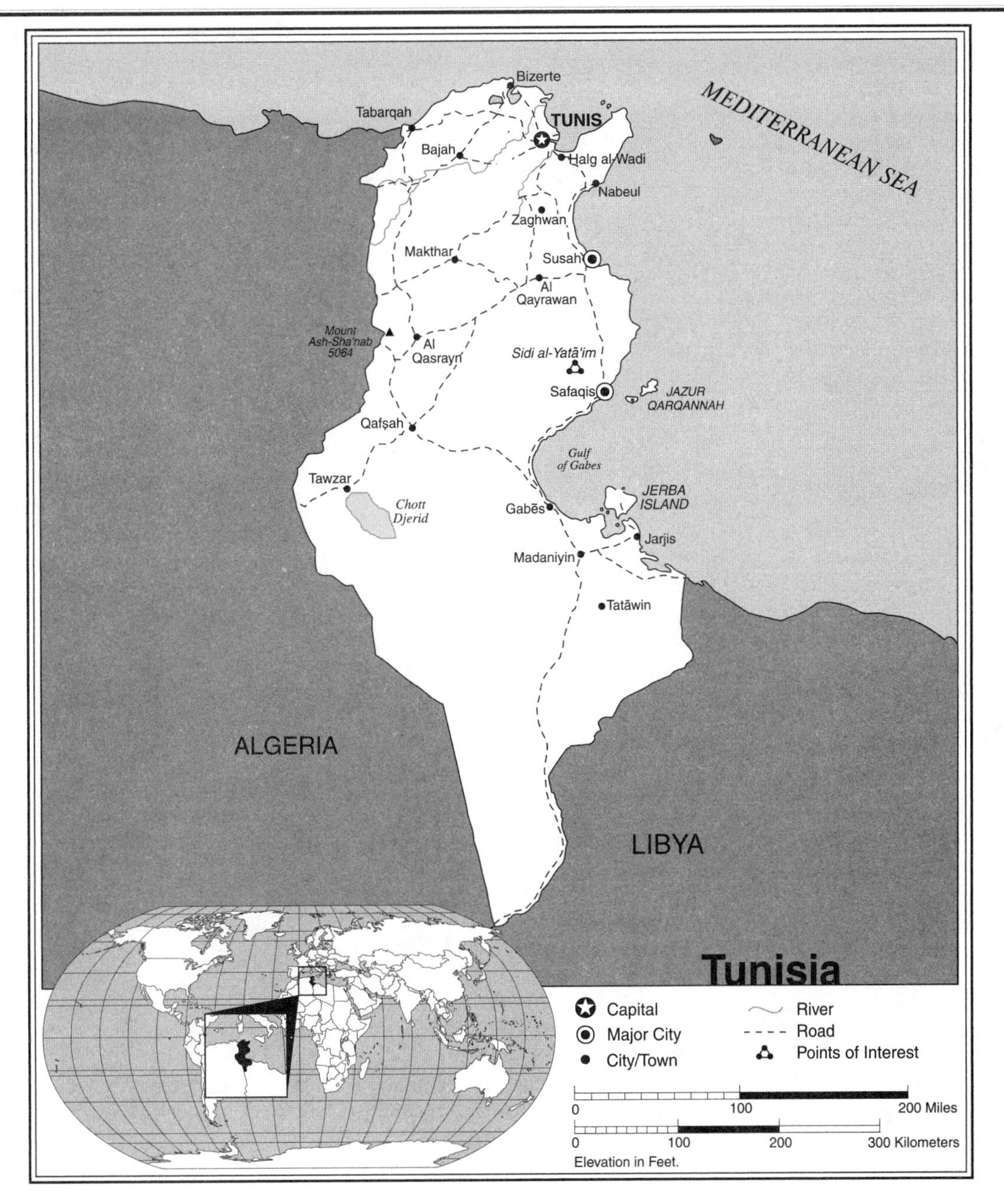

Tunisia Statistics

GEOGRAPHY

Area in Square Miles (Kilometers): 63,153 (163,610) (about the size of Georgia)

Capital (Population): Tunis (675,000)

Environmental Concerns: hazardous-waste disposal; water pollution; limited fresh water resources; deforestation; overgrazing; soil erosion; desertification

Geographical Features: mountains in north; hot, dry central plain; semiarid south merges into Sahara

Climate: hot, dry summers; mild, rainy winters; desert in the south, temperate in the north

PEOPLE

Population

Total: 9,815,644

Annual Growth Rate: 1.12%

Rural/Urban Population Ratio: 37/63

Major Languages: Arabic; French

Ethnic Makeup: 98% Arab–Berber; 1% European; 1% others
Religions: 98% Muslim; 1% Christian; less than 1% Jewish

Health

Life Expectancy at Birth: 72.5 years (male); 74 years (female)
Infant Mortality Rate: 28/1,000 live births
Physicians Available (Ratio): 1/1,640 people

Education

Adult Literacy Rate: 70.8%
Compulsory (Ages): 6–16

COMMUNICATION

Telephones: 1,313,000 main lines
Daily Newspaper Circulation: 45 per 1,000 people
Televisions: 156 per 1,000 people
Internet Service Provider: 1 (2000)

TRANSPORTATION

Highways in Miles (Kilometers): 14,345 (23,100)
Railroads in Miles (Kilometers): 1,403 (2,260)
Usable Airfields: 32
Motor Vehicles in Use: 531,000

GOVERNMENT

Type: republic
Independence Date: March 20, 1956 (from France)
Head of State/Government: President Zine El Abidine Ben Ali; Prime Minister Mohammed Ghannouchi
Political Parties: Constitutional Democratic Rally (RCD), ruling party; others are Al-Tajdid Movement, Liberal Socialist Party (PSL), Movement of Democratic Socialists (MDS), Popular Unity Party, Unionist Democratic Union (Al-Nahda "Resistance"), Islamic fundamentalist party (currently outlawed)
Suffrage: universal at 20

MILITARY

Military Expenditures (% of GDP): 1.5%
Current Disputes: none

ECONOMY

Currency ($ U.S. Equivalent): 1.26 dinars = $1
Per Capita Income/GDP: $6,500/$62.8 billion
GDP Growth Rate: 4.8%
Inflation Rate: 3%
Unemployment Rate: 15.6%
Labor Force: 6,259,000
Natural Resources: petroleum; phosphates; iron ore; lead; zinc; salt
Agriculture: olives; dates; oranges; almonds; grain; sugar beets; grapes; poultry; beef; dairy products
Industry: petroleum; mining; tourism; textiles; footwear; food; beverages
Exports: $6.1 billion (primary partners Germany, France, Italy)
Imports: $8.4 billion (primary partners France, Germany, Italy)

SUGGESTED WEB SITES

http://www.cia.gov/cia/publications/factbook/index.html
http://www.tunisiaonline.com

Tunisia Country Report

Tunisia, the smallest of the four North African countries, is less than one tenth the size of Libya, its neighbor to the east. However, its population is nearly twice the size of Libya's.

Tunisia's long coastline has exposed it over the centuries to a succession of invaders from the sea. The southern third of the country is part of the Sahara Desert; the central third consists of high, arid plains. Only the northern region has sufficient rainfall for agriculture. This region contains Tunisia's single permanent river, the Medjerda.

DEVELOPMENT

Associate membership in the European Union has resulted in a number of advantages to Tunisia. One important one is favorable terms for its agricultural exports. Privatization of some 140 state-owned industries, a liberal investment code and tax reform have made possible a GDP growth rate averaging 4.5 to 5 percent annually.

The country is predominantly urban. There is almost no nomadic population, and there are no high mountains to provide refuge for independent mountain peoples opposed to central government. The Tunis region and the Sahel, a coastal plain important in olive production, are the most densely populated areas. Tunis, the capital, is not only the dominant city but also the hub of government, economic, and political activity.

HISTORY

Tunisia has an ancient history that is urban rather than territorial. Phoenician merchants from what is today Lebanon founded a number of trading posts several thousand years ago. The most important one was Carthage, founded in 814 B.C. It grew wealthy through trade and developed a maritime empire. Its great rival was Rome; after several wars, the Romans defeated the Carthaginians and destroyed Carthage. Later, the Romans rebuilt the city, and it became great once again as the capital of the Roman province of Africa. Rome's African province was one of the most prosperous in the empire. The wheat and other commodities shipped to Rome from North African farms were vitally needed to feed the Roman population. When the ships from Carthage were late due to storms, lost at sea, or seized by pirates, the Romans suffered hardship. Modern Tunisia has yet to reach the level of prosperity it had under Roman rule.

The collapse of the Roman Empire in the fifth century A.D. affected Roman Africa as well. Cities were abandoned; the irrigation system that had made the farms prosperous fell into ruin. A number of these Roman cities, such as Dougga, Utica, and Carthage itself, which is now a suburb of Tunis, have been preserved as historical monuments of this period.

Arab armies from the east brought Islam to North Africa in the late seventh century. After some resistance, the population accepted the new religion, and from that time on the area was ruled as the Arab–Islamic province of *Ifriqiya*. The Anglicized form of this Arabic word, "Africa," was eventually applied to the entire continent.

The Arab governors did not want to have anything to do with Carthage, since they associated it with Christian Roman rule. They built a new capital on the site of a village on the outskirts of Carthage, named Tunis. The fact that Tunis has been the capital and major city in the country for 14 centuries has contributed to the sense of unity and nationhood among most Tunisians.[1]

The original Tunisian population consisted of Berbers, a people of unknown origin. During the centuries of Islamic rule, many Arabs settled in the country. Other waves of immigration brought Muslims from Spain, Greeks, Italians, Maltese, and many other nationalities. Until recently, Tunisia also had a large community of Jews, most of whom emigrated to the State of Israel when it was founded in 1948. The blending of ethnic groups and nationalities over the years has created a relatively homogeneous and tolerant society, with few of the conflicts that marked other societies in the Islamic world.

FREEDOM

The campaign against Islamic fundamentalists have significantly reduced the civil rights normally observed in Tunisia. Foreign travel is routinely restricted, passports randomly confiscated and journalists arrested for articles that are said to "defame" the state. Tunisia was expelled from the World Press Organization in 1999 for its restrictions on press freedom.

From the late 1500s to the 1880s, Tunisia was a self-governing province of the Ottoman Empire. It was called a regency because its governors ruled as "regents" on behalf of the Ottoman sultan.

In 1881, a French army invaded and occupied all of Tunisia, almost without firing a shot. The French said that they had intervened because the bey's government could not meet its debts to French bankers and capitalists, who had been lending money for years to keep the country afloat. There was concern also about the European population. Europeans from many countries had been pouring into Tunisia, ever since the bey had given foreigners the right to own land and set up businesses.

HABIB BOURGUIBA

Habib Ben Ali Bourguiba, born in 1903, once said he had "invented" Tunisia, not historically but in the sense of shaping its existence as a modern sovereign nation. The Neo-Destour Party, under Bourguiba's leadership, became the country's first mass political party. It drew its membership from shopkeepers, craftspeople, blue-collar workers, and peasants, along with French-educated lawyers and doctors. The party became the vanguard of the nation, mobilizing the population in a campaign of strikes, demonstrations, and violence in order to gain independence. It was a long struggle. Bourguiba spent many years in prison. But eventually the Neo-Destour tactics succeeded. On March 20, 1956, France ended its protectorate and Tunisia became an independent republic, led by Habib Bourguiba.

One of the problems facing Tunisia today is that its political organization has changed very little since independence. A Constitution was approved in 1959 that established a "presidential republic"—that is, a republic in which the elected president has great power. Bourguiba was elected president in 1957.

Bourguiba was also the head of the Neo-Destour Party, the country's only legal political party. The Constitution provided for a National Assembly, which is responsible for enacting laws. But to be elected to the Assembly, a candidate had to be a member of the Neo-Destour Party. Bourguiba's philosophy and programs for national development in his country were often called Bourguibism. It was tailored to the particular historical experience of the Tunisian people. Since ancient Carthage, Tunisian life has been characterized by the presence of a strong central government able to impose order and bring relative stability to the people. The predominance of cities and villages over nomadism reinforced this sense of order. The experience of Carthage, and even more so that of Rome, set the pattern.

In 1961, Bourguiba introduced a new program for Tunisian development that he termed "Destourian Socialism." It combined Bourguibism with government planning for economic and social development. The name of the Neo-Destour Party was changed to the Destour Socialist Party (PSD) to indicate its new direction. Destourian Socialism worked for the general good, but it was not Marxist; Bourguiba stressed national unanimity rather than class struggle and opposed communism as the "ideology of a godless state." Bourguiba took the view that Destourian Socialism was directly related to Islam. He said once that the original members of the Islamic community (in Muhammad's time in Mecca) "were socialists ... and worked for the common good."[2] For many years after independence, Tunisia appeared to be a model among new nations because of its stability, order, and economic progress. Particularly notable were Bourguiba's reforms in social and political life. Islamic law was replaced by a Western-style legal system, with various levels of courts. Women were encouraged to attend school and enter occupations previously closed to them, and they were given equal rights with men in matters of divorce and inheritance.

Bourguiba strongly criticized those aspects of Islam that seemed to him to be obstacles to national development. He was against women wearing the veil; polygyny; and ownership of lands by religious leaders, which kept land out of production. He even encouraged people not to fast during the holy month of Ramadan, because their hunger made them less effective in their work.

The system was provided with a certain continuity by the election of Bourguiba as president-for-life in 1974, when a constitutional amendment was approved specifying that at the time of his death or in the event of his disability, the prime minister would succeed him and hold office pending a general election. One author observed: "Nobody is big enough to replace Bourguiba. He created a national liberation movement, fashioned the country and its institutions."[3] Yet he failed to recognize or deal with changing political and social realities in his later years.

The new generation coming of age in Tunisia is deeply alienated from the old. Young Tunisians (half the population are under age 15) increasingly protest their inability to find jobs, their exclusion from the political decision-making process, the unfair distribution of wealth, and the lack of political organizations. It seems as if there are two Tunisias: the old Tunisia of genteel politicians and freedom fighters; and the new one of alienated youths, angry peasants, and frustrated intellectuals. Somehow the two have gotten out of touch with each other.

The division between these groups has been magnified by the growth of Islamic fundamentalism, which in Bourguiba's view was equated with rejection of the secular, modern Islamic society that he created. The Islamic Tendency Movement (MTI) emerged in the 1980s as the major fundamentalist group. MTI applied for recognition as a political party after Bourguiba had agreed to allow political activity outside of the Destour Party and had licensed two opposition parties. But MTI's application was rejected.

THE END OF AN ERA

In 1984, riots over an increase in the price of bread in 1984 signaled a turning point for the regime. For the first time in the republic's history, an organized Islamic opposition challenged Bourguiba, on the grounds that he had deformed Islam to create a secular society.

However, Bourguiba turned a deaf ear to all proposals for political change. Having survived several heart attacks and other illnesses to regain reasonably good health, he seemed to feel that he was indestructible. His personal life underwent significant change as he became more authoritarian. He had divorced his French wife, apparently in response to criticism that a true Tunisian patriot would not have a French spouse. His second wife, Wassila, a member of the prominent Ben Ammar family, soon became the power behind the throne. As Bourguiba's mental state deteriorated, he divorced her arbitrarily in 1986. At that time the president-for-life assemed direct control over party and government. As he did so, his actions became increasingly irrational. He would appoint a cabinet minister one day and forget the next that he had done so. Opposition became an obsession with him. The two legal opposition parties were forced out of local and national elections by arrests of leaders and a shutdown of opposition newspapers. The Tunisian Labor Confederation (UGTT) was disbanded, and the government launched a massive purge of fundamentalists.

HEALTH/WELFARE

Tunisia has overhauled its school and university curricula to emphasize respect for other monotheistic religions. They require courses on the Universal Declaration of Human Rights, democracy, and the value of the individual. The new curricula are at variance with government repressive policies, but they do stress Islamic ideals of tolerance for the school-age population.

A decision that would prove crucial to the needed change in leadership was made by Bourguiba in September 1987, when he named Ben Ali as prime minister. Six weeks later, Ben Ali carried out a bloodless coup, removing the aging president under the 1974 constitutional provision that allows the prime minister to take over in the event of a president's "manifest incapacity" to govern. A council of medical doctors affirmed that this was the case. Bourguiba was placed under temporary house arrest in his Monastir villa, but he was allowed visitors and some freedom of movement within the city (after 1990).

Habib Bourguiba died in April 2001, at the age of 96. He was buried next door to this villa, in a mausoleum of white marble. The words inscribed on its door—"Liberator of women, builder of modern Tunisia"—seem an appropriate inscription for the "inventor" of his country.

NEW DIRECTIONS

President Ben Ali (elected to a full five-year term in April 1989) initiated a series of bold reforms designed to wean the country away from the one-party system. Political prisoners were released under a general amnesty. Prodded by Ben Ali, the Destour-dominated National Assembly passed laws ensuring press freedom and the right of political parties to form as long as their platforms are not based exclusively on language, race, or religion. The Assembly also abolished the constitutional provision establishing the position of president-for-life, which had been created expressly for Bourguiba. Henceforth Tunisian presidents would be limited to three consecutive terms in office.

Elections in 1988 underscored Tunisia's fixation on the single-party system. RCD candidates won all 141 seats in the Chamber of Deputies, taking 80 percent of the popular vote. Two new opposition parties, the Progressive Socialist Party and the Progressive Socialist Rally, participated but failed to win more than 5 percent of the popular vote, the minimum needed for representation in the Chamber. MTI candidates, although required to run as independents because of the ban on "Islamic" parties under the revised election law, dominated urban voting, taking 30 percent of the popular vote in the cities. However, the winner-take-all system of electing candidates shut them out as well.

Local and municipal elections have confirmed the RCD stranglehold on Tunisian political life; its performance was the exact opposite of that of the National Liberation Front in neighboring Algeria, where the dominant party was discredited over time and finally defeated in open national elections by a fundamentalist party. In the 1995 local and municipal council elections, RCD candidates won 4,084 out of 4,090 contested seats, with 92.5 percent of Tunisia's 1,865,401 registered voters casting their ballots.

ACHIEVEMENTS

With domestic electric demand rising at the reate of 77 percent annually, the country has moved rapidly to add new power plants. The new Rades plant, powered by a combination of diesel fuel and natural gas, went into operation in mid 2003; it increases national output from 2000 megawatts (mw) to 2,480. Some 94 percent of Tunisian homes now have electric power.

After the election the Chamber of Deputies was enlarged from the present 144 to 160 deputies. Twenty seats would be reserved for members from opposition parties. In the presidential election, Ben Ali was reelected for a third term and again in 1999 for a fourth term. In the latter election he faced modest opposition, and as a result his victory margin was a "bare" 99.44 percent.[6]

The Chamber was enlarged again in time for the 1999 elections, this time to 182 seats, to broaden representation for Tunisia's growing population. The results were somewhat different from the previous election. The RCD won 148 seats to 13 for the MDS, the largest opposition party. However, opposition parties all together increased their representation from 19 seats to 34.

THE ECONOMY

The challenge to Ben Ali lies not only in broadening political participation but also in improving the economy. After a period of impressive expansion in the 1960s and 1970s, the growth rate began dropping steadily, largely due to decreased demand and lowered prices for the country's three main exports (phosphates, petroleum, and olive oil). Tunisia is the world's fourth-ranking producer of phosphates, and its most important industries are those related to production of superphosphates and fertilizers.

Problems have dogged the phosphate industry. The quality of the rock mined is poor in comparison with that of other phosphate producers, such as Morocco. The Tunisian industry experienced hard times in the late 1980s with the drop in global phosphate prices; a quarter of its 12,000-member workforce were laid off in 1987. However, improved production methods and higher world demand led to a 29 percent increase in exports in 1990.

Tunisia's oil reserves are estimated at 1.65 billion barrels. The main producing fields are at El Borma and offshore in the Gulf of Gabes. New offshore discoveries and a 1996 agreement with Libya for 50/50 sharing of production from the disputed Gulf of Gabes oil field have improved oil output, currently about 4.3 million barrels annually.

Tunisia became an associate member of the European Union in 1998, the first Mediterranean country to do so. The terms of the EU agreement require the country to remove trade barriers over a 10-year period. In turn, Tunisian products such as citrus and olives receive highly favorable export terms in EU countries. The EU also provides technical support and training for the government's Mise A Nouveau (Upgrading and Improvement) program intended to enhance productivity in business and industry and compete internationally.

Tunisia's political stability—albeit one gained at the expense of human rights—and its economic reforms have made it a favored country for foreign aid over the years. During the period 1970–2000 it received more World Bank loans than any other Arab or African country. Its economic reform program, featuring privatization of 140 state-owned enterprises since 1987, liberalizing of prices, reduction of tariffs and other reforms, is lauded as a model for development by international financial institutions.

(Photo by Wayne Edge)

Even in the age of globalization many North African farmers use donkeys to help them on their land.

The country's political stability and effective use of its limited resources for development have made it a favored country for foreign aid. Since the 1970s it has received more World Bank loans than any other Arab or African country. The funding has been equitably distributed, so that 60 percent of the population are middle class, and 80 percent own their own homes.

THE FUTURE

Tunisia's progress as an economic beacon of stability in an unstable region has been somewhat offset by a decline in its long-established status as a successful example of a secular, progressive Islamic state. Following President Ben Ali's ouster of his predecessor in 1987, he proclaimed a new era for Tunisians, based on respect for law, human rights, and democracy. Tunisia's "Islamic nature" was reaffirmed by such actions as the reopening of the venerable Zitouna University in Tunis, a center for Islamic scholarship, along with its counterpart in Kairouan. But like other Islamic countries, it has not been free from the scourge of militant Islamic fundamentalism. The fundamentalist movement Al-Nahda ("Renaissance"), which advocates a Tunisian government based on Islamic law, attacked RCD headquarters in February 1991. Many of its members, including the leader, were subsequently arrested and it was outlawed as a political party. However, the success of Osama bin Laden's Al-Qaeda movement is attracting young Muslims everywhere, and has emboldened those members still at large. In April 2003 a bomb attack on the ancient Jewish synagogue in Djerba, a popular tourist resort area, killed 17 people. In addition to the damage done to the important tourist industry, the attack provided further evidence of its goal of overthrowing the Ben Ali regime as "impure," one not true to Islamic law and ideals.

In subsequent years, Tunisia has become an increasingly closed society. The press is heavily censored. Bourguiba's death was not even reported in Tunisia, and the obligatory seven days of mourning were countered by instructions to banks and government offices to keep regular hours. Telephones are routinely tapped. More than 1,000 Ennahda members have been arrested and jailed without trial. In December 2000, a dozen members of another Islamic fundamentalist group were given 17-year jail sentences for forming an illegal organization; their lawyers walked out of the trial to protest the court's bias and procedural abuses.

The regime's repression of Islamic groups, even moderate nonviolent ones, has changed its former image as a tolerant, progressive Islamic country. The Tunisian League for Human Rights, oldest in the Arab world, was closed in 1992. Arrest and harassment of intellectuals, journalists, and others for alleged criticism of the regime are routine. Foreign publications are banned, and opposition leaders are pilloried in the state-controlled press as fundamentalists. Almost the only positive step in the human-rights area was a press code amendment in 2001 that precluded use of torture or physical punishment of journalists.

Following his "tainted" election victory in 1999, Ben Ali announced a new program designed to provide full employment by 2004. Called the 21–21 program, it would supplement an earlier 26–26 one that had brought the public and private sectors together to end poverty and increase home ownership, notably among the poor. In his address announcing the new program, Ben Ali stated: "Change comes from anchoring the democratic process in a steady and incremental progress aimed at avoiding setbacks or losing momentum."[7] However, his own lack of charisma (when compared with his illustrious predecessor), and his regime's continued repression of human rights, suggested otherwise. One analyst wondered if Bourguiba's funeral had marked a watershed after all, "galvanizing a generation too frightened to speak."[8]

NOTES

1. Harold D. Nelson, ed., *Tunisia: A Country Study* (Washington, D.C.: American University, Foreign Area Studies, 1979), p. 68.
2. *Ibid.*, p. 42.
3. *Ibid.*, p. 194. What Nelson means, in this case, by "authoritarianism" is that the French brought to Tunisia the elaborate bureaucracy of metropolitan France, with levels of administration from the center down to local towns and villages.
4. *Ibid.*, p. 196.
5. Jim Rupert, in *The Christian Science Monitor* (November 23, 1984).
6. Mamoun Fandy, in *The Christian Science Monitor* (October 25, 1999).
7. Georgie Ann Geyer, in *The Washington Post* (October 23, 1999).
8. Noted in *The Economist* (April 15, 2000).

Timeline: PAST

264–146 B.C.
Wars between Rome and Carthage, ending in the destruction of Carthage and its rebuilding as a Roman city

800–900
The establishment of Islam in Ifriqiya, with its new capital at Tunis A.D.

1200–1400
The Hafsid dynasty develops Tunisia as a highly centralized urban state

1500–1800
Ottoman Turks establish Tunis as a corsair state to control Mediterranean sea lanes

1881–1956
French protectorate

1956
Tunisia gains independence, led by Habib Bourguiba

1974
An abortive merger with Libya

1980s
Bourguiba is removed from office in a "palace coup;" he is succeeded by Ben Ali

1990s
Tunisia's economic picture brightens; Ben Ali seeks some social modernization; women's rights are expanded

PRESENT

2000s
Human-rights abuses continue

The Western Sahara: Whose Desert?

It is a fearsome place, swept by sand-laden winds that sting through layers of clothing, scorched by 120°F temperatures, its flat, monotonous landscape broken occasionally by dried-up *wadis* (river beds). The Spanish called it Rio de Oro, "River of Gold," in a bitter jest, for it has neither. Rainfall averages two to eight inches a year in a region twice the size of Colorado but without mountains, only rolling dunes swept constantly by sand-laden winds that fill tents, clothing and food with gritty particles of sand. The region had no name until the twentieth century; it was simply the "western" part of the Sahara. For a brief period in the eleventh century A.D. a Berber tribal confederation, the Almoravids, rode out of the territory to find a powerful Islamic sultanate in Morocco and Spain. But after they were overthrown by another tribal confederation, the Almohads, the western Sahara reverted to political obscurity.

As a political entity, the Western Sahara resulted from European colonization in Africa in the late nineteenth century. Britain and France had a head start in establishing colonies. Spain was a latecomer. By the time the Spanish joined the race for colonies, little was left for them in Africa. Since they already controlled the Canary Islands, off the West African coast, it was natural for them to claim Rio de Oro, the nearest area on the coast.

In 1884, Spain claimed Rio de Oro and its adjoining region, Saguia al-Hamra, in a note to other European powers. The Spanish claim was based on the principle that "occupation of a territory's coast entitled a colonial power to control over the interior."[1] But Spanish rights to the Saharan interior clashed with French claims to Mauritania and French efforts to control the independent Sultanate of Morocco to the north. After the establishment of a joint Franco–Spanish protectorate over Morocco in 1912, Rio de Oro and Saguia al-Hamra were recognized as a single Spanish colony, with its boundaries fixed with the French colony of Mauritania on the south and east and Morocco to the north. The nomads of the Western Sahara now found themselves living within fixed boundaries defined by outsiders.

The Spanish moved very slowly into the interior. The entire Western Sahara was not "pacified" until 1934. Spain invested heavily in development of the important Western Sahara phosphate deposits but did little else to develop the colony. The Spanish population was essentially a garrison community, living apart from the Sahrawis, the indigenous Saharan population, in towns or military posts. A few Sahrawis went to Spain or other European countries, where they received a modern education; upon their return, they began to organize a Saharan nationalist movement. Other Sahrawis traveled to Egypt and returned with ideas of organizing a Saharan Arab independent state. But a real sense of either a Spanish Saharan or an independent Sahrawi identity was slow to emerge.[2]

Serious conflict over the Spanish Sahara developed in the 1960s. By that time, both Morocco and Mauritania had become independent. Algeria, the third African territory involved in the conflict, won its independence after a bloody civil war. All three new states were highly nationalistic and were opposed to the continuation of colonial rule over any African people, but particularly Muslim peoples. They encouraged the Sahrawis to fight for liberation from Spain, giving arms and money to guerrilla groups and keeping their borders open.

However, the three states had different motives. Morocco claimed the Western Sahara on the basis of historical ties dating back to the Almoravids, plus the oath of allegiance sworn to Moroccan sultans by Saharan chiefs in the nineteenth and twentieth centuries. Kinship was also a factor; several important Saharan families have branches in Morocco, and both the mother and the first wife of the founder of Morocco's current ruling dynasty, Mulay Ismail, were from Sahrawi families.

The Mauritanian claim to the Spanish Sahara was based not on historical sovereignty but on kinship. Sahrawis have close ethnic ties with the Moors, the majority of the population of Mauritania. Also, Mauritania feared Moroccan expansion, since its territory had once been included in the Almoravid state. A Saharan buffer state between Mauritania and Morocco would serve as protection for the Mauritanians.

Algeria's interest in Spanish Sahara was largely a matter of support for a national liberation movement against a colonial power. The Algerians made no territorial claim to the colony. But Algerian foreign policy has rested on two pillars since independence: the right to self-determination of subject peoples and the principle of self-determination through referendum. Algeria consistently maintains that the Saharan people should have these rights.

In the 1960s, Spain came under pressure from the United Nations to give up its colonies. After much hesitation, the Spanish announced in August 1974 that a referendum would be held under UN supervision to decide the colony's future.

The Spanish action brought the conflict to a head. King Hassan declared that 1975 would mark the restoration of Moroccan sovereignty over the territory. The main opposition to this claim came from Polisario (an acronym for the Popular Front for the Liberation of Saguia al-Hamra and Rio de Oro, the two divisions of the Spanish colony). This organization, formed by Saharan exiles based initially in Mauritania, issued a declaration of independence, and Polisario guerrillas began attacking Spanish garrisons, increasing the pressure on Spain to withdraw. In October 1975, Hassan announced that he would lead a massive, peaceful march of civilians, "armed" only with Korans, into the Spanish Sahara to recover sacred Moroccan territory. This "Green March" of half a million unarmed Moroccan volunteers into Spanish territory seemed an unusual, even risky, method of validating a territorial claim, but it worked. In 1976, Spain reached agreement with Morocco and Mauritania to partition the territory into two zones, one-third going to Mauritania and two-thirds to Morocco. The Moroccan Zone included the important phosphate deposits.

The Polisario rejected the partition agreement. It announced formation of the Sahrawi Arab Democratic Republic (S.A.D.R.), "a free, independent, sovereign state ruled by an Arab democratic system of progressive unionist orientation and of Islamic religion."[3]

Polisario tactics of swift-striking attacks from hidden bases in the vast desert were highly effective in the early stages. Mauritania withdrew from the war in 1978, when a military coup overthrew its government. The new Mauritanian rulers signed a peace treaty in Algiers with Polisario representatives. Morocco, not to be outdone, promptly annexed the Mauritanian share of the territory and beefed up its military forces. A fortified "Sand Wall," which was built in stages from the former border with Rio de Oro down to the Moroccan–Mauritanian border and in 1987 extended about 350 miles to the Atlantic Ocean, provided the Moroccan Army with a strong defensive base from which to launch punitive raids against its elusive foe. The new segment also cut off the Polisario's access to the sea; Polisario raiders had begun to intercept and board fishing vessels in attempts to disrupt development of that important Moroccan resource and to bring pressure on foreign countries (such as Spain) that use the fishing grounds to push Morocco toward a settlement.

Although a large number of member states of the Organization of African Unity (OAU) subsequently recognized the Sahrawi Republic, Morocco blocked its admission to the OAU, on the grounds that it was part of Moroccan territory. However, the drain on Moroccan resources of indefinitely maintaining a 100,000-man army in the desert led King Hassan II to soften his obduracy, particularly in relation to Algeria. With both countries affected by severe economic problems and some political instability, a rapprochement became possible in the late 1980s. Diplomatic relations were restored in 1988; and in 1989, Morocco joined Algeria, Libya, Tunisia, and Mauritania in the Arab Maghrib Union (AMU). The AMU charter binds member states not to support resistance movements in one another's territory. As a result, Algeria withdrew its backing for S.A.D.R. and closed Polisario offices in Algiers.

Algeria's preoccupation with internal affairs and the withdrawal of Algerian and Libyan financial aid placed the Polisario in a difficult position. Two of its founders, Omar Hadrami and Noureddine Belali, defected in 1989 and acknowledged Moroccan sovereignty over the territory. A 1990 amnesty offer by King Hassan for all Polisario members and Saharan exiles was accepted by nearly 1,000 persons; these included S.A.D.R.'s foreign minister, Brahim Hakim.

In 1983, Polisario leaders reached agreement in principle with the king to settle the dispute by referendum. Participants in the referendum would be limited to the original inhabitants of the territory. But with the Moroccan Army entrenched behind its Sand Wall and the Polisario in control of the open desert, there was little incentive on either side toward implementation of the referendum.

UN mediation produced a formal cease-fire in 1991. A UN observer force, the Mission for the Referendum in the Western Sahara (MINURSO), proceeded to the territory to supervise voter registration. By that time, thousands of Moroccan settlers had moved there to take advantage of free land, housing, and other inducements offered by the government to help "Moroccanize" the country's newest province. The new residents changed the population balance, thereby complicating registration procedures. Morocco insisted that they should be eligible to vote in the referendum. A further complication arose from the fact that some 140,000 of the original inhabitants included in the 1974 Spanish census were now refugees in Algeria.

In May 1995, the UN Security Council, prodded by Algeria and Sahrawi activists, approved *Resolution 995*, which called for prompt registration of voters in the territory under the supervision of MINURSO. By December 1998, 147,000 voters had been registered. However, the Moroccan government insisted that 85,000 others, belonging to three Sahran tribes resident there in the past, should be included in the registration rolls. Former U.S. secretary of state James Baker was appointed as a "high-profile" envoy to mediate between Morocco and the Polisario to resolve the registration deadlock and promote a final settlement for the territory.

After several unproductive efforts at mediation, Baker submitted a proposal to Moroccan and Polisario representatives to break the deadlock. The "Baker Plan" would postpone the referendum until 2006. In the interim, the resident Sahrawi population would elect an autonomous governing body with powers limited to local and provincial affairs. Given the Western Sahara's economic importance to Morocco, it seemed unlikely that the proposed referendum would ever take place. Not only has the territory acquired a Moroccan majority in its population; its important phosphate rock deposits should give the country's economy a major boost when they are developed. The possibility remains of offshore oil discoveries, and its vast empty spaces could easily absorb settlers from Moroccos's overcrowded cities. The UN has spent more than $600 million on peacekeeping efforts in Western Sahara as it has attempted to resolve the issue over the last 13 years.

On their side the Sahrawi refugees crowded into four refuggee camps near Tindouf, Algeria, have developed, surprisingly, a representative democratic system with an elected parliament, a 95 percent literacy rate and a constitution guaranteeing gender equality and respect for all religions. Elected local councils undergird the parliament, and there is a high degree of volunteerism to take care of needed public services such as trash collection and food rations distribution. "We may well have developed a blueprint for an independent Western Sahara," says a tribal leader. "But we have been landless for so long, I don't know if the UN is just waiting for us to disappear, or what?"[4] Despite its recognition by 75 countries, global collective memory seems thus far to have failed to hold the SADR in its sight.

NOTES

1. John Damis, Conflict in Northwest Africa: The Western Sahara Dispute (Palo Alto, CA: Hoover Institution Press, 1983), p. 110.
2. *Ibid*, p. 13, notes that a tribal assembly (Jama'a) was formed in 1967 for the Sahrawis but that its 43 members were all tribal chiefs or their representatives; it had only advisory powers.
3. John Thorne, "Sahara refugees from a progressive society." *Christian Science Monitor*, March 26, 2004.

Southern Africa

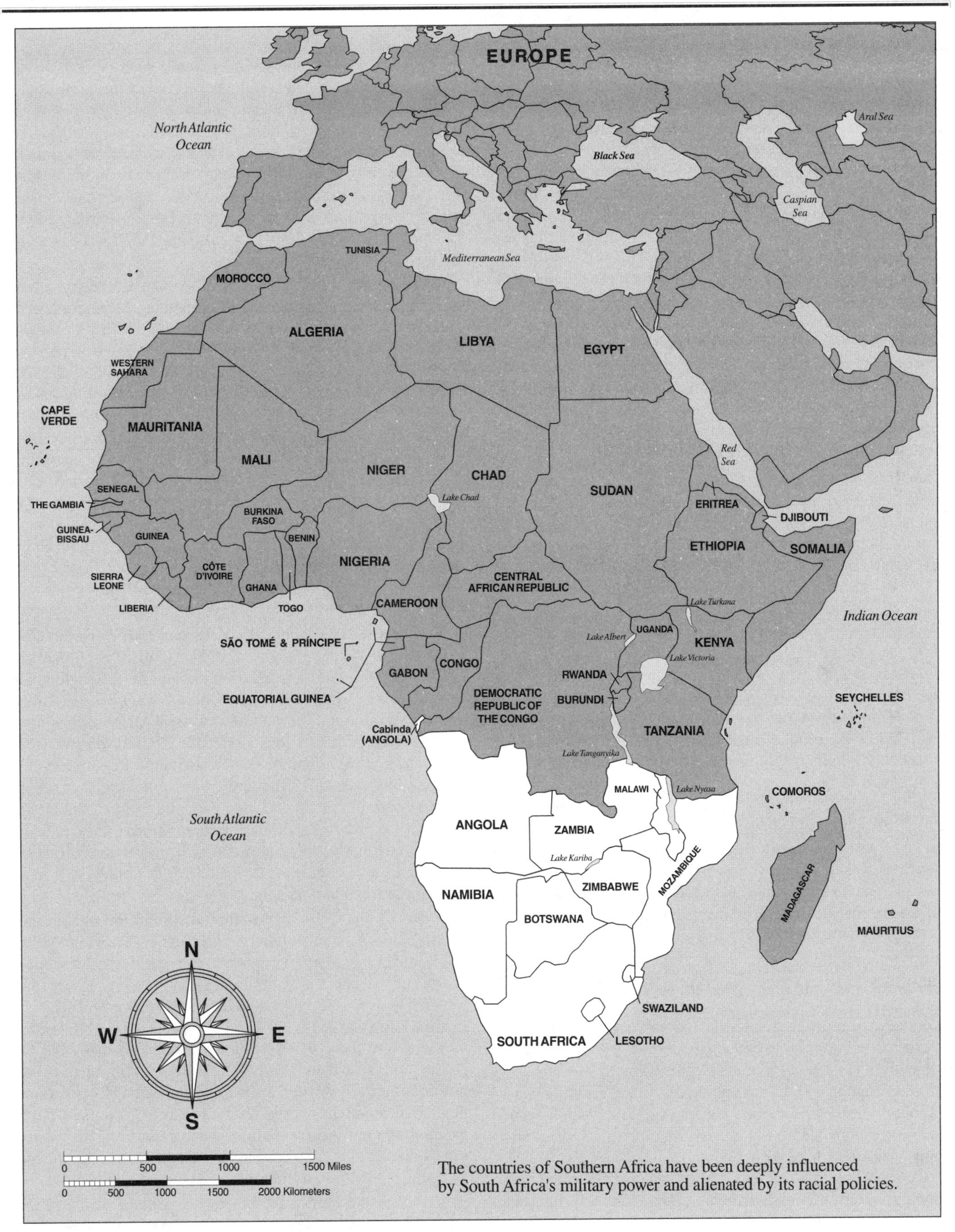

The countries of Southern Africa have been deeply influenced by South Africa's military power and alienated by its racial policies.

Southern Africa: The Continuing Struggle for Self-Determination

Southern Africa—which includes the nations of Angola, Botswana, Lesotho, Malawi, Mozambique, Namibia, South Africa, Swaziland, Zambia, and Zimbabwe—is a diverse region made up of savannas and forest, snow-topped mountains and desert, temperate Mediterranean and torrid tropical climates. Until recently much of the region was marred by armed conflict. But the wars and civil conflicts that have long plagued Angola, Mozambique, Lesotho, Namibia, and South Africa in particular have for now ended. In the case of Angola, this has been facilitated by developments in the Democratic Republic of the Congo (D.R.C., the former Zaire), which for decades had served as a source of, often Western-backed, regional destabilization. The more recent involvement of the governments of Namibia, Angola, and Zimbabwe in the D.R.C.'s internal conflicts appears to be coming to a close with a peace agreement initiated by South Africa. As a result, military complexes can be scaled down in favor of diverting more resources to the challenges of economic development while addressing the very serious challenge of the HIV/AIDS pandemic.

Southern African identity is as much defined by the region's peoples and their past and present interactions as by the region's geographic features. An appreciation of local history is crucial to understanding the forces that both divide and unite the region today. Throughout the twentieth century the dominant theme in the history of Southern Africa was the evolving struggle of the region's indigenous black African majority to free itself of the racial hegemony of white settlers from Europe and their descendants. With the end of the liberation wars, however, the major themes of the twenty-first century are peace and development. It is the pace and style of this development which is the key feature of Southern Africa today. War is the enemy of development. One of the primary themes in Southern Africa today is land, and the domination of the land by various groups of Europeans. Much of the history of Southern Africa has been shaped by the loss of land by indigenous communities to European settlers. The resulting imbalance in land distribution has not been resolved by the establishment of black-majority rule in the former settler-dominated states of Zimbabwe, Namibia, and South Africa. The resulting land shortage for the majority of rural dwellers in those states remains a major constraint to socioeconomic development.

EUROPEAN MIGRATION AND DOMINANCE

A dominant theme in the modern history of Southern Africa has been the evolving struggle of the region's indigenous black African majority to free itself of the racial hegemony of white settlers from Europe and their descendants. By the eighth century A.D., but probably earlier, the southernmost regions of the continent were populated by a variety of black African ethnic groups who spoke languages belonging to the Bantu as well as the Khoisan linguistic classifications. Members of these two groupings practiced both agriculture and pastoralism; archaeological evidence indicates that livestock keeping in Southern Africa pre-dates the time of Christ. Some, such as the BaKongo of northern Angola and the Shona peoples of the Zimbabwean plateaux, had, by the fifteenth century, organized strong states; others, like most Nguni-speakers prior to the early 1800s, lived in smaller communities. Trade networks existed throughout the region, linking various local peoples not only to one another but also to the markets of the Indian Ocean and beyond. For example, porcelains from China have been unearthed in the grounds of the Great Zimbabwe, a stone-walled settlement that flourished in the fifteenth century.

In the 1500s, small numbers of Portuguese began settling along the coasts of Angola and Mozambique. A century later, in 1652, the Dutch established a settlement at Africa's southernmost tip, the Cape of Good Hope. While the Portuguese flag generally remained confined to coastal enclaves until the late 1800s, the Dutch colony expanded steadily into the interior throughout the 1700s, seizing the land of local Khoisan communities. Unlike the colonial footholds of the Portuguese and other Europeans on the continent, which prior to the 1800s were mostly centers for the export of slaves, the Dutch Cape Colony imported slaves from Asia as well as from elsewhere in Africa. Although not legally enslaved, conquered Khoisan were also reduced to servitude. In the process, a new society evolved at the Cape. Much like the American South before the U.S. Civil War, the Cape Colony was racially divided between free white settlers and subordinated peoples of mixed African and Afro-Asian descent.

During the Napoleonic Wars, Britain took over the Cape Colony. Shortly thereafter, in 1820, significant numbers of English-speaking colonists began arriving in the region. The arrival of the British coincided with a period of political realignment throughout much of Southern Africa that is commonly referred to as the "Mfecane." Until recently, the historical literature has generally attributed this upheaval to dislocations caused by the rise of the Zulu state, under the great warrior prince Shaka. However, more recent scholarship on the Mfecane has focused on the disruptive effects of increased traffic in contraband slaves from the interior to the Cape and the Portuguese stations of Mozambique, following the international ban on slave trading.

In the 1830s, the British abolished slavery throughout their empire and extended limited civil rights to nonwhites at the Cape. In response, a large number of white Dutch-descended Boers, or Afrikaners, moved into the interior, where they founded two republics that were free of British control. This migration, known as the Great Trek, did not lead the white settlers into an empty land. The territory was home to many African groups, who lost their farms and pastures to the superior firepower of the early Afrikaners, who often coerced local communities into supplying corvee labor for their farms and public works. But a few African polities, like Lesotho and the western Botswana kingdoms, were able to preserve their independence by acquiring their own firearms.

In the second half of the nineteenth century, white migration and dominance spread throughout the rest of Southern

(United Nations photo 131283 by J. P. Laffont)

Angolan youths celebrated when the nation became independent in 1974.

Africa. The discovery of diamonds and gold in northeastern South Africa encouraged white exploration and subsequent occupation farther north. In the 1890s, Cecil Rhodes's British South Africa Company occupied modern Zambia and Zimbabwe, which then became known as the Rhodesias. British traders, missionaries, and settlers also invaded the area now known as Malawi. Meanwhile, the Germans seized Namibia, while the Portuguese began to expand inland from their coastal enclaves. Thus, by 1900, the entire region had fallen under white colonial control.

With the exception of Lesotho and Botswana, which were occupied as British "protectorates," all of the European colonies in Southern Africa had significant populations of white settlers, who in each case played a predominant political and economic role in their respective territories. Throughout the region, this white supremacy was fostered and maintained through racially discriminatory policies of land alienation, labor regulation, and the denial of full civil rights to nonwhites. In South Africa, where the largest and longest-settled white population resided, the Afrikaners and English-speaking settlers were granted full self-government in 1910—with a Constitution that left the country's black majority virtually powerless.

Yet despite the many changes that have occurred throughout African history, there is one thing that remains constant. In Southern Africa's Kgalahari Desert (shared by South Africa, Namibia, and Botswana), a group of Khoisan individuals, known to the outside world as Bushmen, continue to practice their ancient way of like as hunters and gatherers. Geneticists say the Bushman are the oldest surviving human beings. Within their genetic structure they carry the original genes of the human race. To add majesty to their heritage, contemporary linguist claim that the click language spoken by the Bushmen is also the oldest known language in the world. Hunters and gatherers in the ancient world they survive today with a variety of skills and have intermingled with the more recent arrivals to form a rich blend of diverse cultures.

BLACK NATIONALISM AND SOUTH AFRICAN DESTABILIZATION

After World War II, new movements advocating black self-determination gained ascendancy throughout the region. However, the progress of these struggles for majority rule and independence was gradual. By 1968, the countries of Botswana, Lesotho, Malawi, Swaziland, and Zambia had gained their independence. The area was then polarized between liberated and nonliberated nations. In 1974, a military uprising in Portugal brought statehood to Angola and Mozambique, after long armed struggles by liberation forces in the two territories. Wars of liberation also led to the overthrow of white-settler rule in Zimbabwe, in 1980, and the independence of Namibia, in 1990. Finally, in 1994, South Africa completed a negotiated transition to a nonracial government.

South Africa's liberation has far-reaching implications for the entire Southern African subcontinent as well as the country's own historically oppressed masses. Since the late nineteenth century, South Africa has been the region's economic hub. Today, it accounts for about 80 percent of the total Southern African gross domestic product. Most of the subcontinent's roads and rails also run through South Africa. For generations, the country has recruited expatriate as well as local black African workers for its industries and mines. Today, it is the most economically developed country on the continent, with manufactured goods and agricultural surpluses that are in high demand elsewhere. By the late 1980s, when the imposition of economic sanctions against the then-apartheid regime was at its height, some 46 African countries were importing South African products. With sanctions now lifted, South Africa's economic role on the continent is likely to increase substantially.

A significant milestone was South Africa's admittance in 1994 as the 11th member of the Southern African Development Community (SADC). This organization's ultimate goal is to emulate the European Union (formerly called the European Community or Common Market) by promoting economic integration and political coordination among Southern Africa's states (in-

(United Nations photo 128704 by Jerry Frank)
South Africa's economic and military dominance overshadows the region's planning. Pictured above is Cape Town, South Africa's chief port and the country's legislative capital.

cluding Tanzania). While South Africa is expected to be at the center of the Community, SADC's roots lie in past efforts by its other members to reduce their ties to that country. The organization grew out of the Southern African Development Coordination Conference (SADCC), which was created by the region's then–black-ruled states in 1980 to lessen their dependency on white-ruled South Africa. Each SADCC government assumed responsibility for research and planning in a specific developmental area: Angola for energy, Mozambique for transport and communication, Tanzania for industry, and so on.

In its first decade, SADCC succeeded in attracting considerable outside aid for building and rehabilitating its member states' infrastructure. The organization's greatest success was the Beira corridor project, which enabled the Mozambican port to serve once more as a major regional transit point. Other successes included telecommunications independence of South Africa, new regional power grids, and the upgrading of Tanzanian roads to carry Malawian goods to the port of Dar es Salaam. In 1992, with South Africa's liberation on the horizon, the potential for a more ambitious and inclusive SADC grouping became possible.

In 1996, South African president Nelson Mandela replaced Botswana's president, Sir Ketumile Masire, as SADC chairman, while Pretoria became the headquarters of a new SADC "Organ for Politics Defense and Security." The new South Africa's role as security coordinator within SADC was especially ironic: Before 1990, it had been the violent destabilizing policies of South Africa's military that had sabotaged efforts toward building greater regional cooperation. SADCC members, especially those that were further linked as the so-called Frontline States (Angola, Botswana, Mozambique, Tanzania, Zambia, and Zimbabwe), were then hostile to South Africa's racial policies. To varying degrees, they provided havens for those oppressed by these policies. South Africa responded by striking out against its exiled opponents through overt and covert military operations, while encouraging insurgent movements among some of its neighbors, most notably in Angola and Mozambique.

In Angola, South Africa (along with the United States) backed the rebel movement UNITA, while in Mozambique, it assisted RENAMO. Both of these movements resorted to the destruction of the railways and roads in their operational areas, a tactic that greatly increased the dependence on South African communications of the landlocked states of Botswana, Malawi, Zambia, and Zimbabwe. It is estimated that in the 1980s, the overall monetary cost to the Frontline States of South Africa's destabilization campaign was about $60 billion. (The same countries' combined annual gross national product was only about $25 billion in 1989.) The human costs were even greater: Hundreds of thousands of people were killed; at least equal numbers were maimed; and in Mozambique alone, more than 1 million people became refugees.

In the Southern African context, South Africa remains a military superpower. Despite the imposition of a United Nations arms embargo between 1977 and 1994, the country's military establishment was able to secure both the arms and sophisticated technology needed to develop its own military/industrial complex. Now a global arms exporter, South Africa is nearly self-sufficient in basic munitions, with a vast and advanced arsenal of weapons. Whereas in 1978 it imported 75 percent of its weapons, today that figure is less than 5 percent. By the 1980s, the country had also developed a nuclear arsenal, which it now claims to have dismantled. However, the former embargo was not entirely ineffective—while South African industry produced many sophisticated weapons systems, it found it increasingly difficult to maintain its regional superiority in such high-technology fields as fighter aircraft. By 1989, the increasing edge of Angolan pilots and air-defense systems was a significant factor in the former South African regime's decision to disengage from the Angolan Civil War. The economic costs of South African militarization were also steep. In addition to draining some 20 percent of its total budget outlays, the destabilization campaign contributed to increased international economic sanctions, which between 1985 and 1990 cost its own economy at least $20 billion. Today, both South Africa and its neighbors hope to benefit from a "peace dividend." But after a generation of militarization, progress in shifting resources from lethal to peaceful pursuits will be gradual.

Throughout the 1980s, South Africa justified its acts of aggression by claiming that it was engaged in counterinsurgency operations against guerrillas of the African National Congress (ANC) and Pan Africanist Congress (PAC), which were then struggling for the regime's overthrow. In fact, the various Frontline States took a cautious attitude toward the activities of South African political refugees, generally forbidding them from launching armed attacks from within their borders. In

(United Nations photo 120577)

This sign, once displayed in a park in Pretoria, South Africa, reflected the restrictions of apartheid, formerly the South African government's official policy of racial discrimination.

1984, both Angola and Mozambique formally signed agreements of mutual noninterference with South Africa. But within a year, these accords had repeatedly and blatantly been violated by South Africa.

Drought, along with continued warfare, has resulted in recurrent food shortages in much of Southern Africa in recent decades—again especially in Angola and Mozambique. The early 1980s' drought in Southern Africa neither lasted as long as nor was as widely publicized as those of West Africa and the Horn, yet it was as destructive. Although some countries, such as Botswana, Mozambique, and Zimbabwe, as well as areas of South Africa, suffered more from nature than others, the main features of the crisis were the same: water reserves were depleted; cattle and game died; and crop production declined, often by 50 percent or more.

Maize and cereal production suffered everywhere. South Africa and Zimbabwe, which are usually grain exporters, had to import food. The countries of Angola, Botswana, and Lesotho each had more than half a million people who were affected by the shortfalls, while some 2 million were malnourished in Mozambique. But in 1988, the rains returned to the region, raising cereal production by 40 percent. Zimbabwe was able not only to export but also to provide food aid to other African countries. However, South African destabilization contributed to continuing food scarcities in many parts of Angola and Mozambique.

In 1991–1992, the entire region was once more pushed toward catastrophe, with the onset of the worst single drought year in at least a century. Although most of the region experienced improved rainfall in 1993–1994, many areas are still afflicted by food shortages, while the entire region remains vulnerable to famine. Up to 4.5 million people remain at risk of starvation in Mozambique, and another 3 million in Angola, while Malawi has had to struggle to feed hundreds of thousands of Mozambican refugees along with its own population.

A NEW ERA

Recent events have given rise to hopes for a new era of peace and progress in the region. In 1988, Angola, Cuba, and South Africa reached an agreement, with U.S. and Soviet support, that led to South Africa's withdrawal from Namibia and the removal of Cuban troops from Angola, where they had been supporting government forces. In 1990, Namibia gained its independence under the elected leadership of SWAPO—the liberation movement that had fought against local South African occupation for more than a quarter of a century. In Mozambique, two decades of fighting between the FRELIMO government and South African–backed RENAMO rebels ended in a peace process that has resulted in the successful holding of two multiparty elections. In Zambia and Malawi, multiparty democracy was restored, resulting in the electoral defeat of long-serving authoritarian rulers. The most significant development in the region, however, has been South Africa's transformation. There, the 1990 release of prominent political prisoners, particularly Nelson Mandela, the unbanning of the ANC and PAC, and the lifting of internal state-of-emergency restrictions resulted in extended negotiations that led to an end to white-minority rule.

Southern Africa has also experienced some reversals in recent years. After an on-again, off-again start, direct negotiations between the Angolan government and UNITA rebels led in 1991 to a UN–supervised peace process. But this agreement collapsed in 1992, when UNITA rejected multiparty election results. Although external support for UNITA declined, the movement was able to continue to wreak havoc on much of the Angolan countryside through its trafficking of illegal diamonds and other commodities across the uncontrolled border between Angola and the D.R.C. This in turn encouraged Angola to intervene in the D.R.C. on behalf of the government against UNITA–aligned rebels. By thus effectively encircling UNITA, the Angolan government was ultimately able to achieve military victory where mediation had failed.

Having finally come to the end of its epoch of struggle against white-minority rule, Southern Africa as a whole may be on the threshold of sustained growth. Besides their now-

Angola (Republic of Angola)

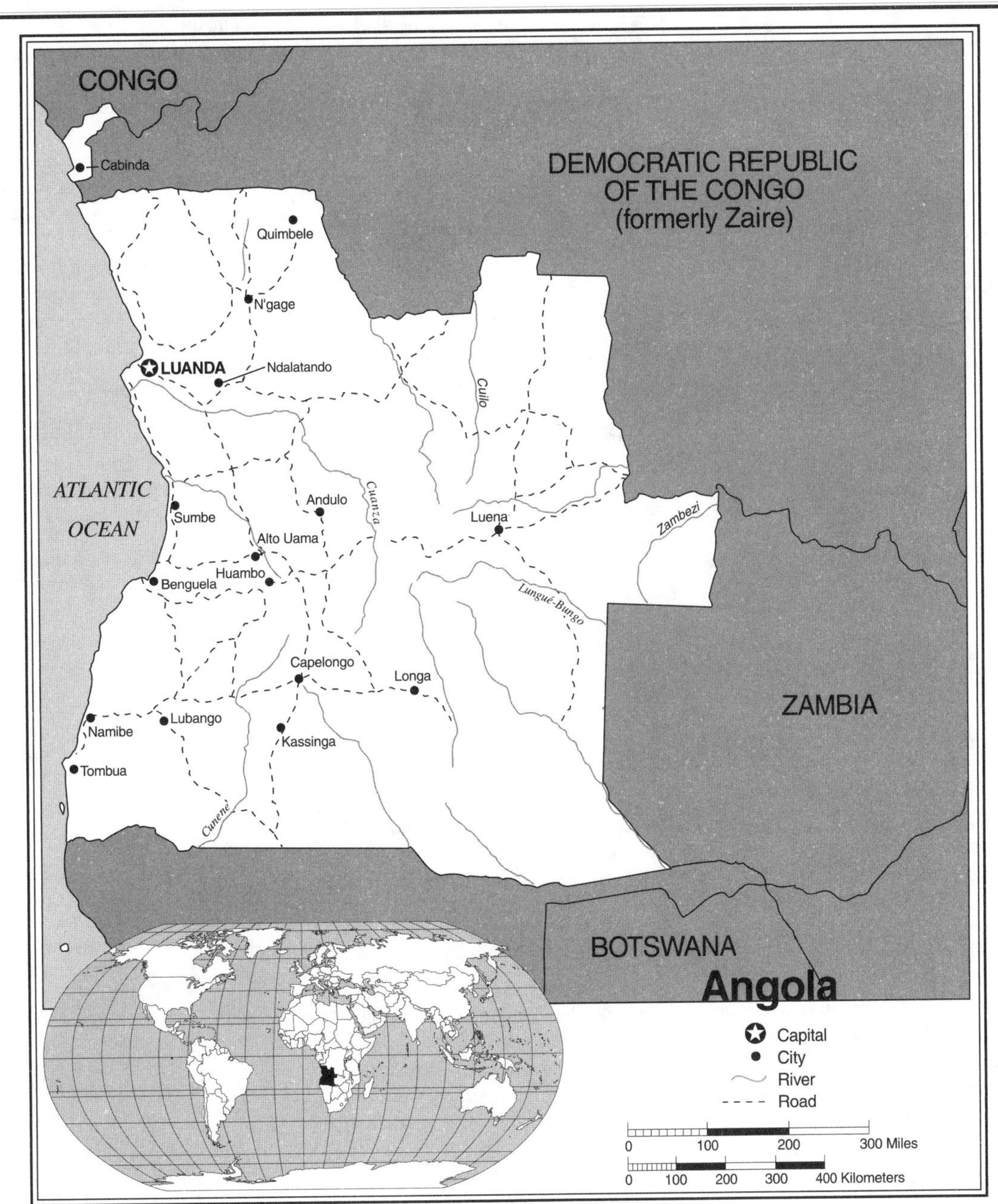

Angola Statistics

GEOGRAPHY

Area in Square Miles (Kilometers): 481,351 (1,246,699) (about twice the size of Texas)

Capital (Population): Luanda (2,819,000)

Environmental Concerns: soil erosion; desertification; deforestation; loss of habitat and biodiversity; water pollution

Geographical Features: a narrow coastal plain rises abruptly to a vast interior plateau

Climate: semiarid in south and along coast to Luanda; the north has a cool, dry season and then a hot, rainy season

PEOPLE

Population

Total: 10,978,552

Annual Growth Rate: 1.93%

Rural/Urban Population Ratio: 66/34

Major Languages: Portuguese; Bantu and other African languages

Ethnic Makeup: 37% Ovimbundu; 25% Kimbundu; 13% Bakongo; 25% others

Religions: 47% indigenous beliefs; 38% Roman Catholic; 15% Protestant

Health

Life Expectancy at Birth: 37 years (male); 40 years (female)

Infant Mortality: 192/1,000 live births

Physicians Available: 1/15,136 people
HIV/AIDS Rate in Adults: 2.78%

Education

Adult Literacy Rate: 42%
Compulsory (Ages): 7–15; free

COMMUNICATION

Telephones: 70,000 main lines
Daily Newspaper Circulation: 11/1,000 people
Televisions: 48/1,000 people
Internet Users: 41,000 (2002)

TRANSPORTATION

Highways in Miles (Kilometers): 47,508 (76,626)
Railroads in Miles (Kilometers): 1,982 (3,189)
Usable Airfields: 244
Motor Vehicles in Use: 223,000

GOVERNMENT

Type: republic
Independence Date: November 11, 1975 (from Portugal)
Head of State/Government: President José Edouardo dos Santos is both head of state and head of government
Political Parties: Popular Movement for the Liberation of Angola; National Front for the Liberation of Angola; others
Suffrage: universal at 18

MILITARY

Military Expenditures (% of GDP): 22%
Current Disputes: apparent end of civil war

ECONOMY

Currency ($ U.S. Equivalent): 74.6 new kwanzas = $1
Per Capita Income/GDP: $1,900/$20.4 billion
GDP Growth Rate: 1.5%
Inflation Rate: 76.6%
Unemployment Rate: unemployment and underemployment affect more than half the population
Labor Force by Occupation: 85% agriculture; 15% industry and services
Natural Resources: petroleum; diamonds; phosphates; iron ore; copper; feldspar; gold; bauxite; uranium
Agriculture: coffee; sisal; cotton; sugarcane; tobacco; vegetables; bananas; plantains; livestock; forest products; fish
Industry: petroleum; minerals; fish processing; brewing; tobacco products; textiles; food processing; construction; others
Exports: $9.6 billion (primary partners United States, European Union, China)
Imports: $4.08 billion (primary partners European Union, South Korea, South Africa)

SUGGESTED WEB SITES

http://www.angola.org
http://www.angolapress-angop.as/index-e.asp
http://www.angolanews.com
http://www.sas.upenn.edu/African_Studies/Country_Specific/Angola.html

Angola Country Report

In April 2002, after the death of Jonas Savimbi, the leader of the rebel Union for Total Independence of Angola (UNITA) forces, and the concurrent decimation of his army, a cease-fire was signed in Luanda between UNITA and the governing Popular Movement for the Liberation of Angola (MPLA). After 27 years, the war in Angola was over. The death of Savimbi in a gunfight with government forces in February 2002 rendered UNITA ineffective.

DEVELOPMENT

Most of Angola's export revenues currently come from oil. There are important diamond and iron mines, but their output has suffered due to the war, which also prevented the exploitation of the country's considerable reserves of other minerals. Angola has enormous agricultural potential, but only about 2% of its arable land is under cultivation.

By early 2002 UNITA had been driven out of the Democratic Republic of the Congo (D.R.C.), its rear base, by the combined forces of Angola, Zimbabwe, Namibia, and the D.R.C. Without its rear base of support, funds derived from diamond sales after the Blood diamonds campaign, and leader Savimbi, UNITA was largely a spent force. In February 2003 the UN mission overseeing the peace process ended, and by June the UNITA had transformed itself into a political party, electing Isaias Samakuva as its new leader.

A peace accord had been signed in 1994, after which the United Nations sent in peacekeepers. But the fighting steadily worsened again, and the MPLA government requested in 1999 that the UN withdraw its peacekeeping forces. The peacekeepers withdrew, and sanctions were instituted against UNITA that froze the European bank accounts for UNITA's leaders and isolated funds used to trade in gems.

Angola, however, remains with large quantities of land mines spread throughout the countryside, and the physical infrastructure has been left in tatters. It will take years to recover from the effects of the civil war, but the country's mineral resources will provide the foundation for development in the future.

In 1996, hopes for a cease-fire had been raised when UNITA agreed in principle to join the MPLA in forming a "Government of National Unity," but a settlement was never implemented. Instead, both sides maintained their military capacities while supporting proxy forces in neighboring states.

FREEDOM

Despite new constitutional guarantees, pessimists note that neither UNITA nor MPLA has demonstrated a strong commitment to democracy and human rights in the past. Within UNITA, Jonas Savimbi's word was law; he was known to have critics within his movement burned as "witches." For a time UNITA had its own Internet address, perhaps the first armed faction to do so.

Since 1975, more than half a million Angolans perished as a result of fighting between the two movements, including many passive victims of land mines. Up to 1 million others fled the country, while another 1 million or so were internally displaced. According to a report by the human-rights organization Africa Watch, tens of thousands of Angolans have lost their limbs "because of the indiscriminate use of landmines by both sides of the conflict." Angola's small and impoverished population could not have perpetuated such carnage were it not for decades of external interference in the nation's internal affairs. The United States, the Soviet Union, South Africa, Cuba, Zaire, and many others helped to create and sustain this tragedy.

(United Nations photo 131277 by J. P. Laffont)

Angola's war for independence from Portugal led to the creation of a one-party state.

After the end of the Cold War and the demise of South Africa's apartheid regime, there was an almost complete cutoff of outside support for the conflict. An agreement in April 1991 between the MPLA and UNITA to participate in United Nations–sponsored elections led to a dramatic decline in violence during 16 months of "phony peace." The successful holding of elections in September 1992 further raised hopes of a new beginning for Angola. While the MPLA appeared to have topped the poll, UNITA and the smaller National Front for the Liberation of Angola (FNLA) secured a considerable vote. But hopes for a new beginning under an all-party government of national unity were quickly dashed by UNITA's rejection of the election result. As a result, the country was plunged into renewed civil war.

THE COLONIAL LEGACY

The roots of Angola's long suffering lie in the area's colonial underdevelopment. The Portuguese first made contact with the peoples of the region in 1483. They initially established peaceful trading contact with the powerful Kongo kingdom and other coastal peoples, some of whom were converted to Catholicism by Jesuit missionaries. But from then to the mid-1800s, the outsiders primarily saw the area as a source of slaves. Angola has been called the "mother of Brazil" because up to 4 million Angolans were carried away from its shores to that land, chained in the holds of slave ships. With the possible exception of Nigeria, no African territory lost more of its people to the trans-Atlantic slave trade.

Following the nineteenth-century suppression of the slave trade, the Portuguese introduced internal systems of exploitation that very often amounted to slavery in all but name. Large numbers of Angolans were pressed into working on coffee plantations owned by a growing community of white settlers. Others were forced to labor in other sectors, such as diamond mines or public-works projects.

HEALTH/WELFARE

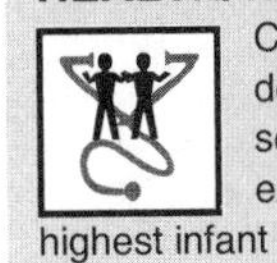

Civil war caused a serious deterioration of Angola's health services, resulting in lower life expectancy and one of the highest infant mortality rates in the world.

Although the Portuguese claimed that they encouraged Angolans to learn Portuguese and practice Catholicism, thus becoming "assimilated" into the world of the colonizers, they actually made little effort to provide education. No more than 2 percent of the population ever achieved the legal status of *assimilado.* The assimilados, many of whom were of mixed race, were concentrated in the coastal towns. Of the few interior Angolans who became literate, a large proportion were the products of Protestant, non-Portuguese, mission schools. Because each mission tended to operate in a particular region and teach from its own syllabus, usually in the local language, an unfortunate by-product of these schools was the reinforcement (the creation, some would argue) of ethnic rivalries among the territory's educated elite.

In the late colonial period, the FNLA, MPLA, and UNITA emerged as the three major liberation movements challenging Portuguese rule. Although all three sought a national following, each built up an ethnoregional core of support by 1975. The FNLA grew out of a movement whose original focus was limited to the northern Kongo-speaking population, while UNITA's principal stronghold was the largely Ovimbundu-speaking south-central plateaux. The MPLA had its strongest following among assimilados and Kimbundu-speakers, who are predominant in Luanda, the capital, and the interior to the west of the city. From the beginning, all three movements cultivated separate sources of external support.

ACHIEVEMENTS

Between 1975 and 1980, the Angolan government claimed that it had tripled the nation's primary-school enrollment, to 76%. That figure subsequently dropped as a result of war.

The armed struggle against the Portuguese began in 1961, with a massive FNLA–inspired uprising in the north and MPLA–led unrest in Luanda. To counter the northern rebellion, the Portuguese resorted to the saturation bombing of villages. In the first year of fighting, this left an estimated 50,000 dead (about half the total number killed throughout the anticolonial struggle). The liberation forces were as much hampered by their own disunity as by the brutality of Portugal's counterinsurgency tactics. Undisciplined rebels associated with the FNLA, for example, were known to massacre not only Portuguese plantation owners but many of their southern workers as well. Such incidents contributed to UNITA's split from the FNLA in 1966. There is also evidence of UNITA forces cooperating with the Portuguese in attacks on the MPLA. Besides competition with its two rivals, the MPLA also encountered some difficulty in keeping its urban and rural factions united.

CIVIL WAR

The overthrow of Portugal's Fascist government in 1974 led to Angola's rapid decolonization. Attempts to create a transitional government of national unity among the three major nationalist movements failed. The MPLA succeeded in seizing Luanda, which led to a loose alliance between the FNLA and UNITA. As fighting between the groups escalated, so did the involvement of their foreign backers. Meanwhile, most of Angola's 300,000 or more white settlers fled the country, triggering the collapse of much of the local economy. With the notable exception of Angola's offshore oil industry, most economic sectors have since failed to recover their preindependence output as a result of the war.

While the chronology of outside intervention in the Angolan conflict is a matter of dispute, it is nonetheless clear that, by October 1975, up to 2,000 South African troops were assisting the FNLA–UNITA forces in the south. In response, Cuba dispatched a force of 18,000 to 20,000 to assist the MPLA, which earlier had gained control of Luanda. These events proved decisive during the war's first phase. On the one hand, collaboration with South Africa led to the withdrawal of Chinese and much of the African support for the FNLA–UNITA cause. It also contributed to the U.S. Congress' passage of the Clarke Amendment, which abruptly terminated the United States' direct involvement. On the other hand, the arrival of the Cubans allowed the MPLA quickly to gain the upper hand on the battlefield. Not wishing to fight a conventional war against the Cubans by themselves, the South Africans withdrew their conventional forces in March 1976.

By 1977, the MPLA's "People's Republic" had established itself in all of Angola's provinces. It was recognized by the United Nations and most of its membership as the nation's sole legitimate government, the United States numbering among the few that continued to withhold recognition. However, the MPLA's apparent victory did not bring an end to the hostilities. Although the remaining pockets of FNLA resistance were overcome following an Angola–Zaire rapprochement in 1978, UNITA maintained its largely guerrilla struggle. Until 1989, UNITA's major supporter was South Africa, whose destabilization of the Luanda government was motivated by its desire to keep the Benguela railway closed (thus diverting traffic to its own system) and harass Angola-based SWAPO forces. Besides supplying UNITA with logistical support, the South Africans repeatedly invaded southern Angola and on occasion infiltrated sabotage units into other areas of the country. South African aggression in turn justified Cuba's maintenance (by 1988) of some 50,000 troops in support of the government. In 1986, the U.S. Congress approved the resumption of "covert" U.S. material assistance to UNITA via Zaire.

An escalation of the fighting in 1987 and 1988 was accompanied by a revival of negotiations for a peace settlement among representatives of the Angolan government, Cuba, South Africa, and the United States. In 1988, South African forces were checked in a battle at the Angolan town of Cuito Cuanavale. South Africa agreed to withdraw from Namibia and end its involvement in the Angolan conflict. It was further agreed that Cuba would complete a phased withdrawal of its forces from the region by mid-1991.

THE FUTURE

With the end of the war, Angola now faces the daunting task of rebuilding its devastated infrastructure and resettling the hundreds of thousands of refugees who fled the fighting. Many Angolans remain dependent on food aid. Although one of Africa's major oil producers, Angola is one of the world's poorest countries, with the average life expectancy among the lowest on the continent. The extreme deprivations of the war have led to problems in Angola's with far-reaching social, political, and economic implications. Angola has denied allegations that oil revenue has been squandered through corruption and mismanagement. However, corruption has long been viewed as a major problem within Angola, with numerous high level military and government officials implicated in illicit activities. In 2003 the International Monetary Fund (IMF) dispatched investigators to track down missing oil money. By April 2004 more than 3,000 people were arrested in a crackdown on illegal diamond mining and trafficking. The government stated that 11,000 people were expelled since December 2003 in a campaign against exploitation of economic resources. The vast majority of these individuals were diamond traders.

Timeline: PAST

1400s
The Kongo state develops

1483
The Portuguese make contact with the Kongo state

1640
Queen Nzinga defends the Mbundu kingdom against the Portuguese

1956
The MPLA is founded in Luanda

1961
The national war of liberation begins

1975
Angola gains independence from Portugal

1976
South African–initiated air and ground incursions into Angola

1979
President Agostinho Neto dies; José dos Santos becomes president

1986
Jonas Savimbi visits the United States; U.S. "material and moral" support for UNITA resumes

1990s
Talks for national reconciliation break down; multiparty elections are held

PRESENT

2000s
Savimbi is killed in a gun battle with government forces

Civil war appears to be over

Botswana (Republic of Botswana)

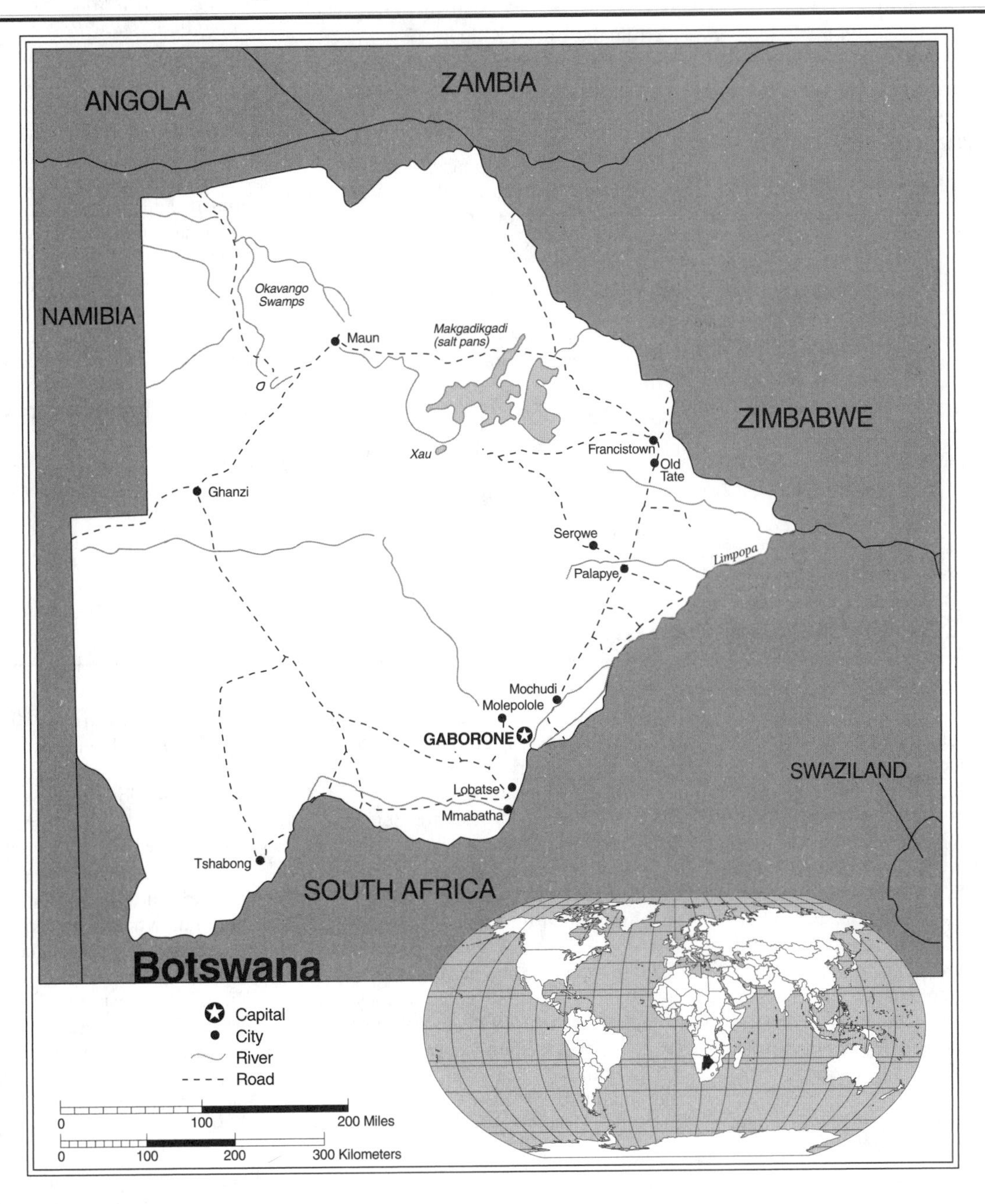

Botswana Statistics

GEOGRAPHY

Area in Square Miles (Kilometers): 231,804 (600,372) (about the size of Texas)
Capital (Population): Gaborone (248,000)
Environmental Concerns: overgrazing; desertification; limited freshwater resources
Geographical Features: mainly flat to gently rolling tableland; Kalahari sandvelt covers 50% of the country; wetlands in the north
Climate: semiarid; warm winters and hot summers

PEOPLE

Population

Total: 1,561,973
Annual Growth Rate: -0.89%
Rural/Urban Population Ratio: 52/48
Major Languages: English; Setswana
Ethnic Makeup: 80% Tswana; 12% Kalanga; 18% others
Religions: 85% indigenous beliefs; 15% Christian

Health

Life Expectancy at Birth: 37 years (male); 40 years (female)

Infant Mortality: 69.98/1,000 live births
Physicians Available: 1/4,395 people
HIV/AIDS Rate in Adults: 35.8%

Education

Adult Literacy Rate: 79.8%

COMMUNICATION

Telephones: 150,000 main lines
Daily Newspaper Circulation: 29/1,000 people
Televisions: 24/1,000 people
Internet Users: 60,000 (2002)

TRANSPORTATION

Highways in Miles (Kilometers): 6,130 (10,217)
Railroads in Miles (Kilometers): 583 (940)
Usable Airfields: 92
Motor Vehicles in Use: 150,000

GOVERNMENT

Type: parliamentary republic
Independence Date: September 30, 1966 (from the United Kingdom)
Head of State/Government: President Festus Mogae is both head of state and head of government
Political Parties: Botswana Democratic Party; Botswana Alliance Movement; Botswana National Front; Botswana Congress Party; others
Suffrage: universal at 18

MILITARY

Military Expenditures (% of GDP): 3.5%
Current Disputes: none

ECONOMY

Currency ($ U.S. Equivalent): 5.25 pulas = $1
Per Capita Income/GDP: $9,000/$14.2 billion
GDP Growth Rate: 7.2%
Inflation Rate: 9.2%
Unemployment Rate: 40%
Population Below Poverty Line: 47%
Natural Resources: diamonds; copper; nickel; salt; soda ash; potash; coal; iron ore; silver
Agriculture: sorghum; maize; pulses; peanuts; cowpeas; beans; sunflower seeds; livestock
Industry: diamonds; copper; nickel; salt; soda ash; potash; tourism; livestock processing
Exports: $2.5 billion (primary partners Europe, Southern Africa)
Imports: $2.1 billion (primary partners Southern Africa, Europe)

SUGGESTED WEB SITES

http://www.gov.bw
http://www.info.bw
http://www.cia.gov/cia/publications/factbook/geos/bc.html

Botswana Country Report

Since its independence in 1966, Botswana has enjoyed one of the highest average economic growth rates in the world. This has resulted in a better life for many of its citizens. But serious challenges remain if the country is to realize the ambitious goals contained within its "Vision 2016" program of national development. Economic growth has not been accompanied by equity. More than 40 percent of households remain below the official poverty line. Another concern is the feeling of many Botswana that their country's economy continues to be dominated by noncitizens, while many locals remain unemployed or underemployed.

As Africa's longest continuous multiparty democracy, Botswana has a long tradition of lively and unimpeded public debate and is among the continent's most stable countries. Relatively free of corruption, and with a good human rights record, the country is known for its fiscally conservative economic policies, utilized to sustain the most profitable diamond mines in the world and shift the income derived from the mines towards social development projects. Botswana protects some of the continent's largest areas of wilderness. The Kalahari Desert, home to the dwindling group of Bushman hunter gatherers, makes up much of the country and most areas are too arid to sustain any agriculture other than cattle.

Perhaps the greatest challenge currently facing Botswana is its sad distinction of having the world's highest recorded rate of HIV/AIDS infection. About one third of all Batswana between the ages of 15 and 40 are believed to be living with HIV/AIDS. Known to some as "Africa for Beginners," Botswana has become a victim of its own economic success, as foreign investment has lead to better transport links which in turn have aided the spread of HIV/AIDS. The disease has left many thousands of children orphaned and has dramatically reduced the national life expectancy rate, from 69 years in 1995 to just under 40 years today. Based on current trends, UNAIDS estimates that overall life expectancy could drop further to as low as 27 years by 2010.

DEVELOPMENT

The United Nations' 1990 Human Development Report singles out Botswana among the nations of Africa for significantly improving the living conditions of its people. In 1989, President Masire was awarded The Hunger Project's leadership prize, based on Botswana's record of improving rural nutritional levels during the 1980s despite 7 years of severe drought.

The man currently leading Botswana in the face of the above challenges is the country's third president, Festus Mogae, who was inaugurated in April 1998 following the retirement of the long-serving Sir Ketumile Masire (1980–1998). In October 1999, Mogae and his Botswana Democratic Party won an easy victory in national elections, against a divided opposition. But his government was subsequently rocked by conflict between several veteran members of the cabinet and the youthful, charismatic vice-president, Ian Khama, a former army commander and the son of the country's first president, Sir Seretse Khama (1966–1980). In August 2000, it was announced that Ian Khama would be given a new coordinating role to assure adequate performance by various ministries, leading many local commentators to proclaim him as Botswana's first "prime minister." President Festus Mogae's party scored a landslide victory in the October 2004 elections, winning a new five-year mandate to rule. Upon being sworn in, President Mogae promised to tackle poverty and unemployment, and pledged to tackle the spread of HIV/AIDS; he has said that he aims to achieve an AIDS-free Botswana by 2016.

Over the past four decades, Botswana has been hailed as a model of postcolonial development in Africa. When the country emerged from 81 years of British colonial occupation, it was ranked as one of the 10 poorest countries in the world, with an annual per capita income of just $69. In the years since, the nation's economy has grown at an average annual rate of 9 per-

(Photo by Wayne Edge)

Three blocks away from the game park with these monkeys you'll find one of the largest shopping centers in South Africa.

cent. Social services have been steadily expanded, and infrastructure has been created. Whereas at independence the country had no significant paved roads, most major towns and villages are now interlinked by ribbons of asphalt and tarmac. The nation's capital, Gaborone, has emerged as a vibrant city. Schools, hospitals, and businesses dot the landscape, while the country's all-digital telecommunications network is among the most advanced in the world. But the failure of such growth to promote greater equity has also led to a rise in social tensions.

FREEDOM

Democratic pluralism has been strengthened by the growth of a strong civil society and an independent press. Concern has been voiced about social and economic discrimination against Khoisan-speaking communities living in remote areas of the Kalahari, who are known to many outsiders as "Bushmen."

Botswana's economic success has come in the context of its unbroken postindependence commitment to political pluralism, respect for human rights, and racial and ethnic tolerance. Freedoms of speech and association have been upheld. In October 1999, the nation held its eighth successive multiparty election. The Botswana Democratic Party, which has ruled since independence, increased its majority in part due to a split in the main opposition party, the left-leaning opposition Botswana National Front.

Most of Botswana's people share Setswana as their first language, which is understood by about 95% of the population. There also exist a number of sizable minority communities—Kalanga, Herero, Kalagari, Khoisan groups, and others—but contemporary ethnic conflict is relatively modest. In 2002, Parliament agreed to amend the Constitution to remove all references to "tribes" while widening the representation within the Ntlo ya Dikgosi, an advisory house of traditional leaders. In the nineteenth century, most of Botswana was incorporated into five Tswana states, each centering around a large settlement of 10,000 people or more. These states, which incorporated non-Tswana communities, survived through agropastoralism, hunting, and their control of trade routes linking Southern and Central Africa.

Lucrative dealing in ivory and ostrich feathers allowed local rulers to build up their arsenals and thus deter the aggressive designs of South African whites. An attempt by white settlers to seize control of southeastern Botswana was defeated in an 1852–1853 war. However, European missionaries and traders were welcomed, leading to a growth of Christian education and the consumption of industrial goods.

HEALTH/WELFARE

After years of being praised as a model of primary health-care delivery, Botswana's public-health service has come under increased criticism for a perceived decline in quality. Botswana's high HIV-positive rate has placed the system under serious stress.

A radical transformation took place after the imposition of British overrule in 1885. Colonial taxes and economic decline stimulated the growth of migrant labor to the mines and industries of South Africa. (In a few regions, migrant earnings remain the major source of income to this day.) Although colonial rule brought much hardship and little benefit, the twentieth-century relationship between the people of Botswana and the British was complicated by local fears of being incorporated into South Africa. For many decades, leading nationalists championed continued rule from London as a shield against their powerful, racially oppressive neighbor.

ECONOMIC DEVELOPMENT

Economic growth since independence has been largely fueled by the rapid expansion of mining activity. Botswana has become one of the world's leading producers of diamonds, which typically account for 80 percent of its export earnings. Local production is managed by Debswana Corporation, an even partnership between the Botswana government and DeBeers, a South Africa–based global corporation; DeBeers' Central Selling Organization has a near monopoly on diamond sales worldwide. The Botswana government has a good record of maximizing the local benefits of Debswana's production.

ACHIEVEMENTS

In 1999, Botswana's Mpule Kwelagobe was crowned as Miss Universe. In July 2000, Botswana launched its first national television service. A UN report ranked Botswana first in Africa in its percentage of women holding middle- and senior-level managerial positions.

The nickel/copper/cobalt mining complex at Selibi-Pikwe is the largest nongovernment employer in Botswana. Falling metal prices and high development costs have reduced the mine's profitability. Despite high operating efficiency, it is expected to close by the end of the decade. Given that mining can make only a modest contribution to local employment, and because of the potential vulnerability of the diamond market, Botswana is seeking to expand its small manufacturing and service sectors. Meat processing is currently the largest industrial activity outside minerals, but efforts are under way to attract overseas investment in both private and parastatal production. Botswana already has a liberal foreign-exchange policy and has established Bedia as an agency to promote foreign direct investment.

Although it has been negatively affected by events in neighboring Zimbabwe (in September 2003 Botswana begin erecting a fence along its border with Zimbabwean illegal immigrants), tourism is of growing importance. Northern Botswana is particularly noted for its bountiful wildlife and stunning scenery. The region includes the Okavango Delta, a vast and uniquely beautiful swamp area, and the Chobe National Park, home of the world's largest elephant herds.

Agriculture is still a leading economic activity for many Batswana. The standard Tswana greeting, *Pula* ("Rain"), reflects the significance attached to water in a society prone to its periodic scarcity. Botswana suffered severe droughts between 1980 and 1987 and again in 1991 and 1992, which—despite the availability of underground water supplies—had a devastating effect on both crops and livestock. Up to 1 million cattle are believed to have perished. Small-scale agropastoralists, who make up the largest segment of the population, were particularly hard hit. However, government relief measures prevented famine. The government provides generous subsidies to farmers, but environmental constraints hamper efforts to achieve food self-sufficiency even in non-drought years.

Commercial agriculture is dominated by livestock. The Lobatse abbatoir, opened in 1954, stimulated the growth of the cattle industry. Despite periodic challenges from disease and drought, beef exports have become relatively stable. Much of the output of the Botswana Meat Commission has preferential access to the European Union. There is some concern about the potential for future reductions in the European quota. Because most of Botswana's herds are grazed in communal lands, questions about the allocation of pasture are a source of local debate. There is also a growing, but largely misinformed, international concern that wildlife are being threatened by overgrazing livestock.

SOUTH AFRICA

Before 1990, Botswana's progress took place against a backdrop of political hostility on the part of its powerful neighbor, South Africa. Since the nineteenth century, Botswana has sheltered many South African refugees fleeing racist oppression. This led to periodic acts of aggression against the country, especially during the 1980s, when Botswana became the repeated victim of overt military raids and covert terrorist operations. The establishment of a nonracial democracy in South Africa has led to a normalization of relations.

Gaborone is the headquarters of the Southern African Development Community, which was originally conceived to reduce the economic dependence of its 10 member nations on the South African apartheid state. With South Africa now a member, SADC now plans to transform itself into a common market. Despite the countries' past political differences, Botswana has maintained a customs union with South Africa that dates back to the colonial era.

Timeline: PAST

700s
Emergence of the Tswana trading center at Toutswemogala

1820s
Kololo and Ndebele invaders devastate the countryside

1830s
Tswana begin to acquire guns through trade in ivory and other game products

1852–1853
Batswana defeat Boer invaders

1885
The British establish colonial rule over Botswana

1966
Botswana regains its independence

1980s
Elections in 1984 and 1989 result in landslide victories for the Democratic Party; the National Front is the major opposition party

1990s
The ruling Democratic Party wins the 1994 and 1999 elections; the opposition National Front makes gains in 1994 but loses seats in 1999

PRESENT

2000s
Despite astounding national economic growth, many Batswana remain poor

HIV/AIDS crisis

Lesotho (Kingdom of Lesotho)

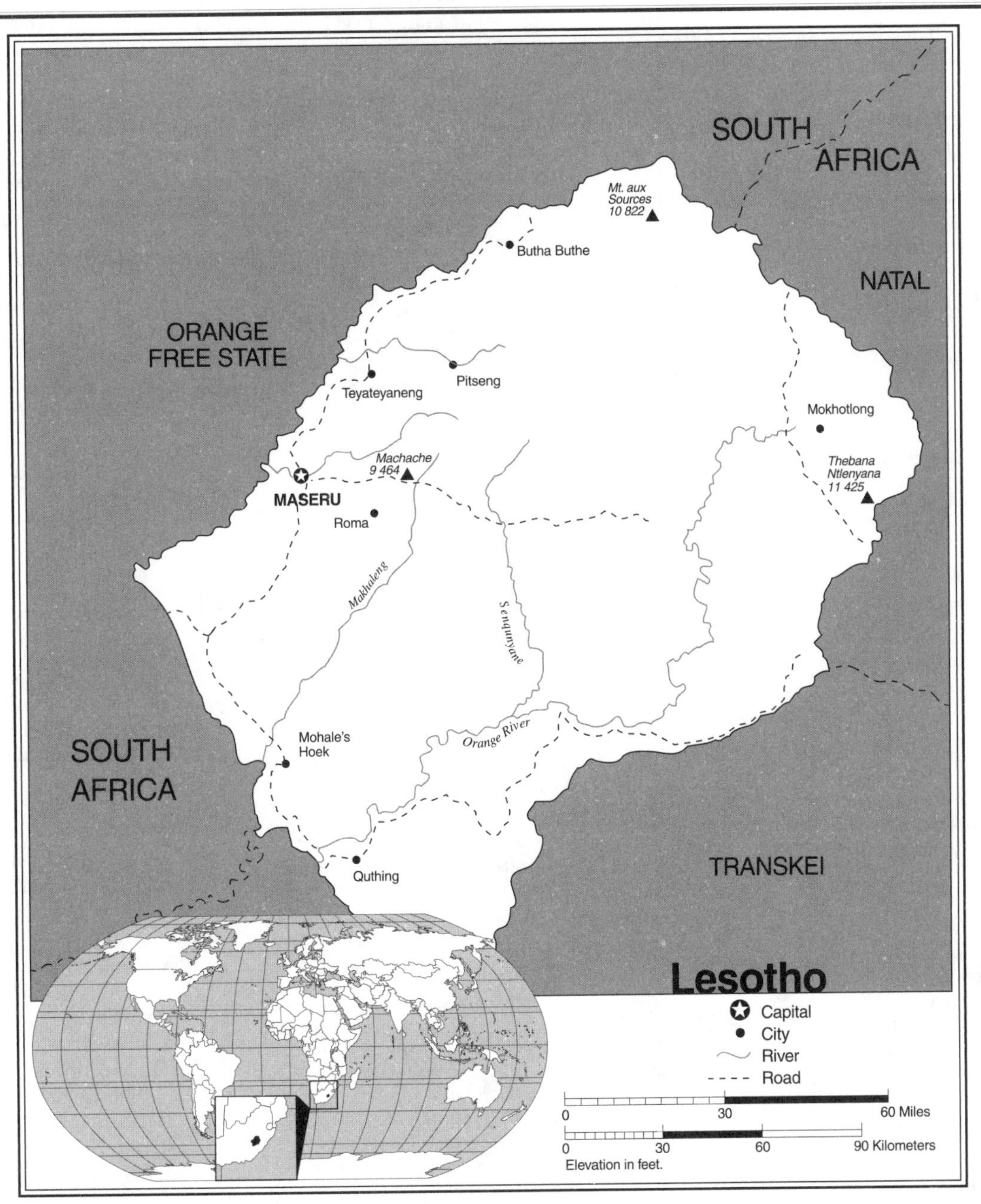

Lesotho Statistics

GEOGRAPHY

Area in Square Miles (Kilometers): 11,716 (30,344) (about the size of Maryland)
Capital (Population): Maseru (271,000)
Environmental Concerns: overgrazing; soil erosion; soil exhaustion; desertification; water pressures
Geographical Features: mostly highland, with plateaus, hills, and mountains; landlocked
Climate: temperate

PEOPLE

Population

Total: 1,865,040
Annual Growth Rate: 0.14%
Rural/Urban Population Ratio: 73/27
Major Languages: English; Sesotho
Ethnic Makeup: 99.7% Sotho
Religions: 80% Christian; 20% indigenous beliefs

Health

Life Expectancy at Birth: 46 years (male); 48 years (female)
Infant Mortality: 83/1,000 live births
Physicians Available: 1/14,306 people
HIV/AIDS Rate in Adults: 28.9%

Education

Adult Literacy Rate: 84.8%

Compulsory (Ages): 6–13; free

COMMUNICATION

Telephones: 28,600 main lines
Daily Newspaper Circulation: 7/1,000 people
Televisions: 7/1,000 people
Internet Users: 21,000 (2002)

TRANSPORTATION

Highways in Miles (Kilometers): 2,973 (4,955)
Railroads in Miles (Kilometers): 1.6 (2.6)
Usable Airfields: 28
Motor Vehicles in Use: 23,000

GOVERNMENT

Type: parliamentary constitutional monarchy
Independence Date: October 4, 1966 (from the United Kingdom)
Head of State/Government: King Letsie III; Prime Minister Pakalitha Mosisili
Political Parties: Lesotho Congress for Democracy; Basotho National Party; Basutoland Congress Party; Marematlou Freedom Party; others
Suffrage: universal at 18

MILITARY

Current Disputes: none

ECONOMY

Currency ($ U.S. Equivalent): 8.28 malotis = $1
Per Capita Income/GDP: $3,000/$5.3 billion
GDP Growth Rate: 4%
Inflation Rate: 6.1%
Unemployment Rate: 45%
Labor Force by Occupation: 86% subsistence agriculture; 35% of male wage earners work in South Africa
Natural Resources: water; agricultural and grazing land; some diamonds and other minerals
Agriculture: corn; wheat; sorghum; pulses; barley; livestock
Industry: food and beverages; textiles; handicrafts; construction; tourism
Exports: $250 million (primary partners Southern Africa, North America)
Imports: $720 million (primary partners Southern Africa, Asia)

SUGGESTED WEB SITES

http://www.lesotho.gov/ls
http://www.mbendi.co.za/cylecy.htm
http://www.cia.gov/cia/publications/factbook/geos/lt.html

Lesotho Country Report

In 2002, the Lesotho Congress for Democracy (LCD) party returned to power after securing a majority of parliamentary seats in elections. The elections, which were endorsed as free and fair by international election observers, were held under a revised Constitution designed to give smaller parties a voice in Parliament.

DEVELOPMENT

Despite an infusion of international aid, Lesotho's economic dependence on South Africa has not decreased since independence; indeed, it has been calculated that the majority of outside funds have actually ended up paying for South African services.

As a result of the election Prime Minister Mosisili was sworn-in for a second five-year term in June 2002. Throughout the independence era Lesotho has had a number of highly visible corruption cases, the most important of which was the May 2002 conviction of Masupha Sole, ex-chief executive of Lesotho Highlands Development Authority. Sole, who was found guilty of accepting bribes from foreign construction companies, was paid millions over a 10-year span in return for business on the Lesotho Highlands Water Project, which supplies water to South Africa. The small farmers throughout Lesotho, who are not supplied with water through irrigation, instead their main source of water was through rain, and the rains in the highlands are sporatic.

During February 2004 Prime Minister Mosisili declared a state of emergency and appealed for food aid. Aid officials say hundreds of thousands face shortages after three-year struggle against drought. Paradoxically in March 2004 the official opening of the first phase of the multi-billion-dollar Lesotho Highlands Water Project, which supplies water to South Africa, took place.

FREEDOM

In Lesotho, basic freedoms and rights are compromised by continuing political instability. Basotho journalists have come out against proposed measures that they say will gag Lesotho's vigorous independent press.

From October 1998 until the new government took over in 2002, Lesotho was ruled by a transitional executive, put into place following the intervention of Botswana and South African troops in support of the previous LCD–led government, which was being threatened by a military coup. Large segments of Lesotho's defense force resisted the intervention, causing scores of deaths on both sides. In the process, many businesses in the capital city, Maseru, and other centers were heavily looted by rioters, who directed much of their anger against Asians.

Listed by the United Nations as one of the world's least-developed nations, each year the lack of opportunity at home causes up to half of Lesotho's adult males to seek employment in neighboring South Africa, where jobs are becoming increasingly scarce. The retrenchment of Basotho mineworkers has led to a local unemployment rate of nearly 50 percent. This dire economic situation is aggravated by Lesotho's chronic political instability.

Lesotho is one of the most ethnically homogeneous nations in Africa; almost all of its citizens are Sotho. The country's emergence and survival were largely the product of the diplomatic and military prowess of its nineteenth-century rulers, especially its great founder, King Moshoeshoe I. In the 1860s, warfare with South African whites led to the loss of land and people as well as an acceptance of British overrule. For nearly a century, the British preserved the country but also taxed the inhabitants and generally neglected their interests. Consequently, Lesotho remained dependent on South Africa. However, despite South African attempts to incorporate the country politically as well as economically, Lesotho's independence was restored by the British in 1966.

Lesotho's politicians were bitterly divided at independence. The conservative Basotho National Party (BNP) had won an upset victory in preindependence elections, with strong backing from the South African government and the local Roman Catholic Church—Lesotho's largest Christian denomination. The opposition, which walked out of the independence talks, was polarized between a pro-royalist faction, the Marematlou Freedom Party (whose regional sympathies largely lay with the African National Congress of South Africa), and the Basotho Congress Party (or BCP, which was allied with the rival Pan-Africanist Congress).

HEALTH/WELFARE

With many of Lesotho's young men working in the mines of South Africa, much of the resident population relies on subsistence agriculture. Despite efforts to boost production, malnutrition, aggravated by drought, is a serious problem.

Soon after independence, the BNP prime minister, Leabua Jonathan, placed the king, Moshoeshoe II, under house arrest. (Later, the king was temporarily exiled.) The BCP won the 1970 elections, but Jonathan, possibly at the behest of South Africa, declared a state of emergency and nullified the results.

In the early 1980s, armed resistance to Jonathan's dictatorship was carried out by the Lesotho Liberation Army (LLA), an armed faction of the BCP. The Lesotho government maintained that the LLA was aided and abetted by South Africa as part of that country's regional destabilization efforts. By 1983, both the South African government and the Catholic Church hierarchy were becoming nervous about Jonathan's establishment of diplomatic ties with various Communist-ruled countries and the growing sympathy within the BNP, in particular its increasingly radical youth wing, for the ANC. South African military raids and terrorist attacks targeting anti-apartheid refugees in Lesotho became increasingly common. Finally, a South African blockade of Lesotho in 1986 led directly to Jonathan's ouster by his military.

Lesotho's new ruling Military Council, initially led by Major General Justinus Lekhanya, was closely linked to South Africa. In 1990, Lekhanya had Moshoeshoe II exiled (for the second time), after he refused to agree to the dismissals of several senior officers. Moshoeshoe's son Letsie was installed in his place. In 1991, Lekhanya was himself toppled by the army. The new leader, General Elias Rameama, promised to hold multiparty elections. In 1992, Moshoeshoe returned, to a hero's welcome, but he was prevented from resuming his role as monarch. His status was uncertain after elections in March 1993 brought the BCP back to power.

ACHIEVEMENTS

Lesotho has long been known for the high quality of its schools, which for more than a century and a half have trained many of the leading citizens of Southern Africa.

Under its aging leader, Ntsu Mokhele, the BCP faced military opposition to its rule. An outbreak of internal fighting within the Royal Lesotho Defense Force (RLDF) culminated, in April 1994, in the assassination of the deputy prime minister and the kidnapping of cabinet members by mutinous soldiers. In August, King Letsie tried to dismiss the government and suspend the constitution. In the face of growing unrest, Botswana, South Africa, and Zimbabwe acted on behalf of the Southern African Development Community as mediators—and subsequently as guarantors—of constitutional rule. The BCP and Moshoeshoe were both restored to power; the latter was killed in January 1996 in an auto accident. In June 1997, a schism in the BCP's ranks led Mokhele to establish the new Lesotho Congress for Democracy party, taking most of the BCP with him. The LCD won a landslide victory in May 1998 elections, but the remnants of the BCP and other opposition parties refused to accept the result, inciting King Letsie and the military to intervene. The resulting mutiny within the RLDF led to the South African/Botswana intervention to restore order.

Timeline: PAST

1820s
Lesotho emerges as a leading state in Southern Africa

1866
Afrikaners annex half of Lesotho

1870–1881
The Sotho successfully fight to preserve local autonomy under the British

1966
Independence is restored

1970
The elections and Constitution are declared void by Leabua Jonathan

1974
An uprising against the government fails

1979
The Lesotho Liberation Army begins a sabotage campaign

1986
South African destabilization leads to the overthrow of Jonathan by the military

1990s
Troops from South Africa and Botswana intervene in Lesotho to avert a coup

PRESENT

2000s
The Lesotho Congress for Democracy is returned to power in parliamentary elections

Malawi (Republic of Malawi)

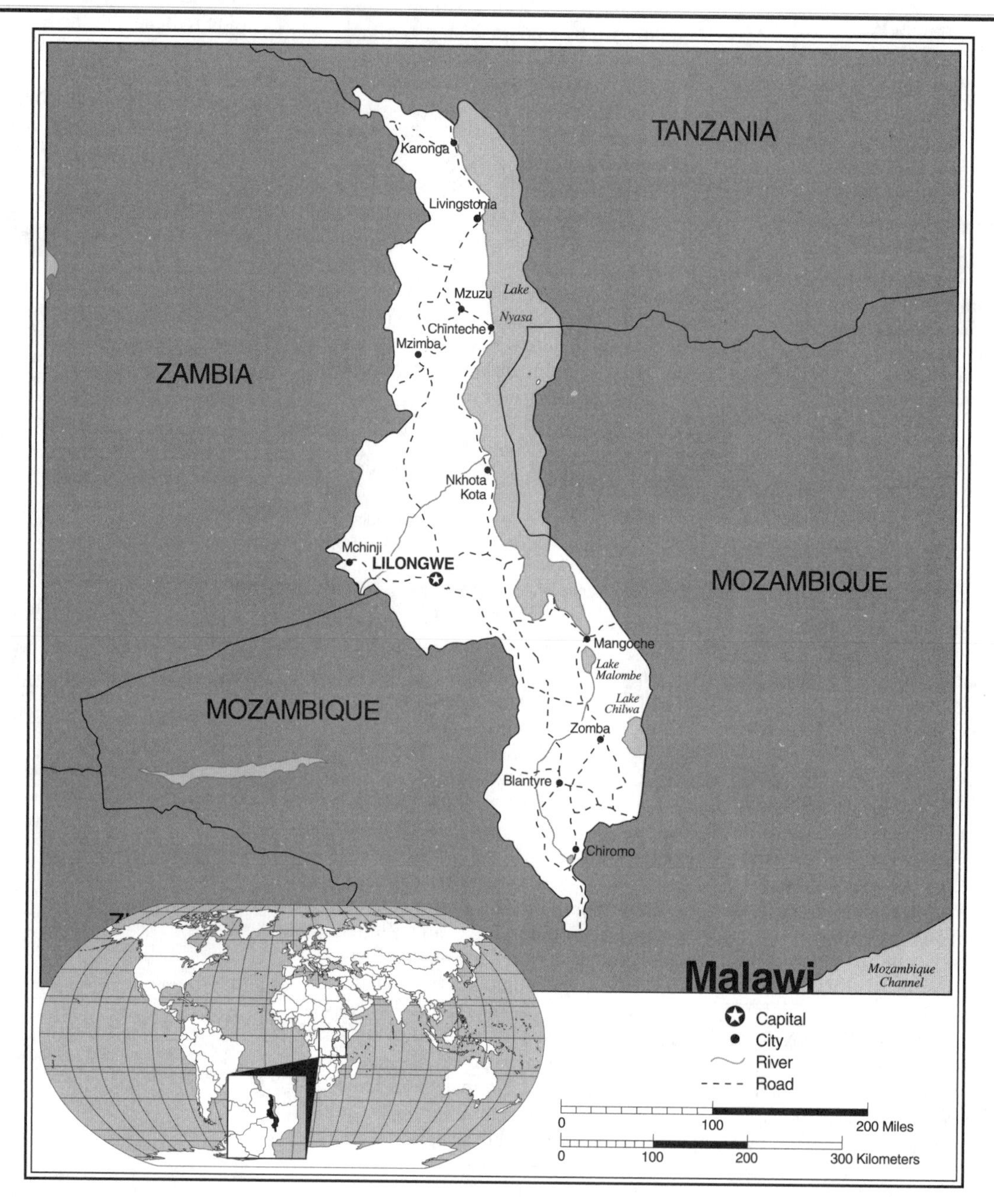

Malawi Statistics

GEOGRAPHY

Area in Square Miles (Kilometers): 45,747 (118,484) (about the size of Pennsylvania)

Capital (Population): Lilongwe (523,000)

Environmental Concerns: deforestation; land degradation; water pollution; siltation of fish spawning grounds

Geographical Features: narrow, elongated plateau with rolling plains, rounded hills, some mountains; landlocked

Climate: subtropical

PEOPLE

Population

Total: 11,906,855

Annual Growth Rate: 2.14%

Rural/Urban Population Ratio: 76/24

Major Languages: Chichewa; English; regional languages

Ethnic Makeup: 90% Chewa; 10% Nyanja, Lomwe, other Bantu groups

Religions: 75% Christian; 20% Muslim; 5% indigenous beliefs

Health

Life Expectancy at Birth: 36 years (male); 37 years (female)

Infant Mortality: 120/1,000 live births

Physicians Available: 1/47,634 people
HIV/AIDS Rate in Adults: 15.96%

Education

Adult Literacy Rate: 58%
Compulsory (Ages): 6–14

COMMUNICATION

Telephones: 38,000 main lines
Internet Users: 36,000 (2003)

TRANSPORTATION

Highways in Miles (Kilometers): 8,756 (14,594)
Railroads in Miles (Kilometers): 489 (789)
Usable Airfields: 44
Motor Vehicles in Use: 55,000

GOVERNMENT

Type: multiparty democracy
Independence Date: July 6, 1964 (from the United Kingdom)
Head of State/Government: President Binguwa Mutharika is both head of state and head of government
Political Parties: United Democratic Front; Malawi Congress Party; Alliance for Democracy; Malawi Democratic Party
Suffrage: universal at 18

MILITARY

Military Expenditures (% of GDP): 0.76%
Current Disputes: boundary dispute with Tanzania

ECONOMY

Currency ($ U.S. Equivalent): 91.36 kwachas = $1
Per Capita Income/GDP: $660/$7 billion
GDP Growth Rate: 1.7%
Inflation Rate: 9.5%
Labor Force by Occupation: 86% agriculture
Population Below Poverty Line: more than half
Natural Resources: limestone; uranium; coal; bauxite; arable land; hydropower
Agriculture: tobacco; tea; sugarcane; cotton; potatoes; cassava; sorghum; pulses; livestock
Industry: tobacco; sugar; tea; sawmill products; cement; consumer goods
Exports: $415.5 million (primary partners South Africa, United States, Germany)
Imports: $463.6 million (primary partners South Africa, United Kingdom, Zimbabwe)

SUGGESTED WEB SITES

http://www.malawi.net
http://www.maform.malawi.net
http://www.cia.gov/cia/publications/factbook/geos/mi.html

Malawi Country Report

In 2003, Malawi was facing its worst famine in more than 50 years. In a land where most of the population engages in farming, the government was struggling to feed the people after the lack of rains in 2001–2002. The harvest in 2002 was 25 percent below that of the last five years. Some 6 million Malawians live below the poverty line, of whom an estimated 70 percent are at serious risk of starvation unless help is found quickly. The situation has been aggravated by Malawi's sale of strategic grain reserves, allegedly on the advice of IMF/World Bank experts. Very little of the money has been collected from the sales. As $40 million worth of maize grain (the staple food) went missing and unpaid for, apparently to local speculators as well as international buyers, people have began asking questions. The scandal could not have come at a worse time for President Bakili Muluzi, who was trying to get Parliament to approve a bill that would allow him to serve a third term in office instead of the normal two terms (which he has already completed). In his drive to remain in power, there have been increasing reports of systematic intimidation directed against Muluzi's opponents.

Parliament, however, refused to accept an amendment to the constitution that would allow president Muluzi to stand for a third term. Thus, Binguwa Mutharika was chosen as the candidate of the ruling United Democratic Front. Viewed as a relative outsider, his nomination surprised many UDF members and led to several party heavyweight defections including Vice President Justin Malewezi, who subsequently contested the presidential election as an independent candidate.

DEVELOPMENT

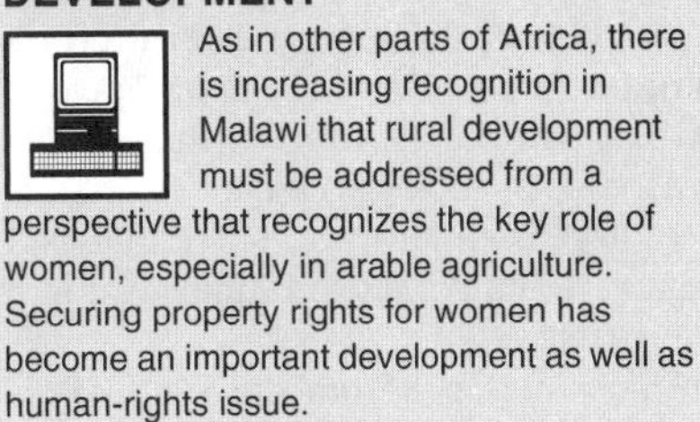

As in other parts of Africa, there is increasing recognition in Malawi that rural development must be addressed from a perspective that recognizes the key role of women, especially in arable agriculture. Securing property rights for women has become an important development as well as human-rights issue.

However, the run-up to the poll was overshadowed by opposition claims of irregularities in the voters' roll. The presidential and parliamentary elections, originally planned for May 18, 2004, were postponed by two days following a High Court appeal by the main opposition Mgwirizano (Unity) coalition. The voting was not delayed for long, though, and on May 24, 2004 candidate Binguwa Mutharika was sworn in as President. European Union and Commonwealth observers said although voting passed peacefully, they were concerned about "serious inadequacies" in the poll.

Malawi's current crisis stands in contrast to the optimism that was generated in 1994 when Muluzi and his United Democratic Front (UDF) ended 30 years of dictatorial rule under Malawi's first president, Dr. Hastings Banda. Muluzi's victory over the then–96-year-old Banda brought an end to what had been one of Africa's most repressive regimes. But since Banda's death in 1997, a degree of nostalgia has emerged regarding his era, perhaps reflecting the failure of political liberalization to bring improvement to the country's weak economy, which is largely dependent on tobacco production. The spread of the HIV/AIDS pandemic and a rising crime rate have also shaken public confidence. While the old political order has been swept aside, the new order has yet to deliver better conditions for most Malawians. In 2004 President Maluzi, after admitting there was a major problem occurring due to the rise of AIDS and that his brother had died of the disease, announced the government would provide anti-viral drugs to AIDS sufferers free of charge.

Since the early months of independence in 1964, when he purged his cabinet and ruling Malawi Congress Party (MCP) of most of the young politicians who had promoted him to leadership in the nationalist struggle, Banda ruthlessly used his secret police and MCP's militia, the Malawi Young Pioneers (MYP), to eliminate potential alternatives to his highly personalized dictatorship. Generations of Malawians, including those living abroad, grew up with the knowledge that voicing critical thoughts about the self-proclaimed "Life President," or *Ngwazi* ("Great Lion"), could prove fatal. Only senior army officers, Banda's long-time "official state hostess" Tamanda Kadzamira, and her uncle John Tembo, the powerful minister of state, survived Ngwazi's jealous exercise of power.

FREEDOM

Although greatly improved since the end of the Banda era, serious human-rights problems remain, including the abuse and death of detainees by police. Prison conditions are poor. Lengthy pretrial detention, an inefficient judicial system, and limited resources have called into question the ability of defendants to receive timely and fair trials. High levels of crime have prompted a rise in vigilante justice.

In 1992, Banda's grip began to weaken. Unprecedented antigovernment unrest gave rise to an internal opposition, spearheaded by clergy, underground trade unionists, and a new generation of dissident politicians. By 1993, this opposition had coalesced into two major movements: the southern-based UDF, and the northern-based Alliance for Democracy (AFORD). The detention of AFORD's leader, Chakufwa Chihana, and others failed to stem the tide of opposition. A referendum in June showed two-to-one support for a return to multiparty politics. In November, while Banda was hospitalized in South Africa, young army officers seized the initiative by launching a crackdown against the MYP while purging a number of senior officers from their own ranks. Thereafter, the army played a neutral role in assuring the success of the election.

HEALTH/WELFARE

Malawi's health service is considered exceptionally poor even for an impoverished country. The country has one of the highest child mortality rates in the world, and more than half of its children under age 5 are stunted by malnutrition.

While the ruthless efficiency of its security apparatus contributed to past perceptions of Malawi's stability, Banda did not survive by mere repression. A few greatly benefited from the regime. Until 1979, the country enjoyed an economic growth rate averaging 6 percent per year. Almost all this growth came from increased agricultural production. The postindependence government favored large estates specializing in exported cash crops. While in the past the estates were almost exclusively the preserve of a few hundred white settlers, today many are controlled by either the state or local individuals.

ACHIEVEMENTS

Although it is the poorest, most overcrowded country in the region, Malawi's response to the influx of Mozambican refugees was described by the U.S. Committee for Refugees as "no less than heroic."

In the 1970s, the prosperity of the estates helped to fuel a boom in industries involved in agricultural processing. Malawi's limited economic success prior to the 1980s came at the expense of the vast majority of its citizens, who survive as small landholders growing food crops. By 1985, some 86 percent of rural households farmed on less than five acres. Overcrowding has contributed to serious soil depletion while marginalizing most farmers to the point where they can have little hope of generating a significant surplus. In addition to land shortage, peasant production has suffered from low official produce prices and lack of other inputs. The northern half of the country, which has almost no estate production, has been relatively neglected in terms of government expenditure on transport and other forms of infrastructure. Many Malawian peasants have for generations turned to migrant labor as a means of coping with their poverty, but there have been far fewer opportunities for them in South Africa and Zimbabwe in recent years.

Under pressure from the World Bank, the Malawian government has since 1981 modestly increased its incentives to the small landholders. Yet these reforms have been insufficient to overcome the continuing impoverishment of rural households, which has been aggravated in recent decades by a decline in migrant-labor remittances. The maldistribution of land in many areas remains a major challenge. On a more positive note, peace in Mozambique has reopened landlocked Malawi's access to the Indian Ocean ports of Beira and Nacala while reducing the burden of dealing with what once numbered 600,000 refugees. Communications infrastructure to the ports, damaged by war, is now being repaired.

Timeline: PAST

1500s
Malawi trading kingdoms develop

1859
Explorer David Livingstone arrives along Lake Malawi; missionaries follow

1891
The British protectorate of Nyasaland (present-day Malawi) is declared

1915
Reverend John Chilembwe and followers rise against settlers and are suppressed

1944
The Nyasaland African Congress, the first nationalist movement, is formed

1964
Independence, under the leadership of Hastings Banda

1967
Diplomatic ties are established with South Africa

1971
"Ngwazi" Hastings Kamuzu Banda becomes president-for-life

1990s
Bakili Muluzi is elected president, ending Banda's 30-year dictatorship; Banda dies in 1997

PRESENT

2000s
Political liberalization fails to improve the economy

Famine threatens millions of Malawians

2002
Railway line linking central Malawi and the Mozambican Port of Nacala reopened after almost 20 years, giving the nation access to the Indian Ocean coast

Mozambique (Republic of Mozambique)

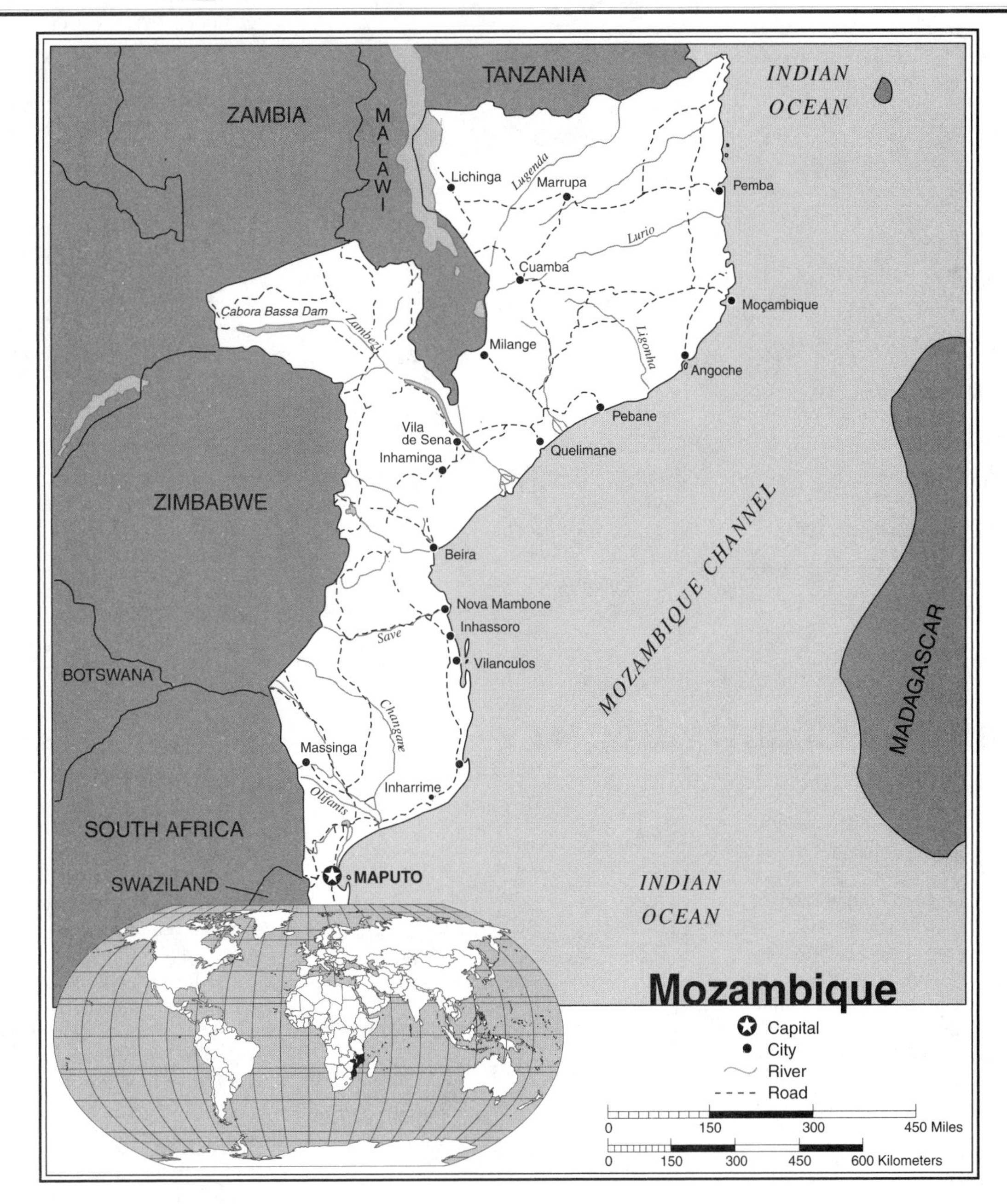

Mozambique Statistics

GEOGRAPHY

Area in Square Miles (Kilometers): 309,494 (801,590) (about twice the size of California)

Capital (Population): Maputo (1,134,000)

Environmental Concerns: civil war and drought have had adverse consequences on the environment; water pollution; desertification

Geographical Features: mostly coastal lowlands; uplands in center; high plateaus in northwest; mountains in west

Climate: tropical to subtropical

PEOPLE

Population

Total: 18,811,731

Annual Growth Rate: 1.22%

Rural/Urban Population Ratio: 61/39

Major Languages: Portuguese; indigenous dialects

Ethnic Makeup: nearly 100% indigenous groups

Religions: 50% indigenous beliefs; 30% Christian; 20% Muslim

Health

Life Expectancy at Birth: 38 years (male); 37 years (female)
Infant Mortality: 138.5/1,000 live births
Physicians Available: 1/131,991 people
HIV/AIDS Rate in Adults: 12.6%–16.4%

Education

Adult Literacy Rate: 47.8%
Compulsory (Ages): 7–14

COMMUNICATION

Telephones: 83,700 main lines
Daily Newspaper Circulation: 8/1,000 people
Televisions: 3.5/1,000 people
Internet Users: 50,000 (2002)

TRANSPORTATION

Highways in Miles (Kilometers): 17,886 (29,810)
Railroads in Miles (Kilometers): 1,879 (3,131)
Usable Airfields: 166
Motor Vehicles in Use: 89,000

GOVERNMENT

Type: republic
Independence Date: June 25, 1975 (from Portugal)
Head of State/Government: President Joaquim Alberto Chissano; Prime Minister Luisa Diogo
Political Parties: Front for the Liberation of Mozambique (Frelimo); Mozambique National Resistance—Electoral Union (Renamo)
Suffrage: universal at 18

MILITARY

Military Expenditures (% of GDP): 1%
Current Disputes: none; cease-fire since 1992

ECONOMY

Currency ($ U.S. Equivalent): 23,467 meticais = $1
Per Capita Income/GDP: $1,200/$21.23 billion
GDP Growth Rate: 7%
Inflation Rate: 14%
Unemployment Rate: 21%
Labor Force by Occupation: 81% agriculture; 13% services; 6% industry
Population Below Poverty Line: 70%
Natural Resources: coal; titanium; natural gas; hydropower
Agriculture: cotton; cassava; cashews; sugarcane; tea; corn; rice; fruits; livestock
Industry: processed foods; textiles; beverages; chemicals; tobacco; cement; glass; asbestos; petroleum products
Exports: $795 million (primary partners South Africa, Zimbabwe, Spain)
Imports: $1.14 billion (primary partners South Africa, Portugal, United States)

SUGGESTED WEB SITES

http://www.mozambique.mz.eindex.htm
http://poptel.org.uk/mozambique-news
http://www.africaindex.africainfo.no/africaindex1/countries/mozambique.html
http://www.cia.gov/cia/publications/factbook/geos/mz.html

Mozambique Country Report

Mozambique has made steady economic and political progress over the past decade in the face of grinding poverty, natural disasters, and the immense burden of overcoming the legacy of a bitter civil war. Although many in the region note that the country has served as a kind of magnet for foreign investment, problems with human-rights issues remain with accusations of torture and harassment of any political opponent of the government. An example of this was seen in November of 2002 when two defendants, on trial for the murder of prominent journalist Carlos Cardoso, alleged that Nymphine Chissano, the son of President Chissano, was linked to the crime. Chissano denied any knowledge of the murder, and was not put on trial. In June 2002, the ruling Mozambique Liberation Front (Frelimo) party chose Armando Guebuza, an independence-struggle veteran, as its presidential candidate for the December 2004 presidential elections, after its long-serving incumbent, Joaquim Chissano, declined to run again.

In February 2000, the eyes of the world focused on devastating floods in Mozambique. In some places the country's two main rivers, the Limpopo and Save, expanded miles beyond their normal banks, engulfing hundreds of villages and destroying property and infrastructure. Many Mozambicans were left homeless. The disaster was a serious setback for the nation, which had been making steady economic progress after three decades of civil war. Even before the floods, Mozambique (which remains one of the world's poorest countries) faced immense economic, political, and social challenges.

DEVELOPMENT

To maintain minimum services and to recover from wartime and flood destruction, Mozambique relies on the commitment of its citizens and international assistance. Western churches have sent relief supplies, food aid, and vehicles.

A 1992 cease-fire agreement, followed by the holding of multiparty elections in 1994 and 1999, has seemingly brought peace to the country. However, the opposition Mozambique National Resistance Movement (Renamo), which narrowly lost both polls, rejected the 1999 electoral outcome.

Since the 1994 elections Frelimo, which has governed the country since independence in 1975, has faced a large opposition bloc in Parliament from Renamo, its old civil-war adversary, which currently controls 117 out of 250 seats. A now-peaceful Renamo is, arguably, less of a challenge to the government than the dictates of international donors, upon whose funding it now depends. While the government must be concerned about improving the dismal living conditions endured by the majority of its people, the donors have insisted on fiscal austerity and a privatization program that has led to retrenchments as well as a loss of government influence.

FREEDOM

While the status of political and civil liberties has improved, the government's overall human-rights record continues to be marred by security-force abuses (including extra-judicial killings, excessive use of force, torture, and arbitrary detention) and an ineffective and only nominally independent judicial system.

Frelimo originally came to power as a result of a liberation war. Between 1964 and 1974, it struggled against Portuguese colonial rule. At a cost of some 30,000 lives, Mozambique gained its independence in 1975 under Frelimo's leadership. Although the new nation was one of the

(United Nations photo 153689 by Kate Truscott)

The drain on natural resources resulting from civil war, the displacement of approximately one fifth of the population, persistent drought, and, most recently, devastating floods, have led to the need for Mozambique to import food to stave off famine.

least-developed countries in the world, many were optimistic that the lessons learned in the struggle could be applied to the task of building a dynamic new society based on Marxist-Leninist principles.

HEALTH/WELFARE

Civil strife, widespread Renamo attacks on health units, and food shortages drastically curtailed health-care goals and led to Mozambique's astronomical infant mortality rate.

Unfortunately, hopes for any sort of postindependence progress were quickly dashed by Renamo, which was originally established as a counterrevolutionary fifth column by Rhodesia's (Zimbabwe) Central Intelligence Organization. More than 1 million people died due to the rebellion, a large proportion murdered in cold blood by Renamo forces. It is further estimated that, out of a total population of 17 million, some 5 million people were internally displaced, and about 2 million others fled to neighboring states. No African nation paid a higher price in its resistance to white supremacy.

Although some parts of Mozambique were occupied by the Portuguese for more than 400 years, most of the country came under colonial control only in the early twentieth century. The territory was developed as a dependency of neighboring colonial economies rather than that of Portugal itself. Mozambican ports were linked by rail to South Africa and the interior colonies of British Central Africa—that is, modern Malawi, Zambia, and Zimbabwe. In the southern provinces, most men, and many women, spent time as migrant laborers in South Africa. The majority of the males worked in the gold mines.

Most of northern Mozambique was granted to three predominantly British concessions companies, whose abusive policies led many to flee the colony. For decades, the colonial state and many local enterprises also relied on forced labor. After World War II, new demands were put on Mozambicans by a growing influx of Portuguese settlers, whose numbers swelled during the 1960s, from 90,000 to more than 200,000. Meanwhile, even by the dismal standards of European colonialism in Africa, there continued to be a notable lack of concern for human development. At independence, 93 percent of the African population in Mozambique were illiterate. Furthermore, most of those who had acquired literacy or other skills had done so despite the Portuguese presence.

ACHIEVEMENTS

Between 1975 and 1980, the illiteracy rate in Mozambique declined from 93% to 72% while classroom attendance more than doubled. Progress slowed during the 1980s due to civil war. Today, the overall literacy rate stands at about 40%.

Although a welcome event in itself, the sudden nature of the Portuguese empire's collapse contributed to the destabilization of postindependence Mozambique. Because Frelimo had succeeded in establishing itself as a unified nationalist front, Mozambique was spared an immediate descent into civil conflict, such as that which engulfed Angola, Portugal's other major African possession. However, the economy was already bankrupt due to the Portuguese policy of running Mozambique on a nonconvertible local currency. The rapid transition to independence compounded this problem by encouraging the sudden exodus of almost all the Portuguese settlers.

Perhaps even more costly to Mozambique in the long term was the polarization between Frelimo and African supporters of the former regime, who included about 100,000 who had been active in its security forces. The rapid Portuguese withdrawal was not conducive to the difficult task of reconciliation. While Frelimo did not subject the "compromised ones" to bloody reprisals, their rights were circumscribed, and many were sent, along with prostitutes and other "antisocial" elements, to "reeducation camps." While the historically pro-Portuguese stance of the local Catholic hierarchy would have complicated its relations with the new state under any circumstance, Frelimo's Marxist rejection of religion initially alienated it from broader numbers of believers.

A TROUBLED INDEPENDENCE

Frelimo assumed power without the benefit or burden of a strong sense of administrative continuity. While it had begun to create alternative social structures in its "liberated zones" during the anticolonial struggle, these areas had encompassed only a small percentage of Mozambique's population and infrastructure. But Frelimo was initially able to fill the vacuum and launch aggressive development efforts. Health care and education were expanded, worker committees successfully ran many of the enterprises abandoned by the settlers, and communal villages coordinated rural development. However, efforts to promote agricultural collectivization as the foundation of a command economy generally led to peasant resistance and economic failure. Frelimo's ability to adapt and implement many of its programs under trying conditions was due largely to its disciplined mass base (the party's 1990 membership stood at about 200,000).

No sooner had Mozambique begun to stabilize itself from the immediate dislocations of its decolonization process than it became embroiled in the Rhodesian conflict. Mozambique was the only neighboring state to impose fully the "mandatory" United Nations economic sanctions against Rhodesia (present-day Zimbabwe). Between 1976 and 1980, this decision led to the direct loss of half a billion dollars in rail and port revenues. Furthermore, Frelimo's decision to provide bases for the fighters of the Patriotic Front led to a state of undeclared war with Rhodesia as well as its Renamo proxies.

Unfortunately, the fall of Rhodesia did not bring an end to externally sponsored destabilization. Renamo had the support of South Africa. By continuing Renamo's campaign of destabilization, the South African regime gained leverage over its hostile neighbors, for the continued closure of Mozambique's ports meant that most of their traffic had to pass through South Africa. In 1984, Mozambique signed a nonaggression pact with South Africa, which should have put an end to the latter's support of Renamo. However, captured documents and other evidence indicate that official South African support for Renamo continued at least until 1989, while South African supplies were still reaching the rebels under mysterious circumstances. In response, Zimbabwe, and to a lesser extent Malawi and Tanzania, contributed troops to assist in the defense of Mozambique.

In its 1989 Congress, Frelimo formally abandoned its commitment to the primacy of Marxist-Leninist ideology and opened the door to further political and economic reforms. Multipartyism was formally embraced in 1991. With the help of the Catholic Church and international mediators, the government opened talks with Renamo. In October 1992, Renamo's leader, Alfonso Dlakama, signed a peace accord that called for UN–supervised elections. The cease-fire finally came into actual effect in the early months of 1993, by which time the UN personnel on the ground reported that some 3 million Mozambicans were suffering from famine.

Besides their mutual distrust, reconciliation between Renamo and Frelimo was troubled by their leaderships' inability to control their armed supporters. With neither movement able to pay its troops, apolitical banditry by former fighters for both sides increased. International financial and military support, mobilized through the United Nations, was inadequate to meet this challenge. In June and July 1994, a number of UN personnel, along with foreign-aid workers, were seized as hostages. The near-complete collapse of the country, however, has so far encouraged Mozambique's political leaders to sustain the peace drive.

Timeline: PAST

1497
Portuguese explorers land in Mozambique

1820s
The Northern Nguni of Shosagaane invade southern Mozambique, establishing the Gaza kingdom

1962
The Frelimo liberation movement is officially launched

1975
The liberation struggle is successful

1980s
Increased Renamo attacks on civilian and military targets; President Samora Machel is killed in a mysterious airplane crash; Joaquim Chissano becomes president

1990s
Renamo agrees to end fighting, participate in multiparty elections

PRESENT

2000s
Floods lead to enormous losses in life, property, and infrastructure

Chissano decides not to run again; elections are scheduled for 2004

Namibia (Republic of Namibia)

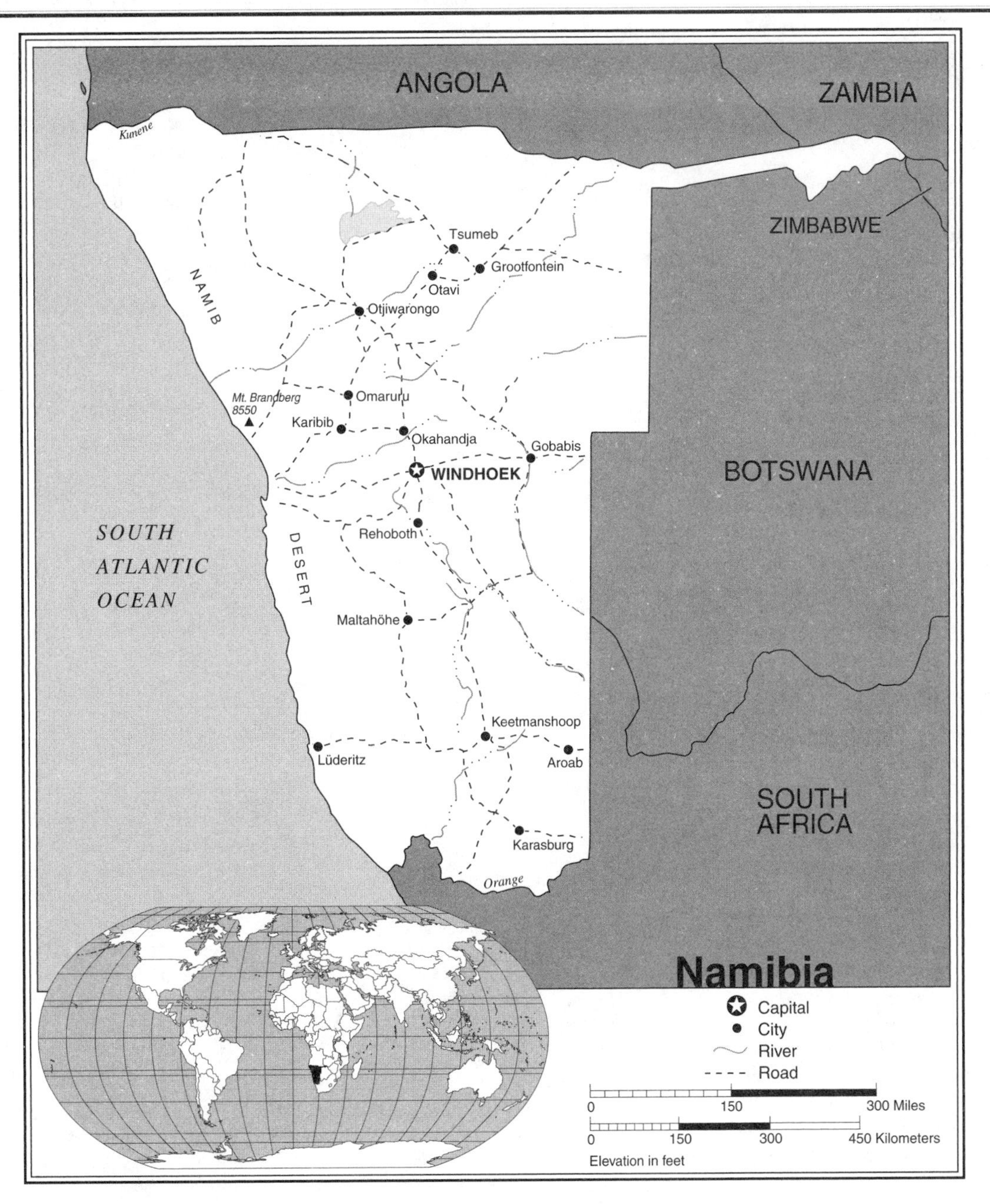

Namibia Statistics

GEOGRAPHY

Area in Square Miles (Kilometers): 318,261 (824,292) (about half the size of Alaska)

Capital (Population): Windhoek (216,000)

Environmental Concerns: very limited natural freshwater resources; desertification

Geographical Features: mostly high plateau; desert along coast and in east

Climate: desert

PEOPLE

Population

Total: 1,954,033

Annual Growth Rate: 1.19%

Rural/Urban Population Ratio: 70/30

Major Languages: English; Ovambo; Kavango; Nama/Damara; Herero; Khoisan; German; Afrikaans

Ethnic Makeup: 50% Ovambo; 9% Kavango; 7% Herero; 7% Damara; 27% others

Religions: 80%–90% Christian; 10%–20% indigenous beliefs

Health

Life Expectancy at Birth: 44 years (male); 41 years (female)
Infant Mortality: 72.4/1,000 live births
Physicians Available: 1/4,594 people
HIV/AIDS Rate in Adults: 19.54%

Education

Adult Literacy Rate: 84%
Compulsory (Ages): 6–16

COMMUNICATION

Telephones: 127,400 main lines
Daily Newspaper Circulation: 27/1,000 people
Televisions: 28/1,000 people
Internet Users: 65,000 (2003)

TRANSPORTATION

Highways in Miles (Kilometers): 39,220 (63,258)
Railroads in Miles (Kilometers): 1,429 (2,382)
Usable Airfields: 137
Motor Vehicles in Use: 129,000

GOVERNMENT

Type: republic
Independence Date: March 21, 1990 (from South African mandate)
Head of State/Government: President Samuel Shafishuna Nujoma; Prime Minister Theo-Ben Gurirab
Political Parties: South West Africa People's Organization; Democratic Turnhalle Alliance of Namibia; United Democratic Front; Monitor Action Group; others
Suffrage: universal at 18

MILITARY

Military Expenditures (% of GDP): 2.6%
Current Disputes: border dispute over at least one island in the Linyanti River

ECONOMY

Currency ($ U.S. equivalent): 7.56 dollars = $1
Per Capita Income/GDP: $7,200/$13.85 billion
GDP Growth Rate: 3.3%
Inflation Rate: 7.3%
Unemployment Rate: 30%–40%
Labor Force by Occupation: 47% agriculture; 33% services; 20% industry
Natural Resources: diamonds; gold; tin; copper; lead; zinc; uranium; salt; cadmium; lithium; natural gas; possible oil, coal reserves; fish; vanadium; hydropower
Agriculture: millet; sorghum; peanuts; livestock; fish
Industry: meat packing; dairy products; fish processing; mining
Exports: $1.58 billion (primary partners United Kingdom, South Africa, Spain)
Imports: $1.7 billion (primary partners South Africa, United States, Germany)

SUGGESTED WEB SITES

http://www.govnet.gov.na/intro.htm
http://www.sas.upenn.edu/African_Studies/Country_Specific/Namibia.html
http://www.cia.gov/cia/publications/factbook/geos/wa.html

Namibia Country Report

The collapse of the UNITA rebel movement in neighboring Angola has brought to a close what had been a perennial source of insecurity along Namibia's northern border. In August 1999, Namibians were shocked when a small group of self-proclaimed separatists launched an armed attack on the town of Katima Mulilo. (Ultimately defeated, the separatists sought refuge in Botswana; most were involuntarily repatriated back to Namibia in 2001–2002.) Subsequent attacks along the border with war-torn Angola and the involvement of Namibian forces in the war in the Democratic Republic of the Congo (Congo-Kinshasa, or D.R.C.) further shook the country's peaceful international image. The incidents, however, did not fundamentally disturb Namibia's steady progress since its liberation from South African rule in 1990.

With the country now fully at peace for the first time in 30 years, the Namibian government is free to concentrate on domestic development. The material resources that went into the war effort can now be put toward peaceful use.

One area earmarked for urgent attention is land reform. Local observers were little surprised in September 2002 when President Samuel Nujoma voiced his support for the Zimbabwean government's seizure of white-held farms. In fact, Nujomo's own ruling party, the South West Africa People's Organization (SWAPO), has committed itself to redistributing to the African farmers half of the land currently owned by white absentee landlords. It is unlikely, however, that land reform in Namibia will be accompanied by the lawless violence that has marred recent developments in Zimbabwe.

DEVELOPMENT

The Nujoma government has instituted English as the medium of instruction in all schools. (Before independence, English was discouraged for African schoolchildren, a means of controlling their access to skills necessary to compete in the modern world.) This effort requires new curricula and textbooks for the entire country.

Yet in a country where arable land is premium, land reform remains an issue. In August 2002 Prime Minister Theo-Ben Gurirab stated that the government viewed land reform as a priority. President Nujoma concurred with his Prime Minister, telling white farmers that they must embrace the reform program. This is not so remarkable except for the fact that the issue of land in Namibia is being worked out on the basis of willing seller willing buyer, with the government acting as the purchaser of the majority population. Prompt government action has had its immediate results. During November 2003 the Union representing black farmworkers called off plans to invade 15 white-owned farms after reaching agreement with white farmers' group. The government says illegal land occupations will not be allowed.

FREEDOM

In line with Namibia's liberal Constitution, human rights are generally respected. There are some problem areas: Demonstrations that do not have prior police approval are banned. The president and other high officials have repeatedly attacked the independent press. There has not been a full accounting of missing detainees who were in SWAPO camps before independence. Security forces have admitted to cases of extra-judicial killing along the Angolan border.

Namibia became independent in 1990, after a long liberation struggle. Its transition from the continent's last colony to a developing nation-state marked the end of a century of often brutal colonization, first by Germany and later South Africa. The German colonial period (1884–1917) was marked by the annihilation of more than 60 percent of the African population in the

(United Nations photo 157251 by J. Isaac)

Developing agricultural production in Namibia is key to the country's economic future. The international economic sanctions that applied before independence have been lifted, and Namibia is now free to enter the potentially profitable markets of Europe and North America. This man working in a cornfield near Grootfontein is part of Namibia's crucial agricultural economy.

southern two thirds of the country, during the uprising of 1904–1907. The South African period (1917–1990) saw the imposition of apartheid as well as a bitter 26-year war for independence between the South African Army (SADF) and SWAPO. During that war, countless civilians, especially in the northern areas of the country, were harassed, detained, and abused by South African–created death squads, such as the *Koevoet* (the Afrikaans word for "crowbar").

Namibia's final liberation was the result of South African military misadventures and U.S.–Soviet cooperation in reducing tensions in the region. In 1987, as it had done many times before, South Africa invaded Angola to assist Jonas Savimbi's UNITA movement. Its objective was Cuito Cuanavale, a small town in southeastern Angola where the Luanda government had set up an air-defense installation to keep South African aircraft from supplying UNITA troops. The SADF met with fierce resistance from the Angolan Army and eventually committed thousands of its own troops to the battle. In addition, black Namibian soldiers were recruited and given UNITA uniforms to fight on the side of the SADF. Many of these proxy UNITA troops later mutinied because of their poor treatment at the hands of white South African soldiers.

HEALTH/WELFARE

The social-service delivery system of Namibia must be rebuilt to eliminate the structural inequities of apartheid. Health care for the black majority, especially those in remote rural areas, will require significant improvements. Public-health programs for blacks, nonexistent prior to independence, must be created.

South Africa failed to capture Cuito Cuanavale, and its forces were eventually surrounded. Faced with military disaster, the Pretoria government bowed to decades of international pressure and agreed to withdraw from its illegal occupation of Namibia. In return, Angola and its ally Cuba agreed to send home troops sent by Havana in 1974 after South Africa invaded Angola for the first time. Key brokers of the cease-fire, negotiations, and implementation of this agreement were the United States and the Soviet Union. This was the first instance of their post–Cold War cooperation.

A plebiscite was held in Namibia in 1989. Under United Nations supervision, more than 97 percent of eligible voters cast their ballots—a remarkable achievement given the vast distances that many had to travel to reach polling stations. SWAPO emerged as the clear winner, with 57 percent of the votes cast. The party's share of the vote increased to 73 percent in the subsequent 1995 elections; support for its main political rival, the Democratic Turnhalle Alliance, dropped to 15 percent.

CHALLENGES AND PROSPECTS

Namibia is a sparsely populated land. More than half of its more than 1.8 million residents live in the northern region known as Ovamboland. Rich in minerals, Namibia is a major producer of diamonds, uranium, copper, silver, tin, and lithium. A large gold mine recently began production, and the end of hostilities has opened up northern parts of the country to further mineral explorations.

Much of Namibia is arid. Until recently, pastoral farming was the primary agricultural activity, with beef, mutton, and goat meat the main products. Independence brought an end to international sanctions applied when South Africa ruled the country, giving Namibian agricultural goods access to the world market. Although some new investment has been attracted to the relatively well-watered but historically neglected northern border regions, most of Namibia's rural majority are barely able to eke out a living, even in nondrought years.

ACHIEVEMENTS

The government of President Sam Nujoma has received high praise for its efforts at racial and political reconciliation after a bitter 26-year war for independence. Nujoma has led these efforts and has impressed many observers with his exceptional political and consensus-building skills.

Despite the economic promise, the fledgling government of Namibia faces severe problems. It inherits an economy structurally perverted by apartheid to favor the tiny white minority. With a glaring division between fabulously wealthy whites and the oppressively poor black majority, the government is faced with the daunting problem of promoting economic development while encouraging the redistribution of wealth. Apartheid ensured that managerial positions were filled by whites, leaving a dearth of qualified and experienced nonwhite executives in the country. This past pattern of discrimination has contributed to high levels of black unemployment today.

The demobilization of 53,000 former SWAPO and South African combatants and the return of 44,000 exiles aggravated this problem. A few former soldiers—notably the Botsotsos, made up of former Koevoet members—turned to organized crime. Having already inherited a civil service bloated by too many white sinecures, the SWAPO administration resisted the temptation of trying to hire its way out of the problem. In 1991–1992, several thousand ex-combatants received vocational training in Development Brigades, modeled after similar initiatives in Botswana and Zimbabwe, but inadequate funding and preparation limited the program's success.

Another major problem lies in Namibia's economic dependence on South Africa. Before independence, Namibia had been developed as a captive market for South African goods, while its resources had been depleted by overexploitation on the part of South African firms. In 1990, all rail and most road links between Namibia and the rest of the world ran through South Africa. But South Africa's March 1, 1994, return of Walvis Bay, Namibia's only port, has greatly reduced this dependence. The port has now been declared a free-trade area. Namibia has also been linked to South Africa through a Common Monetary Area. In 1994, a new Namibian dollar was introduced, replacing the South African rand. But, at least for the time being, the currency's value remains tied to the rand.

The Nujoma government has taken a hard look at these and other economic problems and embarked on programs to solve them. SWAPO surprised everyone during the election campaign by modifying its previously strident socialist rhetoric and calling for a market-oriented economy. Since taking power, it has joined the International Monetary Fund and proposed a code for foreign investors that includes protection against undue nationalizations. Since independence, the Ministry of Finance has pursued conservative policies, which have calmed the country's largely white business community but have been criticized as insufficient to transform the economy for the greater benefit of the impoverished masses.

The government recognizes the need to attract significant foreign investment to overcome the colonial legacy of underdevelopment. In 1993, a generous package of manufacturing incentives was introduced by the Ministry of Trade and Industry. In the same year, the Namibia National Development Corporation was established to channel public investment into the economy. It is too early to assess the success of these initiatives.

NAMIBIA'S FISHING INDUSTRY

Namibia's fishing sector has made an impressive recovery after years of decline. The country's coastal waters had long supported exceptionally high concentrations of sea life due to the upwelling of nutrients by the cold offshore current. But in the years before independence, overfishing, mostly by foreign vessels, nearly wiped out many species. Since then, the government has established a 200-nautical-mile Exclusive Economic Zone along Namibia's coast and passed a Sea Fisheries Act designed to promote the conservation and controlled exploitation of the country's marine resources. These measures have been backed up by effective monitoring on the part of the new Ministry of Fisheries and Marine Resources and the creation of a National Fisheries Research and Information Centre. A rapid recovery in fish stocks has led to an annual growth of 35 percent in the sector's value.

THE ROAD AHEAD

Since 2000, Namibia has shown that regional peace has the capacity to bring about major increases in development. Germany offered a formal apology for the colonial-era killings of tens of thousands of ethnic Hereros, but ruled out compensation for victims' descendants. In November 2004, Hifikepunye Pohamba, President Nujoma's nominee, won the presidential elections as the candidate from the South West African People's Organization (SWAPO). The elections were viewed as free and fair and nonviolent.

Timeline: PAST

1884–1885
Germany is given rights to colonize Namibia at the Conference of Berlin

1904–1907
Herero, Nama, and Damara rebellions against German rule

1966
The UN General Assembly revokes a 1920 South African mandate; SWAPO begins war for independence

1968
Bantustans, or "homelands," are created by South Africa

1971
A massive strike paralyzes the economy

1978
An internal government is formed by South Africa

1980s
Defeat at Cuito Cuanavale leads to a South African agreement to withdraw from Namibia; SWAPO wins UN-supervised elections; a new Constitution is approved

1990s
More than 1,000 refugees flee to Botswana from Namibia's Caprivi regions

PRESENT

2000s
The International Court of Justice awards the disputed Kasikili/Sedudu Island to Botswana

Namibian involvement in the Congo-Kinshasa war

Unrest continues

A bridge across the Zambezi river opens, giving hopes for a boost to regional trade

South Africa (Republic of South Africa)

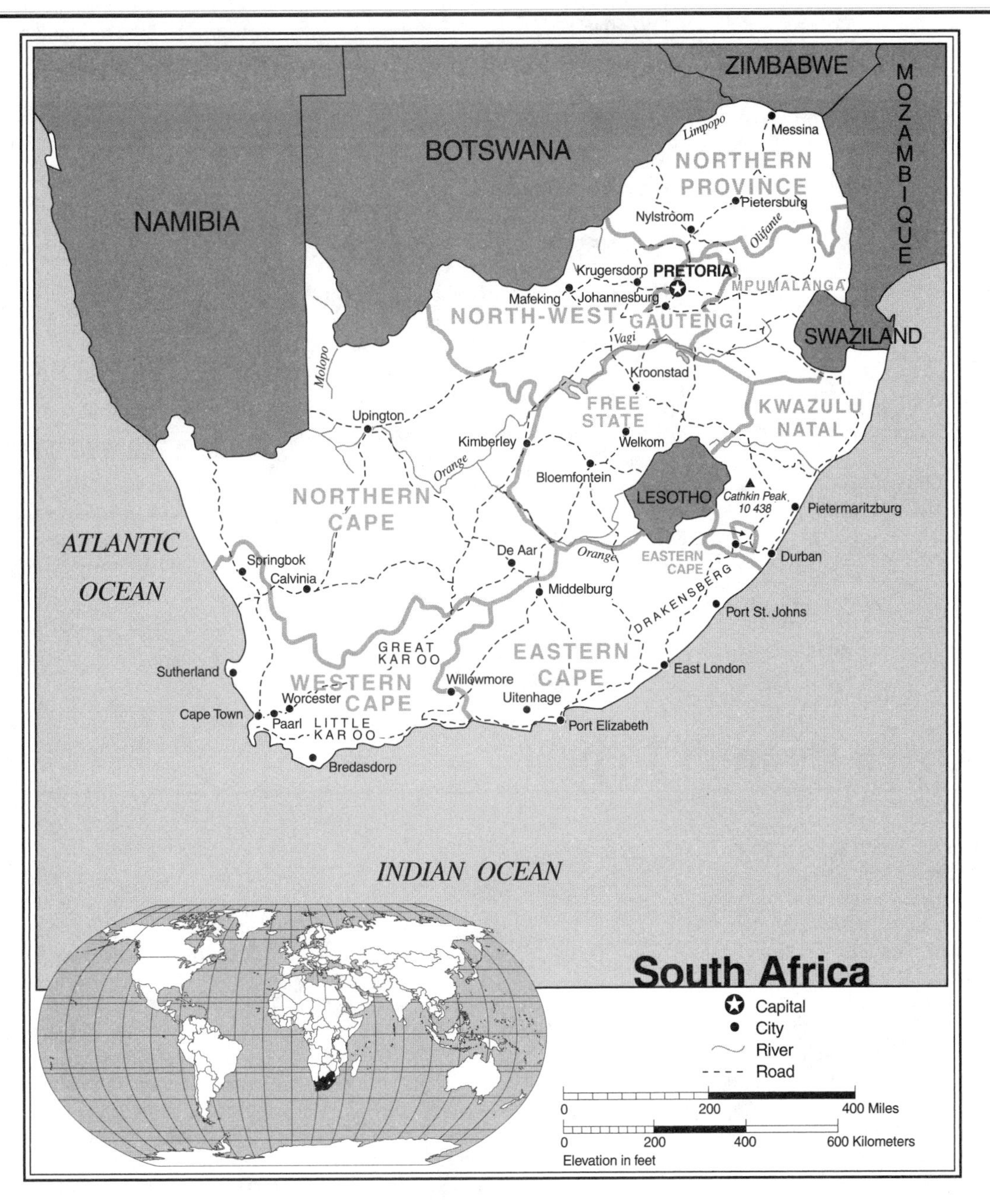

South Africa Statistics*

GEOGRAPHY

Area in Square Miles (Kilometers): 437,872 (1,222,480) (about twice the size of Texas)

Capital (Population): Pretoria (administrative) (1,590,000); Cape Town (legislative) (2,993,000); Bloemfontein (judicial) (364,000)

Environmental Concerns: water and air pollution; acid rain; soil erosion; desertification; lack of fresh water

Geographical Features: vast interior plateau rimmed by rugged hills and a narrow coastal plain

Climate: mostly semiarid; subtropical along the east coast

PEOPLE

Population

Total: 42,718,530
Annual Growth Rate: -0.25%
Rural/Urban Population Ratio: 50/50
Major Languages: Afrikaans; English; Ndebele; Pedi; Sotho; Swati; Tsonga; Tswana; Venda; Xhosa; Zulu

Ethnic Makeup: 75% black; 14% white; 9% Colored; 2% Indian
Religions: 68% Christian; 28.5% indigenous beliefs and animist; 2% Muslim; 1.5% Hindu

Health

Life Expectancy at Birth: 45 years (male); 46 years (female)
Infant Mortality: 61.7/1,000 live births
Physicians Available: 1/1,529 people
HIV/AIDS Rate in Adults: 21.5%

Education

Adult Literacy Rate: 85%
Compulsory (Ages): 7–16

COMMUNICATION

Telephones: 5,000,000+ main lines
Daily Newspaper Circulation: 29/1,000 people
Televisions: 128/1,000 people
Internet Users: 3,100,000 (2002)

TRANSPORTATION

Highways in Miles (Kilometers): 215,157 (358,596)
Railroads in Miles (Kilometers): 12,859 (21,431)
Usable Airfields: 740
Motor Vehicles in Use: 6,000,000

GOVERNMENT

Type: republic
Independence Date: May 31, 1910 (from the United Kingdom)
Head of State/Government: President Thabo Mbeki is both head of state and head of government
Political Parties: African National Congress; New National Party; Inkatha Freedom Party; African Christian Democratic Party; Freedom Front; Pan-Africanist Congress; others
Suffrage: universal at 18

MILITARY

Military Expenditures (% of GDP): 1.5%
Current Disputes: civil unrest; territorial issues with Swaziland

ECONOMY

Currency ($ U.S. Equivalent): 7.56 rands = $1
Per Capita Income/GDP: $10,700/$456.7 billion
GDP Growth Rate: 2.6%
Inflation Rate: 5.8%
Unemployment Rate: 31%
Labor Force by Occupation: 45% services; 30% agriculture; 25% industry
Population Below Poverty Line: 50%
Natural Resources: gold; chromium; coal; antimony; iron ore; manganese; nickel; phosphates; tin; diamonds; others
Agriculture: corn; wheat; sugarcane; fruits; vegetables; livestock products
Industry: mining; automobile assembly; metalworking; machinery; textiles; iron and steel; chemicals; fertilizer; foodstuffs
Exports: $32.3 billion (primary partners Europe, United States, Japan)
Imports: $28.1 billion (primary partners Europe, United States, Saudi Arabia)

SUGGESTED WEB SITES

http://www.gov.za
http://www.southafrica.co.za
http://www.cia/gov/cia/publications/factbook/geos/sf.html

**Note:* When separated, figures for blacks and whites vary greatly.

South Africa Country Report

Since taking over the South African presidency from Nelson Mandela in 1999, Thabo Mbeki has faced a series of major domestic and international challenges. Rising crime, the burgeoning HIV/AIDS pandemic, and the falling value of the country's currency (the rand) against the euro and the U.S. dollar have complicated the immense task of post-apartheid transformation. The economy remains healthy, with the tourist industry doing particularly well despite the international fallout from the 9/11 terrorist attacks in the United States. However, despite the growth of the black middle class, the distribution of wealth in South Africa remains highly skewed, largely reflecting the racial divisions of the old apartheid order. The Mbeki government's basic commitment to market-driven growth, coupled with affirmative-action policies for the previously disadvantaged, has been challenged from both the left and the right. The labor movement, led by the Congress of South African Trade Unions (COSATU), held a general strike against the government's privatization and general economic policies during 2002, notwithstanding COSATU's alliance with Mbeki's political party, the African National Congress (ANC).

DEVELOPMENT

The Government of National Unity's major priority was the implementation of the comprehensive Reconstruction and Development Plan. A major aspect of the plan was a government commitment to build 1 million low-cost houses each year for 5 years.

In April 2004, President Thabo Mbeki was elected by parliament to a second five-year term following the landslide (70% of the votes cast) general election victory of his ruling African National Congress (ANC) party. The election served to verify the popularity of the ANC throughout South Africa, and highlighted the woeful inadequacies of the opposition parties. Mongosuthu Buthelezi, the leader of the Inkatha Freedom Party, who had served as Minister of Home Affairs in the previous government, was dropped from President Mbeki's cabinet. However, even though the ANC won the election by a large margin, the structural conditions of massive poverty and low wages for working class people throughout South African society still plague the nation. Hence, in September 2004 hundreds of thousands of public sector workers went on strike over pay. The government and the unions subsequently came to an agreement, but the threat of the private sector employees taking up the strike banner to increase their pay packages remains.

Moreover the possibilities of violence from white wing forces of the former racist apartheid regime still exist. Bomb explosions in Soweto and a blast near Pretoria during 2002 are thought to be the work of right-wing extremists and, police charged 17 right-wingers with plotting against the state.

In December 2000 local-government elections, the ANC took most of the 237 local councils (59 percent), but the Democratic Alliance—created five months previously from a merger of the Democratic Party (DP), the new National Party (NP), and the Federal Alliance—captured nearly a quarter of the votes. The Inkatha Freedom Party (IFP) won 9 percent. By 2002, however, many of the members from the NP had abandoned the Democratic Alliance in favor of an alliance with the ANC. The ANC also benefited in 2002 from the government being cleared of allegations of official corruption surrounding a large 1999 arms deal.

In the area of HIV/AIDS, Mbeki, along with many of his key associates, courted controversy by expressing skepticism about commonly accepted scientific beliefs about the causal link between the HIV virus and full-blown AIDS, as well as the value of antiretroviral therapies. In December 2001, the country's High Court ruled that pregnant women must be given antiretroviral drugs to help prevent transmission of HIV to their babies. In July 2002, the Constitutional Court ordered the government to provide key anti-AIDS drugs at all public hospitals. The government had argued that such drugs were too costly; but earlier, in April 2001, a group of 39 multinational pharmaceutical companies had withdrawn its legal battle to stop South Africa from importing generic (cheaper) AIDS drugs. The decision to drop the landmark court case was hailed as a major victory for the world's poorest countries.

In hosting the inaugural meeting of the African Union and the World Summit of Sustainable Development in 2002, after the 2001 United Nations Race Conference, South Africa enhanced its diplomatic standing while proving to be an ideal setting for such gatherings. Mbeki's leading efforts to build a "New Economic Partnership for African Development" (NEPAD) were, however, undermined by a lack of consensus with other key African leaders on its blueprint, as well as European and American unease at his inability or unwillingness to take a strong stand against the ruinous policies of Robert Mugabe in neighboring Zimbabwe.

FREEDOM

The government is committed to upholding human rights, which are generally respected. Members of the security forces have committed abuses, however, including torture and excessive use of force. Action has been taken to punish some of those involved. The Truth and Reconciliation Commission, created to investigate apartheid-era human-rights abuses, completed its investigations in 1998. It made recommendations for reparations for victims and granted amnesty for full disclosure of politically motivated crimes.

Mbeki's predecessor, Mandela, remains as a unifying father figure to most South Africans. His retirement ended an extraordinary period in which South Africa struggled to come to terms with its new status as a nonracial democracy, after a long history of white-minority rule.

In April 1994, millions of South Africans turned out to vote in their country's first nonracial elections. Most waited patiently for hours to cast their ballots for the first time. The result was a landslide victory for the African National Congress, which, under the new interim Constitution, would nonetheless cooperate with two of its long-standing rivals, the National Party and the Inkatha Freedom Party, in a "Government of National Unity" (GNU). On May 10, 1994, the ANC's leader was sworn in as South Africa's first black president. Despite the history of often violent animosity between its components, the GNU survived for two years, facilitating national reconciliation. In July 1996, the NP pulled out of the GNU, giving the ANC a freer hand to pursue its ambitious but largely unrealized program of "Reconstruction and Development."

South Africa has decisively turned away from its long, tragic history of racism. For nearly 3 1/2 centuries, the territory's white minority expanded and entrenched its racial hegemony over the nonwhite majority. After 1948, successive NP governments consolidated white supremacy into a governing system known as *apartheid* ("separatehood"). But in a dramatic political about-face, the NP government, under the new leadership of F. W. de Klerk, committed itself in 1990 to a negotiated end to apartheid. Political restrictions inside the country were relaxed through the unbanning of anti-apartheid resistance organizations, most notably the ANC, the Pan-Africanist Congress (PAC), and the South African Communist Party (SACP). Thereafter, three years of on-again, off-again negotiations, incorporating virtually all sections of public opinion, resulted in a 1993 consensus in favor of a five-year, nonracial, interim Constitution.

Notwithstanding its remarkable political progress in recent years, South Africa remains a deeply divided country. In general, whites continue to enjoy relatively affluent, comfortable lives, while the vast majority of nonwhites survive in a state of impoverished deprivation. The boundary between these two worlds remains deep. Under the pre-1990 apartheid system, nonwhites were legally divided as members of three officially subordinate race classifications: "Bantu" (black Africans), "Coloureds" (people of mixed race), or "Asians." (*Note:* Many members of these three groups prefer the common label of "black," which the government now commonly uses in place of Bantu as an exclusive term for black Africans, hereafter referred to in this text as *blacks.*)

THE ROOTS OF APARTHEID

White supremacy in South Africa began with the Dutch settlement at Cape Town in 1652. For 1 1/2 centuries, the domestic economy of the Dutch Cape Colony, which gradually expanded to include the southern third of modern South Africa, rested on a foundation of slavery and servitude. Much like the American South before the Civil War, Cape Colonial society was racially divided between free white settlers and nonwhite slaves and servants. Most of the slaves were Africans imported from outside the local region, although a minority were taken from Asia. The local blacks, who spoke various Khiosan languages, were not enslaved. However, they were robbed by the Europeans of their land and herds. Many were also killed either by European bullets or diseases. As a result, most of the Cape's Khiosan were reduced to a status of servitude. Gradually, the servant and slave populations, with a considerable admixture of European blood, merged to form the core of the so-called Coloured group.

HEALTH/WELFARE

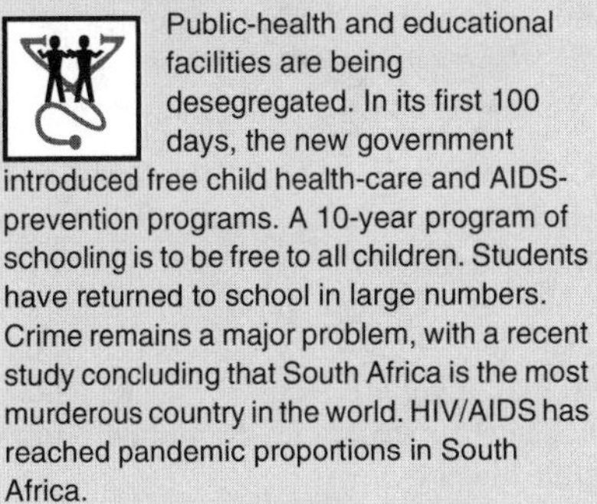

Public-health and educational facilities are being desegregated. In its first 100 days, the new government introduced free child health-care and AIDS-prevention programs. A 10-year program of schooling is to be free to all children. Students have returned to school in large numbers. Crime remains a major problem, with a recent study concluding that South Africa is the most murderous country in the world. HIV/AIDS has reached pandemic proportions in South Africa.

At the beginning of the nineteenth century, the Cape Colony reverted to British control. In the 1830s, the British abolished slavery and extended legal rights to servants. But, as with the American South, emancipation did not end racial barriers to the political and economic advancement of nonwhites. Nonetheless, even the limited reforms that were introduced upset many of the white "Cape Dutch" (or "Boers"), whose society was evolving its own "Afrikaner" identity. (Today, some 60 percent of the whites and 90 percent of the Coloureds in South Africa speak the Dutch-derived Afrikaans language.) In the mid-nineteenth century, thousands of Afrikaners, accompanied by their Coloured clients, escaped British rule by migrating into the interior. They established two independent republics, the Transvaal and the Orange Free State, whose Constitutions recognized only whites as having any civil rights.

The Afrikaners, and the British who followed them, did not settle an empty land. Then, as now, most of the people living in the area beyond the borders of the old Dutch Cape Colony were blacks who

(United Nations photo 1515685)

Resistance groups gained international recognition in their struggle against the South African apartheid regime.

spoke languages linguistically classified as Bantu. While there are nine officially recognized Bantu languages in South Africa, all but two (Tsonga and Venda) belong to either the Sotho-Tswana (Pedi, Sotho, Tswana) or Nguni (Ndebele, Swati, Xhosa, and Zulu) subgroupings of mutually intelligible dialects.

Throughout the 1700s and 1800s, the indigenous populations of the interior and the eastern coast offered strong resistance to the white invaders. Unlike the Khiosan of the Cape, these communities were able to preserve their ethnolinguistic identities. However, the settlers eventually robbed them of most of their land as well as their independence. Black subjugation served the economic interests of white farmers, and later industrialists, who were able to coerce the conquered communities into providing cheap and forced labor. After 1860, many Asians, mostly from what was then British-ruled India, were also brought into South Africa to work for next to nothing on sugar plantations. As with the blacks and Coloureds, the Asians were denied civil rights.

The lines of racial stratification were already well entrenched at the turn of the twentieth century, when the British waged a war of conquest against the Afrikaner republics. During this South African, or Boer, War, tens of thousands of Afrikaners, blacks, and Coloureds died while interned in British concentration camps. The camps helped to defeat the Afrikaner resistance but left bitter divisions between the resistance and pro-British English-speaking whites. However, it was the nonwhites who were the war's greatest losers. A compromise peace between the Afrikaners and the British Empire paved the way for the emergence, in 1910, of a self-governing "Union of South Africa," made up of the former British colonies and Afrikaner republics. In this new state, political power remained in the hands of the white minority.

"GRAND APARTHEID"

In 1948, the Afrikaner-dominated Nationalist Party was voted into office by the white electorate on a platform promising apartheid. Under this system, existing patterns of segregation were reinforced by a vast array of new laws. "Pass laws," which had long limited the movement of blacks in many areas, were extended to apply throughout the country. Black men and women were required to carry "passbooks" at all times to prove their right to be residing in a particular area. Under the Group Areas Act, more than 80 percent of South Africa was reserved for whites (who now make up no more than 14 percent of the population). In this area, blacks were confined to townships or white-owned farms, where, until recently, they were considered to be temporary residents. If they lacked a properly registered job, they were subject to deportation to one of the 10 "homelands."

Under apartheid, the homelands—poor, noncontiguous rural territories that together account for less than 13 percent of South Africa's land—were the designated "nations" of South Africa's blacks, who made up more than 70 percent of the population. Each black was assigned membership in a particular homeland, in accordance with ethnolinguistic criteria invented by the white government. Thus, in apartheid theory, there was no majority in South Africa but, rather, a single white nation—which in reality remained divided among speakers of Afrikaans, English, and other languages, and 10 separate black nations. The Coloureds and the Asians were consigned a never clearly defined intermediate position as powerless communities associated with, but segregated from, white South Africa. The apartheid "ideal" was that each black homeland would eventually become "independent," leaving white South Africa without the "burden" of a black majority. Of course, black "immigrants" could still work for the "white economy," which would remain reliant on black labor. To assure that racial stratification was maintained at the workplace, a

ACHIEVEMENTS

With the end of international cultural and sporting boycotts, South African artists and athletes have become increasingly prominent. In 1993, Nelson Mandela and F. W. de Klerk were awarded the Nobel Peace Prize, following in the footsteps of their countrymen Albert Luthuli and Desmond Tutu.

system of job classification was created that reserved the best positions for whites, certain middle-level jobs for Asians and Coloureds, and unskilled labor for blacks.

Before 1990, the NP ruthlessly pursued its ultimate goal of legislating away South Africa's black majority. Four homelands—Bophutatswana, Ciskei, Transkei, and Venda—were declared independent. The 9 million blacks who were assigned as citizens of these pseudo-states (which were not recognized by any outside country) did not appear in the 1989 South African Census, even though most lived outside of the homelands. Indeed, despite generations of forced removals and influx control, today there is not a single magistrate's district (the equivalent of a U.S. county) that has a white majority.

While for whites apartheid was an ideology of mass delusion, for blacks it meant continuous suffering. In the 1970s alone, some 3.5 million blacks were forcibly relocated because they were living in "black spots" within white areas. Many more at some point in their lives fell victim to the pass laws. Within the townships and squatter camps that ringed the white cities, families survived from day to day not knowing when the police might burst into their homes to discover that their passbooks were not in order.

Under apartheid, blacks were as much divided by their residential status as by their assigned ethnicity. In a relative sense, the most privileged were those who had established their right to reside legally within a township like Soweto. Township dwellers had the advantage of being able to live with their families and seek work in a nearby white urban center. Many of their coworkers lived much farther away, in the peri-urban areas of the homelands. Some in this less fortunate category spent as much as one third of their lives on Putco buses, traveling to and from their places of employment. Still, the peri-urban homeland workers were in many ways better off than their colleagues who were confined to crowded worker hostels for months at a time while their families remained in distant rural homelands. There were also millions of female domestic workers who generally earned next to nothing while living away from their families in the servant quarters of white households. Many of these conditions still persist in South Africa.

Further down the black social ladder were those living in the illegal squatter camps that existed outside the urban areas. Without secure homes or steady jobs, the squatters were frequent victims of night-time police raids. When caught, they were generally transported back to their homelands, from whence they would usually try once more to escape. The relaxation of influx-control regulations eased the tribulations of many squatters, but their lives remained insecure.

Yet even the violent destruction of squatter settlements by the state did not stem their explosive growth. For many blacks, living without permanent employment in a cardboard house was preferable to the hardships of the rural homelands. Nearly half of all blacks live in these areas today. Unemployment there tops 80 percent, and agricultural production is limited by marginal, overcrowded environments.

Economic changes in the 1970s and 1980s tended further to accentuate the importance of these residential patterns. Although their wages on average were only a fraction of those enjoyed by whites, many township dwellers saw their wages rise over several decades, partially due to their own success in organizing strong labor federations. At the same time, however, life in the homelands became more desperate as their populations mushroomed.

Apartheid was a totalitarian system. Before 1994, an array of security legislation gave the state vast powers over individual citizens, even in the absence of a state of emergency, such as existed throughout much of the country between 1985 and 1990. Control was more subtly exercised through the schools and other public institutions. An important element of apartheid was "Bantu Education." Beyond being segregated and unequal, black educational curricula were specifically designed to assure underachievement, by preparing most students for only semiskilled and unskilled occupations. The schools were also divided by language and ethnicity. A student who was classified as Zulu was taught in the Zulu language to be loyal to the Zulu nation, while his or her playmates might be receiving similar instruction in Tsonga or Sotho. Ethnic divisions were also often encouraged at the workplace.

LIMITED REFORMS

In 1982 and 1983, there was much official publicity about reforming apartheid. Yet the Nationalist Party's moves to liberalize the system were limited and were accompanied by increased repression. Some changes were simply semantic. In official publications, the term "apartheid" was replaced by "separate development," which was subsequently dropped in favor of "plural democracy."

A bill passed in the white Parliament in 1983 brought Asian and Coloured representatives into the South African government—but only in their own separate chambers, which remained completely subordinate to the white chamber. The bill also concentrated power in the office of the presidency, which eroded the oversight prerogatives of white parliamentarians. Significantly, the new dispensation completely excluded blacks. Seeing the new Constitution as another transparent attempt at divide-and-rule while offering them nothing in the way of genuine empowerment, most Asians and Coloureds refused to participate in the new political order. Instead, many joined together with blacks and a handful of progressive whites in creating a new organization, the United Democratic Front (UDF), which opposed the Constitution.

In other moves, the NP gradually did away with many examples of "petty" apartheid. In many areas, signs announcing separate facilities were removed from public places; but very often, new, more subtle signs were put up to assure continued segregation. Many gas stations in the Transvaal, for example, marked their bathroom facilities with blue and white figures to assure that everyone continued to know his or her place. Another example of purely cosmetic reform was the legalization of interracial marriage—although it was no longer a crime for a man and a woman belonging to different racial classifications to be wed, until 1992 it remained an offense for such a couple to live in the same house. In 1986, the hated passbooks were replaced with new "identity cards." Unions were legalized in the 1980s, but in the Orwellian world of apartheid, their leaders were regularly arrested. The UDF was not banned but, rather, was forbidden from holding meetings. Although such reforms were meaningless to most nonwhites living within South Africa, some outsiders, including the Reagan administration, were impressed by the "progress."

BLACK RESISTANCE

Resistance to white domination dates back to 1659, when the Khiosan first attempted to counter Dutch encroachments on their pastures. In the first half of the twentieth century, the African National Congress (founded in 1912 to unify what until then had been regionally based black associations) and other political and labor organizations attempted to wage a peaceful civil-rights struggle. An early leader within the Asian community was Mohandas (the Mahatma) Gandhi, who pioneered his strategy of passive resistance in South Africa while resisting the pass laws. In the 1950s, the ANC and associated organizations adopted Gandhian tactics on a massive scale, in a vain attempt to block the enactment of apartheid legislation. Although ANC president Albert Luthuli was awarded the Nobel Peace Prize, the NP regime remained unmoved.

The year 1960 was a turning point. Police massacred more than 60 persons when they fired on a passbook-burning demonstration at Sharpeville. Thereafter, the government assumed emergency powers, banning the ANC and the more recently formed Pan-Africanist Congress. As underground movements, both turned to armed struggle. The ANC's guerrilla organization, *Umkonto we Sizwe* ("Spear of the Nation"), attempted to avoid taking human lives in its attacks. *Poqo* ("Ourselves Alone"), the PAC's armed wing, was less constrained in its choice of targets but proved less able to sustain its struggle. By the mid-1960s, with the capture of such figures as Umkonto leader Nelson Mandela, active resistance had been all but fully suppressed.

A new generation of resistance emerged in the 1970s. Many nonwhite youths were attracted to the teachings of the Black Consciousness Movement (BMC), led by Steve Biko. The BMC and like-minded organizations rejected the racial and ethnic classifications of apartheid by insisting on the fundamental unity of all oppressed black peoples (that is, all nonwhites) against the white power structure. Black consciousness also rejected all forms of collaboration with the apartheid state, which brought the movement into direct opposition with homeland leaders like Gatsha Buthelezi, whom they looked upon as sellouts. In the aftermath of student demonstrations in Soweto, which sparked months of unrest across the country, the government suppressed the BMC. Biko was subsequently murdered while in detention. During the crackdown, thousands of young people fled South Africa. Many joined the exiled ANC, helping to reinvigorate its ranks.

Despite the government's heavy-handed repression, internal resistance to apartheid continued to grow. Hundreds of new and revitalized organizations—community groups, labor unions, and religious bodies—emerged to contribute to the struggle. Many became affiliated through coordinating bodies such as the UDF, the Congress of South African Trade Unions (COSATU), and the South African Council of Churches (SACC). SACC leader Archbishop Desmond Tutu became the second black South African to be awarded the Nobel Peace Prize for his nonviolent efforts to bring about change. But in the face of continued oppression, black youths, in particular, became increasingly willing to use whatever means necessary to overthrow the oppressors.

The year 1985 was another turning point. Arrests and bannings of black leaders led to calls to make the townships "ungovernable." A state of emergency was proclaimed by the government in July, which allowed for the increased use of detention without trial. By March 1990, some 53,000 people, including an estimated 10,000 children, had been arrested. Many detainees were tortured while in custody. Stone-throwing youths nonetheless continued to challenge the heavily armed security forces sent into the townships to restore order. By 1993, more than 10,000 people had died during the unrest.

TOWARD A NEW SOUTH AFRICA

Despite the Nationalist Party's ability to marshall the resources of a sophisticated military–industrial complex to maintain its totalitarian control, it was forced to abandon apartheid along with its four-decade-long monopoly of power. Throughout the 1980s, South Africa's advanced economy was in a state of crisis due to the effects of unrest and, to a lesser extent, of sanctions and other forms of international pressure. Under President P.W. Botha, the NP regime stubbornly refused to offer any openings to genuine reform. However, Botha's replacement in 1989 by F. W. de Klerk opened up new possibilities. The unbanning of the ANC, PAC, and SACP was accompanied by the release of many political prisoners. As many had anticipated, after gaining his freedom in March 1990, ANC leader Nelson Mandela emerged as the leading advocate for change. More surprising was the de Klerk government's willingness to engage in serious negotiations with Mandela and others. By August 1990, the ANC felt that the progress being made justified the formal suspension of its armed struggle.

Many obstacles blocked the transition to a postapartheid state. The NP initially advocated a form of power sharing that fell short of the concept of one person, one vote in a unified state. The ANC, UDF (disbanded in 1991), COSATU, and SACP, which were associated as the Mass Democratic Movement (MDM), however, remained steadfast in their loyalty to the nonracial principles of the 1955 Freedom Charter. Many members of the PAC and other radical critics of the ANC initially feared that the apartheid regime was not prepared to agree to its dismantlement and that the ongoing talks could only serve to weaken black resistance. On the opposite side of the spectrum were still-powerful elements of the white community who remained openly committed to continued white supremacy. In addition to the Conservative Party, the principal opposition in the old white Parliament, there were a number of militant racist organizations, which resorted to terrorism in an attempt to block reforms. Some within the South African security establishment also sought to sabotage the prospects of peace. In March 1992, these far-right elements suffered a setback when nearly 70 percent of white

Timeline: PAST

1000–1500
Migration of Bantu-speakers into Southern Africa

1652
The first settlement of Dutch people in the Cape of Good Hope area

1659
The first Khiosan attempt to resist white encroachment

1815
The British gain possession of the Cape Colony

1820s
Shaka develops the Zulu nation and sets in motion the wars and migrations known as the Mfecane

1899–1902
The Boer War: the British fight the Afrikaners (Boers)

1910
The Union of South Africa gives rights of self-government to whites

1912
The African National Congress is founded

1948
The Nationalist Party comes to power on an apartheid platform

1960
The Sharpeville Massacre: police fire on demonstration; more than 60 deaths result

1976
Soweto riots are sparked off by student protests

1980s
Unrest in the black townships leads to the declaration of a state of emergency; thousands are detained while violence escalates; F. W. de Klerk replaces P. W. Botha as president; anti-apartheid movements are unbanned; political prisoners are released

1990s
Negotiations begin for a nonracial interim Constitution; nonracial elections in May 1994 result in an ANC-led Government of National Unity; Nelson Mandela becomes president

PRESENT

2000s
Thabo Mbeki is inaugurated as president

Mbeki's responses to the HIV/AIDS pandemic draw intense criticism

South African government approved a major program to tackle HIV and AIDS voters approved continued negotiation for democratic reform.

Another troublesome factor was Buthelezi's Inkatha Freedom Party and other, smaller black groups that had aligned themselves in the past with the South African state. Prior to the elections, thousands were killed in clashes between Inkatha and ANC/MDM supporters, especially in the Natal/Kwazulu region. As the positions of the ANC and NP began to converge in 1993, the IFP delegation walked out of the negotiations and formed a "Freedom Alliance" with white conservatives and the leaders of the Bophutatswana and Ciskei homelands. It collapsed in March 1994, following the violent overthrow of the Bophutatswana regime and the defeat of groups of armed white right-wingers that rallied to its defense. Following this debacle, the IFP and more moderate white conservatives—the "Freedom Front"—agreed to participate in national elections. Attempts by more extreme right-wingers to disrupt the elections through a terrorist bombing campaign were crushed in a belated security crackdown.

The elections and the subsequent installation of the GNU were remarkably peaceful, despite organizational difficulties and instances of voting irregularities. In the end, all major parties accepted the result in which the ANC (incorporating MDM) attracted 63 percent of the vote, the NP 20 percent, IFP 10 percent, the Freedom Front 2.2 percent, and the PAC a disappointing 1.2 percent.

Swaziland (Kingdom of Swaziland)

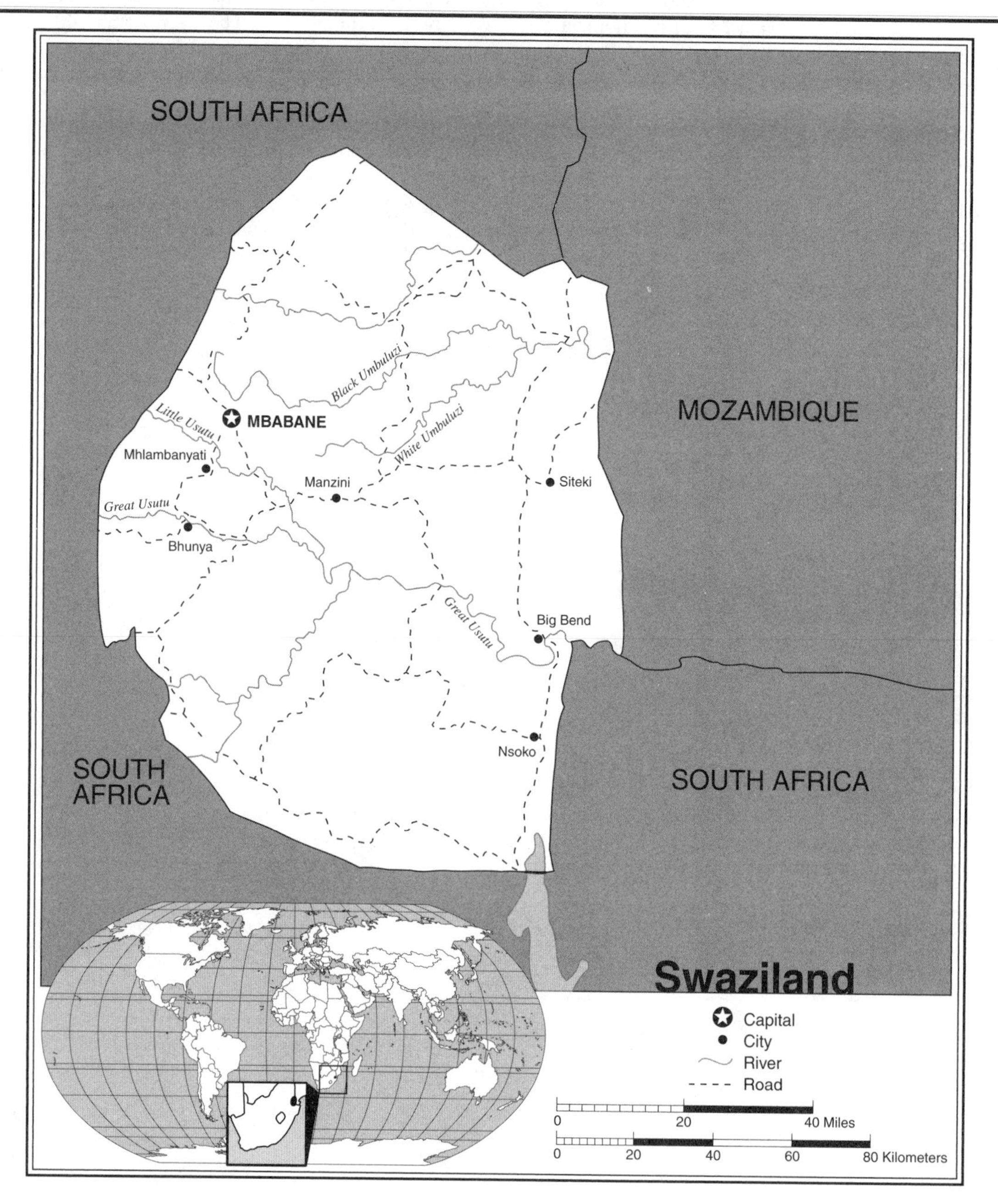

Swaziland Statistics

GEOGRAPHY

Area in Square Miles (Kilometers): 6,704 (17,366) (about the size of New Jersey)

Capital (Population): Mbabane (administrative) (80,000); Lobamba (legislative) (na)

Environmental Concerns: depletion of wildlife populations; overgrazing; soil degradation; soil erosion; limited potable water

Geographical Features: mostly mountains and hills; some sloping plains; landlocked

Climate: from tropical to temperate

PEOPLE

Population

Total: 1,169,241
Annual Growth Rate: 0.55%
Rural/Urban Population Ratio: 74/26
Major Languages: English; SiSwati
Ethnic Makeup: 97% African; 3% European
Religions: 60% Christian; 40% indigenous beliefs

Health

Life Expectancy at Birth: 36 years (male); 38 years (female)
Infant Mortality: 109/1,000 live births
Physicians Available: 1/9,265 people

HIV/AIDS Rate in Adults: 35.6%

Education

Adult Literacy Rate: 78%

COMMUNICATION

Telephones: 46,200 main lines
Televisions: 96/1,000 people
Internet Users: 27,000 (2003)

TRANSPORTATION

Highways in Miles (Kilometers): 1,769 (2,853)
Railroads in Miles (Kilometers): 184 (297)
Usable Airfields: 18
Motor Vehicles in Use: 37,000

GOVERNMENT

Type: monarchy; independent member of the British Commonwealth
Independence Date: September 6, 1968
Head of State/Government: King Mswati III; Prime Minister Absolom Themba Dlamini
Political Parties: banned by the Constitution
Suffrage: n/a

MILITARY

Military Expenditures (% of GDP): 4.7%
Current Disputes: territorial issues with South Africa

ECONOMY

Currency ($ U.S. Equivalent): 7.56 lilangenis = $1
Per Capita Income/GDP: $4,900/$5.7 billion
GDP Growth Rate: 2.5%
Inflation Rate: 7.5%
Unemployment Rate: 34%
Natural Resources: iron ore; asbestos; coal; clay; hydropower; forests; gold; diamonds; quarry stone; talc
Agriculture: corn; livestock; sugarcane; fruits; cotton; rice; sorghum; tobacco; peanuts
Industry: sugar processing; mining; wood pulp; beverages
Exports: $702 million (primary partners South Africa, Europe, Mozambique)
Imports: $850 million (primary partners South Africa, Europe, Japan)

SUGGESTED WEB SITES

http://www.swazi.com/government
http://www.sas.upenn.edu/African_Studies/Country_Specific/Swaziland.html
http://www.cia.gov/cia/publications/factbook/geos/wz.html

Swaziland Country Report

The alleged abduction of a female high school student in October 2002 focused international attention on the struggle between modernist and royal traditionalist forces in Swaziland. The young woman had been removed from her school to be considered for the honor of becoming the 11th wife of King Mswati III. In an unprecedented move, the potential bride's mother accused two of the Swazi king's close associates with kidnapping.

DEVELOPMENT

Much of Swaziland's economy is managed by the Tibiyo TakaNgwana, a royally controlled institution established in 1968 by Sobuza. It is responsible for the financial assets of the communal lands (upon which most Swazis farm) and mining operations.

A small, landlocked kingdom sandwiched between the much larger states of Mozambique and South Africa, casual observers have tended to look upon Swaziland as a peaceful island of traditional Africa that has been immune to the continent's contemporary conflicts. This image, now being increasingly challenged from within is a product of the country's status as the only precolonial monarchy in sub-Saharan Africa to have survived into the modern era. Swazi sociopolitical organization is ostensibly governed in accordance with age-old structures and norms. But below this veneer of timelessness lies a dynamic society that has been subject to internal and external pressures. The holding of restricted, nonparty elections in 1993 and 1998 has not quelled the debate over the country's political future between defenders of the status quo and those who advocate a restoration of multiparty democracy.

FREEDOM

The current political order restricts many forms of opposition, although its defenders claim that local councils, *Tikhudlas*, allow for popular participation in decision making. The leading opposition group is the People's United Democratic Movement.

During the 1993 elections, a "stay-away" campaign in favor of reform, accompanied by quiet diplomacy by neighboring states, helped push the Swazi government toward dialogue on the issue. In 1996, King Mswati announced the appointment of a committee to draw up a new constitution. But progress has since been stalled. The officially banned People's United Democratic Movement (PU-DEMO) and other civil-society groups have long called for a repeal of the 1973 royal decree that abolished constitutional rule, including the guarantee of basic freedoms.

From 1903 until the restoration of independence in 1968, the country remained a British colonial protectorate, despite sustained pressure for its incorporation into South Africa. Throughout the colonial period, the ruling Dlamini dynasty, which was led by the energetic Sobuza II after 1921, served successfully as a rallying point for national self-assertion on the key issues of regaining control of alienated land and opposing union with South Africa. Sobuza's personal leadership in both struggles contributed to the overwhelming popularity of his royalist Imbokodvo Party in the elections of 1964, 1967, and 1972. In 1973, faced with a modest but articulate opposition party, the Ngwane National Liberatory Congress, Sobuza dissolved Parliament and repealed the Westminster-style Constitution, characterizing it as "un-Swazi." In 1979, a new, nonpartisan Parliament was chosen; but authority remained with the king, assisted by his advisory council, the Liqoqo.

HEALTH/WELFARE

Swaziland's low life expectancy and high infant mortality rates have resulted in greater public-health budget allocations. A greater emphasis has also been placed on preventive medicine. However, the extremely high rate of HIV/AIDS poses severe and long-term threats to the nation.

Sobuza's death in 1982 left many wondering if Swaziland's unique monarchist institutions would survive. A prolonged power struggle increased tensions within the ruling order. Members of the Liqoqo seized effective power and appointed a new "Queen Regent," Ntombi. However, palace intrigue continued until Prince Makhosetive, Ntombi's teenage son, was installed as King Mswati III in 1986, at age 18. The new king approved the demotion of the Liqoqo back to its advisory status and has ruled through his appointed prime minister and cabinet.

ACHIEVEMENTS

The University of Swaziland was established in the 1970s and now offers a full range of degree and diploma programs.

One of the major challenges facing any Swazi government is its relationship with South Africa. Under Sobuza, Swaziland managed to maintain its political autonomy while accepting its economic dependence on its powerful neighbor. The king also maintained a delicate balance between the apartheid state and the forces opposing it. In the 1980s, this balance became tilted, with a greater degree of cooperation between the two countries' security forces in curbing suspected African National Congress (ANC) activists. In an abrupt reversal of fortunes, Swaziland's prodemocracy activists now look to the new ANC–led government in South Africa for support.

Swaziland's economy, like its politics, is the product of both internal and external initiatives. Since independence, the nation has enjoyed a high rate of economic growth, led by the expansion and diversification of its agriculture. Success in agriculture has promoted the development of secondary industries, such as a sugar refinery and a paper mill. There has also been increased exploitation of coal and asbestos. Another important source of revenue is tourism, which depends on weekend traffic from South Africa.

Swazi development has relied on capital-intensive, rather than labor-intensive, projects. As a result, disparities in local wealth and dependence on South African investment have increased. Only 16 percent of the Swazi population, including migrant workers in South Africa, were in formal-sector employment by 1989. Until recently the economy was boosted by international investors looking for a politically preferable window to the South African market. An example is Coca Cola's decision to move its regional headquarters and concentrate plant from South Africa to Swaziland; the plant employs only about 100 workers but accounts for 20 percent of all foreign-exchange earnings. The current reform process in South Africa, however, is reducing Swaziland's attraction as a center for corporate relocation and sanctions-busting. It has also increased pressure for greater democracy.

Swaziland is to have a new constitution put into effect in 2004 which provides civil liberties, but critics say the document still leaves ultimate power in the hands of the king. Royalists argue that democracy creates division, and that a king is a strong unifying force.

Timeline: PAST

1800s
Zulu and South African whites encroach on Swazi territory

1900
A protectorate is established by the British

1903
Britain assumes control over Swaziland

1968
Independence is restored

1973
Parliament is dissolved and political parties are banned

1982
King Sobuza dies

1986
King Mswati III is crowned, ending the regency period marked by political instability

1990s
Swaziland's relationship with South Africa shifts; pressures mount for increased democracy

PRESENT

2000s
AIDS is recognized as a formidable threat to the Swazi people

Pressure continues for multipartyism

An alleged kidnapping case draws international attention

Zambia (Republic of Zambia)

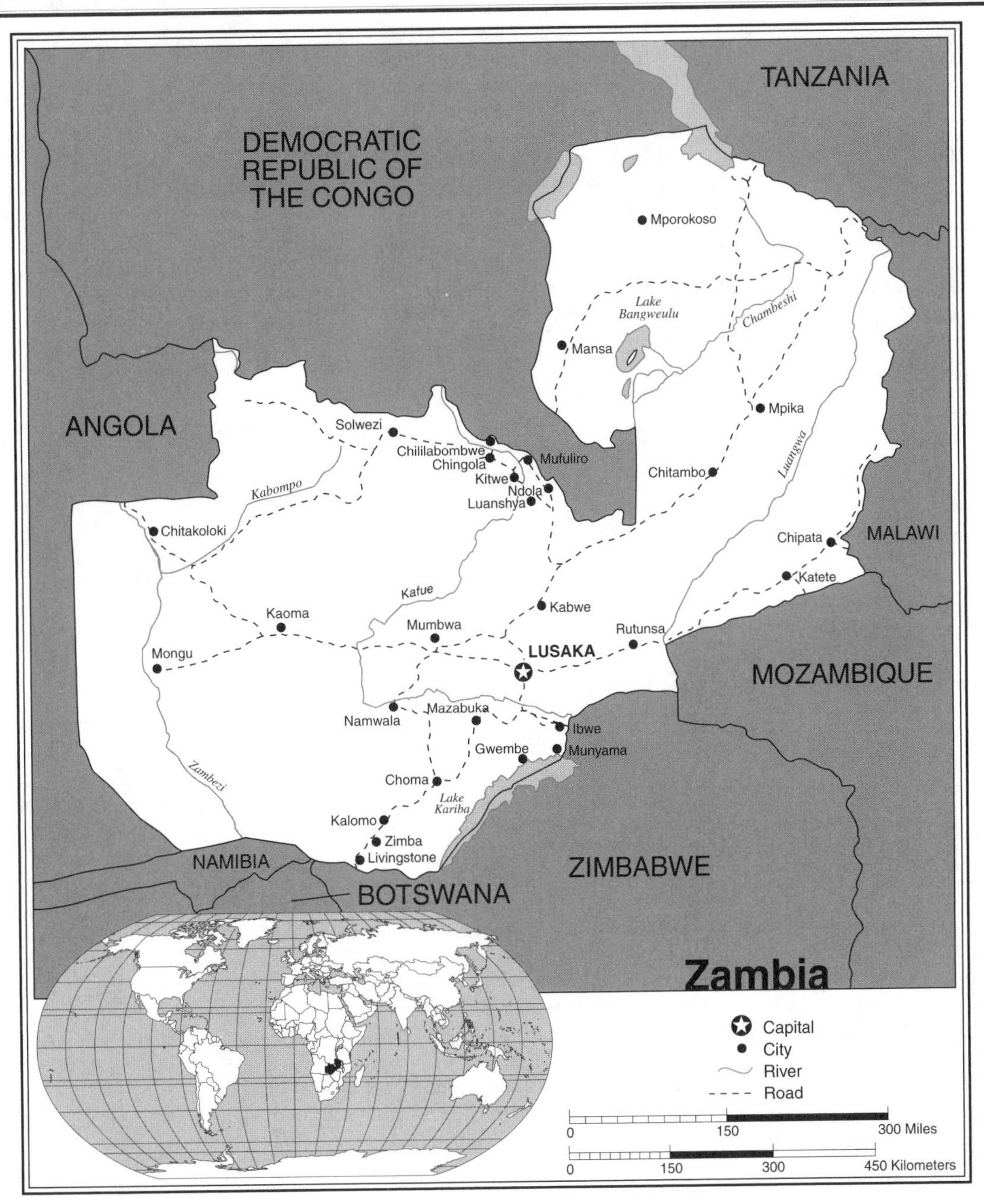

Zambia Statistics

GEOGRAPHY

Area in Square Miles (Kilometers): 290,724 (752,972) (about the size of Texas)

Capital (Population): Lusaka (1,718,000)

Environmental Concerns: air pollution; acid rain; poaching; deforestation; soil erosion; desertification; lack of adequate water treatment

Geographical Features: mostly high plateau with some hills and mountains; landlocked

Climate: tropical; modified by altitude

PEOPLE

Population

Total: 10,462,436

Annual Growth Rate: 1.9%

Rural/Urban Population Ratio: 60/40

Major Languages: English; Bemba; Nyanja; Ila-Tonga; Lozi; others

Ethnic Makeup: 99% African

Religions: 50% Christian; 48% indigenous beliefs; 2% others

Health

Life Expectancy at Birth: 37 years (male); 37 years (female)

Infant Mortality: 89.3/1,000 live births
Physicians Available: 1/10,917 people
HIV/AIDS Rate in Adults: 19.95%

Education

Adult Literacy Rate: 79%
Compulsory (Ages): 7–14

COMMUNICATION

Telephones: 130,000 main lines
Daily Newspaper Circulation: 13/1,000 people
Televisions: 32/1,000 people
Internet Users: 68,200 (2003)

TRANSPORTATION

Highways in Miles (Kilometers): 41,404 (66,781)
Railroads in Miles (Kilometers): 1,294 (2,087)
Usable Airfields: 111
Motor Vehicles in Use: 215,000

GOVERNMENT

Type: republic
Independence Date: October 24, 1964 (from the United Kingdom)
Head of State/Government: President Levy Mwanawasa is both head of state and head of government
Political Parties: Movement for Multiparty Democracy; United National Independence Party; others
Suffrage: universal at 18

MILITARY

Military Expenditures (% of GDP): 0.9%
Current Disputes: none

ECONOMY

Currency ($ U.S. Equivalent): 4,733 kwachas = $1
Per Capita Income/GDP: $870/$8.5 billion
GDP Growth Rate: 3.9%
Inflation Rate: 21.5%
Unemployment Rate: 50%
Labor Force by Occupation: 85% agriculture; 9% services; 6% industry
Natural Resources: copper; zinc; lead; cobalt; coal; emeralds; gold; silver; uranium; hydropower
Agriculture: corn; sorghum; rice; tobacco; cotton; seeds; cassava; peanuts; sugarcane
Industry: livestock; mining; foodstuffs; beverages; chemicals; textiles; fertilizer
Exports: $876 million (primary partners United Kingdom, South Africa, Switzerland)
Imports: $1.2 billion (primary partners South Africa, United Kingdom, Zimbabwe)

SUGGESTED WEB SITES

http://www.zambia.co.zm
http://www.zamnet.zm
http://www.cia.gov/cia/publications/factbook/geos/za.html

Zambia Country Report

After nearly four decades of independence, some three quarters of Zambia's population continue to live below the poverty line, earning less than $1 a day. Up to one in five has also been afflicted by the HIV virus, while many people continue to be struck down by malaria. Floods and drought since 2001 have aggravated Zambia's already poor agricultural output, placing some 2 million people at risk from starvation. The end of the civil war in Angola has, however, brought a welcome relief from border incursions.

DEVELOPMENT

Higher producer prices for agriculture, technical assistance, and rural-resettlement schemes are part of government efforts to raise Zambia's agricultural production. The agricultural sector has shown growth.

The man leading the country in the face of the above challenges is Zambia's third president, Levy Mwanawasa. In January 2002, he was sworn in, replacing his mentor, Frederick Chiluba. Chiluba had tried to alter the Constitution to allow him to run for a third term in office, but members of his own party in Parliament defeated the move. After obtaining the presidency, Mwanawasa stunned the nation by dismissing most of Chiluba's appointees and calling upon Parliament to nullify Chiluba's immunity from prosecution, declaring that the former president was implicated in a series of corruption cases. In July, Parliament voted unanimously to lift the immunity. Although members of the opposition concurred with the ruling party, they further called for a commission of inquiry to also investigate Mwanawasa, who they alleged was implicated in Chiluba-era misdeeds. During September 2004 most of the charges of corruption against former president Frederick Chiluba were dropped, but within hours he was re-arrested on six new charges.

FREEDOM

Under the MMD, police have continued to commit extra-judicial killings and other abuses. The government has continued to try to limit press freedom, while failing to honor its 1991 promise to privatize the public media.

For many, Chiluba's presidency was a disappointment. They had hoped that his election in 1991 as the leader of the Movement for Multiparty Democracy (MMD) was the beginning of a new era of democracy and development, after decades of decline under the one-party rule of his predecessor Kenneth Kaunda. But the Chiluba government largely failed to overcome the effects of high inflation and a shrinking gross domestic product, as well as the mounting challenge of the HIV/AIDS pandemic. Many educated Zambians have left their country in search of opportunities elsewhere.

Zambian politics and society became polarized following the controversial elections of 1996. In that poll, Chiluba and the MMD were reelected amid a boycott by opposition groups, including the former ruling party, the United National Independence Party (UNIP). There were reports of voter-registration irregularities and the enactment of a law barring UNIP's leader, Kaunda, and others from running on account of their being "foreigners." (Kaunda's father had been born before colonial boundaries in what is today Malawi.) The fairness of the elections was also compromised by MMD's use of state resources, especially the public media, in its campaign, and by the charging of Kaunda and nine other UNIP members with treason following a brief bombing campaign by an otherwise still shadowy group calling itself the Black Mambas (after an especially poisonous snake).

In 1997, a Zambian Army captain took control of the national radio station and an-

HEALTH/WELFARE

Life expectancy rates have increased in Zambia since independence, as a result of improved health-care facilities. However, AIDS increasingly looms as a critical problem in Zambia.

ACHIEVEMENTS

Zambia has long played a major role in the fight against white supremacy in Southern Africa. From 1964 until 1980, it was a major base for Zimbabwe nationalists.

nounced a coup. The plot had little support and was quickly crushed. In its aftermath, however, Parliament approved a 90-day state of emergency, during which time prominent critics of the government were detained on allegations of involvement in the coup. Domestic and international human-rights groups reported serious abuses throughout the period.

The deeper roots of Zambia's woes lie in Kenneth Kaunda's 27-year rule. During much of that period, the nation's economy steadily declined. Kaunda consistently blamed his country's setbacks on external forces rather than on his government's failings. There was some justification for his position. The high rate of return on exported copper made the nation one of the most prosperous in Africa until 1975. Since then, fluctuating, but generally depressed, prices for the metal—and the disruption of landlocked Zambia's traditional sea outlets as a result of strife in neighboring states—have had disastrous economic consequences.

Nonetheless, internal factors have also contributed to Zambia's decay. From the early years of Zambia's independence, Kaunda and UNIP showed little tolerance for political opposition. In 1972, the country was legally transformed into a one-party state in which power was concentrated in the hands of Kaunda and his fellow members of UNIP's Central Committee. After 1976, the government ruled with emergency powers. Although Zambia was never as repressive as such neighboring states as Malawi and Zaire/Democratic Republic of the Congo, torture and political detention without trial were common.

Timeline: PAST

A.D. 1889
Rhodes' South African Company is chartered by the British government

1924–1934
Development of the Copperbelt

1953–1963
Federation of Northern Rhodesia, Southern Rhodesia, and Nyasaland is formed; still part of British Empire

1964
Zambia gains independence

1972
Zambia becomes a one-party state under Kenneth Kaunda's United National Independence Party

1980s
South African military raids on Zambia

1990s
Kaunda is defeated in multiparty elections; Frederick Chiluba becomes the nation's second president; a coup attempt is thwarted; the government imposes a state of emergency

PRESENT

2000s
The country grapples with the HIV/AIDS pandemic

Levy Mwanawasa gains the presidency

In its rule, UNIP was supposedly guided by the philosophy of "humanism," a term that became synonymous with the "thoughts of Kaunda." The party also claimed adherence to socialism. But though it was once a mass party that spearheaded Zambia's struggle for majority rule and independence, UNIP came to stand for little other than the perpetuation of its own power.

An underlying economic problem has remained the decline of rural production, despite Zambia's considerable agricultural potential. The underdevelopment of agriculture is rooted in the colonial policies that favored mining to the exclusion of other sectors. Since independence, the rural areas have continued to be neglected in terms of infrastructural investment. Until recently Zambian farmers were paid little for their produce, with government subsidization of imported food. The result has been a continuous influx of individuals into urban areas, despite a lack of jobs for them, and falling food production.

Zambia's rural decline has severely constrained the government's ability to meet the challenge imposed by depressed export earnings from tourism, copper, and other industries. This has led to severe shortages of foreign exchange and chronic indebtedness. After years of relative inertia, the government, during the 1980s, devoted greater attention to rural development. Agricultural production rose modestly in response to increased incentives. But the size and desperate condition of the urban population discouraged the government from decontrolling prices altogether; rising maize prices in 1986 set off riots that left at least 30 people dead. The new MMD government ended the subsidies, but there has since been insufficient investment to revive the collapsed rural economy. Despite such continued neglect, in 2002 the government was adamant in its opposition to allowing the import of genetically modified grains as food aid, supposedly in order to protect domestic crops from the threat of contamination. Meanwhile, famine threatens millions of Zambians.

Zimbabwe (Republic of Zimbabwe)

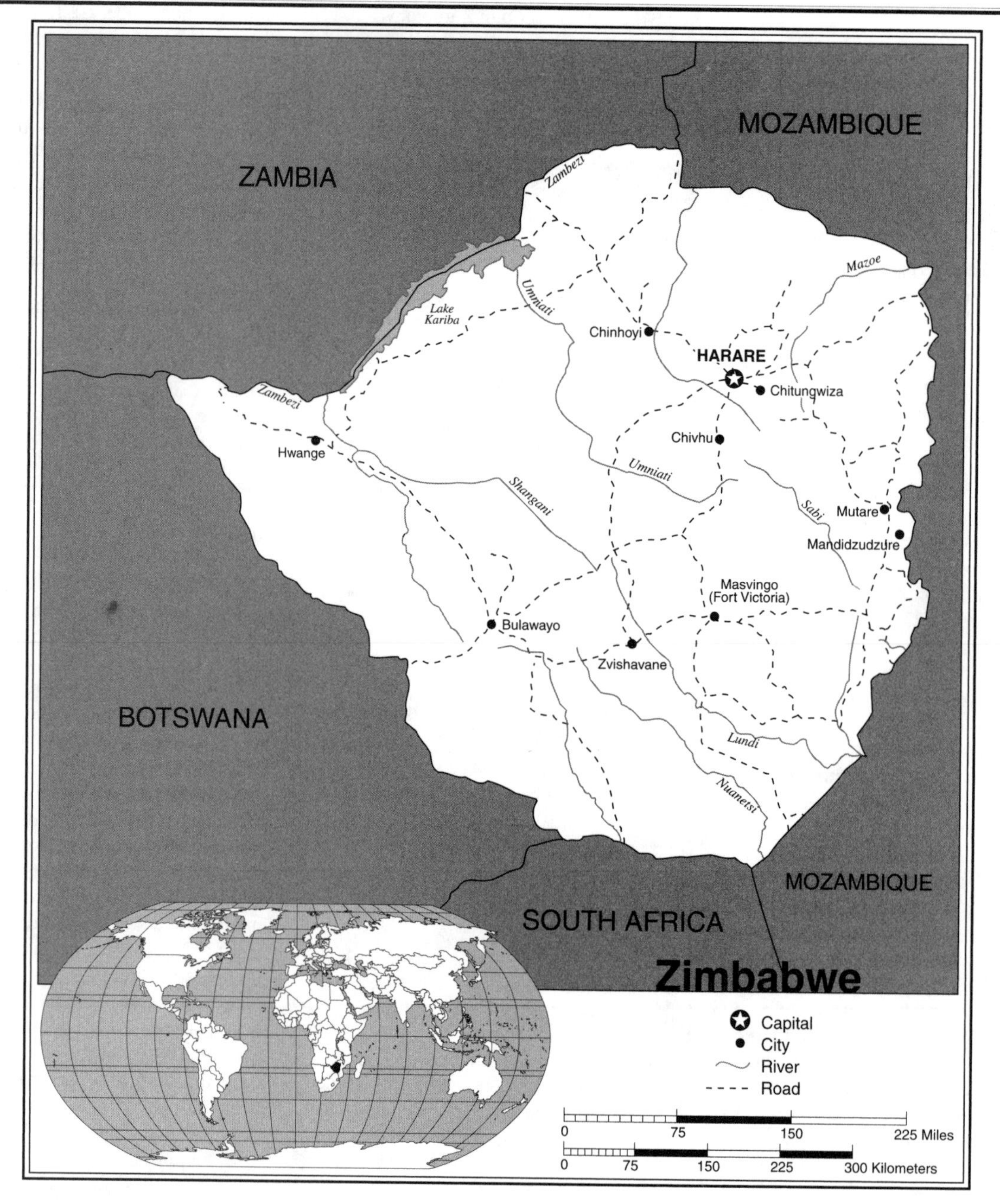

Zimbabwe Statistics

GEOGRAPHY

Area in Square Miles (Kilometers): 150,873 (390,759) (about the size of Montana)

Capital (Population): Harare (1,868,000)

Environmental Concerns: deforestation; soil erosion; land degradation; air and water pollution; poaching

Geographical Features: high plateau with higher central plateau (high veld); mountains in the east; landlocked

Climate: tropical; moderated by altitude

PEOPLE

Population

Total: 12,671,860

Annual Growth Rate: 0.05%

Rural/Urban Population Ratio: 65/35

Major Languages: English; Shona; Ndebele; Sidebele

Ethnic Makeup: 71% Shona; 16% Ndebele; 11% other African; 2% others

Religions: 50% syncretic (part Christian, part indigenous beliefs); 25% Christian; 24% indigenous beliefs; 1% Muslim

Health

Life Expectancy at Birth: 38 years (male); 35 years (female)

Infant Mortality: 62.2/1,000 live births

Physicians Available: 1/6,909 people

HIV/AIDS Rate in Adults: 25.06%

Education

Adult Literacy Rate: 85%
Compulsory (Ages): 6–13

COMMUNICATION

Telephones: 300,900 main lines
Daily Newspaper Circulation: 17/1,000 people
Televisions: 12/1,000 people
Internet Users: 500,000 (2002)

TRANSPORTATION

Highways in Miles (Kilometers): 11,369 (18,338)
Railroads in Miles (Kilometers): 1,655 (2,759)
Usable Airfields: 454
Motor Vehicles in Use: 358,000

GOVERNMENT

Type: parliamentary democracy
Independence Date: April 18, 1980 (from the United Kingdom)
Head of State/Government: Executive President Robert Mugabe is both head of state and head of government
Political Parties: Zimbabwe African National Union–Patriotic Front; Zimbabwe African National Union–Ndonga; Movement for Democratic Change; others
Suffrage: universal at 18

MILITARY

Military Expenditures (% of GDP): 3.8%
Current Disputes: involvement in the war in the Democratic Republic of the Congo

ECONOMY

Currency ($ U.S. Equivalent): 56.86 dollars = $1
Per Capita Income/GDP: $2,500/$28 billion
GDP Growth Rate: -13.6%
Inflation Rate: 384.7%
Unemployment Rate: 70%
Labor Force by Occupation: 66% agriculture; 24% services; 10% industry
Natural Resources: coal; minerals and metals
Agriculture: coffee; tobacco; corn; sugarcane; peanuts; wheat; cotton; livestock
Industry: mining; steel; textiles
Exports: $2.1 billion (primary partners South Africa, United Kingdom, Japan)
Imports: $1.5 billion (primary partners South Africa, United Kingdom, United States)

SUGGESTED WEB SITES

http://www.zimembassy-usa.org
http://www.zcbc.co.zw
http://www.zimweb.com/Dzimbabwe.html
http://www.cia.gov/cia/publications/factbook/geos/Zi.html

Zimbabwe Country Report

Zimbabwe has been in the international spotlight in recent years. The emergence of a strong opposition party, the Movement for Democratic Change (MDC), appears to have been the spark that ignited a series of major initiatives aimed at winning the support of the masses of the people by the country's aging president, Robert Mugabe. In 2000, Mugabe gave his blessing to the occupation of white-owned commercial farms by supposed veterans of Zimbabwe's liberation war, which had ended two decades earlier. The takeover of the farms, and its accompanying campaign of intimidation, played a role in a narrow, disputed victory by Mugabe's Zimbabwean African National Union–Patriotic Front (ZANU–PF) party over the MDC in June 2000 parliamentary elections.

DEVELOPMENT

Peasant production has increased dramatically since independence, creating grain reserves and providing exports for the region. The contribution of communal farmers has been recognized both within Zimbabwe and internationally.

In order to increase popular support for his administration, Mugabe began expropriating white owned land throughout Zimbabwe. The farmers, who did not receive compensation, protested as wave after wave of veterans from the civil war invaded their farms. Some of the states within the international community, particularly Britain, responded by providing aid to MDC, and, in 2002 after highly contested elections which were riddled with inconsistencies and fraud, Zimbabwe was suspended from the Commonwealth. In December 2003, after a decision by the Commonwealth to extend the suspension of Zimbabwe indefinitely, Zimbabwe pulled out of the organization completely.

At the end of November 2002 Agriculture Minister Joseph Made stated that the land-grab was over and the government had seized 35 million acres of land from white farmers. Yet tremendous damage was done to the economy, with inflation raging at a rate of more than 400% a year. The opposition leader Morgan Tsvangirai was subsequently arrested twice and jailed for treason. He was acquitted in 2004 after a lengthy trial.

At the same time, there were growing reports of violence and intimidation directed against suspected MDC supporters, especially in rural areas, as well as attacks on independent journalists. During the presidential campaign, President Mugabe and his supporters blamed western countries, particularly Britain, for being behind a campaign to "recolonize" Zimbabwe. Mugabe described British prime minister Tony Blair as a liar, a scoundrel, and a thief. In the face of credible allegations of massive vote rigging, Mugabe was proclaimed the victor in the elections. Among the international election observers present (a number of European observers were barred) there was a division of opinion as to whether the outcome sufficiently reflected the will of the people. The British Commonwealth ultimately opted to suspend Zimbabwe's membership, while the European Union imposed travel and economic sanctions against Mugabe and his close associates.

FREEDOM

Since the 1990 lifting of the state of emergency that had been in effect since the days of the Federation, Zimbabwe's human-rights record has generally improved. Some government institutions, however, especially the Central Intelligence Organization, are still accused of extra-judicial abuses.

The chaos of the last few years has been accompanied by severe economic deprivation for the masses of Zimbabweans. Those relocated on former white-owned farms have suffered along with others due to the lack of such essential inputs as fertilizers, fuel, and water. To make matters worse, the country has been hit by severe drought. An estimated 1 million people now face the prospect of hunger and starvation in Zimbabwe, and the currency has lost a great deal of its value. Zimbabwe has moved from one of Southern Africa's "breadbaskets" to a dependent regional basket case. Most observers agree that things are going to worsen in the immediate future.

The Mugabe-led ZANU-PF party has dominated Zimbabwe's politics since the country's independence from Britain in 1980. The emergence in 2000 of MDC, led by the former trade unionist Morgan Tsangarai, has posed a formidable foe to ZANU. As a coalition movement, the MDS's primary program has been to replace Mugabe. Having initially failed in this effort, cleavages between various factions of the movement may become more noticeable. In many rural areas, MDC supporters have been driven underground by official repression, but MDC remains strong in the western Matebeleland region as well as major urban areas.

Ideologically, Mugabe belongs to the African liberationist tradition of the 1960s—strong and ruthless leadership, anti-Western, suspicious of capitalism, and deeply intolerant of dissent and opposition. With more than a third of the total land and up to 80 percent of the most productive farming areas in the hands of a few thousand whites before the recent seizure, land redistribution has been a key issue for Zimbabweans. Offering land to landless African farmers is seen by many as an attractive option. In this sense, Robert Mugabe does not stand alone. By adopting mob tactics toward the emotive issue of land—where there was already a widely accepted need for redistribution reform—Mugabe may have bought his regime some additional time. But Zimbabwe's reputation as a law-abiding society now lies in tatters, along with its economy. Many have chosen to leave the country, usually illegally, in search for survival in South Africa or Botswana.

Zimbabwe achieved its formal independence in April 1980, after a 14-year armed struggle by its disenfranchised black African majority. Before 1980, the country was called Southern Rhodesia—a name that honored Cecil Rhodes, the British imperialist who had masterminded the colonial occupation of the territory in the late nineteenth century. For its black African majority, Rhodesia's name was thus an expression of their subordination to a small minority of privileged white settlers whose racial hegemony was the product of Rhodes's conquest. The new name of Zimbabwe was symbolic of the greatness of the nation's precolonial roots.

THE PRECOLONIAL PAST

By the fifteenth century, Zimbabwe had become the center of a series of states that prospered through their trade in gold and other goods with Indian Ocean merchants. These civilizations left as their architectural legacy the remains of stone settlements known as *zimbabwes*. The largest of these, the so-called Great Zimbabwe, lies near the modern town of Masvingo. Within its massive walls are dozens of stella, topped with distinctive carved birds whose likeness has become a symbol of the modern state. Unfortunately, early European fortuneseekers and archaeologists destroyed much of the archaeological evidence of this site, but what survives confirms that the state had trading contacts as far afield as China.

From the sixteenth century, the Zimbabwean civilizations seem to have declined, possibly as a result of the disruption of the East African trading networks by the Portuguese. Nevertheless, the states themselves survived until the nineteenth century. And their cultural legacy is very much alive today, especially among the 71 percent of Zimbabwe's population who speak Shona. Zimbabwe's other major ethnolinguistic community are the Ndebele-speakers, who today account for about 16 percent of the population. This group traces its local origin to the mid-nineteenth-century conquest of much of modern Zimbabwe by invaders from the south under the leadership of Umzilagazi, who established a militarily strong Ndebele kingdom, which subsequently was ruled by his son.

HEALTH/WELFARE

Public expenditure on health and education has risen dramatically since independence. Most Zimbabweans now enjoy access to medical facilities, while primary-school enrollment has quadrupled. Higher education has also been greatly expanded. But the advances are threatened by downturns in the economy, and school fees have been reintroduced.

WHITE RULE

Zimbabwe's colonial history is unique in that it was never under the direct rule of a European power. In 1890, the lands of the Ndebele and Shona were invaded by agents of Cecil Rhodes's British South Africa Company (BSACO). In the 1890s, both groups put up stiff resistance to the encroachments of the BSACO settlers, but eventually they succumbed to the invaders. In 1924, the BSACO administration was dissolved and Southern Rhodesia became a self-governing British Crown colony. "Self-government" was, in fact, confined to the white-settler community, which grew rapidly but never numbered more than 5 percent of the population.

In 1953, Southern Rhodesia was federated with the British colonial territories of Northern Rhodesia (Zambia) and Nyasaland (Malawi). This Central African Federation was supposed to evolve into a "multiracial" dominion; but from the beginning, it was perceived by the black majority in all three territories as a vehicle for continued white domination. As the federation's first prime minister put it, the partnership of blacks and whites in building the new state would be analogous to a horse and its rider—no one had any illusions as to which race group would continue to be the beast of burden.

In 1963, the federation collapsed as a result of local resistance. Black nationalists established the independent "nonracial" states of Malawi and Zambia. For a while, it appeared that majority rule would also come to Southern Rhodesia. The local black community was increasingly well organized and militant in demanding full citizenship rights. However, in 1962, the white electorate responded to this challenge by voting into office the Rhodesia Front (RF), a party determined to uphold white supremacy at any cost. Using already-existing emergency powers, the new government moved to suppress the two major black nationalist movements: the Zimbabwe African People's Union (ZAPU) and the Zimbabwe African National Union (ZANU).

RHODESIA DECLARES INDEPENDENCE

In a bid to consolidate white power along the lines of the neighboring apartheid regime of South Africa, the RF, now led by Ian Smith, made its 1965 Unilateral Declaration of Independence (UDI) of any ties to the British Crown. Great Britain, along with the United Nations, refused to recognize this move. In 1967, the United Nations imposed mandatory economic sanctions against the "illegal" RF regime. But the sanctions were not fully effective, largely due to the fact that they were flouted by South Africa and the Portuguese authorities who controlled most of Mozambique until 1974. The United States continued openly to purchase Rhodesian chrome for a number of years, while many states and individuals engaged in more covert forms of sanctions-busting. The Rhodesian economy initially benefited from the porous blockade, which encouraged the development of a wide range of import-substitution industries.

With the sanctions having only a limited effect and Britain and the rest of the international community unwilling to engage in more active measures, it soon became clear that the burden of overthrowing the RF regime would be borne by the local population. ZANU and ZAPU, as underground movements, began to engage in armed struggle beginning in 1966. The success of their attacks initially was limited; but from 1972, the Rhodesian Security Forces were increasingly besieged by the nationalists' guerrilla campaign. The 1974 liberation of Mozambique from the Portuguese greatly increased the effectiveness of the ZANU forces, who were allowed to infiltrate into Rhodesia from Mozambican territory. Meanwhile, their ZAPU comrades launched attacks from bases in Zambia. In 1976, the two groups became loosely affiliated as the Patriotic Front.

ACHIEVEMENTS

Zimbabwe's capital city of Harare has become an arts and communications center for Southern Africa. Many regional as well as local filmmakers, musicians, and writers based in the city enjoy international reputations. And the distinctive malachite carvings of Zimbabwean sculptors are highly valued in the international art market.

Unable to stop the military advance of the Patriotic Front, which was resulting in a massive white exodus, the RF attempted to forge a power-sharing arrangement that preserved major elements of settler privilege. Although rejected by ZANU or ZAPU, this "internal settlement" was implemented in 1978–1979. A predominantly black government took office, but real power remained in white hands, and the fighting only intensified. Finally, in 1979, all the belligerent parties, meeting at Lancaster House in London, agreed to a compromise peace, which opened the door to majority rule while containing a number of constitutional provisions designed to reassure the white minority. In the subsequent elections, held in 1980, ZANU captured 57 and ZAPU 20 out of the 80 seats elected by the "common roll." Another 20 seats, which were reserved for whites for seven years as a result of the Lancaster House agreement, were captured by the Conservative Alliance (the new name for the RF). ZANU leader Robert Mugabe became independent Zimbabwe's first prime minister.

THE RHODESIAN LEGACY

The political, economic, and social problems inherited by the Mugabe government were formidable. Rhodesia had essentially been divided into two "nations": one black, the other white. Segregation prevailed in virtually all areas of life, with those facilities open to blacks being vastly inferior to those open to whites. The better half of the national territory had also been reserved for white ownership. Large commercial farms prospered in this white area, growing maize and tobacco for export as well as a diversified mix of crops for domestic consumption. In contrast, the black areas, formally known as Tribal Trust Lands, suffered from inferior soil and rainfall, overcrowding, and poor infrastructure. Most black adults had little choice but to obtain seasonal work in the white areas. Black unskilled workers on white plantations, like the large number of domestic servants, were particularly impoverished. But until the 1970s, there were also few opportunities for blacks with higher skills as a result of a de facto "color bar," which reserved the best jobs for whites.

Despite its stated commitment to revolutionary socialist objectives, since 1980, the Mugabe government has taken an evolutionary approach in dismantling the socioeconomic structures of old Rhodesia. This cautious policy is, in part, based on an appreciation that these same structures support what, by regional standards, is a relatively prosperous and self-sufficient economy. Until 1990, the government's hands were also partially tied by the Lancaster House accords, wherein private property, including the large settler estates, could not be confiscated without compensation. In its first years, the government nevertheless made impressive progress in improving the livelihoods of the Zimbabwean majority by redistributing some of the surplus of the still white-dominated private sector. With the lifting of sanctions, mineral, maize, and tobacco exports expanded and import restrictions eased. Workers' incomes rose, and a minimum wage, which notably covered farm employees, was introduced. Rising consumer purchasing power initially benefited local manufacturers. Health and educational facilities were expanded, while a growing number of blacks began to occupy management positions in the civil service and, to a lesser extent, in businesses.

Zimbabwe had hoped that foreign investment and aid would pay for an ambitious scheme to buy out many white farmers and to settle black peasants on their land. However, funding shortfalls resulted in only modest resettlement. Approximately 4,000 white farmers owned more than one third of the land. In 1992, the government passed a bill that allowed for the involuntary purchase of up to 50 percent of this land at an officially set price. While enjoying overwhelming domestic support, this land-redistribution measure came under considerable external criticism for violating the private-property and judicial rights of the large-scale farmers. Others pointed out that, besides producing large surpluses of food in nondrought years, many jobs were tied to the commercial estates. Revelations in 1993–1994 that some confiscated properties had been turned over to leading ZANU politicians gave rise to further controversy.

While gradually abandoning its professed desire to build a socialist society, the Zimbabwean government has continued to face a classic dilemma of all industrializing societies: whether to continue to use tight import controls to protect its existing manufacturing base or to open up its economy in the hopes of enjoying a takeoff based on export-oriented growth. While many Zimbabwean manufacturers would be vulnerable to greater foreign competition, there is now a widespread consensus that limits of the local market have contributed to stagnating output and physical depreciation of local industry in recent years.

POLITICAL DEVELOPMENT

The Mugabe government has promoted reconciliation across the racial divide. Although the reserved seats for whites were abolished in 1987, the white minority (whites now make up less than 2 percent of the population) is well represented within government as well as business. Mugabe's ZANU administration has shown less tolerance of its political opponents, especially ZAPU. ZANU was originally a breakaway faction of ZAPU. At the time of this split, in 1963, the differences between the two movements had largely been over tactics. But elections in 1980 and 1985 confirmed that the followings of both movements have become ethnically based, with most Shona supporting ZANU and Ndebele supporting ZAPU.

Initially, ZANU agreed to share power with ZAPU. However, in 1982, the alleged discoveries of secret arms caches, which ZANU claimed ZAPU was stockpiling for a coup, led to the dismissal of the ZAPU ministers. Some leading ZAPU figures were also detained. The confrontation led to violence that very nearly degenerated into a full-scale civil war. From 1982 to 1984, the Zimbabwean Army, dominated by former ZANU and Rhodesian units, carried out a brutal counterinsurgency campaign against supposed ZAPU dissidents in the largely Ndebele areas of western Zimbabwe. Thousands of civilians were killed—especially by the notorious Fifth Brigade, which operated outside the normal military command structure. Many more fled to Botswana, including, for a period, the ZAPU leader, Joshua Nkomo.

Until 1991, Mugabe's stated intention was to create a one-party state in Zimbabwe. With his other black and white opponents compromised by their past association with the RF and its internal settlement, this largely meant coercing ZAPU into dissolving itself into ZANU. However, the increased support for ZAPU in its core Ndebele constituencies during the 1985 elections led to a renewed emphasis on the carrot over the stick in bringing about the union. In 1987, ZAPU formally merged into ZANU, but their shotgun wedding made for an uneasy marriage.

With the demise of ZAPU, new forces have emerged in opposition to Mugabe and the drive for a one-party state. Principal among these is the Zimbabwe Unity Movement (ZUM), led by former ZANU member Edger Tekere. In the 1990 elections, ZUM received about 20 percent of the vote, in a poll that saw a sharp drop in voter participation. The elections were also marred by serious restrictions on opposition activity and blatant voter intimidation. The deaths of ZUM supporters in the period before the elections reinforced the message of the government-controlled media that a vote for the opposition was an act of suicide. A senior member of the Central Intelligence Organization and a ZANU activist were subsequently convicted of the murder of ZUM organizing secretary Patrick Kombayi. However, they were pardoned by Mugabe.

Mugabe initially claimed that his 1990 victory was a mandate to establish a one-party state. But in 1991, the changing international climate, the continuing strength of the opposition, and growing opposition within ZANU itself caused him to shelve the project. Under 1992 election law, however, ZANU alone was made eligible for state funding.

The survival of political pluralism in Zimbabwe reflects the emergence of a civil society that is increasingly resistant to the concentration of power. Independent non-governmental organizations have successfully taken up many social human-rights issues. Less successful have been attempts to promote an independent press, which has remained almost entirely in government/ZANU hands.

In 1992, the Forum Party, a new opposition movement, was launched, under the leadership of former chief justice Enoch Dumbutshena. But it failed to break the mold of Zimbabwean politics due to its own internal splits and failure to unite with other groups. As a result, Mugabe was easily reelected in March 1996 in a poll with low voter turnout (it was ultimately boycotted by the entire opposition).

Notwithstanding its continuing electoral success, public confidence in the ZANU government has been greatly eroded by its relative failure in handling the 1992 drought crisis. Despite warning signs of the impending catastrophe, little attempt was made to stockpile food. This failure resulted in widespread hunger and dependence on expensive food imports. Long-neglected waterworks, especially those serving Bulawayo, the country's second-largest city, proved to be inadequate. The government also lost support due to its seeming insensitivity to the plight of ordinary Zimbabweans suffering from high rates of unemployment and inflation. With inflation at 22 percent, a civil servants strike was sparked in August 1996 by an across-the-board 6 percent raise for ordinary workers as compared to a 130 percent raise for members of Parliament.

While the welfare of ordinary Zimbabweans may have improved since 1980, popular frustration with the status quo is increasing. In 1998, resentment against the government, resulting from continued economic decline, was aggravated in some quarters by Mugabe's decision to dispatch nearly 3,000 troops to the Democratic Republic of the Congo (D.R.C., the former Zaire) to defend the embattled regime of Laurent Kabila. With formal-sector unemployment in excess of 50 percent and inflation approaching 200 percent, it was a foreign adventure that the country could ill afford.

Timeline: PAST

1400s–1500s
Heyday of the gold trade and Great Zimbabwe

1840s
The Ndebele state emerges in Zimbabwe

1890
The Pioneer Column: arrival of the white settlers

1895–1897
Chimurenga: rising against the white intruders, ending in repression by whites

1924
Local government in Southern Rhodesia is placed in the hands of white settlers

1965
Unilateral Declaration of Independence

1966
Armed struggle begins

1980
ZANU leader Robert Mugabe becomes Zimbabwe's first prime minister

1990s
ZANU and ZAPU merge and win the 1990 elections; elections in 1995 result in a landslide victory for the ruling ZANU-PF

PRESENT

2000s
Mugabe pushes redistribution of white-owned commercial farms

West Africa

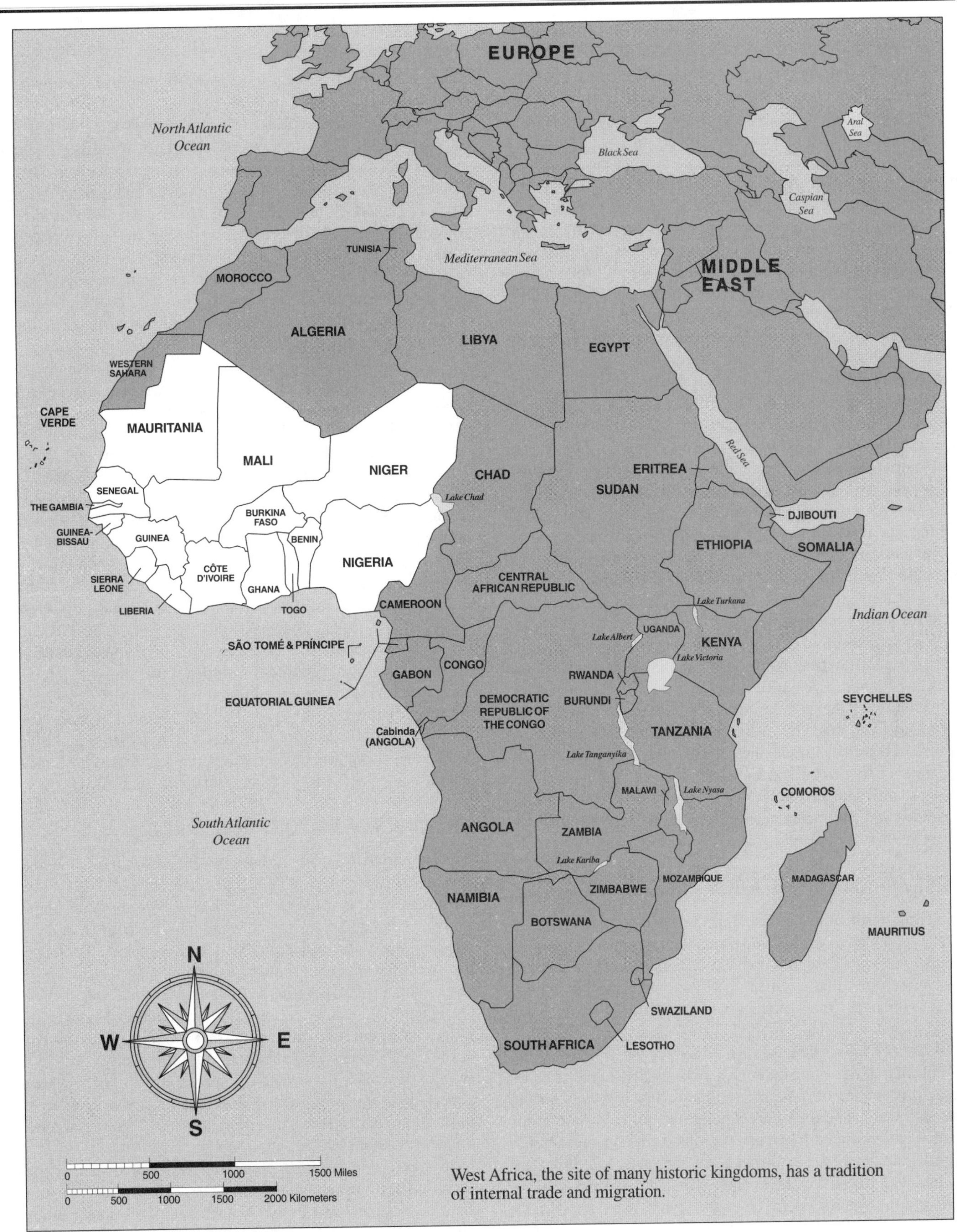

West Africa, the site of many historic kingdoms, has a tradition of internal trade and migration.

West Africa: Seeking Unity in Diversity

Anyone looking at a map of Africa can identify West Africa as the great bulge on the western coast of the continent. It is a region bound by the Sahara Desert to the north, the Atlantic Ocean to the south and west, and, in part, by the Cameroonian Mountains to the east. Each of these boundaries has historically been a bridge rather than a barrier, in that the region has long been linked through trade to the rest of the world.

At first glance, what is more striking is West Africa's great variety, rather than any of its unifying features. It contains the environmental extremes of desert and rain forest. While most of its people rely on agriculture, every type of occupation can be found, from herders to factory workers. Hundreds of languages are spoken; some are as different from one another as English is from Arabic or Japanese. Local cultural traditions and the societies that practice them are also myriad.

Yet the more closely one examines West Africa, the more one is impressed with the features that give the nations of the region a degree of coherence and unity. Some of the common characteristics and features of West Africa as a whole include the vegetation belts that stretch across the region from west to east, creating a similar environmental mix among the region's polities; the constant movement of peoples across local and national boundaries; and efforts being made by West African governments toward greater integration in the region, primarily through economic organizations. West Africans also share elements of a common history.

With the exception of Liberia, all the contemporary states of West Africa were the creations of competing European colonial powers—France, Germany, Great Britain, and Portugal—that divided most of the area during the late 1800s. Before this partition, however, much of the region was linked by the spread of Islam and patterns of trade, including the legacy of intensive involvement between the sixteenth and nineteenth centuries in the trans-Atlantic slave trade. From ancient times, great kingdoms expanded and contracted across the West African savanna and forest, giving rise to sophisticated civilizations.

WEST AFRICAN VEGETATION AND CLIMATE ZONES

Traveling north from the coastlines of such states as Nigeria, Ghana, and Côte d'Ivoire, one encounters tropical rain forests, which give way first to woodland savanna and then to more arid, more open plains. In Mali, Niger, and other landlocked areas to the north, the savanna gives way to the still drier Sahel environment, and finally to the Sahara Desert itself.

Whatever their ethnicity or nationality, the peoples living within each of these vegetation zones generally share the benefits and problems of similar livelihoods. For instance, cocoa, coffee, yams, and cassava are among the cash and food crops planted in the cleared forest and woodland zones, which stretch from Guinea to Nigeria. Groundnuts, sorghum, and millet are commonly harvested in the savanna belt that runs from Senegal to northern Nigeria. Herders in the Sahel, who historically could not go too far south with their cattle because of the presence of the deadly tsetse fly in the forest, continue to cross state boundaries in search of pasture.

People throughout West Africa have periodically suffered from drought. The effects of drought have often been aggravated in recent years by population pressures on the land. These factors have contributed to environmental changes and degradation. The condition of the Sahel in particular has deteriorated through a process of desertification, leading to large-scale relocations among many of its inhabitants. The eight Sahelian countries—Cape Verde, The Gambia, Burkina Faso, Mali, Senegal, Niger, Chad (in Central Africa), and Mauritania—have consequently formed the Committee for Struggle Against Drought in the Sahel (CILSS).

Farther to the south, large areas of woodland savanna have turned into grasslands as their forests have been cut down by land-hungry farmers. Drought has also periodically resulted in widespread brushfires in Ghana, Côte d'Ivoire, Togo, and Benin, fires that have transformed forests into savannas and savannas into deserts. Due to the depletion of forest, the Harmattan (a dry wind that blows in from the Sahara during January and February) now reaches many parts of the coast that in the recent past did not feel its breath. Its dust and haze have become a sign of the new year—and of new agricultural problems—throughout much of West Africa.

The great rivers of West Africa, such as The Gambia, Niger, Senegal, and Volta, along with their tributaries, have become increasingly important both as avenues of travel and trade and for the water they provide. Countries have joined together in large-scale projects designed to harness their waters for irrigation and hydroelectric power through regional organizations, like the Mano River grouping of Guinea, Liberia, and Sierra Leone and the Organization for the Development of the Senegal River, composed of Mali, Mauritania, and Senegal.

THE LINKS OF HISTORY AND TRADE

The peoples of West Africa have never been united as members of a single political unit. Yet some of the precolonial kingdoms that expanded across the region have great symbolic importance for those seeking to enhance interstate cooperation. The Mali empire of the thirteenth to fifteenth centuries, the Songhai empire of the sixteenth century, and the nineteenth-century Fulani caliphate of Sokoto, all based in the savanna, are widely remembered as examples of past supranational glory. The kingdoms of the southern forests, such as the Asante Confederation, the Dahomey kingdom, and the Yoruba city-states, were smaller than the great savanna empires to their north. Although generally later in origin and different in character from the northern states, the forest kingdoms are, nonetheless, sources of greater regional identity.

The precolonial states of West Africa gave rise to great urban centers, interlinked through extensive trade networks. This development was probably the result of the area's agricultural productivity, which supported a relatively high population

(Photo by Wayne Edge)

As this teenage girl in her uniform indicates, education remains a prominent theme for African youth.

density from early times. Many modern settlements have long histories. Present-day Timbuctu and Gao, in Mali, were important centers of learning and commerce in medieval times. Some other examples include Ouagadougou, Ibadan, Benin, and Kumasi, all in the forest zone. These southern centers prospered in the past by sending gold, kola, leather goods, cloth—and slaves—to the northern savanna and southern coast.

The cities of the savannas linked West Africa to North Africa. Beginning in the eleventh century, the ruling groups of the savanna increasingly turned to the universal vision of Islam. While Islam also spread to the forests, the southernmost areas were ultimately more strongly influenced by Christianity, which was introduced by Europeans, who became active along the West African coast in the fifteenth century. For centuries, the major commercial link among Europe, the Americas, and West Africa was the trans-Atlantic slave trade; during the 1800s, however, legitimate commerce in palm oil and other tropical products replaced it. New centers such as Dakar, Accra, and Freetown emerged—resulting either from the slave trade or from its suppression.

THE MOVEMENT OF PEOPLES

Despite the (incorrect) view of many who see Africa as being a continent made up of isolated groups, one constant characteristic of West Africa has been the transregional migration of its people. Herders have moved east and west across the savanna and south into the forests. Since colonial times, many professionals as well as laborers have sought employment outside their home areas.

Some of the peoples of West Africa, such as the Malinke, Fulani, Hausa, and Mossi, have developed especially well-established heritages of mobility. In the past, the Malinke journeyed from Mali to the coastal areas in Guinea, Senegal, and The Gambia. Other Malinke traders made their way to Burkina Faso, Liberia, and Sierra Leone, where they came to be known as Mandingoes.

The Fulani have developed their own patterns of seasonal movement. They herd their cattle south across the savanna during the dry season and return to the north during the rainy season. Urbanized Fulani groups have historically journeyed from west to east, often serving as agents of Islamization as well as promoters of trade. More recently, many Fulani have been forced to move southward as a result of the deterioration of their grazing lands. The Hausa, who live mostly in northern Nigeria and Niger, are found throughout much of West Africa. Indeed, their trading presence is so widespread that some have suggested that the Hausa language be promoted as a lingua franca, or common language, for West Africa.

Millions of migrant laborers are regularly attracted to Côte d'Ivoire and Ghana from the poorer inland states of Burkina Faso, Mali, and Niger, thus promoting continuing economic interdependence among these states. Similar large-scale migrations also occur elsewhere. The drastic expulsion of aliens by the Nigerian government in 1983 was startling to the outside world, in part because few had realized that so many Ghanaians, Nigeriens, Togolese, Beninois, and Cameroonians had taken up residence in Nigeria. Such immigration is not new, though its scale into Nigeria was greatly increased by that country's oil boom. Peoples such as the Yoruba, Ewe, and Vai, who were divided by colonialism, have often ignored modern state boundaries in order to maintain their ethnic ties. Other migrations also have roots in the colonial past. Sierra Leonians worked as clerks and craftspeople throughout the coastal areas of British West Africa, while Igbo were recruited to serve in northern Nigeria. Similarly, Beninois became the assistants of French administrators in other parts of French West Africa, while Cape Verdians occupied intermediate positions in Portugal's mainland colonies.

WEST AFRICAN INTEGRATION

Many West Africans recognize the weaknesses inherent in the region's national divisions. The peoples of the region would benefit from greater multilateral political cooperation and economic integration. Yet there are many obstacles blocking the growth of pan-regional development. National identity is probably even stronger today than it was in the days when Kwame Nkrumah, the charismatic Ghanaian leader, pushed for African unity but was frustrated by parochial interests. The larger and more prosperous states, such as Nigeria and Côte d'Ivoire, are reluctant to share their relative wealth with smaller countries, which, in turn, fear being swallowed.

One-party rule and more overt forms of dictatorship have recently been abandoned throughout West Africa. However, for the moment, the region is still politically divided between those states that have made the transition to multiparty constitutional systems of government and those that are still under effective military control. Overlapping ethnicity is also sometimes more a source of suspicion rather than a source of unity between states. Because the countries were under the rule of different colonial powers, French, English, and Portuguese serve today as official languages of the different nations, which also inherited different administrative traditions. Moreover, during colonial times, independent infrastructures were developed in each country; these continue to orient economic activities toward the coast and Europe rather than encouraging links among West African countries.

Political changes also affect regional cooperation and domestic development. Senegambia, the now-defunct confederation of Senegal and The Gambia, was dominated by Senegal and resented by many Gambians. The Liberian Civil War has also led to division between the supporters and opponents of the intervention of a multinational peacekeeping force.

Despite the many roadblocks to unity, a number of multinational organizations have developed in West Africa, stimulated in large part by the severity of the common problems that the countries face. The West African countries have a good record of cooperating to avoid armed conflict and to settle their occasional border disputes. In addition to the multilateral agencies that are coordinating the struggle against drought and the development of various river basins, there are also various regional commodity cartels, such as the five-member Groundnut Council. The West African Examinations Council standardizes secondary-school examinations in most of the countries where English is an official language, and most of the Francophonic states have the same currency.

The most ambitious and broad organization in the region is the Economic Organization of West African States (ECOWAS), which includes all the states incorporated in the West African section of this text. Established in 1975 by the Treaty of Lagos, ECOWAS aims to promote trade, cooperation, and self-reliance. The progress of the organization in these areas has thus far been limited. But ECOWAS can point to some significant achievements. Several joint ventures have been developed; steps toward tariff reduction are being taken; and ECOWAS members have agreed in principle to eventually establish a common currency.

The ECOWAS states have shown an increasing willingness and capacity to play a leading role in collectively resolving their regional conflicts. Over the past decade, through its multinational peacekeeping force, ECOMOG, members of ECOWAS have jointly intervened to assist in the settlement of internal conflicts in Liberia and Sierra Leone. More recently, ECOWAS has also acted as a mediator in the ongoing civil conflict in Côte d'Ivoire. The modest success of these initiatives points to the pivotal role that must be played by Nigeria in any move toward greater regional cooperation. With about half of West Africa's population and economic output, a revitalized Nigeria has already demonstrated its potential for regional leadership. But Nigeria's own progress, as well as that of the region as a whole, is dependent on its making further progress toward overcoming its own internal political and economic weaknesses.

Benin (Republic of Benin)

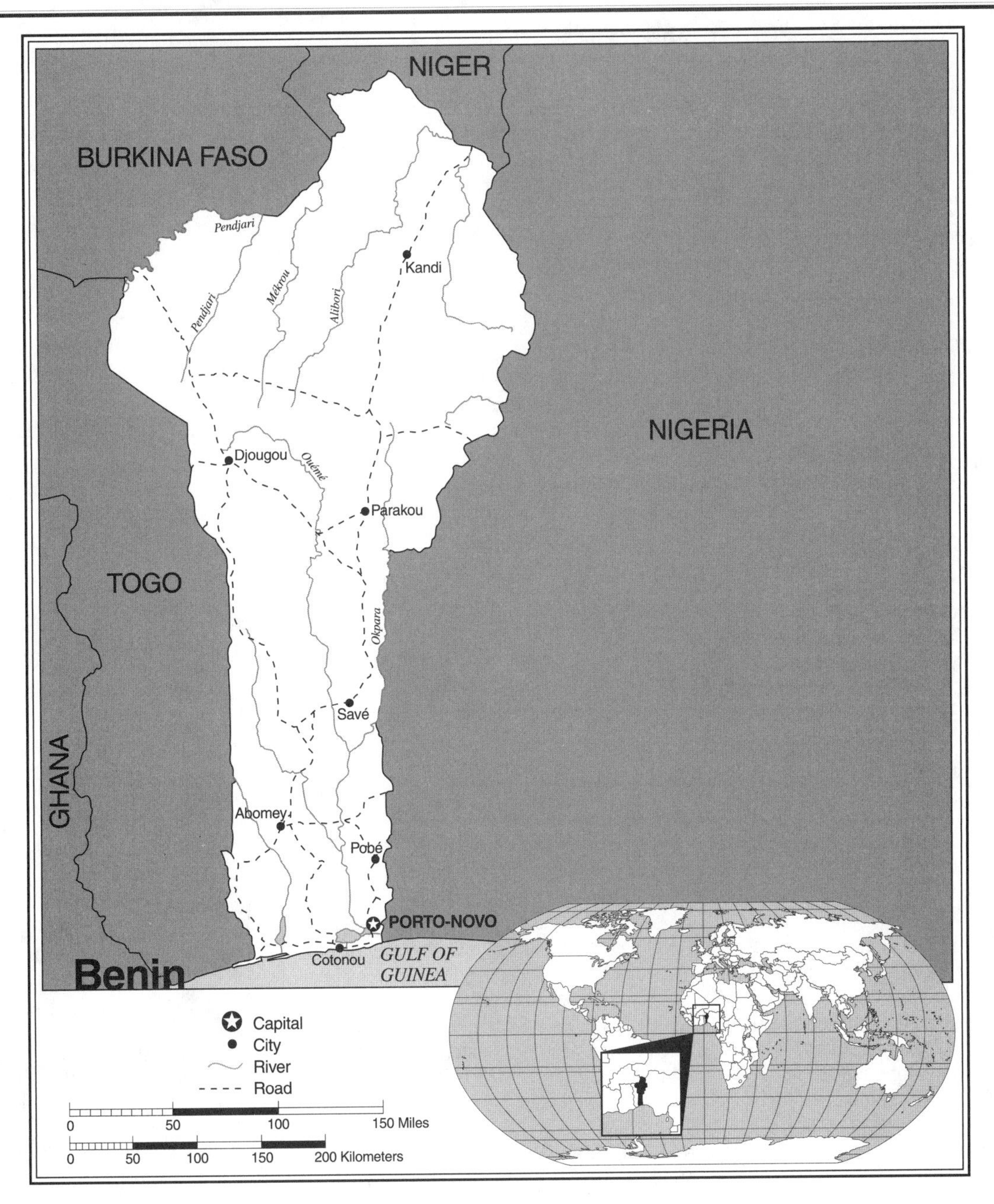

Benin Statistics

GEOGRAPHY

Area in Square Miles (Kilometers): 43,483 (112,620) (about the size of Pennsylvania)

Capital (Population): official: Porto-Novo (225,000); de facto: Cotonou (750,000)

Environmental Concerns: drought; insufficient potable water; poaching; deforestation; desertification

Geographical Features: mostly flat to undulating plain; some hills and low mountains

Climate: tropical to semiarid

PEOPLE

Population

Total: 7,250,033

Annual Growth Rate: 2.91%

Rural/Urban Population Ratio: 58/42

Major Languages: French; Fon; Yoruba; others

Ethnic Makeup: 99% African (most important groupings Fon, Adja, Yoruba, and Bariba); 1% European

Religions: 70% indigenous beliefs; 15% Muslim; 15% Christian

Health

Life Expectancy at Birth: 49 years (male); 51 years (female)
Infant Mortality: 88.5/1,000 live births
Physicians Available: 1/14,216 people
HIV/AIDS Rate in Adults: 4.1%

Education

Adult Literacy Rate: 37%
Compulsory (Ages): 6–12; free

COMMUNICATION

Telephones: 66,500 main lines
Daily Newspaper Circulation: 2/1,000 people
Televisions: 4/1,000 people
Internet Users: 70,000 (2003)

TRANSPORTATION

Highways in Miles (Kilometers): 4,208 (6,787)
Railroads in Miles (Kilometers): 360 (578)
Usable Airports: 5
Motor Vehicles in Use: 55,000

GOVERNMENT

Type: republic
Independence Date: August 1, 1960 (from France)
Head of State/Government: President Mathieu Kérékou is both head of state and head of government
Political Parties: Alliance for Democracy and Progress; Front for Renewal and Development; African Movement for Democracy and Progress; many others
Suffrage: universal at 18

MILITARY

Military Expenditures (% of GDP): 1.2%
Current Disputes: territorial disputes with Niger and Nigeria

ECONOMY

Currency ($ U.S. Equivalent): 581 CFA francs = $1
Per Capita Income/GDP: $1,100/$7.7 billion
GDP Growth Rate: 5.4%
Inflation Rate: 1.5%
Population Below Poverty Line: 37%
Natural Resources: small offshore oil deposits; limestone; marble; timber
Agriculture: palm products; cotton; corn; rice; yams; cassava; beans; sorghum; livestock
Industry: textiles; construction materials; food production; chemical production
Exports: $35.3 million (primary partners Brazil, France, Indonesia)
Imports: $437 million (primary partners France, China, United States)

SUGGESTED WEB SITES

http://www.benindaily.com
http://www.cia.gov/cia/publications/factbook/geos/bn.html

Benin Country Report

Over the past decade, Benin has emerged as one of one of Africa's most stable and democratic states. This has coincided with improved economic growth, though the country remains among the world's poorest in terms of both per capita income and human development. Since gaining its independence from France in 1960, Benin has experienced a series of shifts in political and economic policy that have so far failed to lift most Beninois out of chronic poverty. In this respect, the country's ongoing struggle for development can be seen as a microcosm of the challenges facing much of the African continent.

DEVELOPMENT

Palm-oil plantations were established in Benin by Africans in the mid-nineteenth century. They have continued to be African-owned and capitalist-oriented. Today, there are some 30 million trees in Benin, and palm-oil products are a major export used for cooking, lighting, soap, margarine, and lubricants.

Politically, Benin has been in the forefront of those nations on the continent making the transition away from an authoritarian centralized state toward greater democracy and market reforms. This process has not as yet been accompanied by a decisive shift toward a new generation of leadership. In March 1996, former president Mathieu Kérékou returned to power with 52 percent of the vote, defeating incumbent Nicephore Soglo in Benin's second ballot since the 1990 restoration of multiparty democracy. Five years earlier, Soglo had defeated Kérékou, who had ruled the country as a virtual dictator for 17 years before agreeing to a democratic transition. In the past, Kérékou styled himself as a Marxist Leninist and presided over a one-party state. Today, he presents himself as a "Christian Democrat," affirming that there can be no turning back to the old order. In Parliament, his Popular Revolutionary Party of Benin (PRPB) shares power with other groupings whose existence is primarily a reflection of ethnoregional rather than ideological divisions.

Kérékou's restoration did not result in any significant moves away from his predecessor's economic reforms, which had resulted in a modest rise in gross domestic product, increased investment, reduced inflation, and an easing of the country's debt burden. He is under pressure, however, to raise the living standards of Benin's impoverished masses.

THE OLD ORDER FALLS

Kérékou's first reign began to unravel in late 1989. Unable to pay its bills, his government found itself increasingly vulnerable to mounting internal opposition and, to a lesser extent, to external pressure to institute sweeping political and economic reforms.

FREEDOM

Since 1990, political restrictions have been lifted and prisoners of conscience freed. More recently, however, a number of citizens have been arrested for supposedly inciting people against the government and encouraging them not to pay taxes.

A wave of strikes and mass demonstrations swept through Cotonou, the country's largest city, in December 1989. This upsurge in prodemocracy agitation was partially inspired by the overthrow of Central/Eastern Europe's Marxist-Leninist regimes; ironically, the Stalinist underground Communist Party of Dahomey (PCD) also played a role in organizing much of the unrest. Attempts to quell the demonstrations with force only increased public anger toward the authorities.

In an attempt to defuse the crisis, the PRPB's state structures were forced to give up their monopoly of power by allowing a representative gathering to convene with the task of drawing up a new constitution. For 10 days in February 1990, the Beninois

(United Nations photo)

Benin is one of the least-developed countries in the world. Beninois must often fend for themselves in innovative ways. The peddler pictured above moves among the lake dwellings of a fishing village, selling cigarettes, spices, rice, and other commodities.

gathered around their television sets and radios to listen to live broadcasts of the "National Conference of Active Forces of the Nation." The conference quickly turned into a public trial of Kérékou and his PRPB. With the eyes and ears of the nation tuned in, critics of the regime, who had until recently been exiled, were able to pressure Kérékou into handing over effective power to a transitional government. The major task of this new, civilian administration was to prepare Benin for multiparty elections while trying to stabilize the deteriorating economy.

The political success of Benin's "civilian coup d'etat" placed the nation in the forefront of the democratization process then sweeping Africa. But liberating a nation from poverty is a much more difficult process.

A COUNTRY OF MIGRANTS

Benin is one of the least-developed countries in the world. Having for decades experienced only limited economic growth, in recent years the nation's real gross domestic product has actually declined.

HEALTH/WELFARE

One third of the national budget of Benin goes to education, and the percentage of students receiving primary education has risen to 50% of the school-age population. College graduates serve as temporary teachers through the National Service System, but more teachers and higher salaries are needed.

Emigration has become a way of life for many. The migration of Beninois in search of opportunities in neighboring states is not a new phenomenon. Before 1960, educated people from the then-French colony of Dahomey (as Benin was called until 1975) were prominent in junior administrative positions throughout other parts of French West Africa. But as the region's newly independent states began to localize their civil-service staffs, most of the Beninois expatriates lost their jobs. Their return increased bureaucratic competition within Benin, which, in turn, led to heightened political rivalry among ethnic and regional groups. Such local antagonisms contributed to a series of military coups between 1963 and 1972. These culminated in Kérékou's seizure of power.

While Beninois professionals can be found in many parts of West Africa, the destination of most recent emigrants has been Nigeria. The movement from Benin to Nigeria is facilitated by the close links that exist among the large Yoruba-speaking communities on both sides of the border. After Nigeria, the most popular destination has been Côte d'Ivoire. This may change, however, as economic recession in both of those states has led to heightened hostility against the migrants.

THE ECONOMY

Nigeria's urban areas have also been major markets for food exports. This has encouraged Beninois farmers to switch from cash crops (such as cotton, palm oil, cocoa beans, and coffee) to food crops (such as yams and cassava), which are smuggled across the border to Nigeria. The emergence of this parallel export economy has been encouraged by the former regime's practice of paying its farmers among the lowest official produce prices in the region. Given that agriculture, in terms of both employment and income generation, forms the largest sector of the Beninois economy, the rise in smuggling activities has inevitably contributed to a growth of graft and corruption.

ACHIEVEMENTS

Fon appliquéd cloths have been described as "one of the gayest and liveliest of the contemporary African art forms." Formerly these cloths were used by Dahomeyan kings. Now they are sold to tourists, but they still portray the motifs and symbols of past rulers and the society they ruled.

Benin's small industrial sector is primarily geared toward processing primary products, such as palm oil and cotton, for export. It has thus been adversely affected by the shift away from producing these cash crops for the local market. Small-scale manufacturing has centered around the production of basic consumer goods and construction materials. The biggest enterprises are state-owned cement plants. One source of hope is that with privatization and new exploration, the country's small oil industry will undergo expansion.

Transport and trade are other important activities. Many Beninois find legal as well as illegal employment carrying goods. Due to the relative absence of rain forest (an impediment to travel), Benin's territory has historically served as a trade corridor between the coastal and inland savanna regions of West Africa. Today the nation's roads are comparatively well developed, and the railroad carries goods from the port at Cotonou to northern areas of the country. An extension of the railroad will eventually reach Niamey, the capital of Niger. The government has also tried, with little success, to attract tourists in recent years, through such gambits as selling itself as the "home of Voodoo."

POLITICS AND RELIGION

Kérékou's narrow victory margin in 1996 amid charges and countercharges of electoral fraud underscored the continuing north–south division of Beninois politics and society. Although he is now a self-proclaimed Christian, Kérékou's political base remains the mainly Muslim north, while Soglo enjoyed majority support in the more Christianized south. Religious allegiance in Benin is complicated, however, by the prominence of the indigenous belief system known as Voodoo. Having originated in Benin, belief in Voodoo spirits has taken root in the Americas, especially Haiti, as well as elsewhere in West Africa. During his first presidency, Kérékou sought to suppress Voodoo, which he branded as "witchcraft." Soglo, on the other hand, publicly embraced Voodoo, which was credited with helping him recover from a serious illness in 1992. On the eve of the 1996 election, Soglo recognized Voodoo as an official religion, proclaiming January 10 as "Voodoo National Day." (This move may have politically backfired, however, as it was condemned by the influential Catholic archbishop of Cotonou.)

A more decisive factor in Soglo's fall was the failure of his free-market reforms to revive the Beninois economy. Modest initial growth was seriously undermined in 1994 by the massive devaluation of the CFA franc, while privatization led to the retrenchment of 10,000 public workers. It was unlikely, however, that Kérékou would be tempted to return to his own failed policies of "Marxist-Beninism."

Timeline: PAST

1625
The kingdom of Dahomey is established

1892
The French conquer Dahomey and declare it a French protectorate

1960
Dahomey becomes independent

1972
Mathieu Kérékou comes to power in the sixth attempted military coup since independence

1975
The name of Dahomey is changed to Benin

1990s
Kérékou announces the abandonment of Marxism-Leninism as Benin's guiding ideology; multiparty elections are held; Kérékou loses power to Nicephore Soglo; Kérékou is reelected 5 years later

PRESENT

2000s
Benin marks its 40th year of independence

Poverty remains an overwhelming problem

Burkina Faso

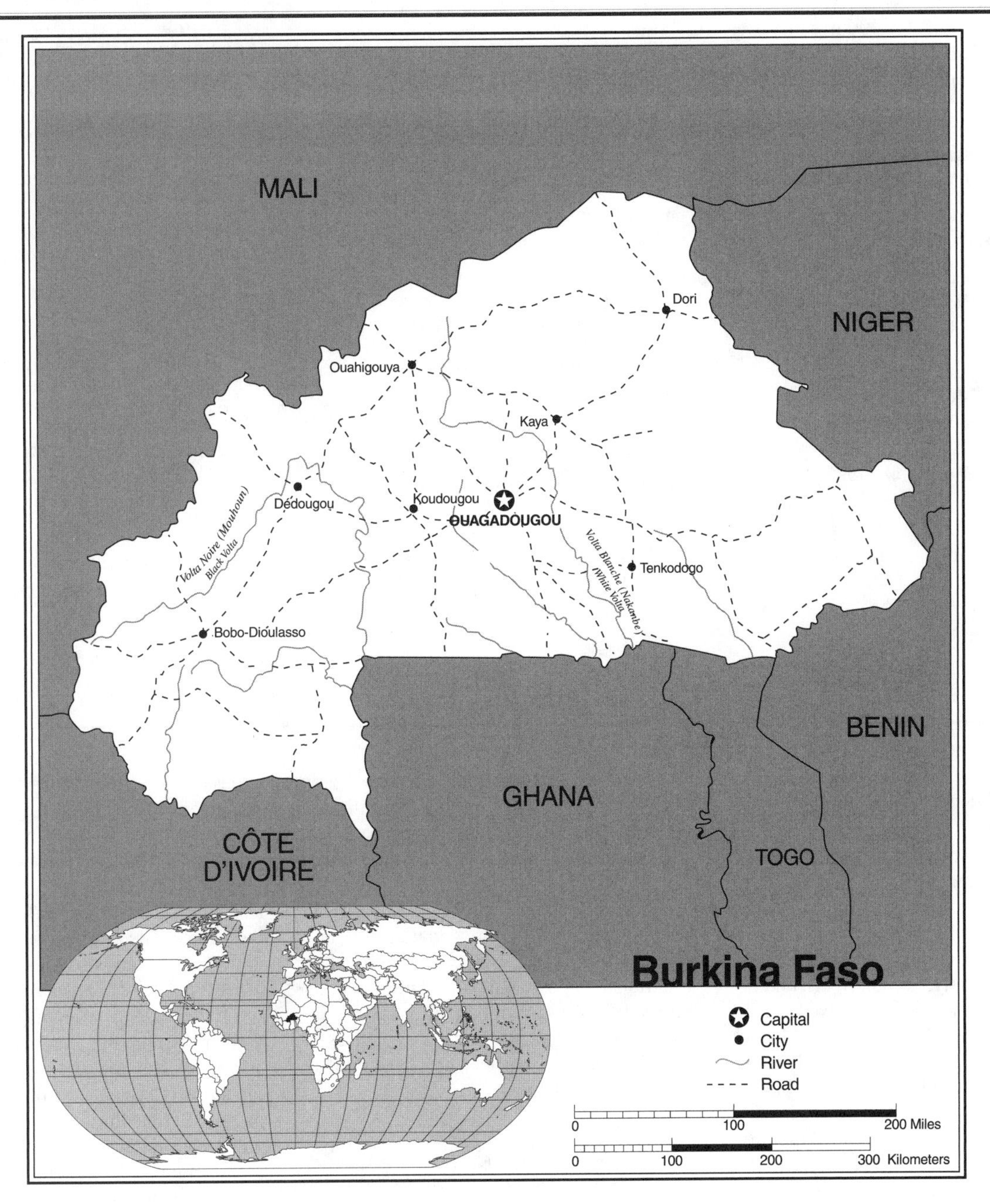

Burkina Faso Statistics

GEOGRAPHY

Area in Square Miles (Kilometers): 106,000 (274,500) (about the size of Colorado)
Capital (Population): Ouagadougou (862,000)
Environmental Concerns: drought; desertification; overgrazing; soil erosion; deforestation
Geographical Features: mostly flat to dissected, undulating plains; hills in west and southeast; landlocked
Climate: tropical; semiarid

PEOPLE

Population

Total: 13,574,820
Annual Growth Rate: 2.64%
Rural/Urban Population Ratio: 82/18
Major Languages: French; Mossi; Senufo; Fula; Bobo; Mande; Gurunsi; Lobi
Ethnic Makeup: about 40% Mossi; Gurunsi; Senufo; Lobi; Bobo; Mande; Fulani
Religions: 50% Muslim; 40% indigenous beliefs; 10% Christian

Health

Life Expectancy at Birth: 45 years (male); 47 years (female)
Infant Mortality: 105.3/1,000 live births
Physicians Available: 1/27,158 people
HIV/AIDS Rate in Adults: 6.44%

Education

Adult Literacy Rate: 36%
Compulsory (Ages): 7–14; free

COMMUNICATION

Telephones: 65,400 main lines
Televisions: 4.4/1,000 people
Internet Users: 48,000 (2003)

TRANSPORTATION

Highways in Miles (Kilometers): 7,504 (12,506)
Railroads in Miles (Kilometers): 385 (622)
Usable Airfields: 33
Motor Vehicles in Use: 55,000

GOVERNMENT

Type: parliamentary
Independence Date: August 5, 1960 (from France)
Head of State/Government: President Blaise Compaoré; Prime Minister Ernest Paramanga Yonli
Political Parties: Congress for Democracy and Progress; African Democratic Rally—Alliance for Democracy and Federation; others
Suffrage: universal

MILITARY

Military Expenditures (% of GDP): 1.4%
Current Disputes: two villages are in a dispute with Benin

ECONOMY

Currency ($ U.S. Equivalent): 581 CFA francs = $1
Per Capita Income/GDP: $1,100/$14.5 billion
GDP Growth Rate: 5.2%
Inflation Rate: 1.9%
Labor Force by Occupation: 90% agriculture
Population Below Poverty Line: 45%
Natural Resources: manganese; limestone; marble; gold; antimony; copper; bauxite; nickel; lead; phosphates; zinc; silver
Agriculture: peanuts; shea nuts; cotton; sesame; millet; sorghum; corn; rice; livestock
Industry: cotton lint; beverages; agricultural processing; soap; cigarettes; textiles; gold
Exports: $265 million (primary partners Venezuela, Benelux, Italy)
Imports: $580 million (primary partners Côte d'Ivoire, Venezuela, France)

SUGGESTED WEB SITES

http://burkinaembassy-usa.org
http://www.sas.upenn.edu/African_Studies/Country_Specific/Burkina.html

Burkina Faso Country Report

Notwithstanding some notable achievements, especially in the utilization of the Volta River and in the promotion of indigenous culture, Burkina Faso (formerly called Upper Volta) remains an impoverished country searching for a governing consensus. Recently its government has faced both domestic and external criticism over the state of the economy, human rights, and allegations that it has been involved in the smuggling of arms for diamonds ("blood diamonds") to the now-defeated rebel movements in Sierra Leone and Angola. In response to the latter allegation, in 2001 a UN–supervised body was set up to monitor the country's trade in weapons. Since falling gold prices forced the closure of its biggest gold mine, Burkina Faso has had little in the way of legitimate exports, leaving the landlocked, semiarid country with few economic prospects, and causing many of its citizens to seek opportunities elsewhere.

Much of Burkina Faso's four decades of independence has been characterized by chronic political instability, with civilian rule being interrupted by the military on seven different occasions. The restoration of multiparty democracy in 1991 under the firm guidance of former military leader Blaise Compaoré seemed to usher in an era of greater political stability. Compaoré's reelection in November 1998 was accepted as legitimate by international observers. But the assassination a month later of independent journalist Robert Zongo touched off a wave of violent strikes and protests. Since then, there have been sustained calls for more fundamental political and social reform from an emerging generation of activists within civil society, including the traditionally powerful trade unions. An umbrella body known as the Collective of Democratic Organizations for the Masses and Political Parties has been formed to challenge the status quo.

DEVELOPMENT

Despite political turbulence, Burkina Faso's economy has recorded positive, albeit modest, annual growth rates for more than a decade. Most of the growth has been in agriculture. New hydroelectric projects have significantly reduced the country's dependence on imported energy.

But power has remained in the hands of Compaoré's hands. Along with his party, the Popular Democratic Organization–Worker's Movement (ODP–MT), he won elections against fragmented opposition in 1991 and 1995, as well as 1998.

Before adopting the mantle of democracy, Compaoré rose to power through a series of coups, the last of which resulted in the overthrow and assassination of the charismatic and controversial Thomas Sankara. A man of immense populist appeal for many Burkinabé, Sankara remains as a martyr to their unfulfilled hopes. By the time of its overthrow, his radical regime had become the focus of a great deal of external as well as internal opposition.

Of the three men directly responsible for Sankara's toppling, two—Boukari Lingani and Henri Zongo—were executed following a power struggle with the third—Compaoré. It is in this context of sanguinary political competition that the assassination of a prominent media critic has once more called into question the government's commitment to political pluralism.

DEBILITATING DROUGHTS

At the time of its independence from France, in 1960, the landlocked country then named the Republic of Upper Volta inherited little in the way of colonial infrastructure. Since independence, progress has been hampered by prolonged periods of severe drought. Much of the country has been forced at times to depend on international food aid. To counteract some of the negative effects of this circumstance, efforts have been made to integrate relief donations into local development schemes.

(United Nations photo 154792 by John Isaac)

Since Burkina Faso gained its independence from France, its progress has been hampered by prolonged periods of drought. Local cooperatives have been responsible for small-scale improvements, such as the construction of the water barrage or barricade pictured above.

Of particular note have been projects instituted by the traditional rural cooperatives known as *naam,* which have been responsible for such small-scale but often invaluable local improvements as new wells and pumps, better grinding mills, and distribution of tools and medical supplies.

FREEDOM

There has been a surprisingly strong tradition of pluralism in Burkina Faso despite the circumscribed nature of human rights under successive military regimes. Freedoms of speech and association are still curtailed, and political detentions are common. The Burkinabé Movement for Human Rights has challenged the government.

Despite such community action, the effects of drought have been devastating. Particularly hard-hit has been pastoral production, long a mainstay of the local economy, especially in the north. It is estimated that a recent drought destroyed about 90 percent of the livestock in Burkina Faso.

To counteract the effects of drought while promoting greater development, the Burkinabé government has developed two major hydroelectric and agricultural projects over the past decade. The Bagre and Kompienga Dams, located east of Ouagadougou, have significantly reduced the country's dependence on imported energy, while also supplying water for large-scale irrigation projects. This has already greatly reduced the need for imported food.

Most Burkinabé continue to survive as agriculturalists and herders, but many people are dependent on wage labor. In the urban centers, there exists a significant working-class population that supports the nation's politically powerful trade-union movement. The division between this urban community and rural population is not absolute, for it is common for individuals to combine wage labor with farming activities. Another population category—whose numbers exceed those of the local wage-labor force—are individuals who seek employment outside of the country. At least 1 million Burkinabé work as migrant laborers in other parts of West Africa. This is part of a pattern that dates back to the early 1900s. Approximately 700,000 of these Burkinabé regularly migrate to Côte d'Ivoire. Returning workers have infused the rural areas with consumer goods and a working-class consciousness.

HEALTH/WELFARE

The inadequacy of the country's public health measures is reflected in the low Burkinabé life expectancy. Mass immunization campaigns have been successfully carried out, but in an era of structural economic adjustment, the prospects for a dramatic improvement in health appear bleak.

UNIONS FORCE CHANGE

As is the case in much of Africa, it is the salaried urban population (at least, next to the army) who have exercised the greatest influence over successive Burkinabé regimes. Trade-union leaders representing these workers have been instrumental in forcing changes in government. They have spoken out vigorously against government efforts to ban strikes and restrain unions. They have also demanded that they be shielded from downturns in the local economy. Although many unionists have championed various shades of Marxist-Leninist ideology, they, along with their natural allies in the civil service, arguably constitute a conservative element within the local society. During the mid-1980s, they became increasingly concerned about the dynamic Sankara's efforts to promote a nationwide network of grassroots Committees for the Defense of the Revolution (CDRs) as vehicles for empowering the nation's largely rural masses.

ACHIEVEMENTS

In 1997, a record total of 19 feature films competed for the Etalon du Yennenga award, the highest distinction of the biannual Pan-African Film Festival, hosted in Ouagadougou. Over the past 3 decades, this festival has contributed significantly to the development of the film industry in Africa. Burkina Faso has nationalized its movie houses, and the government has encouraged the showing of films by African filmmakers.

To many unionists, the mobilization and arming of the CDRs was perceived as a direct challenge to their own status. This threat seemed all the more apparent when Sankara began to cut urban salaries, in the name of a more equitable flow of revenue to the rural areas. When several union leaders challenged this move, they were arrested on charges of sedition. Sankara's subsequent overthrow thus had strong backing from within organized labor and the civil service. These groups, along with the military, remain the principal supporters of Compaoré's ODP–MT and its policy of "national rectification." Yet despite this support base, the government has moved to restructure the until recently all-encompassing public sector of the economy by reducing its wage bill. This effort has impressed international creditors.

Beyond its core of support, the ODP–MT government has generally been met with sentiments ranging from hostility to indifference. While Compaoré claimed—with some justification—that Sankara's rule had become too arbitrary and that he had resisted forming a party with a set of rules, many people mourned the fallen leader's death. In the aftermath of the coup, the widespread use of a new cloth pattern, known locally as "homage to Sankara," became an informal barometer of popular dissatisfaction. Compaoré has also been challenged by the high regard that has been accorded Sankara outside Burkina Faso, as a symbol of a new generation of African radicalism.

Compaoré, like Sankara, has sometimes resorted to sharp anti-imperialist rhetoric. However, his government has generally sought to cultivate good relations with France (the former colonial power) and other members of the Organization for Economic Cooperation and Development, as well as the major international financial institutions. But he has alienated himself from some of his West African neighbors, as well as the Euro–North American diplomatic consensus, through his close ties to Libya and past military support for Charles Taylor's National Patriotic Front in Liberia. Along with Taylor, Compaoré has more recently been accused of, but denies, providing support for the Revolutionary United Front rebels in Sierra Leone. To many outsiders, as well as the Burkinabé people themselves, the course of Compaoré's government remains ambiguous.

Since October 2002 the Ivory Coast has continually accused Burkina Faso of sheltering dissident Ivorian soldiers, many of whom are descendents of individuals who first arrived from Burkina Faso. In turn Burkina Faso raised concerns about attacks on Burkinabes in the Ivory Coast after the September 2002 Ivorian military uprising.

Timeline: PAST

1313
The first Mossi kingdom is founded

1896
The French overcome Mossi resistance and claim Upper Volta

1932
Upper Volta is divided among adjoining French colonies

1947
Upper Volta is reconstituted as a colony

1960
Independence under President Maurice Yameogo

1980s
Captain Thomas Sankara seizes power and changes the country's name to *Burkina* (Mossi for "land of honest men") *Faso* (Dioula for "democratic republic"); Sankara is assassinated in a coup; Blaise Compaoré succeeds as head of state

1990s
Compaoré introduces multipartyism, but his critics are skeptical

PRESENT

2000s
The country marks 4 decades of independence

Burkina Faso is believed to be involved in the "blood diamonds" trade

Cape Verde (Republic of Cape Verde)

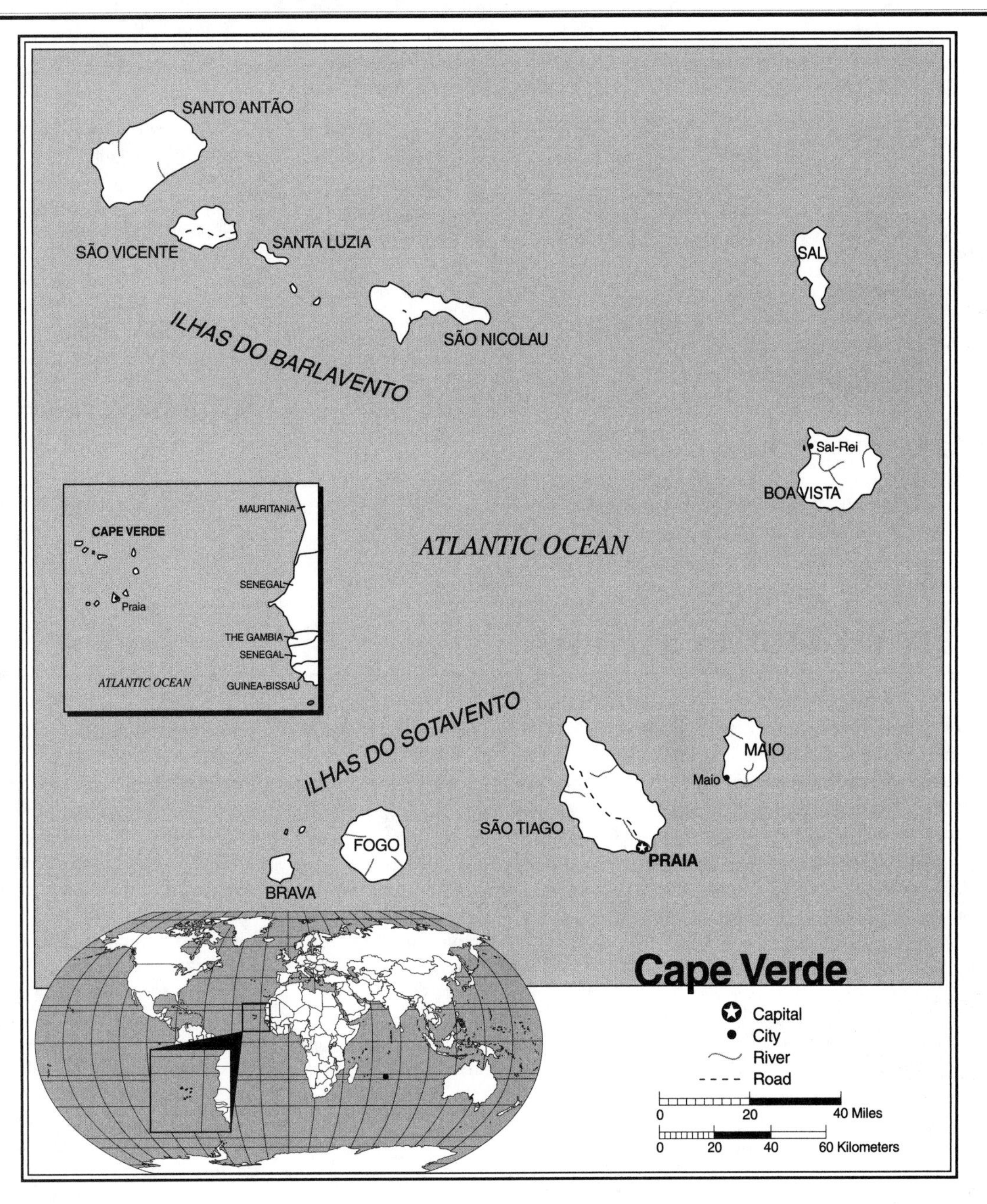

Cape Verde Statistics

GEOGRAPHY

Area in Square Miles (Kilometers): 1,557 (4,033) (about the size of Rhode Island)

Capital (Population): Praia (82,000)

Environmental Concerns: soil erosion; overgrazing; deforestation; desertification; threats to wildlife populations; overfishing

Geographical Features: steep; rugged; rocky; volcanic

Climate: temperate; precipitation meager and very erratic

PEOPLE

Population

Total: 415,294

Annual Growth Rate: 0.73%

Rural/Urban Population Ratio: 39/61

Major Languages: Portuguese; Kriolu

Ethnic Makeup: 71% Creole (mixed); 28% African; 1% European

Religions: Roman Catholicism fused with indigenous beliefs

Health

Life Expectancy at Birth: 66 years (male); 73 years (female)

Infant Mortality: 51.8/1,000 live births
Physicians Available: 1/4,208 people
HIV/AIDS Rate in Adults: 0.04%

Education

Adult Literacy Rate: 72%
Compulsory (Ages): 7–11

COMMUNICATION

Telephones: 71,700 main lines
Televisions: 2.6/1,000 people
Internet Users: 20,400 (2003)

TRANSPORTATION

Highways in Miles (Kilometers): 686 (1,100)
Railroads in Miles (Kilometers): none
Usable Airfields: 6
Motor Vehicles in Use: 18,000

GOVERNMENT

Type: republic
Independence Date: July 5, 1975 (from Portugal)
Head of State/Government: President Pedro Pires; Prime Minister José Maria Neves
Political Parties: Movement for Democracy; African Party for Independence of Cape Verde; Party for Democratic Convergence; Party of Work and Solidarity; others
Suffrage: universal at 18

MILITARY

Military Expenditures (% of GDP): 1.6%
Current Disputes: none

ECONOMY

Currency ($ U.S. Equivalent): 97.7 escudos = $1
Per Capita Income/GDP: $1,500/$600 million
GDP Growth Rate: 4%
Inflation Rate: 3%
Unemployment Rate: 21%
Population Below Poverty Line: 30%
Natural Resources: salt; basalt rock; pozzuolana; limestone; kaolin; fish
Agriculture: bananas; corn; beans; sweet potatoes; sugarcane; coffee; fish
Industry: fish processing; salt mining; shoes and garments; ship repair; food and beverages
Exports: $27.3 million (primary partners Portugal, United Kingdom, Germany)
Imports: $218 million (primary partners Portugal, Germany, France)

SUGGESTED WEB SITES

http://www.sas.upenn.edu/African_Studies/Country_Specific/C_Verde.html
http://virtualcapeverde.net

Cape Verde Country Report

On July 5, 2000, Cape Verdeans proudly celebrated a quarter-century of economic, political, and social progress since independence from Portugal. Despite a late start and unfavorable environmental conditions, the country has emerged as one of postcolonial Africa's tangible success stories.

Over the past decade this has been accompanied by a change in political direction. In 1992, Cape Verde adopted a new flag and Constitution, reflecting the country's transition to political pluralism. After 15 years of single-party rule by the African Party for the Independence of Cape Verde (PAICV), rising agitation led to the legalization of opposition groups in 1990. In January 1991, a quickly assembled antigovernment coalition, the Movement for Democracy (MPD), stunned the political establishment by gaining 68 percent of the votes and 56 out of 79 National Assembly seats. A month later, the MPD candidate, Antonio Mascarenhas Monteiro, defeated the long-serving incumbent, Aristides Pereira, in the presidential elections. It is a credit to both the outgoing administration and its opponents that this dramatic political transformation occurred without significant violence or rancor.

Parliamentary elections in December 1995 resulted in the MPD retaining power, albeit with a reduced majority. In 2001, the political pendulum swung back to PAICV in both the legislative and presidential elections, with Pedro Pires becoming the country's new leader.

The Republic of Cape Verde is an archipelago located about 400 miles west of the Senegalese Cape Verde, or "Green Cape," after which it is named. Unfortunately, green is a color that is often absent in the lives of the islands' citizens. Throughout its history, Cape Verde has suffered from periods of prolonged drought, which before the twentieth century were often accompanied by extremely high mortality rates (up to 50 percent). The last severe drought lasted from 1968 to 1984. Even in normal years, though, rainfall is often inadequate.

DEVELOPMENT

In a move designed to attract greater investment from overseas, especially from Cape Verdean Americans, the country has joined the International Finance Corporation. Efforts are under way to promote the islands as an offshore banking center for the West African (ECOWAS) region.

When the country gained independence, in 1975, there was little in the way of nonagricultural production. As a result, the new nation had to rely for its survival on foreign aid and the remittances of Cape Verdeans working abroad, but the post-independence period has been marked by a genuine improvement in the lives of most Cape Verdeans.

Cape Verde was ruled by Portugal for nearly 500 years. Most of the islanders are the descendants of Portuguese colonists, many of whom arrived as convicts, and African slaves who began to settle on the islands shortly after their discovery by Portuguese mariners in 1456. The merging of these two groups gave rise to the distinct Cape Verdean Kriolu language (which is also spoken in Guinea-Bissau). Under Portuguese rule, Cape Verdeans were generally treated as second-class citizens, although a few rose to positions of prominence in other parts of the Portuguese colonial empire. Economic stagnation, exacerbated by cycles of severe drought, caused many islanders to emigrate elsewhere in Africa, Western Europe, and the Americas.

FREEDOM

The new Constitution should entrench the country's recent political liberalization. Opposition publications have emerged to complement the state- and Catholic Church-sponsored media

In 1956, the African Party for the Independence of Guinea-Bissau and Cape Verde (PAIGC) was formed under the dynamic leadership of Amilcar Cabral, a Cape Verdean revolutionary who hoped to see the two Portuguese colonies form a united nation. Between 1963 and 1974, PAIGC waged a successful war of liberation in Guinea-Bissau that led to the independence of both territories. Although Cabral was assassinated by the Portuguese in 1973, his vision was preserved during the late 1970s by his successors, who, while ruling the two countries separately, maintained the unity of the PAIGC. This arrangement, however, began to break down in the aftermath of a 1980 coup in Guinea-Bissau and resulted in the party's division along national lines. In 1981, the Cape Verdean PAIGC formally renounced its Guinean links, becoming the PAICV.

HEALTH/WELFARE

Greater access to health facilities has resulted in a sharp drop in infant mortality and a rise in life expectancy. Clinics have begun to encourage family planning. Since independence, great progress has taken place in social services. Nutrition levels have been raised, and basic health care is now provided to the entire population.

After independence, the PAIGC/CV government was challenged by the colonial legacy of economic underdevelopment, exacerbated by drought. Massive famine was warded off through a reliance on imported foodstuffs, mostly received as aid. The government attempted to strengthen local food production and assist the 70 percent of the local population engaged in subsistence agriculture. Its efforts took the forms of drilling for underground water, terracing, irrigating, and building a water-desalinization plant with U.S. assistance. Major efforts were also devoted to tree-planting schemes as a way to cut back on soil erosion and eventually make the country self-sufficient in wood fuel.

ACHIEVEMENTS

Cape Verdean Kriolu culture has a rich literary and musical tradition. With emigrant support, Cape Verde bands have acquired modest followings in Western Europe, Lusophone Africa, Brazil, and the United States. Local drama, poetry, and music are showcased on the national television service.

With no more than 15 percent of the islands' territory potentially suitable for cultivation, the prospect of Cape Verde developing self-sufficiency in food appears remote. The few factories that exist on Cape Verde are small-scale operations catering to local needs. Only textiles have enjoyed modest success as an export. Another promising area is fishing.

Timeline: PAST

1462
Cape Verdean settlement begins

1869
Slavery is abolished

1940s
Thousands of Cape Verdeans die of starvation during World War II

1956
The PAIGC is founded

1973
Warfare begins in Guinea-Bissau; Amilcar Cabral is assassinated

1974
A coup in Lisbon initiates the Portuguese decolonization process

1975
Independence

1990s
The PAICV is defeated by the MPD in the country's first multiparty elections; Cape Verde adopts a new Constitution

PRESENT

2000s
Cape Verdeans celebrate 25 years of independence

Power shifts back to PAICV, with Pedro Pires becoming president

Côte d'Ivoire (Republic of Côte d'Ivoire)

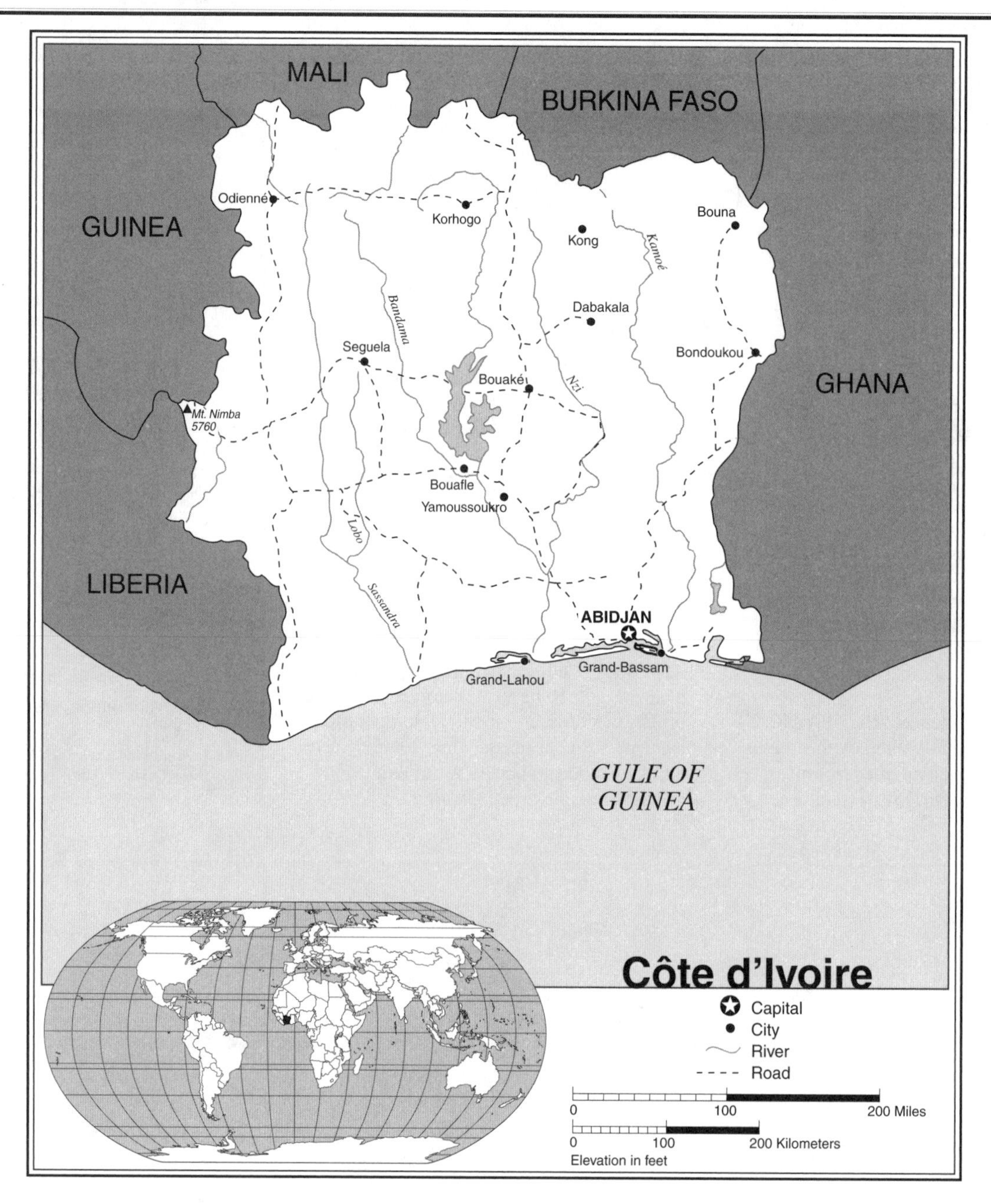

Côte d'Ivoire Statistics

GEOGRAPHY

Area in Square Miles (Kilometers): 124,503 (323,750) (about the size of New Mexico)

Capital (Population): Abidjan (administrative) (3,956,000); Yamoussoukro (political) (120,000)

Environmental Concerns: water pollution; deforestation

Geographical Features: mostly flat to undulating plains; mountains in the northwest

Climate: tropical to semiarid

PEOPLE

Population

Total: 17,327,724

Annual Growth Rate: 2.11%

Rural/Urban Population Ratio: 54/46

Major Languages: French; Dioula; many indigenous dialects

Ethnic Makeup: 42% Akan; 18% Voltaics or Gur; 11% Krous; 16% Northern Mandes; 10% Southern Mandes; 3% others

Religions: 60% Muslim; 22% Christian; 18% indigenous

Health

Life Expectancy at Birth: 43 years (male); 46 years (female)
Infant Mortality: 92.2/1,000 live births
Physicians Available: 1/11,745 people
HIV/AIDS Rate in Adults: 10.76%

Education

Adult Literacy Rate: 50.9%
Compulsory (Ages): 7–13; free

COMMUNICATION

Telephones: 328,000 main lines
Televisions: 57/1,000 people
Internet Users: 90,000 (2002)

TRANSPORTATION

Highways in Miles (Kilometers): 30,240 (50,400)
Railroads in Miles (Kilometers): 408 (660)
Usable Airfields: 36
Motor Vehicles in Use: 255,000

GOVERNMENT

Type: republic
Independence Date: August 7, 1960 (from France)
Head of State/Government: President Laurent Gbagbo; Prime Minister Seydou Diarra
Political Parties: Democratic Party of Côte d'Ivoire; Ivoirian Popular Front; Rally of the Republicans; Ivoirian Workers' Party; others
Suffrage: universal at 18

MILITARY

Military Expenditures (% of GDP): 1.3%
Current Disputes: civil war

ECONOMY

Currency ($ U.S. Equivalent): 581 CFA francs = $1
Per Capita Income/GDP: $1,400/$24 billion
GDP Growth Rate: -1.9%
Inflation Rate: 3.4%
Unemployment Rate: 13%
Natural Resources: petroleum; diamonds; manganese; iron ore; cobalt; bauxite; copper; hydropower
Agriculture: coffee; cocoa beans; bananas; palm kernels; corn; rice; manioc; sweet potatoes; sugar; cotton; rubber; timber
Industry: foodstuffs; beverages; oil refining; wood products; textiles; automobile assembly; fertilizer; construction materials; electricity
Exports: $3.6 billion (primary partners France, the Netherlands, United States)
Imports: $2.4 billion (primary partners France, Nigeria, China)

SUGGESTED WEB SITES

http://www.sas.upenn.edu/African_Studies/Country_Specific/Cote.html
http://www.cia.gov/cia/publications/factbook/geos/iv.html

Côte d'Ivoire Country Report

Once considered an island of political stability and a model of economic growth in West Africa, since the death in 1993 of its first president Félix Houphouët-Boigny, Côte d'Ivoire (previously known by its English name, Ivory Coast) has been shaken by a series of military crises as well as sustained economic decline. In September 2002, a military mutiny sparked fighting that has left the country divided along regional and sectarian lines. The predominantly Muslim northern half of the country has come under the control of rebel soldiers, while the mainly Christian southern half has remained under the rule of the embattled government of Laurent Ghagbo. Although a partial truce between the two sides, upheld by French troops, was negotiated in October, intensive mediation efforts by neighboring states failed to reconcile the two sides. By the end of 2002, the opportunity for a quick end to the crisis appeared to be fading.

In fact the October ceasefire collapsed in November 2002 as armed groups clashed with government forces in a battle for the key cocoa-industry town of Daloa. Moving desperately towards reconciliation in January 2003, President Gbagbo accepted a peace deal at talks in Paris which proposed a power-sharing government. By March 2003 political parties and the rebels agreed on a new government to include nine members from rebel ranks. "Consensus" prime minister, Seydou Diarra, was tasked with forming cabinet. In May 2003 the armed forces signed a "full" ceasefire with rebel groups in May 2003 to end almost eight months of rebellion.

DEVELOPMENT

It has been said that Côte d'Ivoire is "power hungry." The Soubre Dam, being developed on the Sassandra River, is the sixth and largest hydroelectric project in Côte d'Ivoire. It will serve the eastern area of the country. Another dam is planned for the Cavalla River, between Côte d'Ivoire and Liberia.

There are three rebel groups: the New Forces Movement (formerly the Ivory Coast Patriotic Movement (MPCI)), the Movement for Justice and Peace (MJP), and the Ivoirian Popular Movement of the Far West (MPIGO). In July 2003 at a ceremony held in the presidential palace, the military chiefs and rebels declared that the war was over. The rebels pulled out of the unity government in September 2003, after accusing President Gbagbo of failing to honor the peace agreement. During December 2003 the rebels attacked the state run and owned TV station. The govern

Although the rebels rejoined the government in December, by March the government faced new challenges from within its own ranks. The major opposition parties staged a demonstration in the capital city of Abidjan in March 2004. Government troops tried to disperse the demonstration and began shooting indiscriminately into the crowd. The former ruling party—the Ivory Coast Democratic Party (PDCI)—immediately pulled out of the government, accusing President Gbagbo of "destabilising the peace process." A damning report written by the United Nations in May 2004 noted that the opposition rally was used as a pretext for a planned operation by security forces. The report says that more than 120 people were killed and alleges summary executions and torture. Attempting to halt the violence the UN deplored a contingent of peacekeepers into the country in May 2004. After a six-month period of initial success in curtailing the violence, the civil war was re-ignited in November 2004 when the Ivorian Air Force attacked the rebels. France was drawn into the conflict after nine of their soldiers were killed in an air strike. In retaliation for the attack against their troops, the French staged a counter attack on the Ivorian Air Force, destroying all of their planes and killing an undetermined number of Ivorian government soldiers. Violent anti-French protests ensued in Abidjan. The UN imposed an arms embargo against all the combatants in the Ivory Coast in November 2004.

POLITICAL POLARIZATION

Religious and ethnic divisions among Ivoirians in recent years have been aggravated by growing xenophobia against immigrants, who make up at least one third of the country's total population. Under Houphouët-Boigny, people from other African states were allowed to settle and even vote in Côte d'Ivoire. For more than a half-century, Ivoirians and non-Ivoirians alike lived under the certainty of Houphouët-Boigny's leadership. Known by friend and foe alike in his latter years as *Le Vieux* ("The Old Man"), he was a dominant figure not only in Côte d'Ivoire but also throughout Francophone Africa. A pioneering Pan-Africanist, he had served for three years as a French cabinet minister before leading his country to independence in 1960. During his subsequent 33-year rule, the Côte d'Ivoire was seemingly conspicuous for its social harmony as well as economic growth. But his paternalistic autocracy, exercised through the Democratic Party of Côte d'Ivoire (PDCI), had begun to break down before his death.

FREEDOM

Former president Konan Bédié showed little tolerance for dissent, within either the PDCI or society as a whole. Journalists by the score were jailed for such "offenses" as writing "insulting" articles. Six Ivoirian gendarmes were charged in connection with a mass grave discovered near Abidjan in 2000.

Political life entered a new phase in 1990. Months of mounting prodemocracy protests and labor unrest had led to the legalization of opposition parties, previously banned under the country's single-party government; and to the emergence within the PDCI itself, of a reformist wing seemingly committed to the liberalization process. Although the first multiparty presidential and legislative elections in October–November 1990 were widely regarded as having been less than free and fair by outside observers as well as the opposition, many believed that the path was open for further reform. But as Le Vieux's health declined, the reform process was increasingly held hostage by tensions within the PDCI as well as between it and opposition movements, of which the most prominent was Laurent Gbagbo's Ivoirian Popular Front (FPI). Gbagbo and others were briefly jailed in 1992 on charges of inciting violence after mass demonstrations turned to rioting.

Houphouët-Boigny was succeeded by Konan Bédié, a Christian southerner who came out ahead in a power struggle with Allassane Ouatarra, a northern technocrat who had occupied the post of prime minister. Once in power, Bédié stirred up ethnic discord and xenophobia against Muslim northerners. In 1995, he retained the presidency in elections boycotted by supporters of Ouatarra, who was banned from running for office due to a new law mandating that both parents of any presidential candidate must have been born in Côte d'Ivoire. The boycott enjoyed widespread support in the north. For the first time, immigrants were also banned from voting. Violent protests prior to the poll were met with repression.

Bédié's increasingly unpopular rule came to an abrupt end on December 24, 1999, when General Robert Guei assumed power following the country's first coup d'état. Initial international condemnation of the end of 39 years of uninterrupted civilian rule was muted by the obvious jubilation with which many greeted Bédié's overthrow. There was hope that the divisions of the Bédié era might be laid to rest. Guei reached out to Ouatarra and his supporters as well as other opposition and PDCI members. A new Constitution was drafted and accepted in a referendum. But political goodwill evaporated when, in a move designed to assure his own election, Guei excluded Ouatarra from running for president by reintroducing the provision that both parents of candidates must be Ivoirian. An attempted second coup by northern officers was then crushed, further increasing tensions.

With Bédié and others barred, Gbagbo was the only serious contender allowed to run against Guei in the October 2000 elections. Having lost the ballot, Guei's further attempts to rig the election results were frustrated by a popular uprising that led to Gbagbo's assumption of power. Many of Ouattara's supporters were killed following the rejection of their leader's call for new elections. Opposition boycotts of the December 2000 legislative elections resulted in Gbagbo's Ivoirian Popular Front emerging as the biggest single party in Parliament, with a turnout of only 33 percent. This was followed by another failed coup attempt in January 2001 and subsequent security clapdowns. Nevertheless, there were further calls for fresh presidential and legislative elections in March after Ouattara's party gained a majority at local polls.

In a move toward reconciliation, Gbagvo set up a "National Reconciliation Forum" in October 2001. This resulted in Outtara's return from a year-long exile and a subsequent, January 2002, meeting between the country's "big four"—President Gbagbo, Ouatarra, Guei, and Bédié. But the goodwill created by the talks has since collapsed in the wake of fighting, which in its first days resulted in Guei's death, in disputed circumstances.

ECONOMIC DOWNTURN

The on-going cycle of reform, repression, and increasingly violent political conflict has been taking place against the backdrop of a prolonged deterioration in Côte d'Ivoire's once-vibrant economy. The primary explanation for this downturn is the decline in revenue from cocoa and coffee, which have long been the country's principal export earners. This has led to mounting state debt, which in turn has pressured the government to adopt unpopular austerity measures.

The economy's current problems and prospects are best understood in the context of its past performance. During its first two decades of independence, Côte d'Ivoire enjoyed one of the highest economic growth rates in the world. This growth was all the more notable in that, in contrast to many other developing-world "success stories" during the same period, it had been fueled by the expansion of commercial agriculture. The nation had become the world's leading producer of cocoa and third-largest coffee producer.

HEALTH/WELFARE

Côte d'Ivoire has one of the lowest soldier-to teacher ratios in Africa. Education absorbs about 40% of the national budget. The National Commission to Combat AIDS has reported significant success in its campaign to promote condom use, by targeting especially vulnerable groups.

Although prosperity gave way to recession during the 1980s, the average per capita income of the country remained one of Africa's highest. Statistics also indicated that, on average, Ivoirians lived longer and better than people in many neighboring states. But the creation of a productive, market-oriented economy did not eliminate the reality of widespread poverty, leading some to question whether the majority of Ivoirians have derived reasonable benefit from their nation's wealth.

To the further dismay of many young Ivoirians struggling to enter the country's tight job market, much of the political and economic life of Côte d'Ivoire is controlled by its large and growing expatriate population, largely comprised of French and Lebanese. The size of these communities has multiplied since independence. Many foreigners are now quasi-permanent residents who have thrived while managing plantations, factories, and commercial enterprises. Others can be found in senior civil-service positions.

Another group who have until recently prospered are the commercial farmers, who include millions of medium- and small-scale producers. About two thirds of the workforce are employed in agriculture, with coffee alone being the principal source of income for some 2.5 million people. In addition to coffee, Ivoirian planters grow cocoa, bananas, pineapples, sugar, cotton, palm oil, and other cash crops for export. While some of these farmers are quite wealthy, most have only modest incomes.

ACHIEVEMENTS

Ivoirian textiles are varied and prized. Block printing and dyeing produce brilliant designs; woven cloths made strip by strip and sewn together include the white Korhogo tapestries, covered with Ivoirian figures, birds, and symbols drawn in black. The Ivoirian singer Alpha Blondy has become an international superstar as the leading exponent of West African reggae.

In recent years, the circumstance of Ivoirian coffee and cocoa planters has become much more precarious, due to fluctuations in commodities prices. In this respect, the growers, along with their colleagues elsewhere, are to some extent victims of their own success. Their productivity, in response to international demand, has been a factor in depressing prices through increased supply. In 1988, Houphouët-Boigny held cocoa in storage in an attempt to force a price rise, but the effort failed, aggravating the nation's economic downturn. The civil war in the country seriously disrupted the exportation of the cocoa crop in 2004, and threatened the supply of chocolates to western consumers during the Christmas season. As a result, the government has taken a new approach, scrapping plans for future expansion in cocoa production in favor of promoting food crops such as yams, corn, and plantains, for which there is a regional as well as a domestic market.

Until recently, Ivoirian planters continued to hire low-paid laborers from other West African countries. At any given time, there have been about 2 million migrant laborers in Côte d'Ivoire, employed throughout the economy. Their presence is not a new phenomenon but goes back to colonial times. Many laborers come from Burkina Faso, which was once a part of Côte d'Ivoire. A good road system and the Ivoirian railroad (which extends to the Burkinabé capital of Ouagadougou) facilitated the travel of migrant workers to rural as well as urban areas.

DEBT AND DISCONTENT

Other factors may determine how much an Ivoirian benefits from the country's development. Residents of Abidjan, the capital, and its environs near the coast receive more services than do citizens of interior areas. Professionals in the cities make better salaries than do laborers on farms or in small industries. Yet persistent inflation and recession have made daily life difficult for the middle class as well as poorer peasants and workers.

The nonagricultural sectors of the national economy have also been experiencing difficulties. Many state industries are unable to make a profit due to their heavy indebtedness. Serious brush fires, mismanagement, and the clearing of forests for cash-crop plantations have put the nation's once-sizable timber industry in jeopardy. Out of a former total of 12 million hectares of forest, 10 1/2 million have been lost. Plans for expansion of offshore oil production have not been implemented due to an inability to raise investment capital.

Difficulty in raising capital for oil development is a reflection of the debt crisis that has plagued the country since the collapse of its cocoa and coffee earnings. Finding itself in the desperate situation of being forced to borrow to pay interest on its previous loans, the government suspended most debt repayments in 1987. Subsequent rescheduling of negotiations with international creditors resulted in a Structural Adjustment Plan (SAP). This plan has resulted in a reduction in the prices paid to farmers and a drastic curtailment in public spending, leading to severe salary cuts for public and parastatal workers. Recent pressure on the part of the international lending agencies for the Ivoirian government to cut back further on its commitment to cash crops is particularly ironic, given the praise that they bestowed on the same policies in the not-too-distant past. Many also see the imposition of such conditions as hypocritical given the heavy rate of agricultural subsidies within the European Union and United States. Those subsidies have further disadvantaged Ivoirian, along with other African, commercial farmers. With the country now on the brink of full-scale civil war, for most Ivoirians the harsh economic conditions are likely to continue to get worse before they get better.

THE SEARCH FOR STABILITY

The ability of various Côte d'Ivoire governments to gain acceptance for austerity measures has been compromised by corruption and extravagance at the top. A notorious example of the latter was the basilica that was constructed at Yamoussoukro, the home village of Houphouët-Boigny (which also at great cost was made the nation's new capital city before his death). The air-conditioned basilica, patterned after the papal seat of St. Peter's in Rome, is the largest Christian church building in the world. Supposedly a personal gift of Houphouët-Boigny to the Vatican—a most reluctant recipient—its three-year construction is believed to have cost hundreds of millions of U.S. dollars.

The course of domestic conflict in Côte d'Ivoire is being watched elsewhere. For decades Houphouët-Boigny was the doyen of the more conservative, pro-Western leaders in Africa. Hostile to Libya and receptive to both Israel and to "dialogue" with South Africa, Côte d'Ivoire has remained especially close to France, which continues to maintains a military presence in the country. During the recent fighting, U.S. marines also intervened to evacuate foreign nationals.

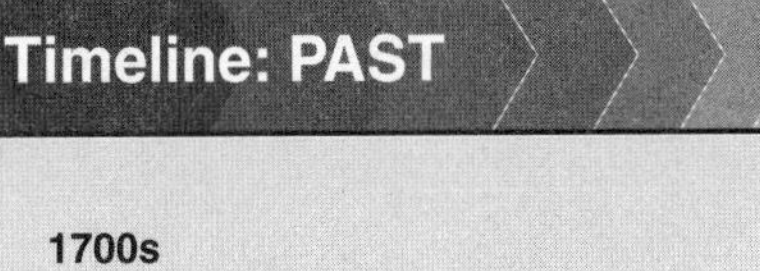

1700s
Agni and Baoulé peoples migrate to the Ivory Coast from the East

1893
The Ivory Coast officially becomes a French colony

1898
Samori Touré, a Malinke Muslim leader and an empire builder, is defeated by the French

1915
The final French pacification of the country takes place

1960
Côte d'Ivoire becomes independent under Félix Houphouët-Boigny's leadership

1980s
The PDCI approves a plan to move the capital from Abidjan to Houphouët-Boigny's home village of Yamoussoukro

1990s
Prodemocracy demonstrations lead to multiparty elections; Houphouët-Boigny dies

PRESENT

2000s
Côte d'Ivoire adjusts after the startling coup in late 1999

Laurent Gbagbo becomes president

A mass grave of 57 bullet-ridden bodies is discovered near Abidjan

The Gambia (Republic of The Gambia)

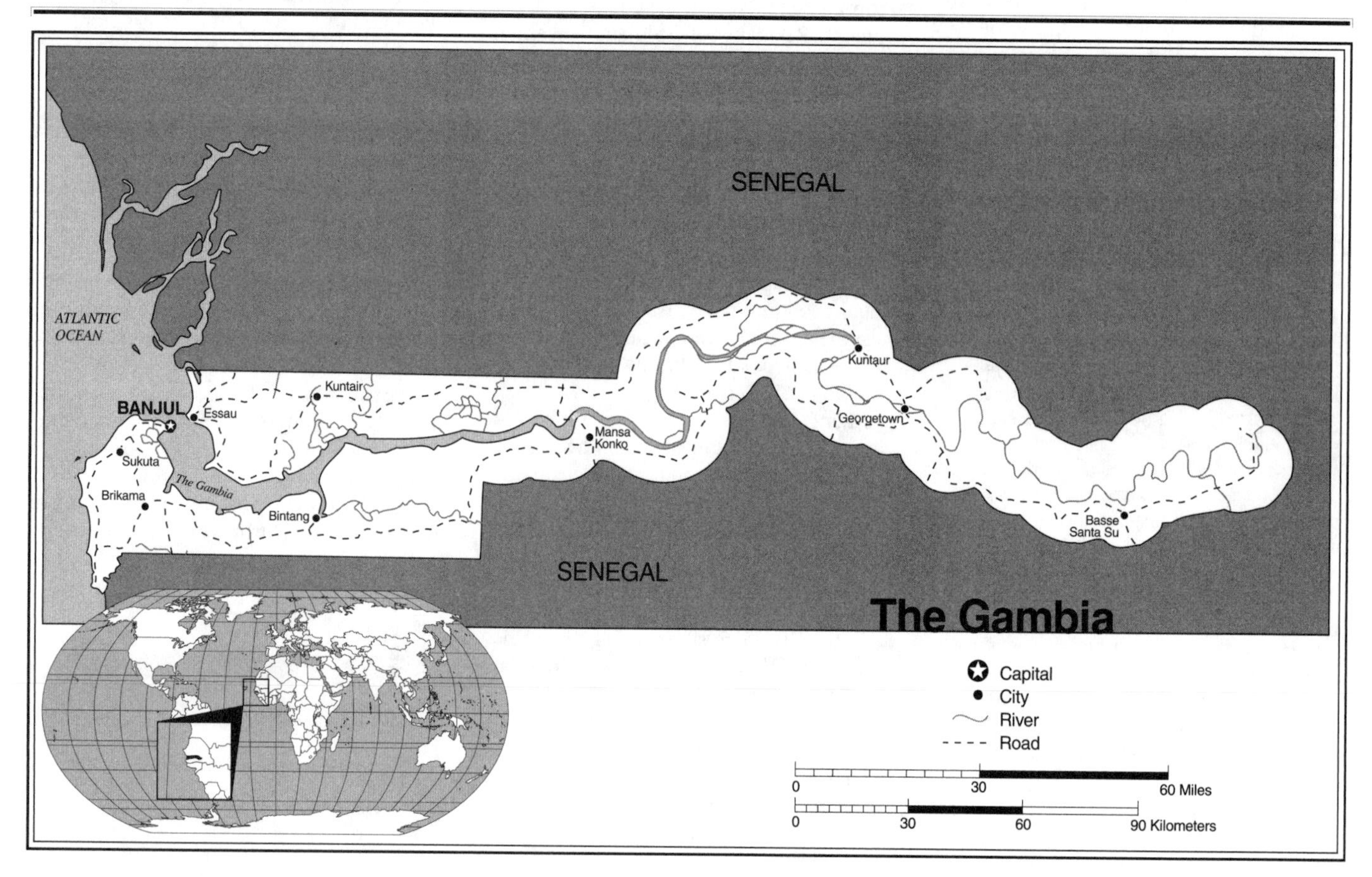

The Gambia Statistics

GEOGRAPHY

Area in Square Miles (Kilometers): (4,361) (11,295) (about twice the size of Delaware)

Capital (Population): Banjul (418,000)

Environmental Concerns: deforestation; desertification; water-borne diseases

Geographical Features: floodplain of The Gambia River flanked by some low hills

Climate: tropical; hot rainy season, cooler dry season

PEOPLE

Population

Total: 1,546,848

Annual Growth Rate: 2.98%

Rural/Urban Population Ratio: 68/32

Major Languages: English; Mandinka; Wolof; Fula; Sarakola; Diula; others

Ethnic Makeup: 42% Mandinka; 18% Fula; 16% Wolof; 24% others (99% African; 1% non-Gambian)

Religions: 90% Muslim; 9% Christian; 1% indigenous beliefs

Health

Life Expectancy at Birth: 52 years (male); 56 years (female)

Infant Mortality: 76.3/1,000 live births

Physicians Available: 1/14,536 people

HIV/AIDS Rate in Adults: 1.95%

Education

Adult Literacy Rate: 47.5%

Compulsory (Ages): 7–13; free

COMMUNICATION

Telephones: 38,000 main lines

Internet Users: 25,000 (2002)

TRANSPORTATION

Highways in Miles (Kilometers): 1,584 (2,640)

Railroads in Miles (Kilometers): none

Usable Airfield: 1

Motor Vehicles in Use: 9,000

GOVERNMENT

Type: republic

Independence Date: February 18, 1965 (from the United Kingdom)

Head of State/Government: President Yahya Jammeh is both head of state and head of government

Political Parties: Alliance for Patriotic Reorientation and Construction; National Reconciliation Party; People's Democratic Organization for Independence and Socialism; others

Suffrage: universal at 18

MILITARY

Military Expenditures (% of GDP): 0.3%
Current Disputes: internal conflicts; boundary dispute with Senegal

ECONOMY

Currency ($ U.S. Equivalent): 19.91 dalasis = $1
Per Capita Income/GDP: $1,770/$2.5 billion
GDP Growth Rate: 3%
Inflation Rate: 14%
Labor Force by Occupation: 75% agriculture; 19% industry and services; 6% government

Natural Resources: fish

Agriculture: peanuts; millet; sorghum; rice; corn, cassava; livestock; fish and forest resources

Industry: processing peanuts, fish, and hides; tourism; beverages; agricultural machinery assembly; wood- and metalworking; clothing

Exports: $139 million (primary partners Benelux, Japan, United Kingdom)
Imports: $200 million (primary partners China, Hong Kong, United Kingdom, the Netherlands)

SUGGESTED WEB SITES

http://www.gambianews.com
http://www.gambianet.com
http://www.gambia.com/gambia.html
http://www.cia.gov/cia/publications/factbook/geos/ga.html

The Gambia Country Report

Since his seizure of power in a 1994 coup, Yahya Jammeh has dominated politics in The Gambia. In 2001, he was reelected president in what international election monitors generally viewed as a free and fair poll. The government called for the elections, however, only shortly after Jammeh lifted a ban on politicians whom he had ousted from power.

DEVELOPMENT

Since independence, The Gambia has developed a tourist industry. Whereas in 1966 only 300 individuals were recorded as having visited the country, the figure for 1988–1989 was over 112,000. Tourism is now the second-biggest sector of the economy. Still, tourism has declined since 2000. After the February 2004 announcement of the discovery of large oil reserves, there is expected to be a major upturn in economic activity.

Subsequent parliamentary elections, in January 2002, were boycotted by most of the opposition, allowing Jammeh's Alliance for Patriotic Reorientation and Construction to win by a landslide. The Gambian opposition has otherwise been weakened by its internal divisions, all attempts to bring about a united front having failed. Opposition to Jammeh's rule has become more open.

In April 2000, Gambians were shocked when student protests in the capital city, Banjul, resulted in the killing of 14 people and the wounding of many more by government security forces. Many interpreted the violence as an ominous official response to the re-emergence of independent voices within the media and civil society, which have been pushing for greater openness and accountability in government.

Jammeh came to power in July 1994, after The Gambia's armed forces overthrew the government of Sir Dawda Jawara, bringing to an abrupt end what had been postcolonial West Africa's only example of uninterrupted multiparty democracy. Under international pressure, elections were held in September 1996 and January 1997, resulting in victories for Jammeh and the Alliance for Patriotic Reorientation and Construction. But the process was marred by the regime's continuing intolerance of genuine opposition. Since the failure of an alleged coup attempt in January 1995, critical voices have been largely silenced by an increasingly powerful National Intelligence Agency. Meanwhile, The Gambia's already weak economy has suffered from reduced revenues from tourism and foreign donors.

FREEDOM

Despite the imposition of martial law in the aftermath of the 1981 coup attempt, The Gambia has had a strong record of respect for individual liberty and human rights. Under its current regime, The Gambia has forfeited its model record of respect for freedoms of speech and association.

The Gambia is Africa's smallest noninsular nation. Except for a small seacoast, it is entirely surrounded by its much larger neighbor, Senegal. The two nations' separate existence is rooted in the activities of British slave traders who, in 1618, established a fort at the mouth of The Gambia River, from which they gradually spread their commercial and, later, political dominance upstream. Gambians have much in common with Senegalese. The Gambia's three major ethnolinguistic groups—the Mandinka, Wolof, and Fula (or Peul)—are found on both sides of the border. The Wolof language serves as a lingua franca in both the Gambian capital of Banjul and the urban areas of Senegal. Islam is the major religion of both countries, while each also has a substantial Christian minority. The economies of the two countries are also similar, with each being heavily reliant on the cultivation of ground nuts as a cash crop.

Timeline: PAST

1618
The British build Fort James at the current site of Banjul, on the Gambia River

1807
The Gambia is ruled by the United Kingdom through Sierra Leone

1965
Independence

1970
Dawda Jawara comes to power

1980s
An attempted coup against President Dawda Jawara; the rise and fall of the Senegambia Confederation

1990s
Jawara is overthrown by a military coup; Yahya Jammeh becomes head of state

PRESENT

2000s
Government security forces kill 14 people during student protests

Jammeh is reelected president, but the opposition gains in parliamentary elections

Discovery of large oil reserves

In 1981, the Senegalese and Gambian governments were drawn closer together by an attempted coup in Banjul. While Jawara was in London, dissident elements within his Paramilitary Field Force joined in a coup attempt with members of two small,

(Photo by Wayne Edge)

Small general stores provide innumerable materials for everyday life including cloth, tools, and utensils.

self-styled revolutionary parties. Based on a 1965 mutual-defense agreement, Jawara received assistance from Senegal in putting down the rebels. Constitutional rule was restored, but the killing of 400 to 500 people during the uprising and the subsequent mass arrest of suspected accomplices left Gambians bitter and divided.

HEALTH/WELFARE

Forty percent of Gambian children remain outside the primary-school setup. Economic Recovery Program austerity has made it harder for the government to achieve its goal of education for all.

In the immediate aftermath of the coup, The Gambia agreed to join Senegal in a loose confederation, which some hoped would lead to a full political union. But from the beginning, the Senegambia Confederation was marred by the circumstances of its formation. The continued presence of Senegalese soldiers in their country led Gambians to speak of a "shotgun wedding." Beyond fears of losing their local identity, many believed that proposals for closer economic integration, through a proposed monetary and customs union, would be to The Gambia's disadvantage. Underlying this concern was the role played by Gambian traders in providing imports to Senegal's market. Other squabbles, such as a long-standing dispute over the financing of a bridge across The Gambia River, finally led to the Confederation's formal demise in 1989. But the two countries still recognize a need to develop alternative forms of cooperation.

The Gambia was modestly successful in rebuilding its politics in the aftermath of the 1981 coup attempt. Whereas the 1982 elections were arguably compromised by the detention of the main opposition leader,

ACHIEVEMENTS

Gambian *griots*—hereditary bards and musicians such as Banna and Dembo Kanute—have maintained a traditional art. Formerly, griots were attached to ruling families; now, they perform over Radio Gambia and are popular throughout West Africa.

Sherif Mustapha Dibba, on (later dismissed) charges of complicity in the revolt, the 1987 and 1992 polls restored most people's confidence in Gambian democracy. In both elections, opposition parties significantly increased their share of the vote, while Jawara's People's Progressive Party retained majority support.

Instances of official corruption had compromised the Jawara government's ability to use its electoral mandate to implement an Economic Recovery Program (ERP), which included austerity measures. The

Gambia has always been a poor country. During the 1980s, conditions worsened as a result of bad harvests and falling prices for groundnuts, which usually account for half of the nation's export earnings. The tourist industry was also disrupted by the 1981 coup attempt. Faced with mounting debt, the government submitted to International Monetary Fund pressure by cutting back its civil service and drastically devaluing the local currency. The latter step initially led to high inflation, but prices have become more stable in recent years, and the economy as a whole has begun to enjoy a gross domestic product growth rate of up to 5 percent per year. As elsewhere, the negative impact of Structural Adjustment has proved especially burdensome to urban dwellers.

Ghana (Republic of Ghana)

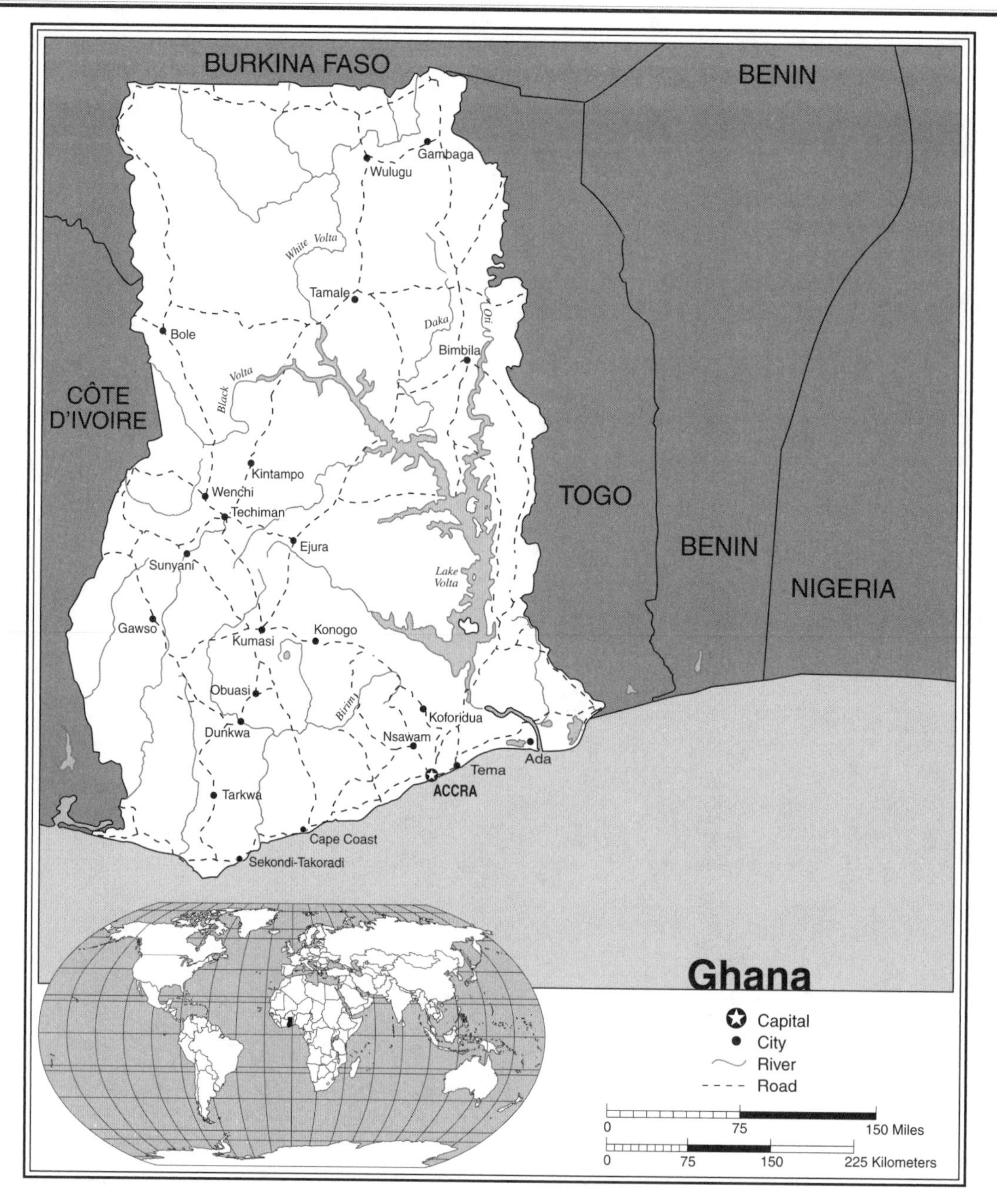

Ghana Statistics

GEOGRAPHY

Area in Square Miles (Kilometers): 92,100 (238,536) (about the size of Oregon)
Capital (Population): Accra (1,925,000)
Environmental Concerns: drought; deforestation; overgrazing; soil erosion; threatened wildlife populations; water pollution; insufficient potable water
Geographical Features: low plains with dissected plateau
Climate: tropical

PEOPLE

Population

Total: 20,757,032
Annual Growth Rate: 1.36%
Rural/Urban Population Ratio: 62/38
Major Languages: English; Akan; Ewe; Ga
Ethnic Makeup: nearly 100% African
Religions: 63% Christian; 21% indigenous beliefs; 16% Muslim

Health

Life Expectancy at Birth: 56 years (male); 59 years (female)
Infant Mortality: 55.6/1,000 live births
Physicians Available: 1/22,452 people
HIV/AIDS Rate in Adults: 3.6%

Education

Adult Literacy Rate: 64.5%
Compulsory (Ages): 6–16

COMMUNICATION

Telephones: 302,300 main lines
Daily Newspaper Circulation: 64/1,000 people
Televisions: 15/1,000 people
Internet Users: 200,000 (2002)

TRANSPORTATION

Highways in Miles (Kilometers): 24,428 (39,400)
Railroads in Miles (Kilometers): 592 (953)
Usable Airfields: 12
Motor Vehicles in Use: 135,000

GOVERNMENT

Type: constitutional democracy
Independence Date: March 6, 1957 (from the United Kingdom)
Head of State/Government: President John Agyekum Kufuor is both head of state and head of government
Political Parties: National Democratic Congress; New Patriotic Party; People's Convention Party; Every Ghanaian Living Everywhere; others
Suffrage: universal at 18

MILITARY

Military Expenditures (% of GDP): 0.7%
Current Disputes: internal conflicts

ECONOMY

Currency ($ U.S. Equivalent): 7,932 cedis = $1
Per Capita Income/GDP: $2,200/$44 billion
GDP Growth Rate: 4.7%
Inflation Rate: 26.7%
Unemployment Rate: 20%
Labor Force by Occupation: 39% services; 36% agriculture; 25% industry
Population Below Poverty Line: 31.4%
Natural Resources: gold; timber; industrial diamonds; bauxite; manganese; fish; rubber; hydropower
Agriculture: cocoa beans; rice; coffee; cassava; peanuts; corn; shea nuts; bananas; timber; fish
Industry: mining; lumbering; light manufacturing; aluminum smelting; food processing
Exports: $1.94 billion (primary partners Togo, United Kingdom, Italy)
Imports: $2.83 billion (primary partners United Kingdom, Nigeria, United States)

SUGGESTED WEB SITES

http://www.ghana.gov.gh
http://www.ghana.com
http://www.ghana-embassy.org/
http://www.cia.gov/cia/publications/factbook/geos/gh.html

Ghana Country Report

In December 2000, John Kufuor defeated then–vice-president John Atta Mills in a ballot that marked the first real transfer of power through elections in Ghana in 4 1/2 decades as an independent republic. The presidential elections also brought to an end two decades of rule by the incumbent Jerry Rawlings, who agreed to step down in accordance with the Constitution. President Kufuor inaugurated a reconciliation commission to look into human-rights violations during military rule. The commission began hearing testimonies in January 2003. Known as the "Gentle Giant," President Kufuor has sought to promote reconciliation without recrimination in a nation with a violent history of political division.

DEVELOPMENT

In the 1960s, Ghana invested heavily in schooling, resulting in perhaps the best-educated population in Africa. Today, hundreds of thousands of professionals who began their schooling under Nkrumah work overseas, annually remitting an estimated $1 billion to the Ghanaian economy. In October 2003 the government approved the merger of two gold-mining firms, ending a bidding war and creating a new gold-mining giant.

In the 1990s, Ghana made gradual but steady progress in rebuilding its economy as well as its political culture, after decades of decline. For many Ghanaians, these gains have been of modest benefit. The country has achieved economic growth, while implementing a socially painful World Bank/International Monetary Fund–sponsored Economic Recovery Program (ERP) that in 2001 was rewarded with a debt-relief package. Yet when adjusted for inflation, per capita income remains below the level that existed in 1957, when Ghana became the first colony in sub-Saharan Africa to obtain independence.

The country is also overcoming the legacy of political instability brought about by revolving-door military coups. In March 1992, Rawlings marked the country's 35th anniversary of independence by announcing an accelerated return to multiparty rule. Eight months later, he was elected by a large majority as the president of what has been hailed as Ghana's "Fourth Republic." Although the election received the qualified endorsement of international monitors, its result was rejected by the main opposition, the New Patriotic Party (NPP). The NPP subsequently boycotted parliamentary elections, allowing an easy victory for Rawlings's National Democratic Congress (NDC), which captured 189 out of 200 seats, with a voter turnout of just under 30 percent. After months of bitter standoff, the political climate has eased since December 1993, when the NPP agreed to enter into a dialogue with the government about its basic demand that the interests of the ruling party be more clearly separated from those of the state.

Ghana's political transformation was a triumph for Rawlings, who ruled since 1981, when he and other junior military officers seized power as the Provisional National Defense Council (PNDC). In the name of ending corruption, they overthrew Ghana's previous freely elected government after it had been in office for less than two years. Rawlings was dismissive of elections at that time: "What does it mean to stuff bits of paper into boxes?" But political success seems to have altered his opinion.

FREEDOM

The move to multipartyism has promoted freedom of speech and assembly. Dozens of independent periodicals have emerged. In April 2002, a state of emergency was declared in the north, after a prominent local leader and more than 30 others were killed in clan violence.

At its independence, Ghana assumed a leadership role in the struggle against colonial rule elsewhere on the continent. Both its citizens and many outside observers were optimistic about the country's future. As compared to many other former colonies, the country seemed to have a sound infrastructure for future progress. Unfortunately, economic development and political democracy have proven to be elusive goals.

HEALTH/WELFARE

In 1991, the African Commission of Health and Human Rights Promoters established a branch in Accra to help rehabilitate victims of human-rights violations from throughout Anglophone Africa. The staff deals with both the psychological and physiological after-effects of abused ex-detainees.

The "First Republic," led by the charismatic Kwame Nkrumah, degenerated into a bankrupt and an increasingly authoritarian one-party state. Nkrumah had pinned his hopes on an ambitious policy of industrial development. When substantial overseas investment failed to materialize, he turned to socialism. His efforts led to a modest rise in local manufacturing, but the sector's productivity was compromised by inefficient planning, limited resources, expensive inputs, and mounting government corruption. The new state enterprises ended up being financed largely from the export earnings of cocoa, which had emerged as Ghana's principal cash crop during the colonial period. Following colonial precedent, Nkrumah resorted to paying local cocoa farmers well below the world market price for their output in an attempt to expand state revenues.

Nkrumah was overthrown by the military in 1966. Despite his regime's shortcomings, he is still revered by many as the leading pan-African nationalist of his generation. His warnings about the dangers of neo-imperialism have proved prophetic.

ACHIEVEMENTS

In 1993, Ghana celebrated the 30th anniversary of the School of the Performing Arts at the University of Legon. Integrating the world of dance, drama, and music, the school has trained a generation of artists committed to perpetuating Ghanaian, African, and international traditions in the arts.

Since Nkrumah's fall, the army has been Ghana's dominant political institution, although there were brief returns to civilian control in 1969–1972 and again in 1979–1981. Both the military and the civilian governments abandoned much of Nkrumah's socialist commitment, but for years they continued his policy of squeezing the cocoa farmers, with the long-term result of encouraging planters both to cut back on their production and to attempt to circumvent the official prices through smuggling. This situation, coupled with falling cocoa prices on the world market and rising import costs, helped to push Ghana into a state of severe economic depression during the 1970s. During that period, real wages fell by some 80 percent. Ghana's crisis was then aggravated by an unwillingness on the part of successive governments to devalue the country's currency, the cedi, which encouraged black-market trading.

RAWLINGS'S RENEWAL

By 1981, many Ghanaians welcomed the PNDC, seeing in Rawlings's populist rhetoric the promise of change after years of corruption and stagnation. The PNDC initially tried to rule through People's Defense Committees, which were formed throughout the country to act as both official watchdogs and instruments of mass mobilization. Motivated by a combination of idealism and frustration with the status quo, the vigilantism of these institutions threatened the country with anarchy until, in 1983, they were reined in. Also in 1983, the country faced a new crisis, when the Nigerian government suddenly expelled nearly 1 million Ghanaian expatriates, who had to be resettled quickly.

Faced with an increasingly desperate situation, the PNDC, in a move that surprised many, given its leftist leanings, began to implement the Economic Recovery Program. Some 100,000 public and parastatal employees were retrenched, the cedi was progressively devalued, and wages and prices began to reflect more nearly their market value. These steps have led to some economic growth, while annually attracting $500 million in foreign aid and soft loans and perhaps double that amount in cash remittances from the more than 1 million Ghanaians living abroad.

The human costs of ERP have been a source of criticism. Many ordinary Ghanaians, especially urban salary-earners, have suffered from falling wages coupled with rising inflation. Unemployment has also increased in many areas (today it is estimated at about 20 percent). Yet a recent survey found surprisingly strong support among "urban lower income groups" for ERP and the government in general. In the countryside, farmers have benefited from higher crop prices and investments in rural infrastructure, while there has been a countrywide boom in legitimate retailing.

ERP continues to have its critics, but it gained substantial support from politicians aligned with Ghana's three principal political tendencies: the Nkrumahists, loyal to the first president's pan-African socialist vision; the Danquah-Busia grouping, named after two past statesmen who struggled against Nkrumah for more liberal economic and political policies; and those loyal to the PNDC. In the November 1992 presidential elections, the NPP emerged as the main voice of the Danquah-Busia camp, while Rawlings's NDC attracted substantial support from Nkrumahists as well as those sympathetic to his own legacy. There was also a body of opinion that was critical of all three historic tendencies, characterizing the NPP and NDC as fronts for power-hungry men fighting yesterday's battles. During the April 1992 referendum to approve the new Constitution, more than half the registered voters (many Ghanaians complained that they were denied registration) failed to participate, despite the government and opposition's joint call for a large "yes" vote. Many also boycotted the November presidential elections. In December 1996, in a poll widely judged to have been fair, Rawlings was reelected. He narrowly defeated his former vice-president, John Kufuor, who enjoyed the backing of both the New Patriotic Party and the Convention People's Party—the same man who eventually would come to succeed him as president.

Timeline: PAST

1482
A Portuguese fort is built at Elmina

1690s
The establishment of the Asante Confederation under Osei Tutu

1844
The "Bonds" of 1844 signed by British officials and Fante chiefs as equals

1901
The British conquer the Asante, a final step in British control of the region

1957
Ghana is the first of the colonial territories in sub-Saharan Africa to become independent

1966
Kwame Nkrumah is overthrown by a military coup

1979
The first coup of Flight Lieutenant Jerry Rawlings

1980s
The second coup by Rawlings; the PNDC is formed

1990s
World Bank and IMF austerity measures; prodemocracy agitation leads to a transition to multiparty rule

PRESENT

2000s
Ghanaians elect John Kufuor of the opposition NPP as president, ending Rawlings's 2-decade rule

Guinea (Republic of Guinea)

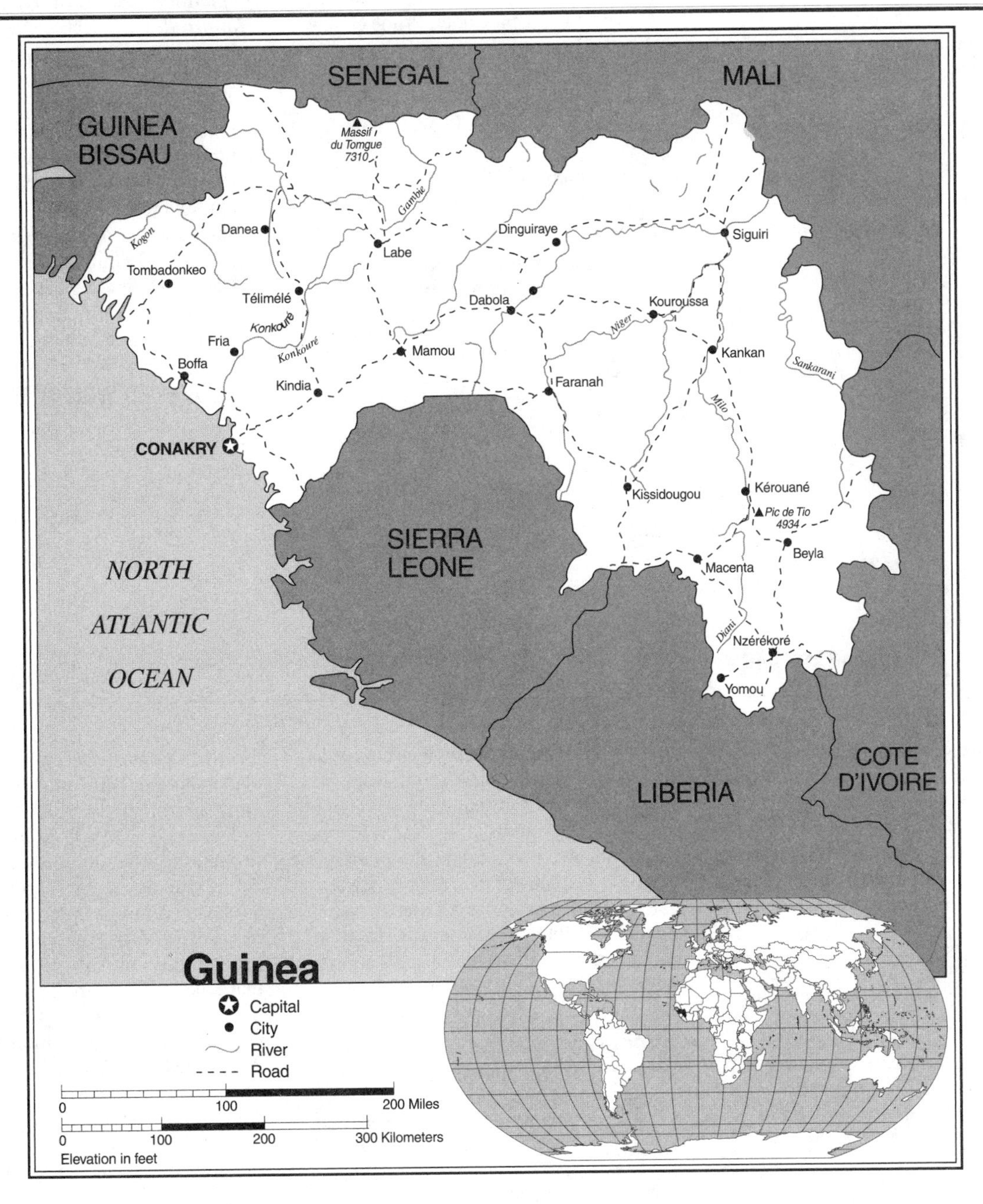

Guinea

GEOGRAPHY

Area in Square Miles (Kilometers): 95,000 (246,048) (about the size of Oregon)
Capital (Population): Conakry (1,272,000)
Environmental Concerns: deforestation; insufficient potable water; desertification; soil erosion and contamination; overfishing; overpopulation
Geographical Features: mostly flat coastal plain; hilly to mountainous interior
Climate: tropical

PEOPLE

Population

Total: 9,246,462
Annual Growth Rate: 2.37%
Rural/Urban Population Ratio: 68/32
Major Languages: French; many tribal languages
Ethnic Makeup: 40% Peuhl; 30% Malinke; 20% Soussou; 10% other African groups
Religions: 85% Muslim; 8% Christian; 7% indigenous beliefs

Health

Life Expectancy at Birth: 43 years (male); 48 years (female)
Infant Mortality: 127/1,000 live births

Physicians Available: 1/9,732 people
HIV/AIDS Rate in Adults: 1.54%

Education

Adult Literacy Rate: 36%
Compulsory (Ages): 7–13; free

COMMUNICATION

Telephones: 37,000 main lines
Televisions: 10/1,000 people
Internet Users: 40,000 (2003)

TRANSPORTATION

Highways in Miles (Kilometers): 18,060 (30,100)
Railroads in Miles (Kilometers): 651 (1,086)
Usable Airfields: 15
Motor Vehicles in Use: 33,000

GOVERNMENT

Type: republic
Independence Date: October 2, 1958 (from France)
Head of State/Government: President (General) Lansana Conté; Prime Minister Francois Lonseny Fall
Political Parties: Party for Unity and Progress; Union for the New Republic; Rally for the Guinean People; many others
Suffrage: universal at 18

MILITARY

Military Expenditures (% of GDP): 3.3%
Current Disputes: refugee crisis as a result of unrest in Sierra Leone and Liberia

ECONOMY

Currency ($ U.S. Equivalent): 1,975 Guinean francs = $1
Per Capita Income/GDP: $2,100/$19 billion
GDP Growth Rate: 3.3%
Inflation Rate: 14.8%
Labor Force by Occupation: 80% agriculture; 20% industry and services
Population Below Poverty Line: 40%
Natural Resources: bauxite; iron ore; diamonds; gold; uranium; hydropower; fish
Agriculture: rice; cassava; millet; sweet potatoes; coffee; bananas; palm products; pineapples; livestock
Industry: bauxite; gold; diamonds; alumina refining; light manufacturing and agricultural processing
Exports: $695 million (primary partners Belgium, United States, Ireland)
Imports: $555 million (primary partners France, United States, Belgium)

SUGGESTED WEB SITES

http://www.sas.upenn.edu/African_Studies/Country_Specific/Guinea.html
htt://www.cia.gov/cia/publications/factbook/geos/gv.html

Guinea Country Report

In recent years, Guinea has managed to maintain internal peace in the face of armed conflict along its borders. But renewed fighting in neighboring Sierra Leone, Liberia, and Côte d'Ivoire has revived fears that the country is being dragged into a wider regional conflict.

DEVELOPMENT

A measure of economic growth in Guinea was reflected in the rising traffic in Conakry harbor, whose volume rose 415% over a 4-year period. Plans are being made to improve the port's infrastructure, but regional conflicts threaten further development.

Since the end of 2000, incursions by rebels along Guinea's border regions with Liberia and Sierra Leone have claimed more than 1,000 lives and caused massive population displacement. The government has accused the governments of Liberia and Burkina Faso, the (now-disarmed) Revolutionary United Front (RUF) of Sierra Leone, and former Guinean Army mutineers of working together to destabilize Guinea. With tensions rising, Alpha Conde, leader of the opposition Guinean People's Rally (RPG), was sentenced in September 2000 to five years in prison, charged with endangering state security and recruiting foreign mercenaries. By February 2001, the government began deploying attack helicopters to the front line to fight with rebels. Meanwhile, the United Nations high commissioner for refugees, Ruud Lubbers, warned that the country's refugee crisis, mostly the result of the conflicts in Sierra Leone and Liberia, was in danger of getting out of control. The country shelters more than half a million (estimates vary widely) cross-border refugees.

FREEDOM

Human rights continue to be restricted in Guinea, with the government's security forces being linked to disappearances, abuse of prisoners and detainees, torture by military personnel, and inhumane prison conditions.

At home, the harassment of journalists and opposition leaders has underscored the government's continued insecurity despite the 1992 transition to multiparty politics. In a constitutional referendum that took place in November 2001, voters endorsed President Lansana Conté's proposal to extend the presidential term from five to seven years. But the opposition boycotted the poll, accusing Conté of trying to stay in office for life. During November 2003 opposition leader Jean-Marie Dore was detained for saying that President Conté was too ill to contest December's presidential election. After being released from jail Mr. Dore and his party boycotted the election in December 2003. With the main opposition parties boycotting the election President Conté won a third term in office.

Since coming to power in 1984, Conté has proven adept at surviving challenges to his authority. In April 1992, he announced that a new Constitution guaranteeing freedom of association would take immediate effect. Within a month, more than 30 political parties had formed. This initiative was a political second chance for a nation whose potential had been mismanaged for decades, under the dictatorial rule of its first president, Sekou Touré.

HEALTH/WELFARE

The life expectancy of Guineans is among the lowest in the world, reflecting the stagnation of the nation's health service during the Sekou Touré years.

From his early years as a radical trade-union activist in the late 1940s until his death in office in 1984, Sekou Touré was Guinea's dominant personality. A descendent of the nineteenth-century Malinke hero Samori Touré, who fiercely resisted the imposition of French rule, Sekou Touré was a charismatic but repressive leader. In

1958, he inspired Guineans to vote for immediate independence from France. At the time, Guinea was the only territory to opt out of Charles de Gaulle's newly established French Community. The French reacted spitefully, withdrawing all aid, personnel, and equipment from the new nation, an event that heavily influenced Guinea's postindependence path. The ability of Touré's Democratic Party of Guinea (PDG) to step into the administrative vacuum was the basis for Guinea's quick transformation into the African continent's first one-party socialist state, a process that was encouraged by the Soviet bloc.

ACHIEVEMENTS

More than 80% of the programming broadcast by Guinea's television service is locally produced. This output has included more than 3,000 movies. A network of rural radio stations is currently being installed.

Touré's rule was characterized by economic mismanagement and the widespread abuse of human rights. It is estimated that 2 million people—at the time about one out of every four Guineans—fled the country during his rule. At least 2,900 individuals disappeared under detention by the government.

By the late 1970s, Touré, pressured by rising discontent and his own apparent realization of his country's poor economic performance, began to modify both his domestic and foreign policies. This shift led to better relations with Western countries but little improvement in the lives of his people. In 1982, Amnesty International publicized the Touré regime's dismal record of political killings, detentions, and torture, but the world remained largely indifferent.

On April 3, 1984, a week after Touré's death, the army stepped in, claiming that it wished to end all vestiges of the late president's dictatorial regime. The bloodless coup was well received by Guineans. Hundreds of political prisoners were released; and the once-powerful Democratic Party of Guinea, which during the Touré years had been reduced from a mass party into a hollow shell, was disbanded. A new government was formed, under the leadership of then-colonel Conté; and a 10-point program for national recovery was set forth, including the restoration of human rights and the renovation of the economy.

Faced with an empty treasury, the new government committed itself to a severe Structural Adjustment Program (SAP). This has led to a dismantling of many of the socialist structures that had been established by the previous government. While international financiers have generally praised it, the government has had to weather periodic unrest and coup attempts. In spite of these challenges, however, it has remained committed to SAP.

Guinea is blessed with mineral resources, which could lead to a more prosperous future. The country is rich in bauxite and has substantial reserves of iron and diamonds. New mining agreements, leading to a flow of foreign investment, have already led to a modest boom in bauxite and diamond exports. Small-scale gold mining is also being developed. These initiatives, however, are being threatened by the conflicts in neighboring states.

Guinea's greatest economic failing has been the poor performance of its agricultural sector. Unlike many of its neighbors, the country enjoys a favorable climate and soils. But, although some 80 percent of Guineans are engaged in subsistence farming, only 3 percent of the land is cultivated, and foodstuffs remain a major import. Blame for this situation largely falls on the Touré regime's legacy of an inefficient, state-controlled system of marketing and distribution. In 1987, the government initiated an ambitious plan of road rehabilitation, which, along with better produce prices, has encouraged farmers to produce more for the domestic market.

Timeline: PAST

1700s
A major Islamic kingdom is established in the Futa Djalon

1898
Samori Touré is defeated by the French

1958
Led by Sekou Touré, Guineans reject continued membership in the French Community; an independent republic is formed

1978
French president Giscard d'Estaing visits Guinea: the beginning of a reconciliation between France and Guinea

1980s
Sekou Touré's death is followed by a military coup; the introduction of SAP leads to urban unrest

1990s
President Lansana Conté begins to establish a multiparty democracy; multiparty elections are held for the presidency; Conté claims victory

PRESENT

2000s
Guinea stays the course of Structural Adjustment despite severe hardships

Fears intensify regarding a regional conflict

Guinea, Sierra Leone, and Liberia agree on measures to secure mutual borders and to tackle insurgency

Prime Minister Lounseny resigns from office

Guinea-Bissau (Republic of Guinea-Bissau)

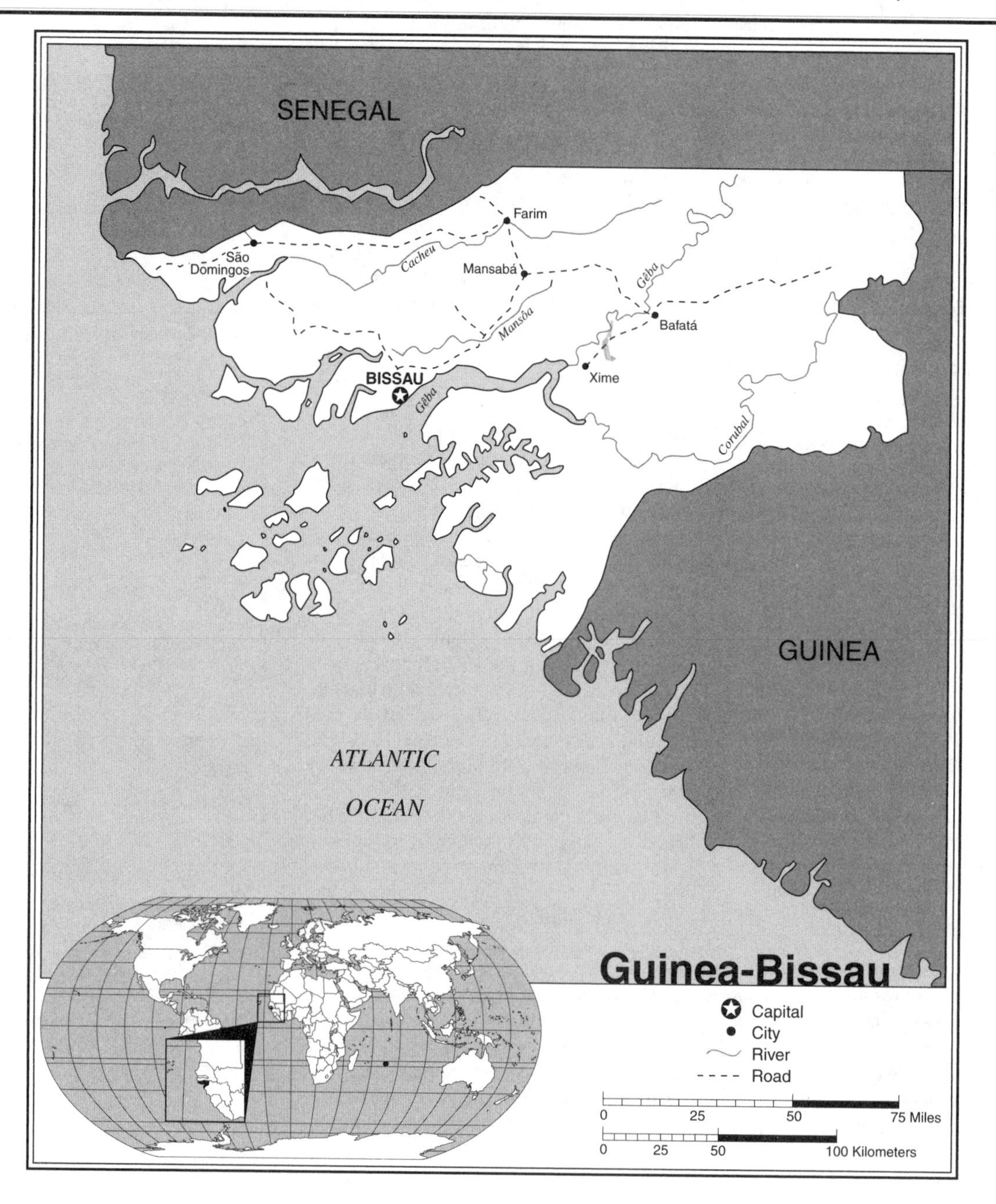

Guinea-Bissau Statistics

GEOGRAPHY

Area in Square Miles (Kilometers): 13,948 (36,125) (about 3 times the size of Connecticut)
Capital (Population): Bissau (292,000)
Environmental Concerns: soil erosion; deforestation; overgrazing; overfishing
Geographical Features: mostly low coastal plain, rising to savanna in the east
Climate: tropical

PEOPLE

Population

Total: 1,388,363
Annual Growth Rate: 1.99%
Rural/Urban Population Ratio: 77/23
Major Languages: Portuguese; Kriolo; various African languages
Ethnic Makeup: 30% Balanta; 20% Fula; 14% Manjaca; 13% Mandinka; 23% others (99% African; 1% others)
Religions: 50% indigenous beliefs; 45% Muslim; 5% Christian

Health

Life Expectancy at Birth: 47 years (male); 52 years (female)
Infant Mortality: 118/1,000 live births
Physicians Available: 1/9,477 people
HIV/AIDS Rate in Adults: 2.5%

Education

Adult Literacy Rate: 54%
Compulsory (Ages): 7–13

COMMUNICATION

Telephones: 10,600 main lines
Internet Users: 19,000 (2003)

TRANSPORTATION

Highways in Miles (Kilometers): 2,610 (4,350)
Railroads in Miles (Kilometers): none
Usable Airfields: 28
Motor Vehicles in Use: 6,000

GOVERNMENT

Type: republic
Independence Date: September 10, 1974 (from Portugal)
Head of State/Government: President Henrique Rosa; Prime Minister Carlos Gomes Junior
Political Parties: African Party for the Independence of Guinea-Bissau and Cape Verde; Front for the Liberation and Independence of Guinea; United Social Democratic Party; Social Renovation Party; Democratic Convergence; others
Suffrage: universal at 18

MILITARY

Military Expenditures (% of GDP): 2.8%
Current Disputes: trouble along the border with Senegal

ECONOMY

Currency ($ U.S. Equivalent): 581 Communaute Financiere Africaine francs (XOF) = $1
Per Capita Income/GDP: $900/$1.2 billion
GDP Growth Rate: -7%
Inflation Rate: 4%
Labor Force by Occupation: 82% agriculture
Natural Resources: fish; timber; phosphates; bauxite; petroleum
Agriculture: corn; beans; cassava; cashew nuts; cotton; fish and forest products; peanuts; rice; palm kernels
Industry: agricultural-products processing; beverages
Exports: $80 million (primary partners India, Italy, South Korea)
Imports: $55.2 million (primary partners Portugal, Senegal, Thailand)

SUGGESTED WEB SITES

http://www.guineabissau.com
http://www.sas.upenn.edu/African_Studies/Country_Specific/G_Bissau.html

Guinea-Bissau Country Report

In February 2000, Kumba Yala of the Social Renovation Party (PRS) took 72 percent of the vote in the second round of presidential elections, defeating the candidate of the former ruling African Party for the Independence of Guinea-Bissau and Cape Verde (PAIGC). Yala's election was the culmination of an 18-month process that brought a measure of political peace to the country, which in 1998 had appeared to be heading toward civil war. Two months of fighting, which had resulted in the displacement of up to one third of the country's population, was brought to an end in August 1998 with the signing of a ceasefire accord between the government and rebel soldiers. This was followed up in November of that year by the establishment of a "Government of National Unity," which presided over a transition to genuine multiparty politics.

DEVELOPMENT

With help from the UN Development Program, Guinea-Bissau has improved the tourism infrastructure of the 40-island Bijagos Archipelago in the hopes of bringing in much-needed revenues.

Yala's reign came to an end when he was ousted from the presidency in a bloodless military coup in September 2003. The military chief who led the coup said the move was, in part, a response to the worsening economic and political situation. Henrique Rosa was chosen by the military authorities to be the head of state pending fresh presidential elections, expected in March 2005. A businessman whose only previous involvement in politics had been his chairing of the National Elections Commission in Guinea-Bissau's first free elections in 1994, Mr. Rosa was at first reluctant to accept the post. He was persuaded to take on the role by the Roman Catholic Bishop of Bissau, Jose Camnate, who headed a committee appointed by the military to help set up a caretaker civilian government. Before taking up his post President Rosa said that he hoped to be a "guarantor of justice, freedom, and peace" for the people of Guinea-Bissau.

The Yala government faced the unenviable challenge of promoting economic development. Since independence, the country has consistently been listed as one of the world's 10 poorest countries. Unfortunately, the period since Yala's installation has been marred by continued political instability. Prior to the 2003 coup, there had been three other attempted coups, hundreds of lives lost in war and political violence, and the pulling out of the ruling coalition by one of the major partners due to lack of consultation. Former president General Mane was killed in 2000 after allegedly trying to stage a coup. Secretary-General Kofi Annan of the United Nations intervened to urge political dialogue in order to defuse domestic tensions. An International Monetary Fund team praised improvements in financial controls, but this came after the country had lost tens of millions of dollars in revenue from corrupt practices of government officials. Two prime ministers and the foreign minister were dismissed by the president for various failings. Meanwhile, the head of the Supreme Court and three judges were dismissed by Yala for allegedly overturning the presidents' decision to expel leaders of a Muslim sect from the country.

FREEDOM

The police have engaged in arbitrary arrests and torture. The fighting between government and rebel troops resulted in some 13,000 civilian casualties during the 1990s.

To many outsiders, the nation has been better known for its prolonged liberation war, from 1962 to 1974, against Portuguese colonial rule. Mobilized by the PAIGC, the freedom struggle in Guinea-Bissau played a major role in the overthrow of the Fascist dictatorship in Portugal itself and the liberation of its other African colonies.

The origins of Portuguese rule in Guinea-Bissau go back to the late 1400s. The area was raided for centuries as a source of slaves, who were shipped to Portugal and its colonies of Cape Verde and Brazil. With the nineteenth-century abolition of slave trading, the Portuguese began to impose forced labor within Guinea-Bissau itself.

In 1956, six *assimilados*—educated Africans who were officially judged to have assimilated Portuguese culture—led by Amilcar Cabral, founded the PAIGC as a vehicle for the liberation of Cape Verde as well as Guinea-Bissau. From the beginning, many Cape Verdeans, such as Cabral, played a prominent role within the PAIGC. But the group's largest following and main center of activity were in Guinea-Bissau. In 1963, the PAIGC turned to armed resistance and began organizing itself as an alternative government. By the end of the decade, the movement was in control of two thirds of the country. In those areas, the PAIGC was notably successful in establishing its own marketing, judicial, and educational as well as political institutions. Widespread participation throughout Guinea-Bissau in the 1973 election of a National Assembly encouraged a number of countries to formally recognize the PAIGC declaration of state sovereignty.

HEALTH/WELFARE

Guinea-Bissau's health statistics remain appalling: an overall 48-year life expectancy, 12% infant mortality, and more than 90% of the population infected with malaria.

INDEPENDENCE

Since 1974, the leaders of Guinea-Bissau have tried to confront the problems of independence while maintaining the idealism of their liberation struggle. The nation's weak economy has limited their success. Guinea-Bissau has little in the way of mining or manufacturing, although explorations have revealed potentially exploitable reserves of oil, bauxite, and phosphates. More than 80 percent of the people are engaged in agriculture, but urban populations depend on imported foodstuffs. This situation has been generally attributed to the poor infrastructure and a lack of incentives for farmers to grow surpluses. Efforts to improve the rural economy during the early years of independence were hindered by severe drought. Only 8 percent of the small country's land is cultivated.

Under financial pressure, the government adopted a Structural Adjustment Program (SAP) in 1987. The peso was devalued, civil servants were dismissed, and various subsidies were reduced. The painful effects of these SAP reforms on urban workers were cushioned somewhat by external aid.

ACHIEVEMENTS

With Portuguese assistance, a new fiber-optic digital telephone system is being established in Guinea-Bissau.

In 1988, in a desperate move, the government signed an agreement with Intercontract Company, allowing the firm to use its territory for five years as a major dump site for toxic waste from Britain, Switzerland, and the United States. In return, the government was to earn up to $800 million. But the deal was revoked after it was exposed by members of the country's exiled opposition; a major environmental catastrophe would have resulted had it gone through.

POLITICAL DEVELOPMENT

Following the assassination of Amilcar Cabral, in 1973, his brother, Luis Cabral, succeeded him as the leader of the PAICG, thereafter becoming Guinea-Bissau's first president. Before 1980, both Guinea-Bissau and Cape Verde were separately governed by a united PAIGC, which had as its ultimate goal the forging of a political union between the two territories. But in 1980, Luis Cabral was overthrown by the military, which accused him of governing through a "Cape Verdean clique." João Vieira, a popular commander during the liberation war who had also served as prime minister, was appointed as the new head of state. As a result, relations between Cape Verde and Guinea-Bissau deteriorated, leading to a breakup in the political links between the two nations.

The PAIGC under Vieira continued to rule Guinea-Bissau as a one-party state for 10 years. The system's grassroots democracy, which had been fostered in its liberated zones during the war, gave way to a centralization of power around Vieira and other members of his military-dominated Council of State. Several coup attempts resulted in increased authoritarianism.

But the government reversed course, and in 1991, the country formally adopted multipartyism. Progress has been slow. An alleged coup attempt in 1993 led to the detention and subsequent trial of a leading opposition figure, João da Costa, on charges of plotting the government's overthrow. Elections finally occurred in July 1994. The vote resulted in a narrow second-round victory for Vieira against a very divided opposition.

Timeline: PAST

1446
Portuguese ships arrive; claimed as Portuguese Guinea; slave trading develops

1915
Portugal gains effective control over most of the region

1956
The African Party for the Independence of Guinea-Bissau and Cape Verde is formed

1963–1973
Liberation struggle in Guinea-Bissau under the PAIGC and Amilcar Cabral

1973
Amilcar Cabral is assassinated; the PAIGC declares Guinea-Bissau independent

1974
Revolution in Portugal leads to recognition of Guinea-Bissau's independence

1980
João Vieira comes to power through a military coup, ousting Luis Cabral

1990s
The country moves toward multipartyism; "Government of National Unity"

PRESENT

2000s
Kumba Yala of the Social Renovation (or Renewal) Party is elected president

Liberia (Republic of Liberia)

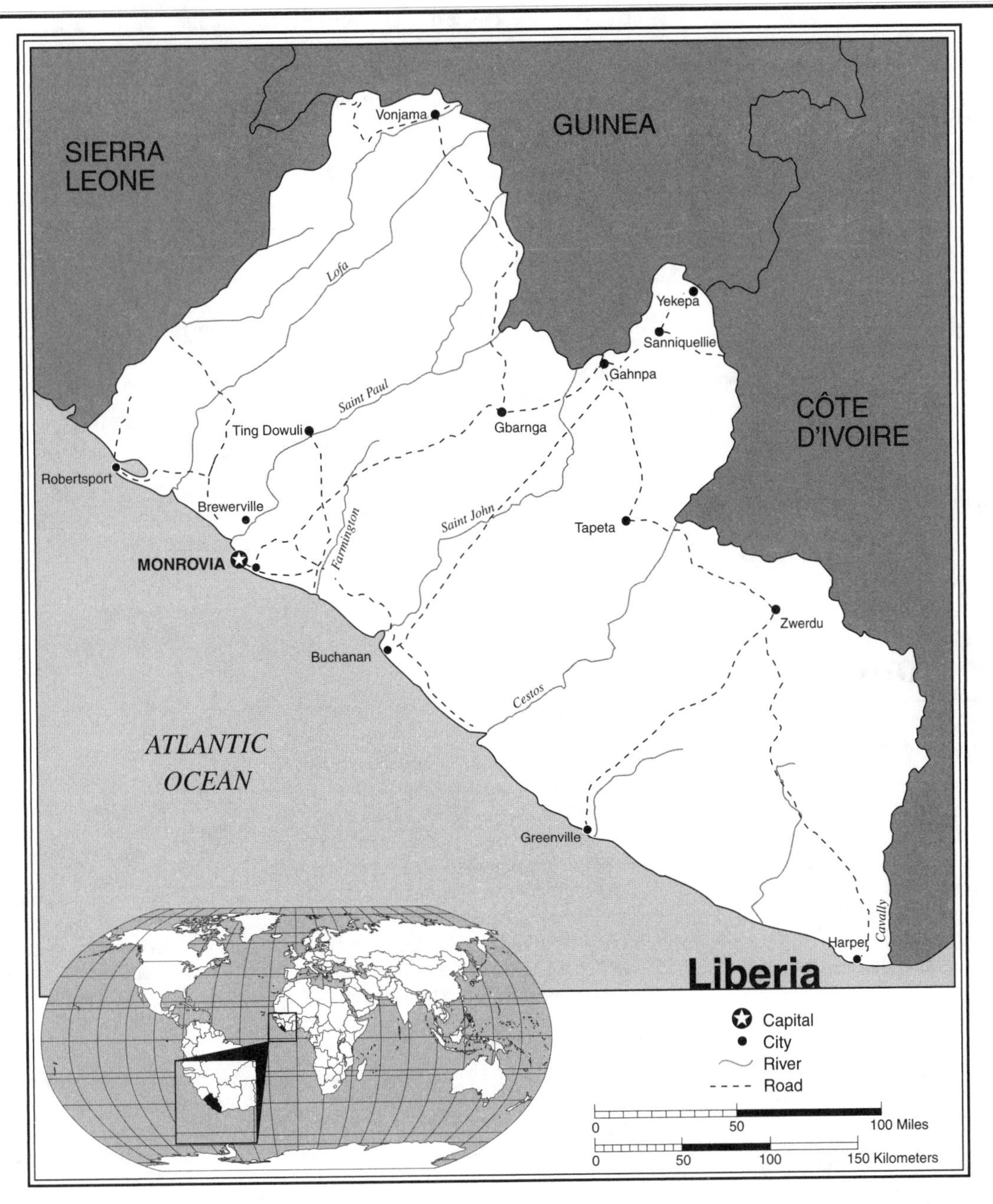

Liberia Statistics

GEOGRAPHY

Area in Square Miles (Kilometers): 43,000 (111,370) (about the size of Tennessee)

Capital (Population): Monrovia (491,000)

Environmental Concerns: soil erosion; deforestation; loss of biodiversity; water pollution

Geographical Features: mostly flat to rolling coastal plains, rising to rolling plateau and low mountains in the northeast

Climate: tropical

PEOPLE

Population

Total: 3,390,635

Annual Growth Rate: 2.7%

Rural/Urban Population Ratio: 56/44

Major Languages: English; Kpelle; Grio; Kru; Krahn; others

Ethnic Makeup: 95% indigenous groups; 5% Americo-Liberian

Religions: 40% indigenous beliefs; 40% Christian; 20% Muslim

Health

Life Expectancy at Birth: 50 years (male); 53 years (female)

Infant Mortality: 130.2/1,000 live births

Physicians Available: 1/8,333 people
HIV/AIDS Rate in Adults: 9%

Education

Adult Literacy Rate: 38.3%
Compulsory (Ages): 7–16; free

COMMUNICATION

Telephones: 7,000 main lines
Daily Newspaper Circulation: 15/1,000 people
Televisions: 20/1,000 people
Internet Users: 1,000 (2002)

TRANSPORTATION

Highways in Miles (Kilometers): 6,180 (10,300)
Railroads in Miles (Kilometers): 306 (490)
Usable Airfields: 47
Motor Vehicles in Use: 28,000

GOVERNMENT

Type: republic
Independence Date: July 26, 1847
Head of State/Government: Chairman Gyude Bryant is both head of state and head of government (note this is an interim position until presidential elections)
Political Parties: National Patriotic Party; National Democratic Party of Liberia; Liberian Action Party; Liberian People's Party; United People's Party; others
Suffrage: universal at 18

MILITARY

Military Expenditures (% of GDP): 1.3%
Current Disputes: civil war; border instabilities with Sierra Leone

ECONOMY

Currency ($ U.S. Equivalent): 61.7 Liberian dollar = $1
Per Capita Income/GDP: $1,100/$3.6 billion
GDP Growth Rate: 3%
Inflation Rate: 15%
Unemployment Rate: 70%
Labor Force by Occupation: 70% agriculture; 22% services; 8% industry
Population Below Poverty Line: 80%
Natural Resources: iron ore; timber; diamonds; gold; hydropower
Agriculture: rubber; rice; palm oil; cassava; coffee; cocoa beans; sugarcane; bananas; sheep; goats; timber
Industry: rubber processing; food processing; diamonds
Exports: $55 million (primary partners Belgium, Germany, Italy)
Imports: $170 million (primary partners France, South Korea, Japan)

SUGGESTED WEB SITES

http://www.fol.org
http://www.blackworld.com/country/liberia.htm
http://www.cia.gov/cia/publications/factbook/geos/li.html

Liberia Country Report

In August 2000, Liberian president Charles Taylor declared a state of emergency as fighting intensified in the north of the country. Since then, rebels calling themselves Liberians United for Reconciliation and Democracy (LURD) have made major advances against government forces. The excesses of President Taylor's troops, who are accused of pillaging the rural countryside, may have strengthened the rebels' hand. There has also been unrest in the urban areas, with Amnesty International acusing government security forces of brutally suppressing civilian protesters. In May 2001, the United Nations Security Council re-imposed an arms embargo on Liberia to punish Taylor for trading weapons for diamonds ("blood diamonds") from rebels in Sierra Leone.

DEVELOPMENT

Liberia's economic and social infrastructure was devastated by the war. Today people are surviving primarily through informal-sector bartering and trading.

Citing a reduced threat from rebels, the state of emergency was lifted in Liberia in September 2002. The rebels initiated a major offensive in March 2003, which ultimately lead to the end of the Taylor administration. The rebels launched battles simultaneously on numerous fronts and moved to within 10 kilometers of the capital Monrovia. Throwing fuel upon the fire in the middle of Liberian peace negotiations taking place in Ghana, the high court in Sierra Leon issued an indictment accusing President Taylor of war crimes over his alleged backing of rebels in Sierra Leone (which had also earned Liberia two years of UN-imposed sanctions). In July, as fighting intensified around the capital, the West African regional group ECOWAS agreed to provide peacekeepers, arriving in August 2003 led by the Nigerians. Surprisingly, once the Nigerians were in the capital, Charles Taylor left the country for good. A transitional government—composed of rebel, government, and civil society groups—assumed control in October 2003. Chariman Gyunde Bryane was sworn in as head of the interim government.

In December 2003 the UN launched a major peacekeeping mission, deploying thousands of troops in rebel-held territory, stating that their aim was to disarm former combatants. During February 2004 international donors pledged more than $500m in reconstruction aid, and the UN Security Council voted to freeze the assets of Charles Taylor.

The LURD forces, who are said to be composed of veterans of past civil strife in Sierra Leone as well as Liberia, have also been accused of human-rights abuses. Their provisioning from among the general population certainly contributed to a rise in hunger in rebel-controlled areas, in northeastern Liberia.

FREEDOM

The Taylor government has tried to reestablish the rule of law. The various government security forces continue to be linked to abuses. In 1998, Taylor accused rivals of plotting a coup and jeopardizing continued efforts to build postwar reconciliation in the country. In 2000, he declared a state of emergency when civil fighting intensified in the north of the country.

Renewed civil war in Liberia has dashed the hopes of those who had believed that the country had achieved peace following the extreme horrors of the previous conflict. Between 1989 and 1996, some 200,000 people were killed. Among the survivors, approximately 750,000 fled the country as refugees, while another 1.2 million were internally displaced, out of a total population of only 2.6 million people. The war ended in July 1997 with Taylor and his National Patriotic Front of Liberia (NPFL) receiving about three quarters of the vote in internationally supervised elections. These elections were the culmination of a year-

(United Nations photo 11233 by N. van Praag)

The Liberian Civil War of 1989–1996 left the country destitute. Political anarchy destroyed much of the infrastructure, economy, and culture. Nearly a tenth of the population were killed, many more displaced from their homes.

long process to restore peace to the country. Taylor's subsequent inauguration brought stability to the country—but it is now under renewed threat.

The restoration of peace in Liberia after July 1997 was overseen by a West African regional military force (ECOMOG) of just over 10,000. The force was deployed throughout the country to provide the security, facilitate the disarmament and demobilization of local combatants, and protect returning refugees. Its relative success in carrying out these functions came about only after years of frustration. When the force finally withdrew in 1998, it left behind 10,000 to 25,000 locally fathered children. ECOMOG's security role was taken over by the reconstituted Armed Forces of Liberia, an uneasy mix of former civil-war rivals.

A major challenge facing Taylor's new government was Liberia's war-ravaged economy, with little formal-sector employment, some $2.2 billion in debt, and a collapsed infrastructure. In December 1997, the president announced that a new currency would be introduced in 1998 to replace the two separate currencies that were in use in different parts of the country in addition to the U.S. dollar, which is also legal tender.

AFRICAN-AMERICAN-AFRICANS

Among the African states, Liberia shares with Ethiopia the distinction of having avoided European rule. Between 1847 and 1980, Liberia was governed by an elite made up primarily of descendants of African-Americans who had begun settling along its coastline two decades earlier. These "Americo-Liberians" make up only 5 percent of the population. But they dominated politics for decades through their control of the governing True Whig Party (TWP). Although the republic's Constitution was ostensibly democratic, the TWP rigged the electoral process.

HEALTH/WELFARE

Outside aid and local self-help efforts were mobilized against famine in Liberia in 1990–1991. But the long and brutal Civil War of the 1990s took a dreadful toll on Liberians' health and well-being.

Most Liberians belong to indigenous ethnolinguistic groups, such as the Kpelle, Bassa, Grio, Kru, Krahn, and Vai, who were conquered by the Americo-Liberians during the 1800s and early 1900s. Some individuals from these subjugated communities accepted Americo-Liberian norms. Yet book learning, Christianity, and an ability to speak English helped an indigenous person to advance within the state only if he or she accepted its social hierarchy by becoming a "client" of an Americo-Liberian "patron." In a special category were the important interior "chiefs," who were able to maintain their local authority as long as they remained loyal to the republic.

During the twentieth century, Liberia's economy was transformed by vast Firestone rubber plantations, iron-ore mining, and urbanization. President William Tubman (1944–1971) proclaimed a "Unification Policy," to promote national integration, and an "Open-Door Policy," to encourage outside investment in Liberia. However, most of the profits that resulted from the modest external investment that did occur left the country, while the wealth that remained was concentrated in the hands of the TWP elite.

During the administration of Tubman's successor, William Tolbert (1971–1980), Liberians became more conscious of the inability of the TWP to address the inequities of the status quo. Educated youths from all ethnic backgrounds began to join dissident associations rather than the regime's patronage system.

As economic conditions worsened, the top 4 percent of the population came to control 60 percent of the wealth. Rural stagnation drove many to the capital city of Monrovia (named after U.S. president James Monroe), where they suffered from high unemployment and inflation. The inevitable explosion occurred in 1979, when the government announced a 50 percent price increase for rice, the national food staple. Police fired on demonstrators, killing and wounding hundreds. Rioting, which resulted in great property damage, led the government to appeal to neighboring Guinea for troops. It was clear that the TWP was losing its grip. Thus Sergeant Samuel Doe enjoyed widespread support when, in 1980, he led a successful coup.

ACHIEVEMENTS

Through a shrewd policy of diplomacy, Liberia managed to maintain its independence when Great Britain and France conquered neighboring areas during the late nineteenth century. It espoused African causes during the colonial period; for instance, Liberia brought the case of Namibia to the World Court in the 1950s.

DOE DOESN'T DO

Doe came to power as Liberia's first indigenous president, a symbolically important event that many believed would herald substantive changes. Some institutions of the old order, such as the TWP and the Masonic Temple (looked upon as Liberia's secret government) were disbanded. The House of Representatives and Senate were suspended. Offices changed hands, but the old administrative system persisted. Many of those who came to power were members of Doe's own ethnic group, the Krahn, who had long been prominent in the lower ranks of the army.

Doe declared a narrow victory for himself in the October 1985 elections, but there was widespread evidence of ballot tampering. A month later, exiled general Thomas Quiwonkpa led an abortive coup attempt. During and after the uprising, thousands of people were killed, mostly civilians belonging to Quiwonkpa's Grio group who were slaughtered by loyalist (largely Krahn) troops. Doe was inaugurated, but opposition-party members refused to take their seats in the National Assembly. Some, fearing for their lives, went into exile.

During the late 1980s, Doe became increasingly dictatorial. Many called on the U.S. government, in particular, to withhold aid until detainees were freed and new elections held. The U.S. Congress criticized the regime but authorized more than $500 million in financial and military support. Meanwhile, Liberia suffered from a shrinking economy and a growing foreign debt, which by 1987 had reached $1.6 billion.

Doe's government was not entirely to blame for Liberia's dire financial condition. When Doe came to power, the Liberian treasury was already empty, in large part due to the vast expenditures incurred by the previous administration in hosting the 1979 Organization of African Unity Conference. The rising cost of oil and decline in world prices for natural rubber, iron ore, and sugar had further crippled the economy. But government corruption and instability under Doe made the bad situation worse.

DOE'S DOWNFALL

Liberia's descent into violent anarchy began on December 24, 1989, when a small group of insurgents, led by Charles Taylor, who had earlier fled the country amid corruption charges, began a campaign to overthrow Doe. As Taylor's NPFL rebels gained ground, the war developed into an increasingly vicious interethnic struggle among groups who had been either victimized by or associated with the regime. Thousands of civilians were thus massacred by ill-disciplined gunmen on both sides; hundreds of thousands began to flee for their lives. By June 1990, with the rump of Doe's forces besieged in Monrovia, a small but efficient breakaway armed faction of the NPFL, under the ruthless leadership of a former soldier named Prince Johnson, had emerged as a deadly third force.

By August, with the United States unwilling to do more than evacuate foreign nationals from Monrovia (the troops of Doe, Johnson, and Taylor had begun kidnapping expatriates and violating diplomatic immunity), members of the Economic Community of West African States decided to establish a framework for peace by installing an interim government, with the support of the regional peacekeeping force known as ECOMOG: the ECOWAS Monitoring Group. The predominantly Nigerian force, which also included contingents from Ghana, Guinea, Sierra Leone, The Gambia, and, later, Senegal, landed in Monrovia in late August. This coincided with the nomination, by a broad-based but NPFL–boycotted National Conference, of Amos Sawyer, a respected academic, as the head of the proposed interim administration.

Initial hopes that ECOMOG's presence would end the fighting proved to be naïve. On September 9, 1990, Johnson captured Doe by shooting his way into ECOMOG headquarters. The following day, Doe's gruesome torture and execution were videotaped by his captors. This "outrage for an outrage" did not end the suffering. Protected by a reinforced ECOMOG, Sawyer was able to establish his interim authority over most of Monrovia, but the rest of the country remained in the hands of the NPFL or of local thugs.

Repeated attempts to get Johnson and Taylor to cooperate with Sawyer in establishing an environment conducive to holding elections proved fruitless. While most neighboring states have supported ECOMOG's mediation efforts, some have provided support (and, in the case of Burkina Faso, troops) to the NPFL, which has encouraged Taylor in his on-again, off-again approach toward national reconciliation.

In September 1991, a new, fiercely anti-NPFL force, the United Liberation Movement of Liberia (ULIMO), emerged from bases in Sierra Leone. The group is identified with former Doe supporters. Subsequent clashes between ULIMO and NPFL on both sides of the Liberian–Sierra Leonean border contributed to the April 1992 overthrow of the Sierra Leonean government as well as the failure of an October 1991 peace accord brokered by the Côte d'Ivoire's then-president, Felix Houphouët-Boigny.

In October 1992, ECOWAS agreed to impose sanctions on the NPFL for blocking Monrovia. ECOMOG then joined ULIMO and remnants of the Armed Forces of Liberia (AFL) in a counteroffensive. In 1993, yet another armed faction, the Liberia Peace Council (LPC), emerged to challenge Taylor for control of southeastern Liberia. In March 1994, an all-party interim State Council, agreed to in principle eight months earlier, was finally sworn in. But it quickly collapsed, while a violent split in ULIMO contributed to further anarchy. In July 2002, UN secretary-general Kofi Annan warned that the Liberian conflict threatens the United Nations' peacekeeping work in neighboring countries.

Timeline: PAST

1500s
The Vai move onto the Liberian coast from the interior

1822
The first African-American settlers arrive from the United States

1871
The first coup exchanges one Americo-Liberian government for another

1931
The League of Nations investigates forced-labor charges

1944
President William Tubman comes to office

1971
William Tolbert becomes president

1980s
Tolbert is assassinated; a military coup brings Samuel Doe to power

1990s
Civil war leads to the execution of Doe, anarchy, and foreign intervention; Charles Taylor and the NPFL win power in internationally supervised elections; the Civil War ends

PRESENT

2000s
President Taylor declares a state of emergency

Civil war resumes

President Taylor steps down—a transitional government is formed

Mali (Republic of Mali)

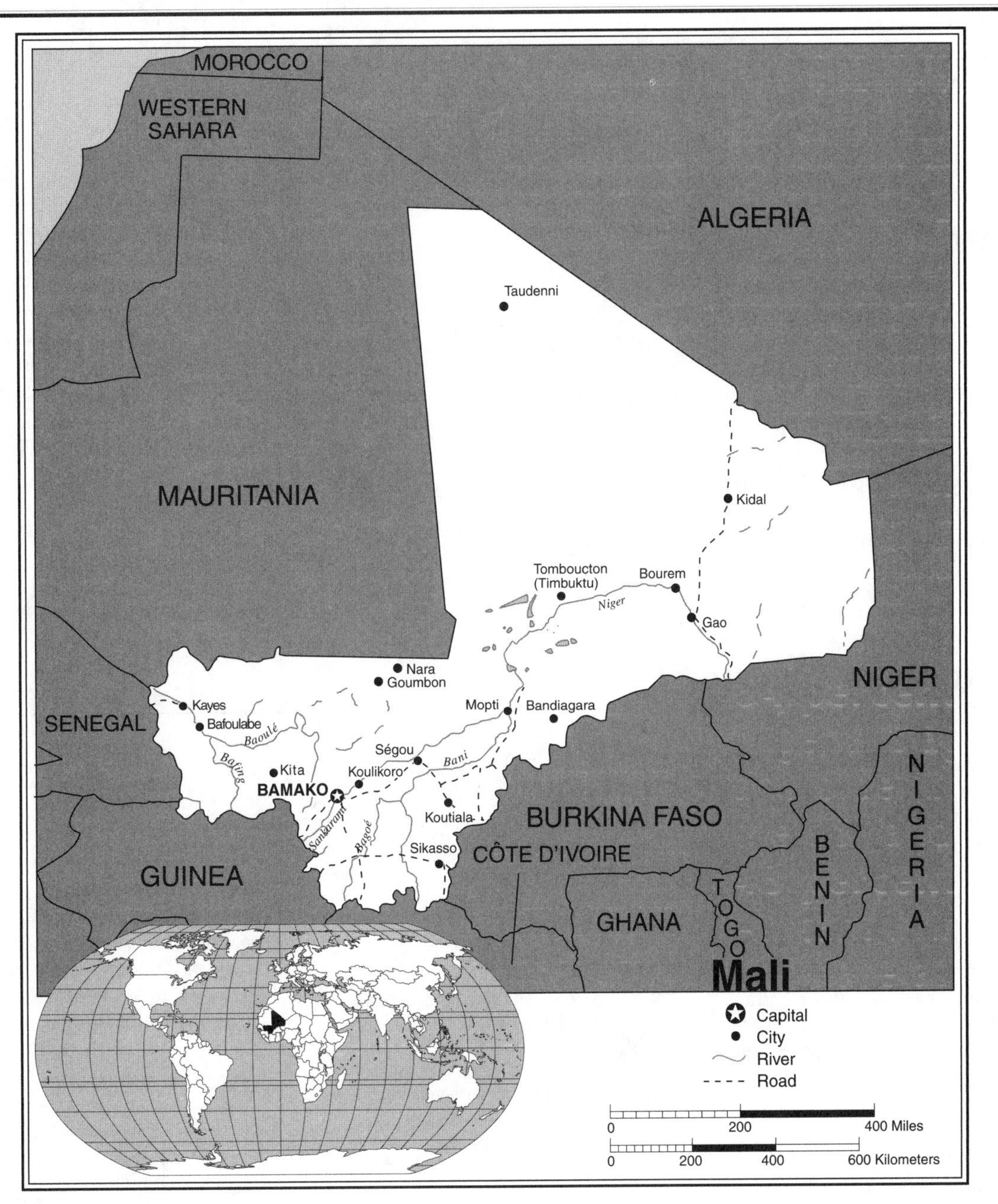

Mali Statistics

GEOGRAPHY

Area in Square Miles (Kilometers): 478,819 (1,240,142) (about twice the size of Texas)

Capital (Population): Bamako (1,161,000)

Environmental Concerns: soil erosion; deforestation; desertification; insufficient potable water; poaching

Geographical Features: mostly flat to rolling northern plains covered by sand; savanna in the south; rugged hills in the northeast

Climate: subtropical to arid

PEOPLE

Population

Total: 11,956,788

Annual Growth Rate: 2.78%

Rural/Urban Population Ratio: 71/29

Major Languages: French; Bambara; numerous African languages

Ethnic Makeup: 50% Mande; 17% Peul; 12% Voltaic; 6% Songhai; 10% Tuareg and Maur (Moor); 5% others

Religions: 90% Muslim; 9% indigenous beliefs; 1% Christian

Health

Life Expectancy at Birth: 46 years (male); 48 years (female)
Infant Mortality: 119.6/1,000 live births
Physicians Available: 1/18,376 people
HIV/AIDS Rate in Adults: 1.7%

Education

Adult Literacy Rate: 38%
Compulsory (Ages): 7–16; free

COMMUNICATION

Telephones: 56,600 main lines
Televisions: 12/1,000 people
Internet Users: 25,000 (2002)

TRANSPORTATION

Highways in Miles (Kilometers): 9,362 (15,100)
Railroads in Miles (Kilometers): 452 (729)
Usable Airfields: 27
Motor Vehicles in Use: 41,000

GOVERNMENT

Type: republic
Independence Date: September 22, 1960 (from France)
Head of State/Government: President Amadou Toumani Touré; Prime Minister Ousmane Issoufi Maiga
Political Parties: Alliance for Democracy; National Congress for Democratic Initiative; Sudanese Union/African Democratic Rally; others
Suffrage: universal at 18

MILITARY

Military Expenditures (% of GDP): 2%
Current Disputes: none

ECONOMY

Currency ($ U.S. Equivalent): 581 CFA francs = $1
Per Capita Income/GDP: $900/$10.3 billion
GDP Growth Rate: 0.5%
Inflation Rate: 4.5%
Unemployment Rate: 14.6% in urban areas
Labor Force by Occupation: 80% agriculture and fishing
Population Below Poverty Line: 64%
Natural Resources: hydropower; bauxite; iron ore; manganese; tin; phosphates; kaolin; salt; limestone; gold; uranium; copper
Agriculture: millet; sorghum; corn; rice; sugar; cotton; peanuts; livestock
Industry: food processing; construction; phosphate and gold mining; consumer-goods production
Exports: $575 million (primary partners Brazil, South Korea, Italy)
Imports: $600 million (primary partners Côte d'Ivoire, France, Senegal)

SUGGESTED WEB SITES

http://www.maliembassy-usa.org
http://www.cia.gov/cia/publications/factbook/geos/ml.html

Mali Country Report

Amadou Toumani Touré, the army general credited with rescuing Mali from military dictatorship and handing it back to its people, won the presidential elections of May 2002. He entered office on a program of anticorruption, peace, and development aimed at alleviation of poverty. This program resonated strongly with the population because corruption was viewed as rampant. In July 2000, an anticorruption commission published a report highlighting embezzlement and mismanagement within a number of state-owned companies and other public bodies.

DEVELOPMENT

In 1989, the government received international funding to overhaul its energy infrastructure. The opening of new gold mines has provided the economy with a boost.

Touré will be building on the legacy of his immediate predecessor, Dr. Alpha Konare, who stepped down from power after two terms in office that moved Mali away from its authoritarian past. The current democratic order was inaugurated a year after a coup led by Touré ended the dictatorial regime of Moussa Traoré. Like his predecessor, Modibo Keita (the first president of Mali), Traoré had governed Mali as a single-party state. True to their word, the coup leaders who seized power in 1991 following bloody antigovernment riots presided over a quick transition to civilian rule.

FREEDOM

The human-rights situation in Mali has improved in recent years, though international attention was drawn to the suppression of opposition demonstrations in the run-up to the 1997 elections.

Konare ruled as an activist scholar who, like many Malians, found political inspiration in his country's rich heritage. But his efforts to rebuild Mali were hampered by a weak economy, aggravated by the 1994 collapse in value of the CFA franc. In 1994 and 1995, violence occurred between security forces and university students protesting against economic hardship. The plight of Malian economic refugees in France gained international attention in 1996, when a number sought sanctuary in a Parisian church and went on a hunger strike in protest against attempts to deport them. The government has enjoyed greater success in reaching a (still fragile) settlement with Tuareg rebels in the country's far north.

AN IMPERIAL PAST

The published epic of Sundiata Keita, the thirteenth-century A.D. founder of the great Mali Empire, is recognized throughout the world as a masterpiece of classical African literature. In Mali itself, Sundiata remains a source of national pride and unity.

HEALTH/WELFARE

About a third of Mali's budget is devoted to education. A special literacy program in Mali teaches rural people how to read and write, by using booklets that concern fertilizers, measles, and measuring fields.

Sundiata's state was one of three great West African empires whose centers lay in modern Mali. Between the fourth and thirteenth centuries, the area was the site of the prosperous, ancient Ghana. The Mali Empire was superseded by that of Songhai, which was conquered by the Moroccans at the end of the sixteenth century. All these empires were in fact confederations. Although they encompassed vast areas united under a single supreme ruler, local communities generally enjoyed a great deal of autonomy.

ACHIEVEMENTS

For centuries, the ancient Malian city of Timbuctu was a leading center of Islamic learning and culture. Chronicles published by its scholars of the Middle Ages still enrich local culture.

From the 1890s until 1960, another form of imperial unity was imposed over Mali (then called French Sudan) and the adjacent territories of French West Africa. The legacy of broader colonial and precolonial unity as well as its landlocked position have inspired Mali's postcolonial leaders to seek closer ties with neighboring countries.

Mali formed a brief confederation with Senegal during the transition period to independence. This initial union broke down after only a few months, but since then the two countries have cooperated in the Organization for the Development of the Senegal River and other regional groupings. The Senegalese port of Dakar, which is linked by rail to Mali's capital city, Bamako, remains the major outlet for Malian exports. Mali has also sought to strengthen its ties with nearby Guinea. In 1983, the two countries signed an agreement to harmonize policies and structures.

ENVIRONMENTAL CHALLENGES

Mali is one of the poorest countries in the world. About 80 percent of the people are employed in (mostly subsistence) agriculture and fishing, but the government usually has to rely on international aid to make up for local food deficits. Most of the country lies within either the expanding Sahara Desert or the semiarid region known as the Sahel, which has become drier as a result of recurrent drought. Much of the best land lies along the Senegal and Niger Rivers, which support most of the nation's agropastoral production. In earlier centuries, the Niger was able to sustain great trading cities such as Timbuctu and Djenne, but today, most of its banks do not even support crops. Efforts to increase cultivation have so far been met with limited overall success.

Mali's frequent inability to feed itself has been largely blamed on locust infestation, drought, and desertification. The inefficient state-run marketing and distribution systems, however, have also had a negative impact. Low official produce prices have encouraged farmers either to engage in subsistence agriculture or to sell their crops on the black market. Thus, while some regions of the country remain dependent on international food donations, crops continue to be smuggled across Mali's borders. Recent policy commitments to liberalize agricultural trading, as part of an International Monetary Funding–approved Structural Adjustment Program (SAP), have yet to take hold.

In contrast to agriculture, Mali's mining sector has experienced promising growth. The nation exports modest amounts of gold, phosphates, marble, and uranium. Potentially exploitable deposits of bauxite, manganese, iron, tin, and diamonds exist. The Manantali Dam in southwestern Mali opened in December 2001. It is expected to provide electricity and jobs for thousands of Malians.

For decades, Mali was officially committed to state socialism. Its first president, Keita, a descendant of Sundiata, established a command economy and one-party state during the 1960s. His attempt to go it alone outside the CFA Franc Zone proved to be a major failure. Under Traoré, socialist structures were modified but not abandoned. Agreements with the IMF ended some government monopolies, and the country adopted the CFA franc as its currency. But the lack of a significant class of private entrepreneurs and the role of otherwise unprofitable public enterprises in providing employment discouraged radical privatization.

Timeline: PAST

1250–1400s
The Mali Empire extends over much of the upper regions of West Africa

late 1400s–late 1500s
The Songhai Empire controls the region

1890
The French establish control over Mali

1960
The Mali Confederation

1968
Moussa Traoré and the Military Committee for National Liberation grab power

1979
Traoré's Democratic Union of the Malian People is the single ruling party

1979–1980
School strikes and demonstrations; teachers and students are detained

1990s
The country's first multiparty elections are held; economic problems stir civic unrest

PRESENT

2000s
The Touré government explores ways to strengthen the economy

Ousmane Issoufi Maiga appointed Prime Minister

Mauritania (Islamic Republic of Mauritania)

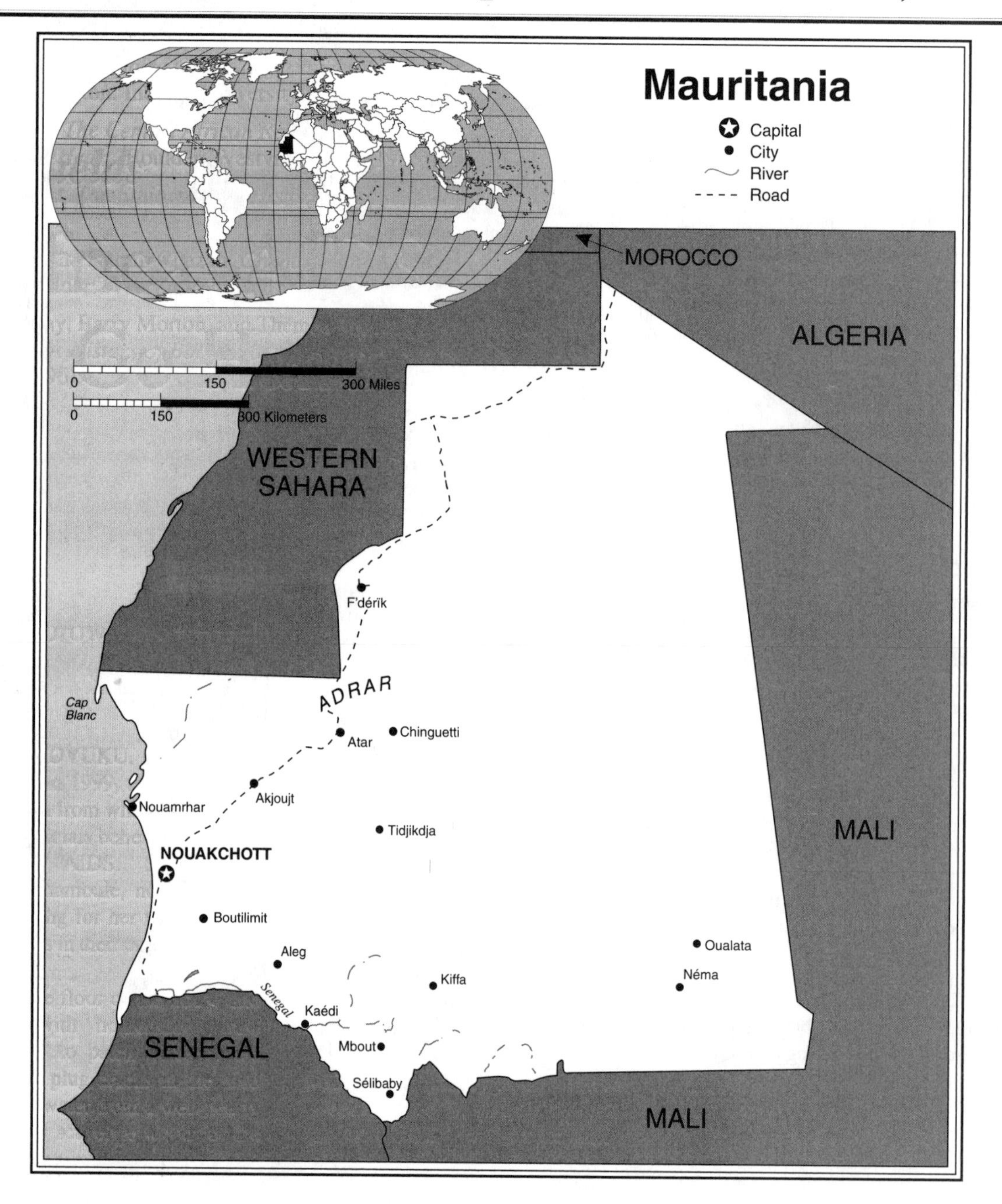

Mauritania Statistics

GEOGRAPHY

Area in Square Miles (Kilometers):

398,000 (1,030,700) (about 3 times the size of New Mexico)

Capital (Population): Nouakchott (626,000)

Environmental Concerns: overgrazing; deforestation; soil erosion; desertification; very limited natural freshwater resources; overfishing

Geographical Features: mostly barren, flat plains of the Sahara; some central hills

Climate: desert

PEOPLE

Population

Total: 2,998,563

Annual Growth Rate: 2.92%

Rural/Urban Population Ratio: 44/56

Major Languages: Hasanixa; Soninke; Arabic; Pular; Wolof

Ethnic Makeup: 40% mixed Maur/ black; perhaps 30% Maur; 30% black
Religion: 100% Muslim

Health

Life Expectancy at Birth: 49 years (male); 54 years (female)
Infant Mortality: 75.2/1,000 live births
Physicians Available: 1/11,085 people
HIV/AIDS Rate in Adults: 1.8%

Education

Adult Literacy Rate: 37.7%
Compulsory (Ages): 6–12

COMMUNICATION

Telephones: 31,500 main lines
Internet Users: 10,000 (2002)

TRANSPORTATION

Highways in Miles (Kilometers): 4,560 (7,600)
Railroads in Miles (Kilometers): 422 (704)
Usable Airfields: 26
Motor Vehicles in Use: 26,500

GOVERNMENT

Type: republic
Independence Date: November 28, 1960 (from France)
Head of State/Government: President Maaouya Ould Sid Ahmed Taya; Prime Minister Sghair Ould M'Bareck
Political Parties: Democratic and Social Republican Party; Union for Democracy and Progress; Popular Social and Democratic Union; others
Suffrage: universal at 18

MILITARY

Military Expenditures (% of GDP): 3.7%
Current Disputes: ethnic tensions

ECONOMY

Currency ($ U.S. equivalent): 276 ouguiyas = $1
Per Capita Income/GDP: $1,800/$5 billion
GDP Growth Rate: 4%
Inflation Rate: 7%
Unemployment Rate: 21%
Labor Force by Occupation: 50% agriculture; 40% services; 10% industry
Population Below Poverty Line: 50%
Natural Resources: iron ore; gypsum; fish; copper; phosphates
Agriculture: millet; sorghum; dates; root crops; cattle and sheep; fish products
Industry: iron-ore and gypsum mining; fish processing
Exports: $359 million (primary partners France, Japan, Italy)
Imports: $335 million (primary partners France, United States, Spain)

SUGGESTED WEB SITE

http://www.cia.gov/cia/publications/ factbook/geos/mr.html

Mauritania Country Report

Since the adoption of its current Constitution in 1991, Mauritania has legally been a multiparty democracy. But in practice, power remains in the hand of President Ould Taya's Republican Social Democratic Party (PRDS). The ruling party won large majorities in the 1992 and 1997 elections, which were boycotted by the leading opposition groupings. Participation in the 2001 elections resulted in some opposition gains but yet another controversial victory for the PRDS. Action for Change, a new party claiming to represent the Haratine (or Harratin), Mauritania's Arab-oriented black underclass of ex-slaves, won four seats. But in the beginning of 2002, the party was banned. Multiparty politics has thus so far failed to assure either social harmony or a respect for human rights. Neither has it resolved the country's severe social and economic problems.

DEVELOPMENT

Mauritania's coastal waters are among the richest in the world. During the 1980s, the local fishing industry grew at an average annual rate of more than 10%. Many now believe that the annual catch has reached the upper levels of its sustainable potential.

Amidst allegations of fraud President Taya won the elections of November 2003 with more than 67% of the votes during the first round. There were attempted coups during June 2003 and August and September 2004. Although each of the three coups failed, there is always the possibility that a future coup might succeed.

For decades, Mauritania has grown progressively drier. Today, about 75 percent of the country is covered by sand. Less than 1 percent of the land is suitable for cultivation, 10 percent for grazing. To make matters worse, the surviving arable and pastoral areas have been plagued by grasshoppers and locusts.

In the face of natural disaster, people have moved. Since the mid-1960s, the percentage of urban dwellers has swelled, from less than 10 percent to 53 percent, while the nomadic population during the same period has dropped, from more than 80 percent to perhaps 20 percent. In Nouakchott, the capital city, vast shantytowns now house nearly a quarter of the population. As the capital has grown, from a few thousand to 626,000 in a single generation, its poverty—and that of the nation as a whole—has become more obvious. People seek new ways to make a living away from the land, but there are few jobs. The best hope for lifting up the economy may lie in offshore oil exploration. A prospecting report in 2002 has attracted the interest of major international oil companies.

FREEDOM

The Mauritanian government is especially sensitive to continuing allegations of the existence of chattel slavery in the country. While slavery is outlawed, there is credible evidence of its continued existence. In 1998, five members of a local advocacy group SOS–Esclaves (Slaves) were sentenced to 13 months' imprisonment for "activities within a non-authorized organization."

Mauritania's heretofore faltering economy has coincided with an increase in racial and ethnic tensions. Since independence, the government has been dominated by the Maurs (or Moors), who speak Hasaniya Arabic. This community has historically been divided between the aristocrats and commoners, of Arab and Berber origin, and the Haratine, who were black African slaves who had assimilated Maurish culture but remain socially segregated. Including the Haratine, the Maurs account for perhaps 60 percent of the citizenry (the government has refused to release comprehensive data from the last two censuses).

The other half of Mauritania's population is composed of the "blacks," who mostly speak Pulaar, Soninke, or Wolof. Like the Maurs, all these groups are Muslim. Thus Mauritania's rulers have stressed Islam as a source of national unity. The country proclaimed itself an Islamic republic at independence, and since 1980 the Shari'a—the Islamic penal code—has been the law of the land.

Muslim brotherhood has not been able to overcome the divisions between the northern Maurs and southern blacks. One major source of friction has been official Arabization efforts, which are opposed by most southerners. In recent years, the country's desertification has created new sources of tension. As their pastures turned to sand, many of the Maurish nomads who did not find refuge in the urban areas moved southward. There, with state support, they began in the 1980s to deprive southerners of their land.

HEALTH/WELFARE

There have been some modest improvements in the areas of health and education since the country's independence, but conditions remain poor. Mauritania has received low marks regarding its commitment to human development.

Oppression of blacks has been met with resistance from the underground Front for the Liberation of Africans in Mauritania (FLAM). Black grievances were also linked to an unsuccessful coup attempt in 1987. In 1989, interethnic hostility exploded when a border dispute with Senegal led to race riots that left several hundred "Senegalese" dead in Nouakchott. In response, the "Moorish" trading community in Senegal became the target of reprisals. Mauritania claimed that 10,000 Maurs were killed, but other sources put the number at about 70. Following this bloodshed, more than 100,000 refugees were repatriated across both sides of the border. Mass deportations of "Mauritanians of Senegalese origin" have fueled charges that the Nouakchott regime is trying to eliminate its non-Maurish population.

ACHIEVEMENTS

There is a current project to restore ancient Mauritanian cities, such as Chinguette, which are located on traditional routes from North Africa to Sudan. These centers of trade and Islamic learning were points of origin for the pilgrimage to Mecca and were well known in the Middle East.

Tensions between Mauritania and Senegal were eased in June 2000 by the newly elected Senegalese president Abdoulaye Wade. This helped to reduce cross-border raids by deported Mauritanians. Genuine peace, however, will require greater reform within Mauritania itself and provision for the return of refugees.

In recent years, the regime in Nouakchott has sent out conflicting signals. Although the government has legalized some opposition parties, it has also continued to pursue its Arabization program and has clamped down on genuine dissent. Maur militias have been armed, and the army has been expanded with assistance from Arab countries.

Timeline: PAST

1035–1055
The Almoravids spread Islam in the Western Sahara areas through conquest

1920
The Mauritanian area becomes a French colony

1960
Mauritania becomes independent under President Moktar Ould Daddah

1978
A military coup brings Khouma Ould Haidalla and the Military Committee for National Recovery to power

1979
The Algiers Agreement: Mauritania makes peace with Polisario and abandons claims to Western Sahara

1980
Slavery is formally abolished

1990s
Multiparty elections are boycotted by the opposition; tensions continue between Mauritania and Senegal

PRESENT

2000s
Desertification takes its toll on the environment and the economy

It is estimated that 90,000 Mauritanians still live in servitude, despite the legal abolishment of slavery

Senegal and Mauritania seek better relations

President Taya re-elected to office

Niger (Republic of Niger)

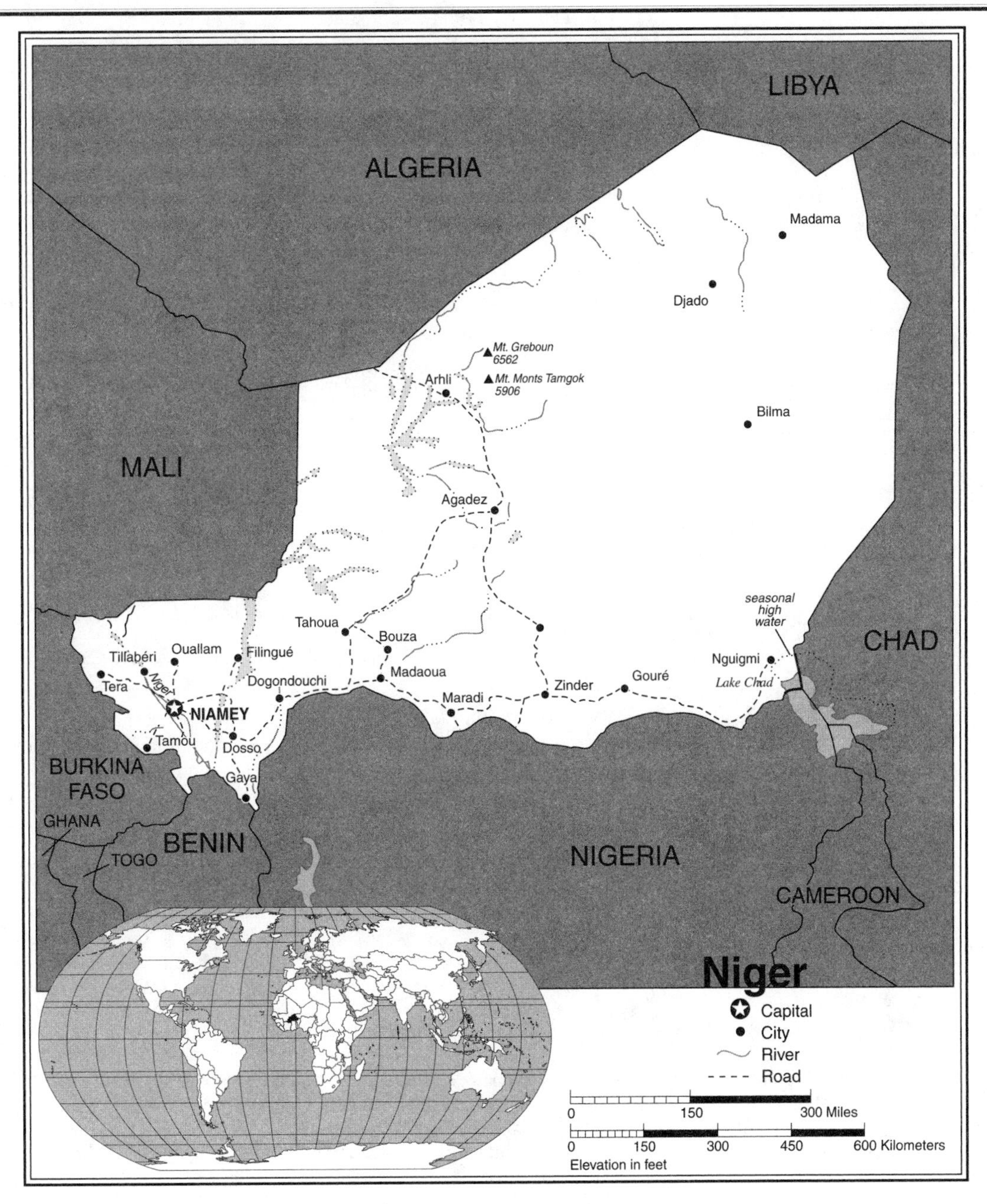

Niger Statistics

GEOGRAPHY

Area in Square Miles (Kilometers): 489,191 (1,267,000) (about twice the size of Texas)

Capital (Population): Niamey (821,000)

Environmental Concerns: overgrazing; deforestation; soil erosion; desertification; poaching and habitat destruction

Geographical Features: mainly desert plains and sand dunes; flat to rolling plains in the south; hills in the north; landlocked

Climate: desert; tropical in the extreme south

PEOPLE

Population

Total: 11,360,538

Annual Growth Rate: 2.67%

Rural/Urban Population Ratio: 80/20

Major Languages: French; Hausa; Djerma

Ethnic Makeup: 56% Hausa; 22% Djerma; 8% Fula; 8% Tuareg; 6% others

Religions: 80% Muslim; 20% indigenous beliefs and Christian

Health

Life Expectancy at Birth: 42 years (male); 42 years (female)

Infant Mortality: 122/1,000 live births
Physicians Available: 1/35,141 people
HIV/AIDS Rate in Adults: 4%

Education

Adult Literacy Rate: 15.3%
Compulsory (Ages): 7–15, free

COMMUNICATION

Telephones: 22,400 main lines
Televisions: 2.8 per 1,000 people
Internet Users: 15,000 (2002)

TRANSPORTATION

Highways in Miles (Kilometers): 6,262 (10,100)
Railroads in Miles (Kilometers): none
Usable Airfields: 27
Motor Vehicles in Use: 51,500

GOVERNMENT

Type: republic
Independence Date: August 3, 1960 (from France)
Head of State/Government: President Mamadou Tandja is both head of state and head of government
Political Parties: National Movement for a Developing Society–Nassara; Democratic and Social Convention–Rahama; Nigerien Party for Democracy and Socialism–Tarayya; Nigerien Alliance for Democracy and Social Progress–Zaman-lahia; others
Suffrage: universal at 18

MILITARY

Military Expenditures (% of GDP): 1.3%
Current Disputes: territorial dispute with Libya; boundary disputes over Lake Chad

ECONOMY

Currency ($ U.S. Equivalent): 581 CFA francs = $1
Per Capita Income/GDP: $820/$9 billion
GDP Growth Rate: 3.8%
Inflation Rate: 3%
Labor Force by Occupation: 90% agriculture; 6% industry and commerce; 4% government
Population Below Poverty Line: 63%
Natural Resources: uranium; coal; iron ore; tin; phosphates; gold; petroleum
Agriculture: millet; sorghum; peanuts; cotton; cowpeas; cassava; livestock
Industry: cement; brick; textiles; chemicals; agricultural products; food processing; uranium mining
Exports: $246 million (primary partners France, Nigeria, Spain)
Imports: $331 million (primary partners France, Côte d'Ivoire, United States)

SUGGESTED WEB SITES

http://www.friendsofniger.org
http://www.cia.gov/cia/publications/factbook/geos/ng.html

Niger Country Report

Niger is ranked by the United Nations as the world's second-poorest country, after war-ravaged Sierra Leone. This circumstance can in part be blamed on poor governance. For most of the past four decades, since it gained independence from France in 1960, Niger has been governed by a succession of military regimes that have left it bankrupt. This has led to chronic instability, as the government has regularly failed to pay its salaries, resulting in strikes by civil servants and mutinies by soldiers. In November 1999, the current president, Mamadou Tandja, was elected under a new Constitution. But ultimate power remains in the hands of the military, which, in January 1996, overthrew Niger's last elected government. In July 1996, the leader of the coup, Colonel Ibrahim Bare Mainassara, claimed victory in elections that were widely condemned as fraudulent. Mainassara's subsequent assassination in April 1999 opened the door to a return to civilian rule, but at this writing, a military committee under strongman Major Daouda Mallam-Wanke continues to wield influence over the government.

DROUGHT AND DESERTIFICATION

Most Nigeriens subsist through small-scale crop production and herding. Yet farming is especially difficult in Niger. Less than 10 percent of the nation's vast territory is suitable for cultivation even during the best of times. Most of the cultivable land lies along the banks of the Niger River. Unfortunately, much of the past four decades has been the worst of times. During this period, Nigeriens have been constantly challenged by recurrent drought and an ongoing process of desertification.

DEVELOPMENT

Nigerien village cooperatives, especially marketing cooperatives, predate independence and have grown in size and importance in recent years. They have successfully competed with well-to-do private traders for control of the grain market. In September 2004 the Nigerien government granted gold mining permits to a number of European nations in an effort to increase gold mining and production.

Drought had an especially catastrophic effect during the 1970s. Most Nigeriens were reduced to dependency on foreign food aid, while about 60 percent of their livestock perished. Some people believe that the ecological disaster that afflicted Africa's Sahel region (which includes southern Niger) during that period was of such severity as to disrupt the delicate long-term balance between desert and savanna. Others, however, have concluded that the intensified desertification of recent years is primarily rooted in human, rather than natural, causes, which can be reversed. In particular, many attribute environmental degradation to the introduction of inappropriate forms of cultivation, overgrazing, deforestation, and new patterns of human settlement.

FREEDOM

Nigeriens have been effectively disenfranchised by the 1996 coup and subsequent fraudulent presidential election. Security forces are known to beat and intimidate opposition political figures. The private media are a target of repression, with a number of journalists having been detained. Opposition meetings and demonstrations are often banned.

Ironically, much of the debate on people's negative impact on the Sahel environment has been focused on some of the agricultural-development schemes that once were perceived as the region's salvation. In their attempts to boost local food production, international aid agencies often promoted so-called Green Revolution programs. These were designed to increase per-acre yields, typically through the intensive planting of new seeds and reliance on imported fertilizers and pesticides. Such projects often led to higher initial local outputs that proved unsustainable, largely due to expensive overhead. In addition, many experts promoting the new agricultural techniques failed to appreciate the value of traditional technologies and forms of social organization in limiting desertification while allowing people to cope with drought. It is now appreciated that patterns of cultivation long championed by Nigerien farmers allowed for soil conservation and reduced the risks associated with pests and poor climate.

HEALTH/WELFARE

A national conference on educational reform stimulated a program to use Nigerien languages in primary education and integrated the adult literacy program into the rural development efforts. The National Training Center for Literacy Agents is crucial to literacy efforts.

The government's recent emphasis has been on helping Niger's farmers to help themselves through the extension of credit, better guaranteed minimum prices, and improved communications. Vigorous efforts have been made in certain regions to halt the spread of desert sands by supporting village tree-planting campaigns. Given the local inevitability of drought, the government has also increased its commitment to the stockpiling of food in granaries. But for social and political as much as economic reasons, government policy has continued to discourage the flexible, nomadic pattern of life that has long characterized many Nigerien communities.

The Nigerien government's emphasis on agriculture has, in part, been motivated by the realization that the nation could not rely on its immense uranium deposits for future development. The opening of uranium mines in the 1970s resulted in the country becoming the world's fifth-largest producer. By the end of that decade, uranium exports accounted for some 90 percent of Niger's foreign-exchange earnings. Depressed international demand throughout the 1980s, however, resulted in substantially reduced prices and output. Although uranium still accounts for 75 percent of foreign-exchange earnings, its revenue contribution in recent years is only about a third of what it was prior to the slump.

The uranium production has in recent years led to some controversy. In January 2003 U.S. President George W. Bush claimed that Iraq tried to acquire uranium from Niger for its nuclear program. This claim was also made in the United Kingdom in a September 2002 dossier on Iraq. Both the United States and the United Kingdom's claims, however, were proven to be unfounded. In March 2003 the Nuclear Regulatory Agency informed the UN that documents relating to the Iraq-Niger uranium claim were forged.

MILITARY RULE

For nearly half of its existence after its independence, Niger was governed by a civilian administration, under President Hamani Diori. In 1974, during the height of drought, Lieutenant Colonel Seyni Kountché took power in a bloodless coup. Kountché ruled as the leader of a "Supreme Military Council," which met behind closed doors. Ministerial portfolios, appointed by the president, were filled by civilians as well as military personnel. In 1987, Kountché died of natural causes and was succeeded by Colonel Ali Saibou.

ACHIEVEMENTS

Niger has consistently demonstrated a strong commitment to the preservation and development of its national cultures through its media and educational institutions, the National Museum, and events such as the annual youth festival at Agades.

The National Movement for the Development of Society (MNSD) was established in 1989 as the country's sole political party, after a constitutional referendum in which less than 4 percent of the electorate participated. But, as was the case in many other countries in Africa, the year 1990 saw a groundswell of local support for a return to multipartyism. In Niger, this prodemocracy agitation was spearheaded by the nation's labor confederation, which organized a widely observed 48-hour general strike. Having earlier rejected the strikers "as a handful of demagogues," in 1991, President Saibou agreed to the formation of a National Conference to prepare a new constitution.

The conference ended its deliberations with the appointment of an interim government, headed by Amadou Cheffou, which led the country to multiparty elections in February–March 1993. After two rounds of voting, the presidential contest was won by Mahamane Ousmane. Ousmane's Alliance of Forces for Change (AFC) opposition captured 50 seats in the new 83-seat National Assembly, while the MNSD became the major opposition party, with 29 seats.

Ousmane's government made a promising start by reaching peace agreements with two rebel movements, the Tuareg Front for the Liberation of Air and Azaouad and the Organization of Army Resistance. But the nation's economic crisis deepened with the 1994 devaluation of the CFA franc. Naturally, Ousmane's political status was weakened. In February 1995, the opposition coalition, led by Hama Amadou, gained control of the National Assembly, resulting in an uneasy government of "cohabitation." Serious student unrest was followed by the military coup in January 1996 that resulted in the installation of Colonel Ibrahim Bare Mainassara as president. Under international pressure, Mainassara agreed to early elections, which were seen to have been fraudulent.

The political turn is likely to further poison interethnic relations in Niger. Since independence, members of the Zarma group have been especially prominent in the government, MNSD, and military. The deposed Ousmane has been Niger's first Hausa leader (the Hausa constitute the country's largest ethnolinguistic group).

Timeline: PAST

1200s–1400s
The Mali Empire includes territories and peoples of current Niger areas

1400s
Hausa states develop in the south of present-day Niger

1800s
The area is influenced by the Fulani Empire, centered at Sokoto, now in Nigeria

1906
France consolidates rule over Niger

1960
Niger becomes independent

1974
A military coup brings Colonel Seyni Kountché and a Supreme Military Council to power

1987
President Kountché dies and is replaced by Ali Saibou

1990s
The Nigerien National Conference adopts multipartyism; President Ibrahim Bare Mainassara is assassinated

PRESENT

2000s
President Mamadou Tandja holds power under the new Constitution

The military retains significant influence

Nigeria (Federal Republic of Nigeria)

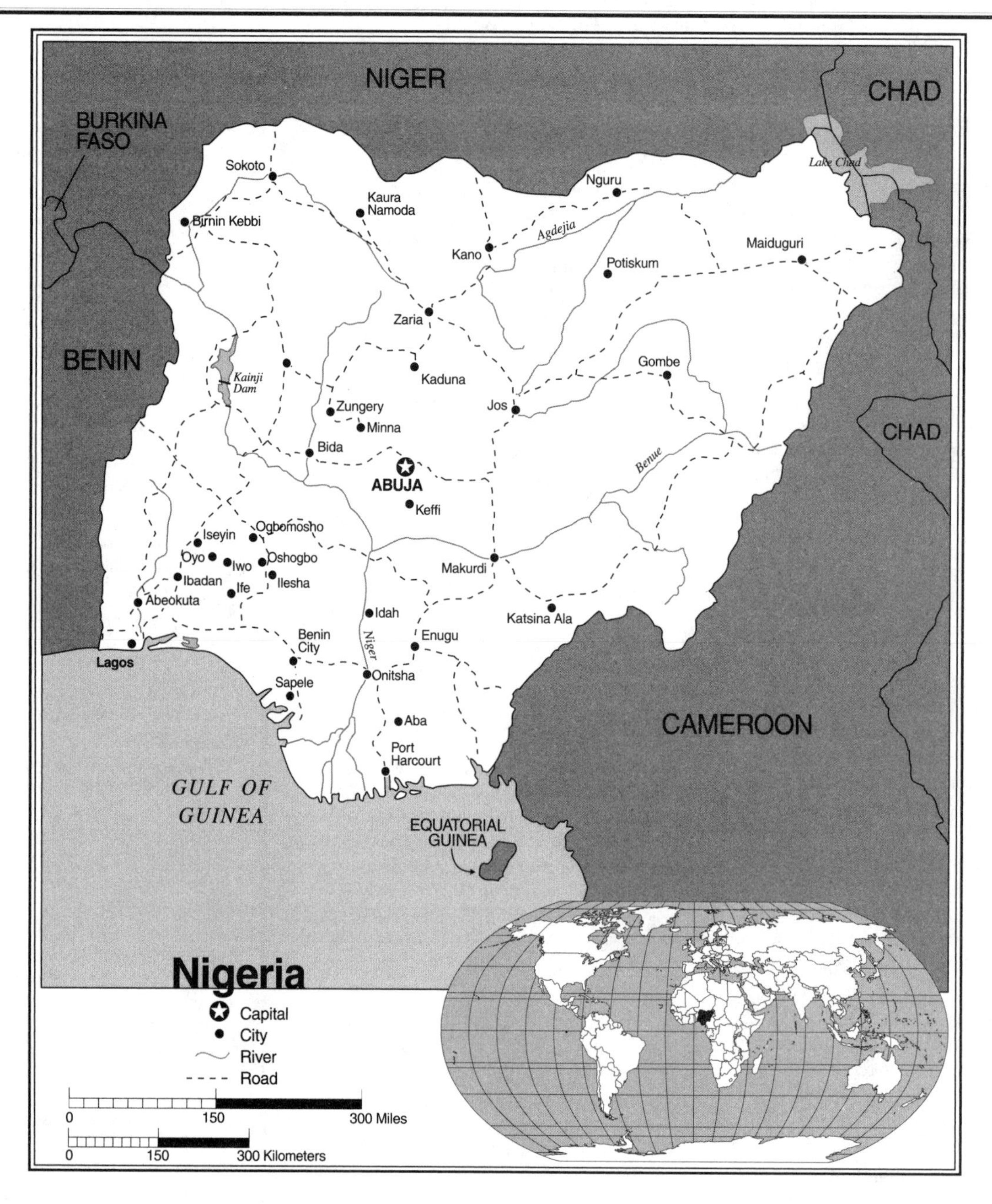

Nigeria Statistics

GEOGRAPHY

Area in Square Miles (Kilometers): 356,669 (923,768) (twice the size of California)

Capital (Population): Abuja (420,000)

Environmental Concerns: soil degradation; deforestation; desertification; drought

Geographical Features: southern lowlands merge into central hills and plateaus; mountains in southeast; plains in north

Climate: varies from equatorial to arid

PEOPLE

Population

Total: 137,253,133

Annual Growth Rate: 2.54%

Rural/Urban Population Ratio: 57/43

Major Languages: English; Hausa; Yoruba; Ibo; Fulani

Ethnic Makeup: about 21% Hausa; 21% Yoruba; 18% Ibo; 9% Fulani; 31% others

Religions: 50% Muslim; 40% Christian; 10% indigenous beliefs

Health

Life Expectancy at Birth: 51 years (male); 51 years (female)

Infant Mortality: 72.5/1,000 live births
Physicians Available: 1/4,496 people
HIV/AIDS Rate in Adults: 5.06%

Education

Adult Literacy Rate: 57%
Compulsory (Ages): 6–15; free

COMMUNICATION

Telephones: 853,100 main lines
Daily Newspaper Circulation: 18 per 1,000 people
Televisions: 38 per 1,000 people
Internet Users: 750,000 (2003)

TRANSPORTATION

Highways in Miles (Kilometers): 120,524 (194,394); but much of the road system is barely usable
Railroads in Miles (Kilometers): 2,226 (3,567)
Usable Airfields: 70
Motor Vehicles in Use: 954,000

GOVERNMENT

Type: republic in transition from military rule
Independence Date: October 1, 1960 (from the United Kingdom)
Head of State/Government: President Olusegun Obasanjo is both head of state and head of government
Political Parties: People's Democratic Party; Alliance for Democracy; All People's Party
Suffrage: universal at 18

MILITARY

Military Expenditures (% of GDP): 1%
Current Disputes: civil strife; various border disputes

ECONOMY

Currency ($ U.S. Equivalent): 128.3 nairas = $1
Per Capita Income/GDP: $900/$114 billion
GDP Growth Rate: 7.1%
Inflation Rate: 13.8%
Unemployment Rate: 28% (1992 est.)
Labor Force by Occupation: 70% agriculture; 20% services; 10% industry
Population Below Poverty Line: 45%
Natural Resources: petroleum; tin; columbite; iron ore; coal; limestone; lead; zinc; natural gas; hydropower
Agriculture: cocoa; peanuts; rubber; yams; cassava; sorghum; palm oil; millet; corn; rice; livestock; timber; fish
Industry: mining; petroleum; food processing; textiles; cement; building materials; chemicals; agriculture products; printing; steel
Exports: $20.3 billion (primary partners United States, Spain, India)
Imports: $13.7 billion (primary partners United Kingdom, United States, France)

SUGGESTED WEB SITES

http://www.nigeria.com
http://www.nigeriatoday.com
http://www.nigeriadaily.com
http://nigeriaworld.com
http://www.africanews.org/west/nigeria/
http://www.cia.gov/cia/publications/factbook/geos/ni.html

Nigeria Country Report

Nigeria, the "sleeping giant of Africa," is showing signs of waking up. Its president, Olusegun Obasanjo, has been a driving force behind the New Partnership for African Development (NEPAD), which commits African states to good governance, respect for human rights, and efforts to end regional conflicts. In return, the continent is seeking more aid and foreign investment, and a lifting of trade barriers that impede African exports. The Nigerian government has also been playing a leading role in attempts to bring peace to West Africa, where President Obasanjo is generally referred to as an honest broker. Nigeria's military has been active in the settlement of regional disputes in Liberia, Côte d'Ivoire, and Sierra Leone. But the country's regional diplomatic standing was compromised in October 2002, when it reneged on a previous agreement to abide by the judgment of the International Court of Justice in a long-running border dispute with Cameroon.

The first legislative elections since the end of military rule in 1999 were held throughout Nigeria in April 2003. Polling was marked by delays and allegations of fraud arising from the opposition. President Obasanjo's People's Democratic Party won a parliamentary majority, with President Obasanjo elected for a second term in office with more than 60% of the votes cast. Although a number of opposition parties rejected the result, and the EU observers say the polling was marred by "serious irregularities," Obasanjo was sworn into office as president, marking the first civilian transfer of power in Nigeria's history.

DEVELOPMENT

Nigeria hopes to mobilize its substantial human and natural resources to encourage labor-intensive production and self-sufficient agriculture. Recent bans on food imports will increase local production, and restrictions on imported raw materials should encourage research and local input for industry.

With a vast population of nearly 130 million, Nigeria's human resources are yet to be fully tapped in the interest of the country. Poverty and inequality between the rich and the poor remain extreme. Nigeria's industrious people hope that the restoration of democracy will allow them to make renewed progress in the face of these challenges. In February 1999, Obasanjo was elected Nigeria's first civilian president in 15 years. His government has since struggled to push forward with the immense task of governing the diverse communities that make up Africa's most populous country. But ethnic/religious tension and corruption have continued to plague Nigeria. Transparency International has ranked the country as the second most corrupt in the world. In January 2002, a blast at a munitions dump in the principal city, Lagos, left more than 1,000 people dead, sparking renewed calls for public accountability.

Since early 2000, attempts to introduce Shari'a (Islamic law) in northern areas of the country have touched off severe violence between Muslim and Christian communities, as well as international condemnation for stoning sentences against single mothers convicted of adultery. The most serious violence occurred in 2001, when some 40,000 people were reportedly displaced following ethnic fighting between the Tiv people and several smaller ethnic groups in the central Nasarawa state. In February 2002, some 100 people were killed and thousands more left temporarily homeless in Lagos, as a result of further ethnic strife, which the city's governor suggested was orchestrated by retired army officers seeking to restore military rule.

Despite such incidents, there are signs that Nigeria's civil society is being rebuilt after almost two decades of military rule. Nigeria's transition to civilian rule followed the unexpected deaths of its last dictator, General Sani Abacha, and his most famous political prisoner, Chief Mashood Abiola. (The latter was considered by many to have won annulled 1993 presidential elections.) In the wake of the deaths, a caretaker administration under General Abdulsalem Abubakar, in cooperation with previously repressed sections of civil society, promised to restore democracy.

The Nigerian government's credibility at home and abroad was enhanced by the freeing of political prisoners and Abubakar's personal paying of respects to the late Abiola. Among those freed was Obasanjo, who had briefly served as Nigeria's military ruler in the 1970s, before handing over power to civilians. Besides the jailing of such figures as Abiola and Obasanjo, the Abacha regime had reduced Nigeria to the status of a pariah state through its execution of dozens of political opponents, including the internationally prominent writer, ecologist, and human-rights activist Ken Saro Wiwa.

FREEDOM

Under Abacha, Nigeria had one of the world's worst human-rights records. In 1998, the Nigerian Advocacy Group for Human Rights joined other international groups in issuing a statement insisting that nothing essentially changed after Abubakar succeeded Abacha. With the transition to civilian rule under former political detainee Obasanjo, the situation should improve.

Since Nigeria's independence, in 1960, its citizens have been through an emotional, political, and material rollercoaster ride. It has been a period marred by interethnic violence, economic downturns, and mostly military rule. But there have also been impressive levels of economic growth, cultural achievement, and human development. To some people, this land of great extremes typifies both the hopes and frustrations of its continent.

Nigeria's hard-working population is responsible for one of Africa's largest economies. But per capita income is still only $840 per year, which is about average for the globe's most impoverished continent but down from Nigeria's estimated 1980 per capita income of $1,500.

A decade ago, it was common to equate Nigeria's wealth with its status as Africa's leading oil producer, but oil earnings have since plummeted. Although hydrocarbons still account for about 90 percent of the country's export earnings and 75 percent of its government revenue, the sector's current contribution to total gross domestic product is a more modest 20 percent.

NIGERIA'S ROOTS

For centuries, the River Niger, which cuts across much of Nigeria, has facilitated long-distance communication among various communities of West Africa's forest and savanna regions. This fact helps to account for the rich variety of cultures that have emerged within the territory of Nigeria over the past millennium. Archaeologists and historians have illuminated the rise and fall of many states whose cultural legacies continue to define the nation.

Precolonial Nigeria produced a wide range of craft goods, including leather, glass, and metalware. The cultivation of cotton and indigo supported the growth of a local textiles industry. During the mid-nineteenth century, southern Nigeria prospered through palm-oil exports, which lubricated the wheels of Europe's Industrial Revolution. Earlier, much of the country was disrupted through its participation in the slave trade. Most African-Americans have Nigerian roots.

Today, more than 250 languages are spoken in Nigeria. Pidgin, which combines an English-based vocabulary with local grammar, is widely used as a lingua franca in the cities and towns. Roughly two thirds of Nigerians speak either Hausa, Yoruba, or Igbo as a home language. During and after the colonial era, the leaders of these three major ethnolinguistic groups clashed politically from their separate regional bases.

HEALTH/WELFARE

Nigeria's infant mortality rate is now believed to have dropped to about 72 per 1,000 live births. (Some estimate it to be as high as 150 per 1,000.) While social services grew rapidly during the 1970s, Nigeria's strained economy since then has led to cutbacks in health and education.

The British, who conquered Nigeria in the late nineteenth and early twentieth centuries, administered the country through a policy of divide-and-rule. In the predominantly Muslim, Hausa-speaking north, they co-opted the old ruling class while virtually excluding Christian missionaries. But in the south, the missionaries, along with their schools, were encouraged, and Christianity and formal education spread rapidly. Many Yoruba farmers of the southwest profited through their cultivation of cocoa. Although most remained as farmers, many of the Igbo of the southeast became prominent in nonagricultural pursuits, such as state employees, artisans, wage workers, and traders. As a result, the Igbo tended to migrate in relatively large numbers to other parts of the colony.

REGIONAL CONFLICTS

At independence, the Federal Republic of Nigeria was composed of three states: the Northern Region, dominated by Hausa speakers; the Western Region, of the Yoruba; and the predominantly Igbo Eastern Region. National politics quickly deteriorated into conflict among these three regions. At one time or another, politicians in each of the areas threatened to secede from the federation. In 1966, this strained situation turned into a crisis following the overthrow by the military of the first civilian government.

In the coup's aftermath, the army itself was divided along ethnic lines; its ranks soon became embroiled in an increasingly violent power struggle. The unleashed tensions culminated in the massacre of up to 30,000 Igbo living in the north. In response, the Eastern Region declared its independence, as the Republic of Biafra. The ensuing civil war between Biafran partisans and federal forces lasted for three years, claiming an estimated 2 million lives. During this time, much of the outside world's attention became focused on the conflict through visual images of the mass starvation that was occurring in rebel-controlled areas under federal blockade. Despite the extent of the war's tragedy, the collapse of Biafran resistance was followed by a largely successful process of national reconciliation. The military government of Yakubu Gowon (1966–1975) succeeded in diffusing ethnic politics, through a restructured federal system based on the creation of new states. The oil boom, which began soon after the conflict, helped the nation-building process by concentrating vast resources in the hands of the federal government in Lagos.

ACHIEVEMENTS

When many of their leading writers, artists, and intellectuals were exiled and the once-lively press was suppressed, Nigerians found some solace in the success of their world-class soccer team and other athletes. In September, Nigeria's first satellite, NigeriaSat-1, was launched by Russian Rocket.

CIVILIAN POLITICS

Thirteen years of military rule ended in 1979. A new Constitution was implemented, which abandoned the British parliamentary model and instead adopted a modified version of the U.S balance-of-powers system. In order to encourage a national outlook, Nigerian presidential candidates needed to win a plurality that included at least one-fourth of the vote in two-thirds of the states.

Five political parties competed in the 1979 elections. They all had similar platforms, promising social welfare for the masses, support for Nigerian business, and a foreign policy based on nonalignment and anti-imperialism. Ideological differences tended to exist within the parties as much as among them, although the People's Redemption Party (PRP) of Aminu Kano became the political home for many Socialists. The most successful party was the somewhat right-of-center National Party of Nigeria (NPN), whose candidate, Shehu Shagari, won the presidency.

New national elections took place in August and September 1983, in which Shagari received more than 12 million of 25.5 million votes. However, the reelected government did not survive long. On December 31, 1983, there was a military coup, led by Major General Muhammad Buhari. The 1979 Constitution was suspended, Shagari and others were arrested, and a federal military government was reestablished. Although no referendum was ever taken on the matter, it is clear that many Nigerians welcomed the coup: this initial response was a reflection of widespread disillusionment with the Second (civilian) Republic.

The political picture seemed very bright in the early 1980s. A commitment to national unity was well established. Although marred by incidents of political violence, two elections had successfully taken place. Due process of law, judicial independence, and press freedom—never entirely eliminated under previous military rulers—had been extended and were seemingly entrenched. But the state was increasingly seen as an instrument of the privileged that offered little to the impoverished masses, with an electoral system that, while balancing the interests of the elite in different sections of the country, failed to empower ordinary citizens. A major reason for this failing was pervasive corruption. People lost confidence as certain officials and their cronies became millionaires overnight. Transparent abuses of power had also occurred under the previous military regime. Conspicuous kleptocracy (rule by thieves) had been tolerated during the oil-boom years of the 1970s, but it became the focus of popular anger as Nigeria's economy contracted during the 1980s.

OIL BOOM—AND BUST

Nigeria, as a leading member of the Organization of Petroleum Exporting Countries (OPEC), experienced a period of rapid social and economic change during the 1970s. The recovery of oil production after the Civil War and the subsequent hike in its prices led to a massive increase in government revenue. This allowed for the expansion of certain types of social services. Universal primary education was introduced, and the number of universities increased from five (in 1970) to 21 (in 1983). A few Nigerians became very wealthy, while a growing middle class was able to afford what previously had been luxuries.

Oil revenues had already begun to fall off when the NPN government embarked upon a dream list of new prestige projects, most notably the construction of a new federal capital at Abuja, in the center of the country. While such expenditures provided lucrative opportunities for many businesspeople and politicians, they did little to promote local production.

Agriculture, burdened by inflationary costs and low prices, entered a period of crisis, leaving the rapidly growing cities dependent on foreign food. Nonpetroleum exports, once the mainstay of the economy, either virtually disappeared or declined drastically.

The oil industry, although responsible for the vast majority of Nigeria's foreign exchange earning, has caused a number of problems throughout the country. In July 2003 a nationwide strike took place for nine days in a successful attempt to get government to reduce the price of fuel. During September 2004 deadly battles between gangs in the oil city of Port Harcourt prompted a strong crackdown by troops. The human-rights group Amnesty International cited the death toll at 500, after the government authorities claimed that only 20 people died. A very successful four-day general strike over fuel prices took place in October 2004, stoking fears about the country's oil exports and driving up the price of oil worldwide.

While gross indicators appeared to report impressive industrial growth in Nigeria, most of the new industry depended heavily on foreign inputs and was geared toward direct consumption rather than the production of machines or spare parts. Selective import bans led merely to the growth of smuggling.

The golden years of the 1970s were also banner years for inappropriate expenditures, corruption, and waste. For a while, given the scale of incoming revenues, it looked as if these were manageable problems. But GDP fell drastically in the 1980s with the collapse of oil prices. As the economy worsened, populist resentment grew.

In 1980, an Islamic movement condemning corruption, wealth, and private property defied authorities in the northern metropolis of Kano. The army was called in, killing nearly 4,000. Similar riots subsequently occurred in the cities of Maiduguri, Yola, and Gombe. Attempts by the government to control organized labor by reorganizing the union movement into one centralized federation sparked unofficial strikes (including a general strike in 1981). In an attempt to placate the growing number of unemployed Nigerians, more than 1 million expatriate West Africans, mostly Ghanaians, were suddenly expelled, a domestically popular but essentially futile gesture.

REFORM OR RETRIBUTION?

Buhari justified the military's return to power on the basis of the need to take drastic steps to rescue the economy, whose poor performance he blamed almost exclusively on official corruption. A "War Against Indiscipline" was declared, which initially resulted in the trial of a number of political leaders, some of whose economic crimes were indeed staggering. The discovery of large private caches of naira (the Nigerian currency) and foreign exchange fueled public outrage. Tribunals sentenced former politicians to long jail terms. In its zeal, the government looked for more and more culprits, while jailing journalists and others who questioned aspects of its program. In 1985, Major General Ibrahim Babanguida led a successful military coup, charging Buhari with human-rights abuses, autocracy, and economic mismanagement.

Babanguida released political detainees. In a clever strategy, he also encouraged all Nigerians to participate in national forums on the benefits of an International Monetary Fund loan and Structural Adjustment Program (SAP). The government turned down the loan but used the consultations to legitimize the implementation of "home-grown" austerity measures consistent with IMF and World Bank prescriptions.

The 1986 budget signaled the beginning of SAP. The naira was devalued, budgets were restricted, and the privatization of many state-run industries was planned. Because salaries remained the same while prices rose, the cost of basic goods rose dramatically, with painful consequences for middle- and working-class Nigerians as well as for the poor.

Although the international price of oil improved somewhat in the late 1980s, there was no immediate return to prosperity. Continued budgetary excesses on the part of the government (which heaped perks on its officer corps and created more state governments to soak up public coffers), coupled with instability, undermined SAP sacrifices. In 1988, the government attempted a moderate reduction in local fuel subsidies. But when, as a result, some transport owners raised fares by 50 to 100 percent, students and workers protested, and bank staff and other workers went on strike. Police killed demonstrators in Jos. Domestic fuel prices have since remained among the lowest in the world, encouraging a massive smuggling of petroleum to neighboring states. This has recently led to the ironic situation of a severe local petroleum shortage.

The Babanguida government faced additional internal challenges while seeking to project an image of stability to foreign investors. Coup attempts were foiled in 1985 and 1991, while chronic student unrest led to the repeated closure of university campuses. Religious riots between Christians and Muslims became endemic in many areas, leading to hundreds, if not thousands, of deaths.

In 1986, Babanguida promised a phased return to full civilian control. But his program of guided democratization degenerated into a farce. Local nonparty elections were held in 1987, and a (mostly elected) Constituent Assembly subsequently met and approved modifications to Nigeria's 1979 Constitution. Despite the trappings of electoral involvement, the Transitional Program was tightly controlled. Many politicians were banned as Babanguida tried to impose a two-party system on what traditionally had been a multiparty political culture. When none of 13 potential parties gained his approval, the general decided to create two new parties of his own: the "a little to the left" Social Democratic Party (SDP) and the "a little to the right" National Republican Convention (NRC).

Doubts about the military in general and Babanguida's grasp on power in particular were raised in 1990, when a group of dissident officers launched yet another coup. In radio broadcasts, the insurrectionists announced the expulsion of five northern states from the federal republic, thus raising the specter of a return to interethnic civil war. The uprising was crushed.

A series of national elections were held in 1992 between the two officially sponsored parties. But public indifference and/or fear of intimidation, institutionalized by the replacement of the (ideally, secret) ballot with a procedure of publicly lining up for one's candidate, compromised the results. Allegations of gross irregularities led to the voiding of first-round presidential primary elections and the banning of all the candidates. After additional delays, accompanied by a serious antigovernment rioting in Lagos and other urban areas, escalating intercommunal violence, and further clampdowns on dissent, a presidential poll was finally held in June 1993 between two government-approved candidates: Mashood Abiola and Bashir Tofa. The result was a convincing 58 percent victory for the SDP's Abiola, though an estimated 70 percent of the electorate refused to participate in the charade.

Babanguida annulled the results before they had been officially counted (the final results were released by local officials in defiance of Babanguida's regime). Instead, in August 1992, he resigned and installed an interim government led by an ineffectual businessman, Ernest Shonekan. Growing unrest—aggravated by an overnight 600 percent increase in domestic fuel prices and a dramatic airline hijacking by a group calling itself the Movement for the Advancement of Democracy (MAD)—led to the interim regime's rapid collapse. In November, the defense minister, General Sani Abacha, reimposed full military rule.

Resistance to military rule steadily increased throughout 1994. Abiola was arrested in June after proclaiming himself president. His detention touched off nationwide strikes, which shut down the oil industry and other key sectors of the economy.

CULTURAL PROMINENCE

Nigeria is renowned for its arts. Contemporary giants include Wole Solyinka, who received the Nobel Prize for Literature for his work—plays such as "The Trials of Brother Jero" and "The Road," novels such as *The Interpreters,* and poems and nonfiction works. Two other literary giants are Chinua Achebe, author of *Things Fall Apart, A Man of the People,* and *Anthills of the Savannah;* and the feminist writer Buchi Emecheta, whose works include *The Joy of Motherhood.* The legendary Fela Anikulado Kuti's "Afro-Beat" sound and critical lyrics have made him a local hero and international music megastar. Also prominent is "King" Sunny Ade, who has brought Nigeria's distinctive Juju music to international audiences.

Timeline: PAST

1100–1400
Ancient life flourishes

1851
The first British protectorate is established at Lagos

1960
Nigeria becomes independent as a unified federal state

1966–1970
Military seizure of power; proclamation of Biafra; civil war

1979
Elections restore civilian government

1980s
Muhammed Buhari's military coup ends the Second Republic; later, Buhari is toppled by Ibrahim Babanguida; lean times

1990s
Babanguida resigns; Sani Abacha takes the reins; civil unrest and violence intensify; military strongman Abdulsalam Abubakar takes power; elections bring civilian Olusegun Obasanjo to power

PRESENT

2000s
Ethnic and religious conflict intensifies

Hopes for democratic pluralism in Nigeria revive

First civilian transfer of power in Nigeria's history

Senegal (Republic of Senegal)

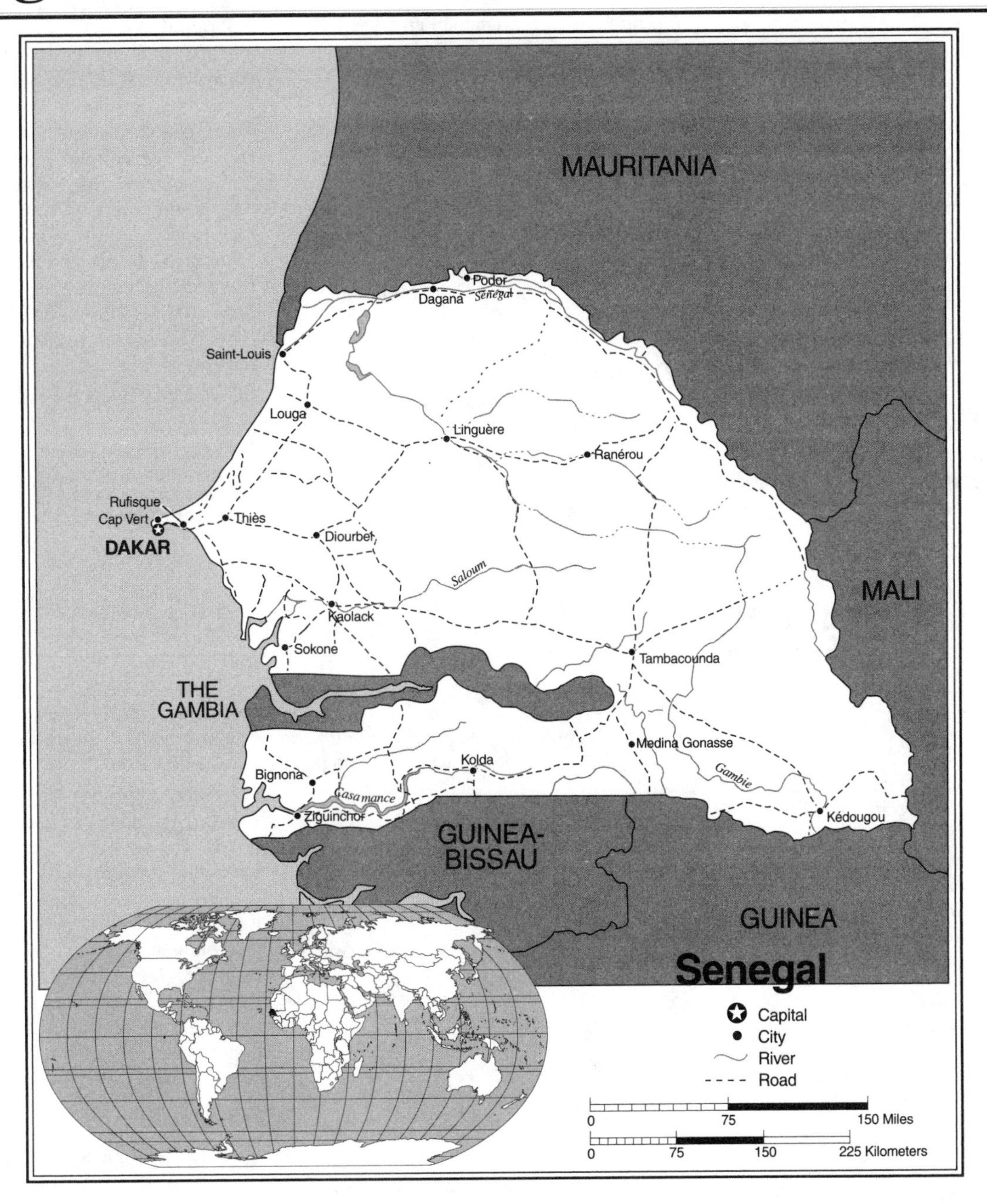

Senegal Statistics

GEOGRAPHY

Area in Square Miles (Kilometers): 76,000 (196,840) (about the size of South Dakota)

Capital (Population): Dakar (2,160,000)

Environmental Concerns: poaching; deforestation; overgrazing; soil erosion; desertification; overfishing

Geographical Features: low, rolling plains, foothills in the southeast; Gambia is almost an enclave of Senegal

Climate: tropical

PEOPLE

Population

Total: 10,852,147

Annual Growth Rate: 2.52%

Rural/Urban Population Ratio: 53/47

Major Languages: French; Wolof; Pulaar; Diola; Mandinka

Ethnic Makeup: 43% Wolof; 24% Pular; 15% Serer; 18% others

Religions: 94% Muslim; 5% Christian; 1% indigenous beliefs

Health

Life Expectancy at Birth: 61 years (male); 65 years (female)

Infant Mortality: 55.4/1,000 live births

Physicians Available: 1/14,825 people

HIV/AIDS Rate in Adults: 1.4%

Education

Adult Literacy Rate: 39%
Compulsory (Ages): 7–13

COMMUNICATION

Telephones: 235,000 main lines
Televisions: 6.9/1,000 people
Internet Users: 225,000 (2003)

TRANSPORTATION

Highways in Miles (Kilometers): 8,746 (14,576)
Railroads in Miles (Kilometers): 565 (905)
Usable Airfields: 20
Motor Vehicles in Use: 160,000

GOVERNMENT

Type: republic
Independence Date: April 4, 1960 (from France)
Head of State/Government: President Abdoulaye Wade; Prime Minister Macky Sall
Political Parties: Socialist Party; Senegalese Democratic Party; Democratic League–Labor Party Movement; Independence and Labor Party; others
Suffrage: universal at 18

MILITARY

Military Expenditures (% of GDP): 1.4%
Current Disputes: civil unrest; issue with The Gambia; tensions with Mauritania and Guinea-Bissau

ECONOMY

Currency ($ U.S. equivalent): 581 CFA francs = $1
Per Capita Income/GDP: $1,600/$17 billion
GDP Growth Rate: 5.7%
Inflation Rate: 0%
Unemployment Rate: 48%
Labor Force by Occupation: 70% agriculture
Population Below Poverty Line: 54%
Natural Resources: fish; phosphates; iron ore
Agriculture: peanuts; millet; sorghum; corn; rice; cotton; vegetables; livestock; fish
Industry: agricultural and fish processing; phosphate mining; fertilizer production; petroleum refining; construction materials
Exports: $1 billion (primary partners France, Italy, Spain)
Imports: $1.3 billion (primary partners France, Nigeria, Germany)

SUGGESTED WEB SITES

http://www.senegal-online.com/anglais/index.html
http://www.sas.upenn.edu/African_Studies/Country_Specific/Senegal.html

Senegal Country Report

The year 2002 was marked by both triumph and tragedy for the people of Senegal. At the World Cup, the national soccer team beat world champions France in the opening game before going to qualify for the quarterfinals. But the nation was subsequently thrown into grief when hundreds of lives were lost in a ferryboat disaster. Both incidents took attention away from the country's broader struggle to pull itself out of chronic poverty.

DEVELOPMENT

The recently built Diama and Manantali Dams will allow for the irrigation of many thousands of acres for domestic rice production. At the moment, large amounts of rice are imported to Senegal, mostly to feed the urban population.

In March 2000, Senegalese politics entered a new era with the electoral victory of veteran opposition politician Abdoulaye Wade over incumbent Abdou Diouf. Wade became the country's third president. Like his predecessors, Wade faces daunting challenges. Much of Senegal's youthful, relatively well-educated population remains unemployed. Widespread corruption and a long-running separatist rebellion in the southern region of Casamance will also test the new regime. Taking a step-by-step approach, Wade has been able to bring about a reduction in the separatist rebellion, and corruption is on the decline. But, also like his predecessors, Wade should be able to draw upon the underlying strength of Senegal's culturally rich multiethnic society, which has maintained its cohesion through decades of adversity.

FREEDOM

Senegal's generally favorable human-rights record is marred by persistent violence in its southern region of Casamance, where rebels are continuing to fight for independence. A 2-year cease-fire broke down in 1995 after an army offensive was launched against the rebel Movement of Democratic Forces of Casamance.

To his great credit President Wade can claim to have reduced tensions, and an end is in sight to the long simmering, low intensity separatist war, going on since independence in the southern Casamance region. Since the start of the war hundreds of people have been killed and thousands have fled to Guinea-Bissau. Rebel fighters remain active, although the leader of the Movement of Democratic Forces of Casamance declared in 2003 that the war was over.

THE IMPACT OF ISLAM

The vast majority of Senegalese are Muslim. Islam was introduced into the region by the eleventh century A.D. and was spread through trade, evangelism, and the establishment of a series of theocratic Islamic states from the 1600s to the 1800s.

Today, most Muslims are associated with one or another of the Islamic Brotherhoods. The leaders of these Brotherhoods, known as marabouts, often act as rural spokespeople as well as the spiritual directors of their followers. The Brotherhoods also play an important economic role. For example, the members of the Mouride Brotherhood, who number about 700,000, cooperate in the growing of the nation's cash crops.

FRENCH INFLUENCE

In the 1600s, French merchants established coastal bases to facilitate their trade in slaves and gum. As a result, the coastal communities have been influenced by French culture for generations. More territory in the interior gradually fell under French political control.

Although Wolof is used by many as a lingua franca, French continues to be the common language of the country, and the educational system maintains a French character. Many Senegalese migrate to France, usually to work as low-paid laborers. The French maintain a military force near the capital, Dakar, and are major investors in the Senegalese economy. Senegal's judiciary and bureaucracy are modeled after those of France.

HEALTH/WELFARE

Like other Sahel countries, Senegal has a high infant mortality rate and a low life expectancy rate. Health facilities are considered to be below average, even for a country of Senegal's modest income, but recent child-immunization campaigns have been fairly successful.

POLITICS

Under Diouf, Senegal strengthened its commitment to multipartyism. After succeeding Leopold Senghor, the nation's scholarly first president, Diouf liberalized the political process by allowing an increased number of opposition parties effectively to compete against his own ruling Socialist Party (PS). He also restructured his administration in ways that were credited with making it less corrupt and more efficient. Some say that these moves did not go far enough, but Diouf, who inclined toward reform, had to struggle against reactionary elements within his own party.

In national elections in 1988, Diouf won 77 percent of the vote, while the Socialists took 103 out of 120 seats. Outside observers believed that the elections had been plagued by fewer irregularities than in the past. However, opposition protests against alleged fraud touched off serious rioting in Dakar. As a result, the city was placed under a three-month state of emergency. Diouf's principal opponent, Abdoulaye Wade of the Democratic Party (PD), was among those arrested and tried for incitement. But subsequent meetings between Diouf and Wade resulted in an easing of tensions. Indeed, in 1991, Wade shocked many by accepting the post of minister of state in Diouf's cabinet. Elections in 1993 were less controversial, with Diouf being reelected with 58 percent of the vote. PS representation dropped to 84 seats.

In March 1995, a new, multiparty "Government of National Unity" was formed, which survived despite the defection of one of its members, the Independent Labor Party, in September 1996. But interparty tension grew in the face of Diouf's failure to appoint an independent elections commission in preparation for elections in November 1996.

THE ECONOMY

Many believe that the *Sopi* (Wolof for "change") riots of 1988 were primarily motivated by popular frustration with Senegal's weak economy, especially among its youth (about half of the Senegalese are under age 21), who face an uncertain future. Senegal's relatively large (47 percent) urban population has suffered from rising rates of unemployment and inflation, which have been aggravated by the country's attempt to implement an International Monetary Fund–approved Structural Adjustment Program (SAP). In recent years, the economy has grown modestly but has so far failed to attract the investment needed to meet ambitious privatization goals. Among rural dwellers, drought and locusts have also made life difficult. Fluctuating world market prices and disease as well as drought have undermined groundnut exports.

ACHIEVEMENTS

Dakar, sometimes described as the "Paris of West Africa," has long been a major cultural center for the region. Senegalese writers such as former president Leopold Senghor were founders of the Francophonic African tradition of Negritude.

Senegal has also been beset by difficulties in its relations with neighboring states. The Senegambia Confederation, which many hoped would lead to greater cooperation with The Gambia, was dissolved in September 1989. Relations with Guinea-Bissau are strained as a result of that nation's failure to recognize the result of international arbitration over disputed, potentially oil-rich waters. Senegalese further suspect that individuals in Guinea-Bissau may be linked to the separatist unrest in Senegal's Casamance region. There some 1,000 people died in an insurgency campaign between the Senegalese Army and the guerrillas of the Movement of Democratic Forces of Casamance. In July 1993, the rebels agreed to a cease-fire, but the cease-fire collapsed in 1995. In August 2000, the rebels agreed to reopen talks with Wade's new administration.

But the major source of cross-border tension has been Mauritania. In 1989, longstanding border disputes between the two countries led to a massacre of Senegalese in Mauritania, setting off widespread revenge attacks against Mauritanians in Senegal. More than 200,000 Senegalese and Mauritanians were repatriated. Relations between the two countries have remained tense, in large part due to the persecution of Mauritania's "black" communities by its Maur-dominated military government. Many Mauritanians belonging to the persecuted groups have been pushed into Senegal, leading to calls for war, but in 1992, the two countries agreed to restore diplomatic, air, and postal links.

Timeline: PAST

1659
The French occupy present-day St. Louis and, later, Gorée Island

1700s
The Jolof kingdom controls much of the region

1848
All Africans in four towns of the coast vote for a representative to the French Parliament

1889
Interior areas are added to the French colonial territory

1960
Senegal becomes independent as part of the Mali Confederation; shortly afterward, it breaks from the Confederation

1980s
President Leopold Senghor retires and is replaced by Abdou Diouf; Senegalese political leaders unite in the face of threats from Mauritania

1990s
Serious rioting breaks out in Dakar protesting the devaluation of the CFA franc

PRESENT

2000s
Tensions remain with Guinea-Bissau and Mauritania

Abdoulaye Wade wins the presidency

Sierra Leone (Republic of Sierra Leone)

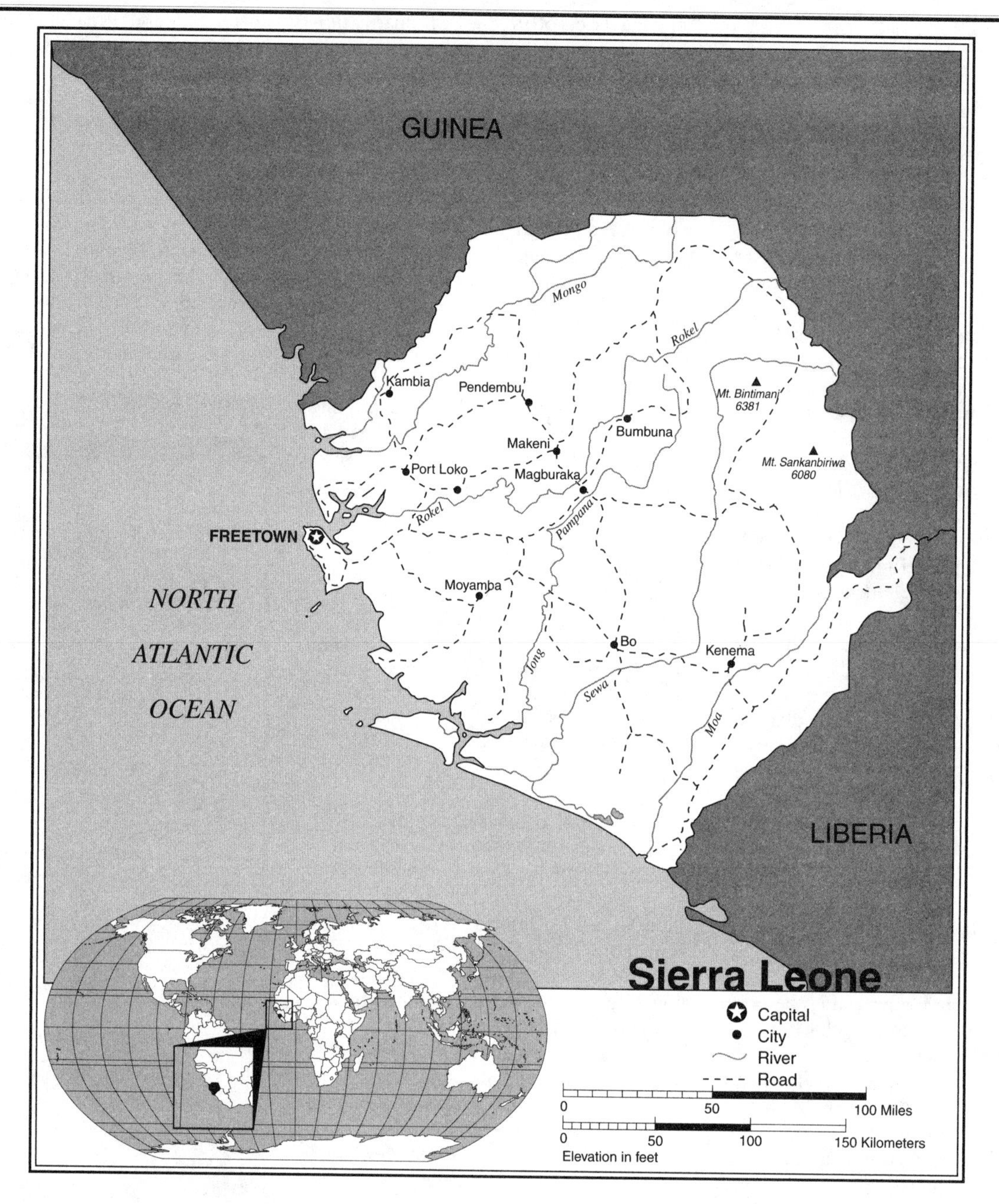

Sierra Leone Statistics

GEOGRAPHY

Area in Square Miles (Kilometers): 27,925 (72,325) (about the size of South Carolina)

Capital (Population): Freetown (837,000)

Environmental Concerns: soil exhaustion; deforestation; overfishing; population pressures

Geographical Features: a coastal belt of mangroves; wooded, hilly country; upland plateau; mountainous east

Climate: tropical; hot, humid

PEOPLE

Population

Total: 5,883,889

Annual Growth Rate: 2.27%

Rural/Urban Population Ratio: 64/36

Major Languages: English, Krio, Temne, Mende

Ethnic Makeup: 30% Temne; 30% Mende; 30% other African; 10% others

Religions: 60% Muslim; 30% indigenous beliefs; 10% Christian

Health

Life Expectancy at Birth: 43 years (male); 49 years (female)

Infant Mortality: 144.3/1,000 live births

Physicians Available: 1/10,832 people

HIV/AIDS Rate in Adults: 2.99%

Education

Adult Literacy Rate: 31.4%

COMMUNICATION

Telephones: 25,000 main lines
Internet Users: 20,000 (2001)

TRANSPORTATION

Highways in Miles (Kilometers): 7,020 (11,700)
Railroads in Miles (Kilometers): 52 (84)
Usable Airfields: 10
Motor Vehicles in Use: 44,000

GOVERNMENT

Type: constitutional democracy
Independence Date: April 27, 1961 (from the United Kingdom)
Head of State/Government: President Ahmad Tejan Kabbah is both head of state and head of government
Political Parties: Sierra Leone People's Party; National Unity Party; others
Suffrage: universal at 18

MILITARY

Military Expenditures (% of GDP): 1.5%
Current Disputes: hopes for a lasting peace after a decade of civil war

ECONOMY

Currency ($ U.S. Equivalent): 2,347 leones = $1
Per Capita Income/GDP: $500/$2.7 billion
GDP Growth Rate: 6.5%
Inflation Rate: 10%
Population Below Poverty Line: 68%
Natural Resources: diamonds; titanium ore; bauxite; gold; iron ore; chromite
Agriculture: coffee; cocoa; palm kernels; rice; palm oil; peanuts; livestock; fish
Industry: mining; petroleum refining; small-scale manufacturing
Exports: $65 million (primary partners New Zealand, Belgium, United States)
Imports: $145 million (primary partners Czech Republic, United Kingdom, United States)

SUGGESTED WEB SITES

http://www.sierra-leone.org
http://www.sierraleonenews.com
http://www.fosalone.org
http://www.africanews.org/west/sierraleone/
http://www.sas.upenn.edu/African_Studies/Country_Specific/S_Leone.html

Sierra Leone Country Report

In 2002, Sierra Leone emerged from a decade of civil war with the help of Britain (its former colonial power), a large United Nations peacekeeping mission, and other international elements. More than 17,000 UN troops disarmed tens of thousands of rebels and militia fighters. It was the biggest UN peacekeeping success in Africa for many years, following debacles in the 1990s in Angola, Rwanda, and Somalia. Currently the country is rebuilding its infrastructure and civil society. President Ahmed Tejan Kabbah won a landslide victory in May elections, in which his Sierra Leone People's Party also secured a majority in Parliament. In July 2002, a "Truth and Reconciliation Commission" was established to help the people of Sierra Leone overcome the trauma of the war, which was characterized by widespread atrocities.

DEVELOPMENT

The recently relaunched Bumbuna hydroelectric project should reduce Sierra Leone's dependence on foreign oil, which has accounted for nearly a third of its imports. In response to threats of boycotting, the country's Lungi International Airport was upgraded. Persistent inflation and unemployment have taken a severe toll on the country's people.

In July 2003 rebel leader Foday Sankoh died in prison of natural causes while waiting to be tried for war crimes. In August 2003 President Kabbah testified to the Truth and Reconciliation commission that he had no say over operations of pre-government militias during the civil war. The much awaited disarmament and rehabilitation of more than 70,000 civil war combatants was officially completed in February 2004. In March 2004 the UN backed War Crimes Tribunal opened a courthouse to try senior militia leaders from both sides of civil war—the trails themselves began in June.

Sierra Leone's period of political instibility began in April 1992, when army Captain Valentine Strasser announced the overthrow of the long-governing All People's Congress (APC). The coup was initially welcomed, as the APC governments of the deposed president Joseph Momoh and is similarly deposed predecessor Siaka Stevens had been renowned for their institutionalized corruption and economic incompetence. But disillusionment grew as the Strasser-led National Provisional Ruling Council postponed holding multiparty elections, while sinking into its own pattern of corruption. The emergence of the RUF insurgency brought further misery, with both the rebels and army being accused of abuses.

Hopes that the (in many quarters unexpected) successful holding of democratic elections in February–March 1996 would lead to peace and reconciliation were dashed in May 1997, when dissident junior officers overthrew the elected government of President Kabbah. An Armed Forces Revolutionary Council (AFRC), led by Major Johnny Paul Koroma (who had been awaiting trial on charges stemming from an earlier coup attempt) banned political parties and all public demonstrations and meetings and announced that all legislation would be made by military decree. The AFRC soon revealed itself to be a vehicle of the rebel Revolutionary United Front (RUF) as well as of elements within the military unwilling to accept a return to civilian control.

FREEDOM

The deposed AFRC/RUF regime unleashed a terror campaign, including extra-judicial killings, torture, mutilation, rape, beatings, arbitrary arrest, and the detention of unarmed civilians. Junta forces killed and/or amputated the arms of detainees. Prior to the coup, RUF was infamous for its murderous attacks on civilians during raids in which children were commonly abducted and forced to commit atrocities against their relatives as a form of psychological conditioning.

The AFRC/RUF regime attracted overwhelming regional condemnation, with the international community sanctioning efforts by the Economic Community of West African States (ECOWAS) to restore Kabbah to power. This was finally achieved in February 1998, when military units of ECOMOG, the Nigerian-led ECOWAS peace-monitoring force, attacked and routed the junta's forces in the capital city, Freetown, after Koroma abandoned his agreement to step down peacefully.

In January 1999, rebels backing the RUF seized parts of the capital city, Freetown, from ECOMOG. After weeks of bitter fighting they were driven out, but 5,000 people had been killed, and the city was devastated. A cease-fire was declared that May, following a further ECOMOG offensive against the RUF. In July, after six weeks of talks in Lomé, Togo, a new peace agreement was signed under which the rebels were to receive posts in government and assurances that they would not be prosecuted for war crimes. In accordance with the agreement, the RUF leader, Foday Sankoh, was brought into a transitional government pledged to restoring democracy, law, and order, with UN peacekeeper assistance.

HEALTH/WELFARE

Life expectancy for both males and females in Sierra Leone is only in the 40s, while the infant mortality rate, 144.3 per 1,000 live births, remains appalling. In 1990, hundreds, possibly thousands, of Sierra Leone children were reported to have been exported to Lebanon on what amounted to slave contracts. The UNEP Human Development Index rates Sierra Leone last, out ot 174 countries.

In November–December 1999, UN forces arrived to police the agreement, but ECOMOG troops continued to be attacked outside Freetown. In April–May 2000, as rebel troops attacked the capital, UN forces came under attack in the eastern part of the country, but far worse was in store when first 50, then several hundred UN troops were abducted. To protect and evacuate British citizens, 800 British paratroopers were sent to Freetown. Working alongside the UN, these troops helped to recapture hostages and secure the airport, while Sankoh was captured. In January 2001, presidential and parliamentary elections were postponed due to continuing strife. But by March the rebel army had begun to surrender, allowing for its forces' disarmament and participation in the elections.

ACHIEVEMENTS

The Sande Society, a women's organization that trains young Mende women for adult responsibilities, has contributed positively to life in Sierra Leone. Beautifully carved wooden helmet masks are worn by women leaders in the society's rituals. Ninety-five percent of Mende women join the Society.

Sierra Leone is the product of a unique colonial history. Freetown was founded by waves of black settlers who were brought there by the British. The first to arrive were the so-called "Black Poor," a group of 400 people sent from England in 1787. Shortly thereafter, former slaves from Jamaica and Nova Scotia arrived; they had gained their freedom by fighting with the British, against their American masters, in the U.S. War of Independence. About 40,000 Africans who were liberated by the British and others from slave ships captured along the West African coast were also settled in Freetown and the surrounding areas in the first half of the nineteenth century.

The descendants of Sierra Leone's various black settlers blended African and British ways into a distinctive *Krio,* or Creole, culture. Besides speaking English, they developed their own Krio language, which has become the nation's lingua franca. Today, the Krio make up only about 5 percent of Sierra Leone's multiethnic population.

As more people were given the vote in the 1950s, the indigenous communities ended Krio domination in local politics. The first party to win broad national support was the Sierra Leone People's Party (SLPP), under Sir Milton Margai, which led the country to independence in 1961. During the 1967 national elections, the SLPP was narrowly defeated by Stevens' APC. From 1968 to 1985, Stevens presided over a steady erosion of Sierra Leone's economy and civil society. The APC's increasingly authoritarian control coincided with the country's economic decline. Although rich in its human as well as natural resources at independence, today Sierra Leone is ranked as one of the world's poorest countries.

Revenues from diamonds (which formed the basis for prosperity during the 1950s) and gold have steadily fallen due to the depletion of old diggings and massive smuggling. The two thirds of Sierra Leone's labor force employed in agriculture have suffered the most from the nation's faltering economy. Poor producer prices, coupled with an international slump in demand for cocoa and robusta coffee, have cut into rural incomes. The promise by Stevens' successor, Momoh, to improve producer prices as part of a "Green Revolution" program went unfulfilled. Like its minerals, much of Sierra Leone's agricultural production has been smuggled out of the country. In 1989, the cost of servicing Sierra Leone's foreign debt was estimated to be 130 percent of the total value of its exports. This grim figure led to the introduction of an International Monetary Fund–supported Structural Adjustment Program (SAP), whose austerity measures made life even more difficult for urban dwellers.

In some ways the situation in Sierra Leone provides lessons for Africa. Disciplined armies can be forces of stability, just as undisciplined armies can give rise to chaos. The final phase of the conflict also demonstrated the potential of obtaining peace through concerted efforts of the regional states accompanied by external powers. Perhaps another lesson was the usefulness of targeted economic sanctions. In the case of Sierra Leone, the RUF survived for many years through profits gained through diamond smuggling. The relative success in separating such "blood diamonds" from legitimate exports has given rise to greater international control over the marketing of the gems, which may prove to be a model for similar situations in the future.

Timeline: PAST

1400–1750
Early inhabitants arrive from Africa's interior

1787
Settlement by people from the New World and recaptured slave ships

1801
Sierra Leone is a Crown colony

1898
Mende peoples unsuccessfully resist the British in the Hut Tax War

1961
Independence

1978
The new Constitution makes Sierra Leone a one-party state

1985
President Siaka Stevens steps down; Joseph Momoh, the sole candidate, is elected

1990s
Debt-servicing cost mounts; SAP; Liberian rebels destabilize Sierra Leone; Momoh is overthrown

PRESENT

2000s
Civil war ends

Ahmed Kabbah wins reelection

Togo (Togolese Republic)

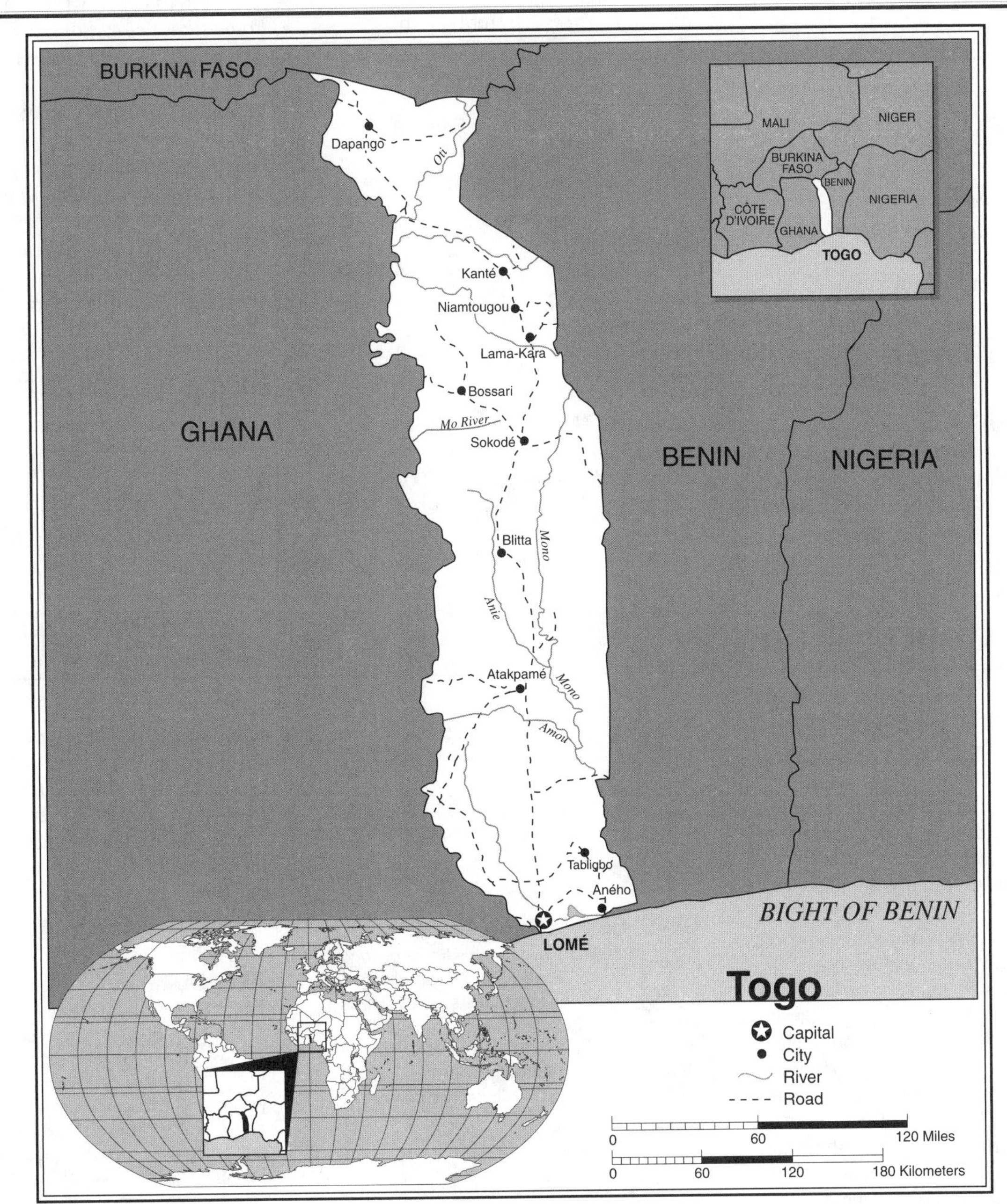

Togo Statistics

GEOGRAPHY

Area in Square Miles (Kilometers): 21,853 (56,600) (about the size of West Virginia)

Capital (Population): Lomé (732,000)

Environmental Concerns: drought; deforestation

Geographical Features: gently rolling savanna in north; central hills; southern plateau; low coastal plain with extensive lagoons and marshes

Climate: tropical to semiarid

PEOPLE

Population

Total: 5,556,812

Annual Growth Rate: 2.27%

Rural/Urban Population Ratio: 67/33

Major Languages: French; Ewe; Mina; Dagomba; Kabye; Dasomsa

Ethnic Makeup: 99% African—Ewe; Mina; Kabye; many others

Religions: 70% indigenous beliefs; 20% Christian; 10% Muslim

Health

Life Expectancy at Birth: 52 years (male); 56 years (female)

Infant Mortality: 69.3/1,000 live births

Physicians Available: 1/11,270 people

HIV/AIDS Rate in Adults: 5.98%

Education

Adult Literacy Rate: 51.7%
Compulsory (Ages): 6–12

COMMUNICATION

Telephones: 60,600 main lines
Televisions: 36/1,000 people
Internet Users: 210,000 (2003)

TRANSPORTATION

Highways in Miles (Kilometers): 4,512 (7,520)
Railroads in Miles (Kilometers): 352 (532)
Usable Airfields: 9
Motor Vehicles in Use: 109,000

GOVERNMENT

Type: republic under transition to multiparty democratic rule
Independence Date: April 27, 1960 (from French-administered UN trusteeship)
Head of State/Government: President Gnassingbé Eyadéma; Prime Minister Koffi Sama
Political Parties: Assembly of the Togolese People; Coordination des Forces Nouvelles; Action Committee for Renewal; Patriotic Pan-African Convergence; Union of Forces for Change; others
Suffrage: universal for adults

MILITARY

Military Expenditures (% of GDP): 1.8%
Current Disputes: civil unrest; tensions with Benin

ECONOMY

Currency ($ U.S. Equivalent): 581 CFA francs = $1
Per Capita Income/GDP: $1,500/$7.6 billion
GDP Growth Rate: 3.3%
Inflation Rate: -1%
Labor Force by Occupation: 65% agriculture; 30% servies; 5% industry
Population Below Poverty Level: 32%
Natural Resources: phosphates; limestone; marble; arable land
Agriculture: coffee; cocoa; yams; cassava; millet; sorghum; rice; livestock; fish
Industry: phosphates mining; textiles; handicrafts; agricultural processing; cement; beverages
Exports: $306 million (primary partners Benin, Nigeria, Ghana)
Imports: $420 million (primary partners Ghana, France, Côte d'Ivoire)

SUGGESTED WEB SITES

http://www.republicoftogo.com/english/index.htm
http://www.republicoftogo.com/
http://www.sas.upenn.edu/African_Studies/Country_Specific/Togo.html

Togo Country Report

In recent years, Togo has become a prime example of the difficulty of achieving democratic reform in the face of determined resistance by a ruling clique that enjoys military backing and a strong ethnic support base. For the past four decades, the country has been politically dominated by supporters of its long-serving president, Gnassingbé Eyadéma. Their grip on power was evident once more in October 2002, when Eyadéma's Assembly (or Rally) of the Togolese People (RPT) party won another landslide victory in legislative elections, even in the face of continuing allegations of vote rigging and human-rights abuses. The true political contest may have occurred earlier in the year, when an apparent power struggle resulted in Eyadéma's sacking his prime minister and long-time ally, Agbeyome Kodjo. The outcome reconfirmed Eyadéma's status as one of Africa's true political survivors.

Emerging from the ranks of the military, Eyadéma first seized power in 1967. This followed a period of instability in the wake of the assassination of the country's first president, Sylvanius Olympio, by the Togolese military. In 1969, Eyadéma institutionalized his increasingly dictatorial regime as a one-party state. All Togolese have been required to belong to the RPT. But in 1991, faced with mass prodemocracy demonstrations in Lomé, the capital city, Eyadéma acquiesced to opposition calls for a "National Conference" that would end the RPT's monopoly of power. Since then, Eyadéma has survived Togo's turbulent return to multiparty politics with characteristic ruthlessness, skillfully taking advantage of the weakness of his divided opponents.

DEVELOPMENT

Much hope for the future of Togo is riding on the recently created Free Trade Zone at Lomé. Firms within the zone are promised a 10-year tax holiday if they export at least three quarters of their output. The project is backed by the U.S. Overseas Private Investment Corporation.

In December 2002, the Parliament altered the constitution by removing a clause which would have barred President Eyadéma from seeking a third presidential term. In June 2003 Eyadéma was reelected president of the country. With his election victory in hand President Eyadéma reinstated the previous Prime Minister Koffi Sama and his government, then began talks on a national unity government in July 2003. A new unity government was announced, but the main opposition parties were not included.

DEMOCRACY VS. DICTATORSHIP

Meeting in July–August 1991, the National Conference turned into a public trial of the abuses of the ruling regime. Resisting the president's attempts to dissolve it, the Conference appointed Kokou Koffigoh as the head of an interim government, charged with preparing the country for multiparty elections. The RPT was to be disbanded, and Eyadéma himself was barred from standing for reelection.

FREEDOM

Togo continues to have a poor human-rights record. Its progovernment security forces have been responsible for extrajudicial killings, beatings, arbitrary detentions, and interference with citizens' rights to movement and privacy. Freedoms of speech and of the press are restricted. Interethnic killings have led to major population displacements.

In November–December 1991, however, soldiers loyal to Eyadéma launched a bloody attack on Koffigoh's residence. The French, whose troops had intervened in the past to keep Eyadéma in power, refused Koffigoh's plea for help. Instead, the coup attempt ended with the now-almost-irrelevant Koffigoh and Eyadéma agreeing to maintain their uneasy cohabitation. Elections were henceforth to include the RPT. Despite the "compromise," there was an upsurge in political violence in 1992, which included the May shooting of Gil-

(United Nations photo 117,584 by Anthony Fisher)

Potable water is not universally available in Togo. While food production in Togo is officially said to be adequate, outside observers contend that the drought-prone north is uncomfortably reliant on the more agriculturally productive southern areas.

christ Olympio (the son of Sylvanus) and other Eyadéma opponents. In September, "rebel" soldiers once more held the government hostage.

A January 1993 massacre of the prodemocracy demonstrators pushed the country even further to the brink. Some 300,000 southern Togolese, mostly Ewe-speakers, fled the country, fearing "ethnic cleansing" by the largely northern, Kabye-speaking army. In 1993–1994, exiled anti-Eyadéma militants—many of whom coalesced as the Front of the National Committee for the Liberation of the Togolese People (FNCL)—began to fight back. The army chief of staff was among those killed during a daring raid on the main military headquarters in Lomé, in which grenades were also thrown into Eyadéma's bedroom.

HEALTH/WELFARE

The nation's health service has declined as a result of austerity measures. Juvenile mortality is 15%. Self-induced abortion now causes approximately 17% of the deaths among Togolese women of child-bearing age. School attendance has dropped in recent years.

In July 1993, Eyadéma and his more moderate opponents signed a peace accord in Burkina Faso, pledging renewed movement toward election. A month later, however, the opposition boycotted a snap presidential poll. Thereafter, Eyadéma gave ground, agreeing to internationally supervised legislative elections in February 1994. After two rounds of voting, amid escalating violence, Eyadéma's RPT and the main opposition—Action Committee for Renewal (CAR), led by Yaovi Agboyibor—each controlled about 35 seats in the 75-seat Assembly. (The situation was clouded by judicial reviews of the results in five constituencies.) The balance of power rested with former Organization of African Unity secretary-general Edem Kodjo's Togo Union for Democracy (UTD), which had entered the election allied with CAR. But in May, Kodjo became prime minister, with Eyadéma's backing. The failure of the moderate opposition to capitalize on its apparent victory in undoubtedly flawed elections strengthened the determination of the militants to carry on by other means.

STRUCTURAL ADJUSTMENT

Togo's political crisis has taken place against a backdrop of economic restructuring. In 1979, Togo adopted an economic-recovery strategy that many consider to have been a forerunner of other Structural Adjustment Programs (SAPs) introduced throughout most of the rest of Africa. Faced with mounting debts as a result of falling export revenue, the government began to loosen the state's grip over the local economy. Since 1982 a more rigorous International Monetary Fund/World Bank–supported program of privatization and other market-oriented reforms has been pursued. Given this chronology, Togo's economic prospects have become a focus of attention for those looking for lessons about the possible effects of SAP. Both proponents and opponents of SAP have grounds for debate.

ACHIEVEMENTS

The name of Togo's capital, Lomé, is well known in international circles for its association with the Lomé Convention, a periodically renegotiated accord through which products from various African, Caribbean, and Pacific countries are given favorable access to European markets.

Supporters of Togo's SAP point out that since 1985, the country has enjoyed an average growth in gross domestic product of 3.3 percent per year. While this statistic is an improvement over the 1.7 percent rate recorded between 1973 and 1980, however, it is well below the 7.2 percent growth that prevailed from 1965 to 1972. (The GDP growth rate in 2002 was estimated at 2.2 percent.) During the 1980s, there was a rise in private consumption, 7.6 percent per year, and a drop in inflation, from about 13 percent in 1980 to an estimated 2 percent in 1989. A rate of 2.3 percent was estimated for 2002.

The livelihoods of certain segments of the Togolese population have materially improved during the past decade. Beneficiaries include some of the two thirds of the workforce employed in agriculture. Encouraged by increased official purchase prices, cash-crop farmers have expanded their outputs of cotton and coffee. This is especially true in the case of cotton production, which tripled between 1983 and 1989. Nearly half the nation's small farmers now grow the crop.

Balanced against the growth of cotton has been a decline in cocoa, which emerged as the country's principal cash crop under colonialism. Despite better producer prices during the mid-1980s, output fell as a result of past decisions not to plant new trees. Given the continuing uncertainty of cocoa prices, this earlier shift may prove to have been opportune. The long-term prospects of coffee are also in doubt, due to a growing global preference for the arabica beans of Latin America over the robusta beans that thrive throughout much of West Africa. As a result, the government had to reverse course in 1988, drastically reducing its prices for both coffee and cocoa, a move that it hopes will prove to be only temporary.

Eyadéma's regime has claimed great success in food production, but its critics have long countered official reports of food self-sufficiency by citing the importation of large quantities of rice, a decline in food production in the cotton-growing regions, and widespread childhood malnutrition. The country's food situation is complicated by an imbalance between the drought-prone northern areas and the more productive south. In 1992, famine threatened 250,000 Togolese, mostly northerners.

There have been improvements in transport and telecommunications. The national highway system, largely built by the European Development Fund, has allowed the port of Lomé to develop as a transshipment center for exports from neighboring states as well as Togo's interior. At the same time, there has been modest progress in cutting the budget deficit. But it is in precisely this area that the cost of Togo's SAP is most apparent. Public expenditures in health and education declined by about 50 percent between 1982 and 1985. Whereas school enrollment rose from 40 percent to more than 70 percent during the 1970s, it has slipped back below 60 percent in recent years.

The ultimate justification for Togo's SAP has been to attract overseas capital investment. In addition to sweeping privatization, a Free Trade Zone has been established. But overseas investment in Togo has always been modest. There have also been complaints that many foreign investors have simply bought former state industries on the cheap rather than starting up new enterprises. Furthermore, privatization and austerity measures are blamed for unemployment and wage cuts among urban workers. One third of the state-divested enterprises have been liquidated.

Whatever the long-term merits of Togo's SAP, it is clear that it has so far resulted in neither a pattern of sustainable growth nor an improved standard of living for most Togolese. For the foreseeable future, the health of Togo's economy will continue to be tied to export earnings derived from three commodities—phosphates, coffee, and cocoa—whose price fluctuations have been responsible for the nation's previous cycles of boom and bust.

Timeline: PAST

1884
Germany occupies Togo

1919
Togo is mandated to Britain and France by the League of Nations following Germany's defeat in World War I

1956–1957
UN plebiscites result in the independence of French Togo and incorporation of British Togo into Ghana

1960
Independence is achieved

1963
Murder of President Sylvanus Olympio; a new civilian government is organized after the coup

1967
The coup of Colonel Etienne Eyadéma, now President Gnassingbé Eyadéma

1969
The Coalition of the Togolese People becomes the only legal party in Togo

1990s
Prodemocracy demonstrations lead to interim government and the promise of multiparty elections; Eyadéma survives escalating violence and controversial elections

PRESENT

2000s
Eyadéma retains power

Eyadéma is named chairman of the OAU

From rubble to revival

A South African man turns a dump into a cultural mecca.

By Megan Lindow

SOWETO, SOUTH AFRICA

The top of an old water tower offers Mandla Mentoor the best point from which to survey the subtle transformations taking place in his community below.

From here, the view is vast and panoramic: Sprawling, haphazard jumbles of shanties and geometric rows of matchbox houses stretch to the horizon, divided by pockets of barren land. They are Soweto's black townships, created more than 50 years ago by the former apartheid government.

"We call this place Somoho, the Soweto Mountain of Hope," says Mr. Mentoor, a wiry man with a thin goatee and a broad grin, gesturing to the open hillside on top of which the water tower rests. "It has seen both rain and storm in terms of trouble."

Just a few years ago, this 45-acre space that divides Mentoor's township of Tshiawelo in half was strewn with garbage. During the 1980s and early '90s, residents protested apartheid by refusing to pay local taxes, so uncollected garbage soon piled high in Soweto's open spaces. Criminals frequented the area, women were raped, and local people sometimes found abandoned babies and dead bodies in the rubble, Mentoor recalls.

Today, however, the trash is gone, and patches of dusty hillside have been planted with trees and vegetable gardens. Residents have built makeshift theaters and cooking huts, and walls of rock have been piled up to form "dialogue circles"—spaces for meetings, parties, and performances.

Projects like this reflect a "greening" movement that is slowly spreading in neglected urban townships and degraded rural settlements, where most South Africans live. While communities improve themselves for a variety of reasons, Soweto's changes were spearheaded by one individual, Mentoor, on a mission to bring culture and employment to his home. "Through the development of this mountain, the young people are having fun and giving back to their communities.... They are becoming changed people," says Mentoor. "They have ownership of it."

Turning weeds into gardens

In South Africa, there are several areas that are either under development or ripe for revival. "So many parks and open spaces here you find have no purpose. They're not giving direct benefits to people," says John Nzira of Food and Trees for Africa, a nonprofit organization that works with local governments and communities to turn derelict areas into food gardens and other amenities.

But in the small, northern town of Vryburg, residents planted olive trees along a crime-ridden strip of land between two roads, turning a previous eyesore into a lucrative community venture. Finding productive uses for discarded bits of land builds communities, generates income, improves nutrition, and reduces crime, Mr. Nzira says.

In the densely packed settlement of Diepsloot in the hills north of Johannesburg, with the help of Food and Trees for Africa, unemployed men and women have planted a garden of tomatoes, peppers, spinach, carrots, pumpkins, and medicinal plants on what had been an overgrown plot of land.

"Before, I had no knowledge of farming," says Pepu Mashele, a thin woman who lives in a nearby squatter settlement. She works in the garden nearly every day and earns money selling the produce. "Before, I was just staying at home, but now it's different. I'm generating something."

In Soweto, Mentoor began working out of his house near "the mountain" in 1991 to address problems of unemployment, crime, and environmental degradation in his community.

At first, he says, he funded his ventures and supported his family by selling his artwork and "stealing from relatives," he admits with a grin, adding, "the entire project was started with 50 cents."

Mentoor formed his passion for art and the environment as a child participating in Boy Scouts. It was only during the early '90s, as South Africa began its transition to democracy, that he saw an opportunity to make a difference in the community.

"Through the development of this mountain, the young people are having fun and giving back to their communities.... They are becoming changed people."
—Mandla Mentoor

At first, he says, he recruited young people and unemployed women to salvage paper, cans, and other waste materials to sell, but he quickly realized this was not the best way to make money.

So he developed Amandla Waste Creations and began teaching people to use these materials to make low-cost building materials and crafts such as papier-mâché and wire sculptures to sell to tourists.

"I was a student of '76," says Mentoor, referring to the notorious 1976 student riots in Soweto. "I've grown up with the tag of 'lost generation,' and that never jelled with me as a youth."

The organization's first real grant money came when Mentoor won the World Wilderness Forum's Green Trust Award in 2002. Mentoor's group voted to use the prize money ($1,500) to buy rakes and masks needed to clean up "the mountain."

"We turned into a serious laughing-stock in the community," Mentoor recalls. "But I kept reminding them of the starfish story: One day, a young person found all the fish had washed out of the sea and onto the shore, and didn't know what to do to save them. So he started throwing them back one by one. His friends said, 'You're mad, how are you going to save them by throwing them in one by one?' And he said, 'at least I'm saving those ones. At least I've made my contribution.' "

The following year, Mentoor won a fellowship from Ashoka, a US aid organization. Ever since, he's relied on networking with environmental groups to gain piecemeal funding to support Somoho.

Climbing 'the mountain'

Mentoor's house, once a modest four-room bungalow, is now crammed with visitors. He has added an office. A model of a house built using tires and glass bottles sits in his backyard.

On the blocks surrounding his house, and at the foot of Somoho, shop spaces are filled with art studios, women's baking and sewing enterprises, and film and recording studios—all offshoots of Mentoor's organization. Hundreds of young people have become involved in sports, music, and dance in Somoho since restoration efforts began.

Eventually, Mentoor says, these projects will all be moved up to the "mountain." Mentoor also hopes the space will include a craft market, an environmental education center, and even an African restaurant in the water tower.

Sydney Cindi, who runs the waste-art section of the program, says he's trying to get young people involved so they won't make the mistakes he did. He learned to work with clay in prison, where he served four years for robbery.

"To me, Somoho is not just a project, it's a school of learning," he says. "When we started on the mountain it was a dumping place. Now it's a place where people sit under the trees."

The West helps, and harms, as Southern Africa seeks food

By Danna Harman

Last May, with signs of a food crisis in Southern Africa growing, the United Nations World Food Program (WFP) rented out a two-story building in suburban Johannesburg, South Africa. Gray carpeting was rolled out, cubicles were erected, coffeemakers wheeled in, computers hooked up, and ID badges handed out. The emergency center sprung into high gear.

Some 50 WFP employees were flown in from other stations across the world. Tim Smith, who used to work in the Sierra Leone office, was brought in to the Johannesburg headquarters in July and put in charge of ports and shipping logistics. His job: to track incoming food donations and coordinate the movement of the cargo to distribution points around the continent.

Mr. Smith's mobile phone rings nonstop. The distribution agents in Maputo, Mozambique, want to know when exactly the Liberty Grace—a US freighter filled with 50,000 tons of yellow corn—will be arriving. Someone calls to say there is a storm coming down the coast which could hamper offloading. An overland logistics officer tells him that some trucks have broken down and a transport agent wants more money.

"There are always hitches," he says calmly. "But nothing more or less than the usual. We just keep it all moving."

Smith and his team are working hard to help mitigate a food shortage affecting 14.5 million people in six Southern African countries. Little rain has fallen on the region this growing season, so people here are relying on international donations. But the food that Smith and the WFP are distributing—sent here by the US, Europe, and Japan—might not even be needed, say observers, if those same countries would open up their markets to African goods. Some argue that Africa is in the position it is in now—unable to remedy the food shortage itself—because rich countries have put up trade barriers that have kept Africa poor and reliant on the West.

A rich man's business

Eliam Diamond lives on the shores of Lake Malawi in the small fishing village of Chiwawala, some 100 miles from the capital, Lilongwe. But he is not a fisherman.

"Fishing," says this father of six, "is a rich man's business." Fishing requires a net and rope and bait, he patiently explains, and, most significantly, a canoe. He has none of these.

He is not a farmer, either. He tried for some time, digging in the soil with a handmade hoe and lovingly sprinkling in corn seeds. But his plot was too small, his production too meager, and the market too saturated.

"Farming used to be the poor man's business," says Mr. Diamond. "But these days, it is just like fishing—for the rich man."

Diamond is a weaver. He makes mats out of dried palm leaves. A six-foot sleeping mat takes him four days to make and sells for as little as 4 cents, not enough to buy what little food there is here. So he relies on handouts.

A few days ago, Diamond picked up his monthly ration of donated US corn from the WFP at the Ngodzi distribution center near his village, carrying home the 110-pound bags tied to his old bicycle. Throughout Malawi, there are an estimated 3.3 million people like Diamond—too poor to buy food, living on monthly distributions from international aid organizations.

But, say critics, the problem of terrible poverty will not be solved with handouts.

Rich donors—the US, the European Union, and Japan, in particular—would be doing much more good in the long term if they concentrated on trade, instead of aid, goes the argument.

Critics also say that the large farm subsidies paid by wealthy countries to their own farmers hurt farmers in developing countries. The subsidized farmers can flood the market and sell their goods for less than it costs to grow them; the governments make up the difference. Poor farmers can't compete with such low prices, keeping them out of international markets, and keeping their countries from gaining economically.

According to the World Trade Organization, rich countries gave $57 billion in development aid in 2001 but paid more than $350 billion to their own farmers. World Bank figures suggest that giving developing countries more access to rich markets could earn them about $150 billion a year.

"We're subsidizing farmers in the north to the tune of $1 billion a day to preserve the quality of life of those in the north," says Pedro Sanchez, a professor at the College of Natural Resources at the University of California, Berkeley. "Whereas there is only negligible support for poor farmers in Africa and the rest of the developing world. The already strong in this case are only getting stronger."

A sweet deal

Sarah Lowe of Oxfam International, a confederation of 12 development agencies working to find solutions to poverty, says that while many poor countries were convinced by international lending organizations to open up trade, wealthier countries did not do the same.

"The IMF [International Monetary Fund] pressured countries in this region to liberalize their economies—reduce their emergency grain reserves, dismantle their state marketing boards, and remove farm subsidies," Ms. Lowe says. "And what happened? Grain production slumped because farmers no longer received guaranteed prices for their crops, and many could no longer afford inputs like fertilizers, so the price of staples like corn skyrocketed. Meanwhile, the rich countries continued along with their own farm subsidies."

In a recent report, Oxfam illustrates the damage being done by subsidies by looking at handouts given by the EU to sugar farmers in Europe. These farmers then produce surpluses, which flood the world market at artificially low prices.

Meanwhile, South Africa, Malawi, Swaziland, and Zambia—all low-cost sugar producers—are unable to tap potential markets in North America or the Middle East because they are outbid by the subsidized European producers, according to the Oxfam report.

"This means that an agricultural commodity that could play a real part in poverty alleviation in Southern Africa does not do so," the Oxfam report said. "European consumers are paying to destroy livelihoods in some of the world's poorest countries."

Western countries are beginning to see the harm done by subsidies and trade barriers. At the August World Summit on Sustainable Development in Johannesburg, debate over the contentious issue of farm subsidies raged, pitting developing nations against their wealthy counterparts.

Efforts to remove subsidies

The US and the EU affirmed their commitment to helping agricultural practices in the developing world through projects aimed at encouraging sustainable farming, diversifying crops, and expanding winter harvesting. Western countries committed to "move toward phasing out export subsidies and reduce trade-distorting domestic support," though it was generally seen as nonbinding and weak.

Just last week, the Bush Administration proposed the creation of the Southern African Free Trade and Development Agreement with South Africa, Botswana, Lesotho, Namibia, and Swaziland in an effort to open US markets to African goods.

Diamond shakes his head when asked where the corn he has received as food aid came from, or how it got to the distribution center, or whether he will always be able to get free handouts.

"I think it might be from America," he guesses. "In America there are no poor people," he continues, gaining confidence in his story and directing his comments at some neighbors gathered around, who nod in agreement. "The farmers are so rich they can give us their corn for nothing." Diamond stops and ponders the statement. "That is what I think, at least," he admits. "But I don't know for sure. I have never been there.... I have not really been anywhere."

Analysis: Defining genocide

Human rights campaigners accuse Sudan's pro-government Arab militia of carrying out genocide against black African residents of the Darfur region.

BBC News

The militia groups are accused of forcing some one million people from their homes and killing up to 50,000.

Many thousands more are at risk of starving due to a shortage of food in the camps where they have fled.

For months the US government declined to declare whether the killings in Darfur constitute genocide, saying there was not enough information.

But US Secretary of State Colin Powell on 9 September told the US Senate Foreign Relations Committee:

"We concluded that genocide has been committed in Darfur and that the government of Sudan and the Janjaweed bear responsibility and genocide may still be occurring."

United Nations Secretary General Kofi Annan has so far refused to use the term genocide, which would carry a legal obligation to act.

But what is genocide and when can it be applied? Some argue that the definition is too narrow and others that the term is devalued by misuse.

UN definition

The term was coined in 1943 by the Jewish-Polish lawyer Raphael Lemkin who combined the Greek word "genos" (race or tribe) with the Latin word "cide" (to kill).

After witnessing the horrors of the Holocaust—in which every member of his family except his brother and himself was killed—Dr Lemkin campaigned to have genocide recognised as a crime under international law.

> Genocide is ... both the gravest and greatest of the crimes against humanity
>
> Alain Destexhe

His efforts gave way to the adoption of the UN Convention on Genocide in December 1948, which came into effect in January 1951.

Article Two of the convention defines genocide as "any of the following acts committed with the intent to destroy, in whole or in part, a national, ethnical, racial or religious group, as such:

- Killing members of the group
- Causing serious bodily or mental harm to members of the group
- Deliberately inflicting on the group conditions of life calculated to bring about its physical destruction in whole or in part
- Imposing measures intended to prevent births within the group
- Forcibly transferring children of the group to another group

The convention also imposes a general duty on states that are signatories to "prevent and to punish" genocide.

Since its adoption, the UN treaty has come under fire from different sides, mostly by people frustrated with the difficulty of applying it to different cases.

'Too narrow'

Some analysts argue that the definition is so narrow that none of the mass killings perpetrated since the treaty's adoption would fall under it.

The objections most frequently raised against the treaty include:

- The convention excludes targeted political and social groups
- The definition is limited to direct acts against people, and excludes acts against the environment which sustains them or their cultural distinctiveness
- Proving intention beyond reasonable doubt is extremely difficult

- UN member states are hesitant to single out other members or intervene, as was the case in Rwanda
- There is no body of international law to clarify the parameters of the convention (though this is changing as UN war crimes tribunals issue indictments)
- The difficulty of defining or measuring "in part", and establishing how many deaths equal genocide

But in spite of these criticisms, there are many who say genocide is recognisable.

In his book *Rwanda and Genocide in the 20th Century*, former secretary-general of Medecins Sans Frontieres, Alain Destexhe says: "Genocide is distinguishable from all other crimes by the motivation behind it.

"Genocide is a crime on a different scale to all other crimes against humanity and implies an intention to completely exterminate the chosen group.

"Genocide is therefore both the gravest and greatest of the crimes against humanity."

Loss of meaning

Mr Destexhe believes the word genocide has fallen victim to "a sort of verbal inflation, in much the same way as happened with the word fascist".

Because of that, he says, the term has progressively lost its initial meaning and is becoming "dangerously commonplace".

Michael Ignatieff, director of the Carr Centre for Human Rights Policy at Harvard University, agrees.

"Those who should use the word genocide never let it slip their mouths. Those who unfortunately do use it, banalise it into a validation of every kind of victimhood," he said in a lecture about Raphael Lemkin.

"Slavery for example, is called genocide when—whatever it was, and it was an infamy—it was a system to exploit, rather than to exterminate the living."

In the Democratic Republic of Congo, a renegade commander said he captured the town of Bukavu earlier this month to prevent a genocide of Congolese Tutsis—the Banyamulenge.

It later transpired that fewer than 100 people had died.

The differences over how genocide should be defined, lead also to disagreement on how many genocides actually occurred during the 20th Century.

History of genocide

Some say there was only one genocide in the last century—the Holocaust.

Other experts give a long list of what they consider cases of genocide, including the Soviet man-made famine of Ukraine (1932-33), the Indonesian invasion of East Timor (1975), and the Khmer Rouge killings in Cambodia in the 1970s.

Former Yugoslav leader Slobodan Milosevic is on trial in The Hague, charged with genocide in Bosnia from 1992-5.

However, some say there have been at least three genocides under the 1948 UN convention:

- The mass killing of Armenians by Ottoman Turks between 1915-1920—an accusation that the Turks deny
- The Holocaust, during which more than six million Jews were killed
- Rwanda, where an estimated 800,000 Tutsis and moderate Hutus died in the 1994 genocide
- In the case of Bosnia, many believe that massacres occurred as part of a pattern of genocide, though some doubt that intent can be proved in the case of Mr Milosevic

The first case to put into practice the convention on genocide was that of Jean Paul Akayesu, the Hutu mayor of the Rwandan town of Taba at the time of the killings.

In a landmark ruling, a special international tribunal convicted him of genocide and crimes against humanity on 2 September 1998.

Twenty-one ringleaders of the Rwandan genocide have now been convicted by the International Criminal Tribunal for Rwanda.

Earlier this year, the war crimes tribunal for the former Yugoslavia widened the definition of what constitutes genocide.

General Radislav Krstic had appealed against his conviction for his role in the killing of more than 7,000 Muslim men and boys in Srebrenica in 1995.

But the court rejected his argument that the numbers were "too insignificant" to be genocide—a decision likely to set an international legal precedent.

US envoy for war crimes Pierre Prosper has already started to compile a list of those associated with the Janjaweed Arab militia in Sudan.

For the moment, these are threatened with sanctions but in the future, they may be charged with genocide, like those in Rwanda and the former Yugoslavia.

A new generation of African leaders

A graceful concession

EARLY ON Christmas morning 2002, Uhuru Kenyatta, his cheery necktie and plastered-on smile failing to make him look any less exhausted, stood sweating under the lights in Nairobi's Serena Hotel ballroom, slowly reading out the most important speech of his young life.

"These elections were a glowing tribute to the great nation of Kenya and freedom of choice," he began. "I accept the choice of the people and now concede that Mwai Kibaki will be the third president of the Republic of Kenya." There was a desire for change afoot, he continued, "but we were not perceived by the people as the change they were looking for."

Not a particularly notable address by Western standards, but practically revolutionary for Africa. The atmosphere in Nairobi that morning was drum-taut with tension. Riot police slapped their batons in anticipation. Newspaper editors had canceled their correspondents' vacations, expecting anger and violent ethnic clashes—that's what had happened after every other election in the country's history.

But Mr. Kenyatta's grace in defeat caught everyone by surprise and helped defuse the situation. The moment was more than just Kenya's first peaceful end to an election cycle. It marked a new maturity in African leadership.

The lanky Kenyatta is the son of Kenya's first president, Jomo Kenyatta, one of Africa's "big men"—those who wrested power from European overlords. The elder Kenyatta was part of what was supposed to be a new day for the continent—Africa run by Africans.

Yet no sooner had the Europeans left than new overlords took control, this time with African names: Robert Mugabe in Zimbabwe, Idi Amin in Uganda, Mobutu Sese Seko in Congo, to name a few. For the next four decades, Africa was pockmarked by war, corruption, coups, and countercoups. The continent became a front line in the cold war, with the world's superpowers propping up some of the most despicable men. These big men often confused their own interests with those of the countries they ruled over, handing out favors and hoarding their nation's wealth in offshore bank accounts. The new day had faded to dusk.

In the 1990s, hopes hung on the next generation of leaders—men like Rwanda's Paul Kagame, Uganda's Yoweri Museveni, and Ethiopia's Meles Zenawi. They were, again, supposed to usher in a new era of African democracy. And while surveys show more pplitical freedom in Africa now than when the decade began, that change often isn't felt on the ground. Mr. Kagame, for example, won re-election in Rwanda last month with 95 percent of the vote in a poll many saw as less than free and fair. Recent elections in Uganda and Ethiopia have gone the same way.

Now Africa watchers are forced to look to yet another generation. While some are pessimistic, others see another dawn approaching in young African leaders like Kenyatta. His self-effacing concession represented a transition of sorts—from those who did anything to gain power to those who want to embrace democracy, sound business practices, and the rule of law.

"We might not be seeing dramatic and sweeping change yet, but there are a number of people rising up who are able to see what the right thing to do is—and who want to try that," says Ted Dagne, an Eritrean-American specialist at the United States Congressional Research Service in Washington. "Systemic problems loom large, and it's going to take time for the new, independent African-born leaders to change this, but there are some good signs."

Sticking to his guns

AYISI MAKATIANI is often mentioned as one of young Africa's up-and-comers. But it's taken him the better part of a decade to gain that recognition. It was back when he was still studying for his degree by the placid Charles River in Cambridge, Mass., a decade ago that Mr. Makatiani first came up with the idea of starting an Internet business: an online chat room that would link Kenyans living in the US who missed home and wanted to keep in touch.

It didn't take long for Makatiani, an electrical-engineering student at the Massachusetts Institute of Technology and his Kenyan friends to learn that they were on to something. A fast-growing subscription base and demands for hard news from home led them to a more ambitious goal: an Africa-based Internet service provider, complete with African content.

Most people back home still did not have electricity, true, but those were the early days of the dotcom craze. With the sense that anything could happen, they gave it a go. They knew navigating the waters of Kenyan corruption would be daunting, but they weren't prepared for the class 5 rapids they encountered.

Shortly after the opening of the Africa Online offices in downtown Nairobi, Makatiani's competitors, who had ties to high-level government officials, "convinced" the national telephone company to shut down his company's phone lines—

leaving the main server unable to dial out. Customers began canceling subscriptions. "We were offering dialup service, and we had no dial tones," he recalls. "It was not fair. Not easy."

For every **100 professionals** sent abroad from Africa for training, **35 fail to return.**

Source: Science in Africa Magazine

But today, Kenya's first commercial Internet service provider is operating in 10 countries and is considered one of the continents best-run businesses. Makatiani has been named one of the World Economic Forum's leaders of tomorrow and recently started a promising new venture capital firm—Gallium Capital Partners—to fund tech companies in the region. Gallium has already been flagged by Fortune magazine as a model fund for companies in Africa.

Africa needs the kind of economic boost Makatiani's venture-capital fund can provide. Sub-Saharan Africa now is poorer, sicker, and more devastated by war than it was when the colonialists departed. At the start of the 21st century, it has the largest concentration of people in the world living on less than $1 a day, the greatest number of civil wars ongoing, and the highest number of refugees. AIDS has cut life expectancy to 47 years, and only 12 percent of the roads are paved. Corruption still abounds.

But Makatiani always knew two things, he says, as he maneuvers his car along the highways of his current hometown, Johannesburg, South Africa, between meetings: He was going to become a major business player in Africa, and he was going to do it the fair, ethical way.

"That was a time where you simply could not do business without having to pay someone," says Makatiani, a handsome one-time track-and-field champion who today favors dapper suits and conservative ties. "But we didn't want to pay someone. We didn't want to join that club. It was like being part of the mafia."

But he also knew that he could not yell and scream and demand things in Nairobi that were par for the course in Cambridge—like getting a working phone line if he paid his bills. With the help of a colleague's influential father, Makatiani created a politically, well-connected board of directors that began lobbying on Africa Online's behalf, protecting it from unfair demands and steering it toward helpful partners.

"What we had to do was educate [government bureaucrats and suppliers]. There are a lot of people around who have power but who are poor—trying to get a piece of the action. But we refused to cut corners," he says. If his group had started handing out bribes, he says, they would never have seen the end of it.

"Perhaps we were a little bit naive in those early days at Africa Online," he chuckles. "We wanted to stick to our guns. We might have been richer quicker, but I am not in the business of short-term advantages. And I have always been able to sleep at night."

"It will be men and women like Makatiani who will create the wealth that pulls Africa into the developed world," Red Herring, the respected technology magazine, wrote last year. "His company is treading where diplomacy has failed, confronting problems that have thwarted powerful international agencies, and slowly progressing toward its goal of creating a single market out of Africa's 800 million people."

Who knows division like Rwandans?

FOR EVERY AFRICAN who goes abroad and returns with professional expertise and grand visions, there are many more who don't come back. According to the International Organization for Migration and the United Nations Economic Commission for Africa, brain drain has been steadily increasing. Between 1960 and 1975 an estimated 27,000 highly qualified African professionals left their home countries. Between 1985 and 1990, the number was up to 60,000—and Africa has been losing an average of 20,000 annually ever since. These figures do not include the sizable number of students who leave to study overseas—and haven't yet decided whether they will ever return.

If they do, many can be quickly defeated. Corruption bankrupts some. Others are knocked down by poverty, entrenched traditions, AIDS, or tribal warfare.

In 1994, Chris Kayomba was a refugee in Uganda, halfheartedly studying journalism, watching dead bodies flow into Lake Victoria, and dreaming of the day he would go home to Rwanda and make a fresh start.

More than 50 of Mr. Kayomba's relatives—brothers, sisters, aunts, cousins—were killed during the 100 days of genocide in Rwanda. In all, 800,000 people died in the ethnic cleansing.

When the genocide finally ended, Kayomba took a taxi back home to Kigali and, just shy of age 30, got together with some friends to try to do something about repairing the country. They started Umuseso, Rwanda's first daily opposition paper. Umuseso, derived from the Kinyarwanda word for "daybreak," was going to be something fresh, they told themselves. In a land scarred by Tutsi and Hutu tribal hatred, their paper was going to offer straight talk about ethnicity and government—"and Manchester United," adds McDowell Kalisa, a senior editor who also moonlights as the British soccer team's Rwanda fan-club director.

The paper would heal, challenge, bring up new ideas. That's what they thought. But in Africa, good intentions, sometimes can take one only so far.

The team became discouraged. As government harassment grew, one fled to the Netherlands, two were jailed, and others left the country. Kayomba clung to his ideals, getting a scholarship for a masters degree in peace and conflict studies at the University of Londonberry in Northern Ireland. Ireland was something else, he recalls. At first he was not sure they were even speaking English—and he's sure many of them had never ever seen a Rwandan. But soon he began to love it, made friends with everyone—Catholics and Protestants alike. He even became a go-between for them.

"Think about it," he says with a grin. "Who has better experience in evils of division than Rwandans? I know what that's like." In class they studied Israel and the disputed Asian territory of Kashmir, and in the evenings they held debates on different ways of resolving ethnic and religious strife.

When Kayomba came back from Ireland, he had big plans: He would write powerful commentaries in Umuseso about postwar reconciliation between tribes; he would advocate for overcoming the lingering animosities in the country without limiting freedoms; he would organize lectures on how other postconflict societies have dealt with their pasts; and maybe he would even run for office.

So far, he has done none of this. He needed to make money first, he admits. He grew tired of the infighting at the paper, the bureaucracy in and around the government, the prohibition of any real talk about ethnicity, and the mild but persistent harassment of anyone saying anything controversial at all. So he went to teach journalism at the University of Butare and do research on democracy for a Dutch nongovernmental organization (NGO).

He's married now, and makes four times as much money working for the NGO than he would working at his old newspaper. Sometimes he even writes a column for the government paper.

"There is no real independent media here," he says, defending his choice. "No one really addresses the issues anyway."

Umuseso is still around, though these days its just Mr. Kalisa and his friend Robert Sebufirira, writing the stories, doing the editing, delivering the newspaper in a van. The focus has changed, too. There are more sports pages and far less talk about ethnicity. It's prohibited by the government—an extreme measure taken, they say, to prevent a repeat of the horrors that were born out of the combination of free speech and simmering ethnic tension that led, in part, to the genocide. In private, critics argue that the newly reelected President Kagame is using these laws to stifle freedoms and actually stirring ethnic divisions by shoving them under the carpet.

Kayomba's colleagues at Umuseso can understand his choices. "Rwanda is a poor country," says Mr. Sebufirira, "and it's hard to remain courageous when you need to make ends meet.... Don't be surprised if you meet someone with good ideas and you come back five years later and they are speaking the opposite."

"If you study or move out, you get new ideas—you are not confined in a certain cycle and as a result, you look at things so differently," explains Kalisa. "The problem is that you find you can't implement those good ideas back home.... Kayomba got so many good ideas [in Ireland]. But when he wanted to exercise them, he found this was not good ground to work on."

To serve the nation, not just the tribe

KENYANS HAVEN'T FACED genocide like their neighbors, but tribalism is no less of a divisive force. Kenyatta's speech last Christmas—and the way he campaigned—was noteworthy for its relative lack of tribalism.

In Africa, the man with the tribe behind him is expected to take care of his people at the expense of everyone else. National pride or unity is not a concept that comes easily to a continent where colonialists unceremoniously split up rivers, mountains, tribes, and families as they divvied up the land among themselves. Tribalism has been the order of the day ever since. Everything, it seems, takes a backseat to ethnicity. Take Kenya's exalted long-distance runners. When a Kenyan wins the New York marathon, the media in Nairobi hail it as a Kalenjin or a Luhya victory—not a win for Kenya.

Kenyatta's father, as president, gave members of the Kikyuyu tribe—Kenya's largest and most influential—a disproportionate share of political and economic power. Afterward, President Daniel arap Moi exploited distrust of the Kikyuyus for his own ends and handed out favors to his tribesmen, the Kalenjin, as well as to other supportive ethnic groups.

But in last year's elections, both Kenyatta and , current President Kibaki—also a Kikyuyu—campaigned on platforms to stop this cycle. The peaceful elections, with voting patterns less ethnically based than before, may be an example of an emerging national spirit that weaken old ethnic cleavages.

"We are not fighting for liberation anymore," says Kenyatta. "Now it's time to rediscover what sort of leadership we want. We need to design and build systems and create institutions that will serve—not just individuals or this generation—but posterity. America has done this, and this is why it is still standing firm after 200 years."

Fighting 'brain drain'

AMERICA WAS HOME for Ernest Darkoh. He had a nice apartment in the New York borough of Brooklyn, he was making good money, his social life was thriving, and his first nephew had just been born.

But something was gnawing at him.

"I could see my life stretching ahead of me in the States," says the 33-year-old American-born son of Ghanaian parents. "I would be ... just an other professional."

It's the end of a long day at work in his stuffy office in Gabarone, Botswana's tiny capital city, and he sways slightly on a swivel chair. "What I wanted to do was follow my heart," he says. "Go somewhere where my input was really needed."

More than 15 million people have died of AIDS in sub-Saharan Africa, and to date, 11 million have been orphaned. In Botswana, 38 percent of adults are HIV-positive and life expectancy has plummeted to below 40 from over 65. By 2010, it could sink to 29, predicts the United Nations Program on HIV/AIDS—a level not seen in developed nations since the Middle Ages.

Outside input here is needed, and Darkoh—with a medical degree and a master's in public health from Harvard, an MEA from Oxford, and a several years' experience working at McKinsey Company in New York—wanted to give it.

One of his projects at the consulting firm was a study, the first of its kind, of the feasibility of launching HIV/AIDS antiretroviral therapy in Botswana. Soon after, he was recruited by Botswana's government to head its AIDS-drug rollout efforts. It is a groundbreaking project into which private US companies and foundations have poured millions.

The program distributes the drugs free of charge to anyone who needs them. It is generally regarded as the developing world's most comprehensive assault on AIDS and a model for fighting the epidemic elsewhere.

64 percent of Nigerians aged 25 and older living in the US (and **43 percent** of all Africans living in the US) have at least a bachelor's degree.

Source: United Nations

Even so, Darkoh hesitated before accepting. "I had certain criteria in my head that needed to be fulfilled," he says. "I wanted to make sure I knew what I was heading into."

He wanted to make sure he could be effective and had a clear mandate, he says, and he wanted to have independence within the public sector. "Because you can really get bogged down by a system and get nothing done," he explains. "Especially in this part of the world."

He's up early every day and spends most of his time in the office. He complains, only half kiddingly, that he would prefer to be more hands-on with patients, but that someone has to do the administrative stuff. Still, he travels in pretty rarefied circles: He met with President Bush during his trip to Africa this summer, as well as Microsoft founder Bill Gates, who has poured some $50 million into the project Darkoh is spearheading.

Darkoh initially had to overcome the perception that he was too young for the job. "I knew that the key to gaining trust was to show that I was sensitive to the politics and that I could deliver results," he says. "I had to work almost 20 hours a day for the first year of the program."

Getting qualified Africans who study abroad to come back to a place where they will make less money, face more frustration, and often not be able to put into practice some of the advanced techniques they learn in Western schools, can be a challenge, say many here.

"Parents pay a lot of money for their children to get the sort of training I did," says Ibou Thior of Senegal, another Harvard graduate who today is director of the Botswana Harvard AIDS Institute. "And the expectation is that not only will you make a difference—you will also make a living." A person returning from study overseas, argues Mr. Thior, needs to be rewarded, not frustrated. "The government needs to provide good working conditions and opportunities so one can apply what has been learned.... Otherwise, you might not want, or be able, to return."

No one challenges the system

AS AN UNDERGRADUATE at Amherst College in western Massachusetts, Kenyatta would set off to see America during weekends or breaks. He loved the freedom. "The best time of my life," he remembers.

Once, he and his roommates took a road trip to Florida. Another time they caught a cheap charter flight to Los Angeles and drove to San Diego, just to see something new. He switched majors several times, in the end settling on a double major of economics and political science. He dated different women, partied late, and audited random classes on slow afternoons. Everyone knew who he was, says an old schoolmate, but no one cared.

When he graduated, he was ready to apply for an MBA. The idea was for him to run the family's vast business empire. That's what was expected of him as the son of one of Africa's big men.

Kenyatta was born in 1961, just as Kenya was shaking off its British colonial masters. (In Swahili, his name literally means "independence" or freedom.") His father, who helped bring Kenya this independence, dominated the political scene for more than 20 years until his death. Almost automatically, power then passed to the elder Kenyatta's deputy, Daniel arap Mof, who proceeded to rule for another two decades.

But Kenyatta didn't get his MBA. He went home and chose not to run the family's vast enterprises—five-star hotels, airlines, banks, and giant farms—that his father had amassed. Instead, public service called. "It was always there, my interest in politics," he protests, defensive against the charges of nepotism and a life of privilege. "But I brought a lot back from the US which really helped me decide. I left Kenya thinking one way. But then I was able to sit back and see it all in context. It was the first time I saw clearly."

"To be sure, study in the West does not automatically bestow perspective, integrity, or a penchant for democratic principles. Zimbabwe's President Robert Mugabe, for example, has six degrees from prestigious Western universities. Few today would consider Mr. Mugabe at the vanguard of democratic reforms.

Some say study overseas can be counterproductive, imbuing ideals that do not suit the real world back home. But many of those interviewed say that overseas exposure made them "global citizens," giving them a perspective that they wouldn't have been able to get without leaving Africa for a time.

Kenyatta was sheltered growing up, he admits today with a lopsided grin. The people around him did not encourage any real challenge to the system.

"Things were gone one way, and that was the only way," he shrugs, resisting a cigarette—he is trying to quit—and smoothing down his smart gray suit. He owns traditional African garb—a colobus monkey skin and hat, and a fly whisk, for example—but they come out only on special occasions. He prefers his designer clothes.

Kenyatta certainly benefited from Kenya's corruption. But unlike many other sons and daughters of privilege across the continent, he claims to want to fix what has gone wrong. He came back "not exactly to make amends," he says, fumbling as he tries to formulate carefully the delicate sentence, "but, well, began seeing there were a lot of things not necessarily right with the order of things in Kenya."

'Are you trying to be white?'

IF YOU ARE an African, says Darkoh in Botswana, and you leave Africa and come back, people more often than not regard you with suspicion.

"They think you have tried too hard to Westernize," he says. "They ask: 'Are you trying to be white?'" New ideas and dynamic people are not welcomed with open arms, he says.

So governments can become filled with the also-rans. "A crisis like HIV/AIDS comes along and everyone looks to the government to address it—but they can't handle it,'' he complains. "Most of the systemic institutional inadequacies we are currently experiencing with HIV/AIDS existed long before the disease came knocking on our door. HIV/AIDS did not create these systemic deficits—it' has simply exacerbated them."

The numbers bear out Darkoh's concerns. According to statistics from the International Organization for Migration, more African scientists and engineers work in the US than in all of Africa. A few years ago, Zambia had 1,600 doctors; now only 400 practice there. More than 21,000 doctors from Nigeria are working in the US. Sixty percent of Ghana's doctors left during the 1980s, placing the healthcare system in critical condition. An estimated 20 percent of skilled South Africans have left the country in the past 10 years, and in Zimbabwe the professional workforce has shrunk by two thirds in just five years.

In order to replace those who have left the continent for greener pastures, Africa spends an estimated $4 billion annually on recruiting some 100,000 skilled expatriates.

The solution, says Darkoh, is for African governments to invest in getting the right people. "Major corporations do not get the results they do by hiring weak talent," he explains. "The right people in the right place at the right time will deliver the right results." It is time for donors and recipient countries to insist on results and institute accountability frameworks, he says. "In the 1980s, development aid was based on cold war needs, but today, it's about accountahility. That, coupled with African leaders realizing that they themselves have to be more responsible ... those are already improvements," he says.

The New Partnership for africa's Development (NEPAD) is seen as part of this shift in approach. Last year, NEPAD was initiated by African governments themselves, whereby they agreed to become more accountable for gold governance in return for billions of dollars in annual investment, aid, and debt reduction from wealthy donor countries.

From 1985 to 1990, **Africa lost 60,000 professionals,** among them doctors, lecturers, and engineers. The continent has been losing **an average of 20,000** every year since.
Source: United Nations

Africa Online's Makatiani, a NEPAD advocate, already sees signs of progress in Kenya. "During Moi's time, corruption was the norm and no one was ever punished for being corrupt. But time has passed and the new government is changing that." In the Nine months since Kibaki came to power, the government collected more in taxes than in any similar period before, and corrupt businessmen, as Makatiani puts it, "are running for their lives." Makatiani says these changes can be found all over the continent. Corruption "is becoming much less acceptable," he says.

Darkoh adds, "Now, the governments need to further shape up and woo back their Diaspora communities, instead of making it hard for them."

He would rather be in Africa, he says, than anywhere else in the world. He just started his own healthcare services company—BroadReach Healthcare—which assists developing countries, donors, and assistance agencies achieve better outcomes on investments made in healthcare, particularly for HIV/AIDS treatment. He is able to make a. meaningful and tangible difference in Africa, he says, which he might not be able to do in the US. He knows others, Africans and Americans born to African parents, who would come back as well—if conditions were right.

"But they worry," he says, about everything from respect and good working conditions to security, healthcare, and civil liberties. "You might want to be a hero," he suggests, "but when you start thinking about actually moving, your mind begins wandering to questions such as whether there's a health clinic to go to when your kid gets an asthma attack in the middle of the night and what your bank account is looking like."

A call for young people to serve

KENYATTA DOESN'T WORRY about his bank account. But 10 months after his concession speech, his pace has not slackened. He can be found in his office until 11 p.m., his crumpled suit jacket tossed over a chair.

He meets daily with NGOs, visits constituencies across the country, works on restructuring his party, and, from the benches of the opposition in the old assembly hall downtown, raises questions on every issue of the day—from constitutional reform to anticorruption legislation. He embraces the democratic principle of the "loyal opposition" in a country that has never really allowed such a thing.

"I get fed up a lot," admits Kenyatta. "Most of us do."

But, he stresses, the problem is that most young Africans assume leadership is a game of others, and not about them. "You tend to lose the best minds and best assets because young people don't want to engage in the rough and tumble. But that is the wrong mentality. You need to engage," he says.

Makatiani says the politics of Africa are going into "Phase 2": The older generation of leaders were revolutionaries, freedom fighters like Kenyatta's father, accustomed to taking big leaps and getting things fast, he explains. "But the new generation like myself is more realistic and is ready to take smaller steps," he says. "We are ready to work hard for incremental, but real, success."

Kenyatta agrees, and says that he embodies that shift. "I believed, as a child, it was the right of others to be there and set up the rules of the game—and neither I nor anyone else could challenge that. But, you know, with more exposure you begin to think more: Actually, I can do it, too—and differently."

Empty Fields

In Africa, AIDS and Famine Now Go Hand in Hand

As Farmers Die in Swaziland, Their Plots Lie Fallow; Bush Visits the Continent

Five Orphans in a Mud Shack

By Roger Thurow

MAPHATSINDVUKU, Swaziland—Their father died in 1999, their mother in 2000, both of them from what social workers and village officials believe were complications from AIDS. Since then, Makhosazane Nkhambule, now 16 years old, has been caring for her four younger brothers and sisters in their one-room mud-brick shack.

They sweep the floor of the house and the dirt yard with homemade straw brooms. They try to patch holes in the thatched roof and plug cracks in the mud walls, They fetch water from a well nearly a mile away. They scavenge wood for the fire. They go to an informal school in a neighbor's house.

Makhosazane says they can do everything they need to do, except feed themselves. "I would like to plant corn and vegetables, but we have no money to buy seed's or tools," she says. Her parents' cattle could have helped with plowing, but they have also died. The garden beside the hut and the two-acre field behind it haven't been planted since their mother died.

For two years, the orphans scrounged what they could, asking neighbors for scraps of food and waiting for relatives in distant villages to bring something to eat. Last year, the United Nations' World Food Program came to Swaziland to distribute food to those suffering from the drought that has gripped southern Africa. Although the Nkhambule children had no crops to be killed by drought, they began receiving the food aid. So, too, did thousands of other households where the adults who had been tending the fields have died. Most of the victims likely died of HIV/AIDS, which, according to government estimates, infects more than one-third of adults in this tiny, hilly kingdom.

The Nkhambule siblings, barefoot and wearing dirty, shabby clothes, embody what is being called an entirely new variety of famine. It breaks the historical mold of food crises, according to people who are studying it. It isn't caused by weather, war, failed government policy or crop disease, all of which prevent or discourage farmers from bringing in a harvest. Rather, this is a food shortage caused by a disease that kills the farmers themselves. Recovery won't come with weather improvement, new government policies, a peace treaty or improved hybrid crops. Once the farmers die, there is no rain that will make their empty fields grow.

Across southern Africa, the region of the world hardest hit by AIDS, some seven million farmers have died from the epidemic, according to estimates of governments and relief agencies. This has left many families with no means or experience to do the farming. The continuing AIDS crisis threatens to create chronic food shortages and leave large populations "reliant for their survival on a long-term program of international social welfare," says Alex de Waal, an official with the U.N. commission on AIDS and governance in Africa.

President Bush yesterday began a five-day trip to Africa, during which he will confront the continent's AIDS abyss. In South Africa, Botswana and Uganda, he will see the human tragedies and the wider drag on national development, as well as local and international efforts to control the epidemic. He will also promote his own $15 billion initiative to fund various AIDS-treatment programs and prevention strategies in Africa and the Caribbean.

What is emerging in Swaziland has researchers ratcheting up the costs of AIDS. While nearby Botswana and South Africa have great mineral wealth and considerable industrial development, Swaziland is a largely agrarian country. In places such as this, AIDS and food shortages are combining to unravel societies and destabilize populations in new ways. "A drought is usually in certain areas of a country, but AIDS is all over. It is an unbelievable impoverishment agent," says Derek von Wissell, the national director of Swaziland's Emergency Response Council on HIV/AIDS.

He notes that the percentage of women at prenatal clinics diagnosed with HIV skyrocketed to more than 38% in 2002, from just under 4% in 1992. Only Botswana has a higher rate. Over the same period, per capita agricultural production fell by a third, even before the drought took hold last year. The government reports a 54% drop in agricultural production in households where at least one adult member died because of AIDS or other reasons.

In traditional famine, the first to die are usually the weakest, particularly children and the elderly. In the new variety, AIDS strikes at adults in the prime of life, leaving the children and elderly to cope. In Swaziland, the government estimates that more than 15% of children under the age of 15 are orphaned. By 2010, the country forecasts the number of orphans could increase to 120,000, or 12% of the entire population of one million. International aid organizations say that 10% of the nation's households are headed by children, and even more are headed by grandparents too old and weak to work the fields.

Mr. von Wissell says aid workers have found children who haven't had any adult contact for months. He tells of encountering four siblings, led by an 8-year-old, walking naked down a road. Their mother had died, and they had walked 15 miles in search of their grandmother. Mr. von Wissell helped them look and then took them back to their home village, where they were put under the care of the local chief. "We never did find the grandmother," he says.

To help feed and nurture the orphans, the national AIDS council is trying to revive old Swazi social structures. Last year, it persuaded about half of the roughly 350 chiefdoms to set aside a few acres of arable land each to grow food communally for the children. And it is establishing village social centers where the orphans can be fed and have more contact with adults and other children.

In Swaziland, as elsewhere in Africa, the HIV virus has been spread by populations that move frequently between rural homes and jobs in cities, on plantations or in mines. Poor health networks hinder diagnosis, and a scarcity of affordable drugs impedes treatment. Initial efforts at prevention often failed because of ignorance about the disease and how it is spread, as well as a powerful taboo about acknowledging its presence.

The new variety of famine is forcing international relief organizations to retool their strategies. Food relief is usually considered to be temporary aid, until a country recovers from its shortages. But with AIDS, and the prospect of no quick recovery, food relief becomes long-term care. As the AIDS crisis spreads, food shortages increase. As food shortages increase, so does malnutrition, which makes people more susceptible to diseases stemming from a weakened immune system. Although the drought in southern Africa has eased somewhat since last year, the U.N.'s World Food Program has recently launched another appeal to help feed the region. It is asking donor countries and organizations for $308 million to buy close to 540,000 tons of food, enough to feed 6.5 million people until next June.

In Swaziland, "the drought combined with illness has pushed people over the edge," says Sarah Laughton, the WFP's emergency coordinator in the country. "Even if the rains come, they won't recover."

At a WFP food-distribution center in the southern village of Ngologolweni, women and children gather under a big Mopani tree in a schoolyard to wait for their rations. They are "widows, women abandoned by their men, orphans—all those who can't plant, who have no resources," says Dudu Ndlangamandla, a member of the local relief committee.

Bags of food from around the world are lined up in neat rows: rice from Algeria, peas from Japan, a corn-soy blend from the U.S. A 14-year-old boy walks over from the school to register for his rations; he lives with his 17-year-old sister. An older woman with her left big toe sticking out through a hole in her tennis shoe also comes down from the school and joins the line. She helps to cook for the schoolchildren, she says, but she can't grow enough to feed her own family of six children since her husband died. She says he was "poisoned." Despite the high prevalence of HIV here, it is rare for someone to admit having it, or for people to acknowledge that relatives died of AIDS. Instead, they use various euphemisms.

At a small community center close to the capital of Mbabane, a kettle of porridge cooks on an outside fire, and a big pot of vegetable soup heats up on a stove inside. This is one of a network of neighborhood locations that offer a warm meal to orphans every afternoon after school, with support from U.N. agencies. "We had been seeing the children scavenging for food in trash cans, and we said we needed to give them a meal to eat," says Janet Aphane, one of 20 local women—most of them retired teachers, nurses and civil servants—who prepare the food.

They began a year ago with 30 children a day and are now up to 80. "We have to turn away a lot of children, because our resources limit us," she says. "About 500 children would come if we threw open the doors."

The orphaned children also keep flocking to 73-year-old Mandathane Ndzima in the rural village of Mpathni. They are her grandchildren, 20 in all, who come under her care as her children die one by one. Her youngest son died first, of tuberculosis, a common illness of those weakened by AIDS. Then her oldest son died in a car accident. Then the middle son died, and his wife, too, of tuberculosis. In total, they left 12 children behind. Now a fourth child, a 40-year-old daughter also suffering from tuberculosis, has returned home with her eight children.

"When my sons were alive, we had enough to eat," Mrs. Ndzima says. They filled the family's plot of several acres with corn and sweet potatoes. After the sons died, the family's few oxen, which pulled the plow, were stolen. This past year, the grandmother and her grandchildren planted only a small portion, without fertilizer, and most of that was lost to the drought. Tall grass grows over the rest of the field. Mrs. Ndzima hopes to cut some and bundle it up to sell as thatch. "If I get enough, maybe I can pay someone to assist us in plowing," she says.

Mrs. Ndzima and her daughter, Moana, sit on a couple of cinder blocks in the middle of their cluster of little houses. Behind them are two cylindrical grain storage bins. At this time of the year, right after the harvest, when her sons were alive and farming, the metal bins would be full to the brim, Mrs. Ndzima says. Now they are empty.

So every month, the grandmother and her grandchildren carry home their ration of food aid: 165 pounds of corn, 11 pounds of beans, a gallon of cooking oil and 11 pounds of the corn-soy blend. In order to get the grandchildren through the month, Mrs. Ndzima says she and her daughter have cut back to two meals a day.

In Mandela's shadow

Thabo Mbeki has great ambitions for the future of his country. But his fiercest critic is not to be shaken off

LAST January, when the African National Congress (ANC) celebrated its 90th anniversary, Thabo Mbeki, South Africa's president, made a long speech. To general astonishment, he made no mention of the ANC's most famous figure, Nelson Mandela. The feeling is mutual. When Mr Mbeki entered parliament at the state opening in February, all stood to respectful attention except Mr Mandela, who sat stock-still in his chair despite gentle tugs from his wife.

At 84, Mr Mandela has lost none of his charisma. On his visits to dusty townships, people stand for hours waiting for him and scream with joy when he appears. In parliament, MPs leap to their feet, sway, dance and sing in adoration. At dinners and private discussions, even close friends flutter with affection and respect. As he takes an armchair at a lunch in Johannesburg a roomful of businessmen, lawyers and family friends sink to their knees and form an attentive circle at his feet. He calls Queen Elizabeth by her first name ("well, she calls me Nelson"), and goes hand-in-hand with Bill Clinton, beaming fo r the cameras. Mr Mbeki, by contrast, can be painfully shy. He admits he is an awkward showman, fondest of his own company. He has solitary interests: reading, the Internet, computer chess, landscape photography, poetry. On presidential tours, known as *imbizos*, he is stiff and formal. With journalists he is often impatient, bristling at ignorant questions. Mr Mandela is widely known by an affectionate nickname, Madiba. Mr Mbeki has none.

It is an open secret that another man, Cyril Ramaphosa, was Mr Mandela's first choice as his successor. But Mr Ramaphosa was out-manoeuvred in the early 1990s, and by 1999 Mr Mbeki was president. Were it not for Mr Mandela, he would be riding high and virtually unopposed. At the ANC 's five-yearly conference, held next week in the vineyard town of Stellenbosch, Mr Mbeki will be picked again as party leader; his close allies are almost certain to romp to all the high party positions; and victory is more or less assured at the general election in 2004.

His party dominates politics, holding two-thirds of the seats in parliament against a divided and feeble opposition. And his country is becoming more and more of a presence in the continent. Mr Mbeki presides over the African Union, Africa's putative answer to the EU , and is also the brains behind the New Partnership for Africa's Development (Nepad), an ambitious plan to attract more capital investment. Last year (at Mr Mandela's request) he sent his deputy-president, Jacob Zuma, and a battalion of soldiers to help keep the peace in Burundi, and he has offered 1,500 more soldiers as UN peacekeepers for eastern Congo. Under Mr Mandela, South Africa shed its isolation in Africa; under Mr Mbeki, it is actively engaged.

Causes for complaint

Yet all is not well within Mr Mbeki's administration, not least because Mr Mandela has let it be widely known that he is unhappy with it. The old man, *tata*, as Mr Mbeki sometimes calls him dismissively, has started to meddle.

Some of his recent unhappiness stems from local politics. In the ANC's heartland of the Eastern Cape, where Mr Mbeki and Mr Mandela were both born, the ANC provincial government is in such disarray that Mr Mbeki has had to send people from national government to sort it out. This has caused enormous discontent within the party. In KwaZulu-Natal, a truce between the ANC and the Zulu nationalist Inkatha Freedom Party, brokered under Mr Mandela but largely negotiated by Mr Mbeki, is collapsing as the two parties fight for local control.

Yet most of the animosity comes from Mr Mbeki's handling of South Africa's AIDS crisis, which Mr Mandela believes has been badly bungled. The disease has already killed hundreds of thousands of South Africans and is set to claim the lives of at least 4.5m more, over 11% of the population. Already, 300,000 households are headed by orphaned children. Unsurprisingly, a recent survey showed that 96% of South Africans consider the disease to be a "very big" problem for the country.

AIDS is already striking hard at the professions, notably teachers and nurses. Some analysts worry that the disease has weakened the capacity of the army (with an infection rate of well over 23%) and the police. Life expectancy is slumping as child mortality rises. Ill-health is also entrenching poverty: for a middle-income country, surprisingly large numbers of people report being short of food.

A year ago, Mr Mandela warned that leaders and their wives must do more to fight AIDS. That was an explicit reference to Mr Mbeki's inactivity, though also an admission of his own negligence as president between 1994 and 1999. All year Mr Mandela has raised the profile of AIDS, drawing a stark contrast with Mr Mbeki's wriggles and denials.

The national government and some provinces have done a lot to boost primary health care, train nurses, fund education-and-prevention campaigns and give out more free condoms. The government also helps to fund research into a vaccine and has opened 18 pilot sites to test the effectiveness of anti-retroviral drugs, which are commonly used in rich countries to keep those infected with HIV healthy. But the president himself has often frustrated these efforts.

Mr Mbeki questions figures that show the epidemic has taken hold in South Africa. He argues that anti-AIDS drugs may be more dangerous than AIDS itself. He refuses to single out AIDS as a special threat, preferring to talk of general "diseases of poverty", and will rarely speak about it publicly. Peter Piot, the head of UNAIDS, was once brought secretly to meet Mr Mbeki in Cape Town, in an effort to persuade him that the human immuno-deficiency virus was the cause of AIDS. The two men sat late into the night drinking whisky and fruitlessly arguing the point.

Mr Mbeki also refuses to encourage people to know their HIV status and to lessen stigma around the disease. He will not take a public AIDS test, and has only once been pictured holding an infected child. Taking their lead from the president, no members of his government and very few MPs, civil servants, or public figures of any sort admit the obvious when their colleagues die of the disease.

Mr Mbeki himself seems personally affronted by the attention given to AIDS. He resents any prejudice against Africans as diseased, in part because apartheid scientists tried to create and spread viruses that would kill only black Africans. In the early 1990s the then ruling National Party even alleged that ANC leaders returning from exile in other parts of Africa (such as Mr Mbeki himself) were bringing AIDS into the country.

Appalled by Mr Mbeki's obtuseness on the subject, Mr Mandela has been trying to force him to do more about it. He openly backs the Treatment Action Campaign (TAC), one of the most aggressive and effective opposition movements in the country. Earlier this year the TAC sued the government in the Constitutional Court and forced it to set up a national programme to give anti-AIDS drugs to infected pregnant women. The group's leader, Zackie Achmat, who himself has AIDS, refuses to use anti-retroviral drugs until the government makes them widely available. Madiba has visited and hugged Mr Achmat, and promised to lobby the president on his behalf. Last week the two men again appeared together, to launch a privately-funded plan to get anti-AIDS drugs to thousands more of the poor.

In September at Orange Farm, a township just south of Johannesburg, Mr Mandela launched a particularly sharp attack. He demanded general provision of anti-retroviral drugs by the government, pushing aside worries about cost, toxicity and capitalist imperialism. Then he ordered two of his grandchildren to join him on stage, wagged his fingers at them, and told the crowd:

> I have 29 grandchildren and six great-grandchildren. They are very naughty. They tell me I have lost power and influence, that I am a has-been. They tell me to sit down. That I must stop pretending I am still the president. Now, you have heard all these important people here today, you have heard what they say about me. So, now you must stop telling me to sit down!

That message was aimed at Mr Mbeki.

Earlier this month, Mr Mandela launched an independent report on AIDS that concluded: "We must manage this disease or it will manage us." After perfunctory praise for government research, he said again that too little is being done, officially, to fight AIDS and the stigma that surrounds it. "I have often said to the government there is a perception that we don't care that thousands are dying of this disease. If you look at the pages of the *Sowetan*, at the death notices, it is clear that our young people are dying."

While many private companies—among them AngloAmerican, Coca-Cola and De Beers—have decided that it is cost-effective to provide free anti-AIDS drugs to infected workers, the government is reluctant to make similar promises to other South Africans. Mr Mandela wants such drugs to be universally available, and ordinary South Africans agree. A recent survey of nearly 10,000 people revealed that over 95% want the government to provide anti-AIDS drugs now.

Not Madiba alone

Mr Mbeki often refuses phone calls from his predecessor. When he does, Mr Mandela is told that he misunderstands the complexity of AIDS. The old man has been asked by senior ANC leaders to stop undermining the government, and to tell the press that the only problem with official AIDS policy is one of communication.

Will his interventions in fact have any effect? Possibly, since he is not a lone voice. Many heads of state and former presidents have privately tried to persuade Mr Mbeki to do more on AIDS, if only for the sake of his own reputation. Mr Clinton, who was beside Mr Mandela at Orange Farm, confided his view of the president's position on AIDS: "In his own mind, he thinks he has moved a long way." The challenge now is to persuade him to do as much as leaders in poorer African countries.

In Nigeria, said Mr Clinton, politicians talk openly about the disease. "It has kept its [infection] rate low because there was no organised, systematic and official denial of the problem there. In South Africa I see systemic obstacles still, to doing what has worked elsewhere. Leaving people to their own devices is just not good enough." In particular, Mr Clinton criticises the failure to make mass use of anti-retrovirals, drugs that are already widely used in Brazil and India.

In September, Mr Piot protested that the World Summit on Sustainable Development, held that month in Johannesburg, paid almost no attention to AIDS. Delegates conceded that the disease was given a low priority to avoid embarrassing the host. Mr Piot and Carol Bellamy, the head of UNICEF, have both complained that Mr Mbeki's Nepad plan for African recovery makes no substantial mention of the fight against AIDS.

Mr Mbeki now seems to be slightly shifting his position, if only to persuade Madiba, Bishop Desmond Tutu and others to stop "backseat driving". In April the government agreed to set out a plan for provincial governments to give nevirapine, an anti-retroviral drug, to infected pregnant women and rape victims. Mr Mbeki's chief spokesman, Joel Netshitenzhe, even suggested that factories might be built to produce generic anti-AIDS drugs. And the government has been talking to firms and health activists about a national treatment plan for the disease.

In the past few months Mr Mbeki has kept silent on the subject, while trying to cultivate a more friendly persona (he has taken up golf so that he can bond with George Bush, who is expected to pay a visit next month). But South Africa's president will not be pushed into a position he opposes, especially not by westerners, the media, the UN or activists.

Ultimately, then, the most effective pressure will come from within his own party. Next week's ANC meeting in Stellenbosch is a chance for members to push Mr Mbeki to act decisively. And one member, the only world-famous one, will go on pushing hardest of all, no matter how often his calls are not returned.

WORLD IN REVIEW

Disconnected Continent

The Difficulties of the Internet in Africa

MAGDA KOWALCZYKOWSKI, Staff Writer, *Harvard International Review*

While the Internet has become an integral part of the Western world, it has only just arrived in Africa. In the United States and northern Europe, an average of one out of three people uses the Internet. In Africa, as in much of the developing world, Internet usage rates below one percent are the norm. Fortunately, this situation is changing rapidly, as initiatives by the United Nations and private corporations attempt to bridge the global digital divide.

According to official statistics, the future of connectivity in African countries is bright. The number of telephone lines is growing at a rate of 10 percent per year, and all of the main lines in Botswana and Rwanda are digital, compared to just under half of all lines in the United States. Cellular phone service, limited to six African countries a decade ago, is available in 42 countries today. Columbia Technology's Africa-One project is expected to complete an optical fiber network for the entire continent this year at a cost of US$1.6 billion.

Support from other private corporations looks promising as well. In cooperation with the Harvard Center for International Development, commercial technology giants such as Sun Microsystems, AOL-Time Warner, and Hewlett-Packard have pledged US$10 million over the next two years toward technology designed to improve the quality of life in 12 developing nations. With figures like these, it is easy to gloss over the real problems in the implementation of this vast network.

WorldTeacher: Namibia

In July 2001, I stepped into a fully equipped computer lab at the teacher resource center in Ongwediva, northern Namibia, beginning my part in a project to help spread computer and Internet literacy to the developing world as a WorldTeach volunteer. With video cameras, CD burners, and high-speed connections, the lab could easily have been in the United States. Although this teacher resource center in Ongwediva is state-of-the-art, it must serve the technological needs of all of northern Namibia. Because of large distances between towns, a lack of vehicles, and limited awareness, many teachers do not exploit this resource to the fullest extent possible.

WorldTeach is designed to eliminate these problems. The 16-teacher contingency traveled to Namibia to teach computer and Internet literacy to students at various primary and secondary schools throughout the country An experiment of sorts, it was the first time that a program of this magnitude had been implemented in the developing world. A partnership with Schoolnet, an England-based company that equips schools with computers and technical support, made the project possible: Schoolnet provided the network, WorldTeach provided the teachers. After a period of training, each WorldTeacher was sent to a Schoolnet-sponsored site where he or she gave lessons on computer and Internet use.

Various development reports and meetings in the past few years have pointed to the explosive potential of information technology in Africa. Statistics on the increasing numbers of telephone lines, Internet Service Providers, and Internet connections have shed light on the vast potential of expanding markets and opportunities in this sector. Expectations are not reality, however. From my viewpoint on the ground in Namibia, I found the actual situation to be much more complex than the large companies and idealistic groups wishing to help these countries would like to pretend.

The enormous project at hand faces many glitches that must be smoothed out before the Internet can become a regular and integral part of the Namibian culture. One difficulty has been financial. Although the first school I visited in Edundja had a special budget allotted for the use of the Internet, it simply was not enough to sustain the program after I had left. I was there five days a week for three weeks, leaving the dial-up connections on all day in the hopes of getting as many students as possible to explore cyberspace. I taught the eighth, ninth, and tenth graders how to open email accounts, how to search the web, and even how to make their own website. The phone bill amassed during this time proved too hefty for the administration to allow Internet use to continue after my departure. At my second school in Uis, I made sure to use the Internet more sparingly. Even so, this school was poorer than the first and the Internet did not thrive there either.

Another issue tied closely to the monetary situation is technical support and resources. Both of the schools in which I taught had many technical problems with the computers. With problems ranging from missing files to insufficient memory, the computers needed to be maintained and fixed on a regular basis, an expensive and time-consuming proposition. When students and teachers have not had a chance to familiarize themselves with computers, they can potentially damage them. Some problems are relatively small and can be fixed quickly and easily, but when the caretakers responsible for the computers do not understand even the basics of troubleshooting, salvageable computers are left to gather dust. Because the teachers and students do not want to break the computers, they are often reluctant to use them without supervision from an expert. At both schools, the teachers were initially very nervous about touching the keyboard. They wanted and needed someone to help them navigate through this technological endeavor.

Once the teachers and students overcame their initial hesitations, they were very enthusiastic about the possibilities of how the Internet could serve them. I attended the students' other classes to see what was being taught, and I supplemented their lessons with extra information and media from the Internet. The students were enthralled by an animation of blood flow in the heart after learning about blood cells in their biology class. The teachers were excited to find lesson plans and activity ideas online and wanted to learn more about this teaching tool.

Barriers to Change

Despite the initial interest of the teachers and students, the situation soon returned to the status quo. The allotted time of three weeks spent at each school was just not enough to change everyone's thinking and integrate the Internet into the education of Namibian students. That transformation would have required many more resources and a concerted effort over a longer period of time. Admitting the Internet into their community entails a recognition of the outside world. The Internet is an innovation of global proportions, but to introduce the ideas behind it to someone who has never worked on a computer before is quite a formidable task.

Even if the students were to understand the power and potential of the Internet, it would not do them much good in the short run. In the rural areas, socio-economic problems prevented students from making the most of their education. Only the rich can attend the University of Namibia or the Namibian Polytechnic University. Many young girls drop out of school due to teenage pregnancy. Parents often want their children at home, tending to the oxen and harvesting the mahangu, a crop used to make the porridge Namibians eat daily. When even the best students end up working as cashiers in the local markets after graduation, the benefits of a good education can be hard to see. Students cannot understand the link between studying hard and a bright future.

Political shortcomings also obstruct the integration of the Internet into society. The Namibian government does not have computers for all of its offices and workers. It is difficult to see how a government can fully embrace the technological movement when it does not understand how to use the Internet in its own affairs. The few schools with computers obtain them from nongovernmental organizations. Without a push from the government for these types of programs, technology cannot be fully integrated into schools and communities. The government is responsible for phone lines, roads, and the other infrastructure vital to the fledgling Internet movement. It alone has the power to break up the monopoly of Telecom, the only communications company in the country, which would make dial-up Internet access much more affordable. In a place where the public treasury is a precious resource that must be parceled out based on national priorities, the Internet takes a back seat to more immediately pressing matters such as controlling the AIDS epidemic.

The programs that are implemented at the local level must be thoroughly researched and carefully planned while leaving enough flexibility to adapt to special conditions. The 2001 WorldTeach program was structured for breadth rather than depth in its first year. To make sure as many schools as possible were connected during one short summer, each teacher covered two schools in a six-week period. The schools spanned the geographic and demographic spectrum from urban areas where students were aware of the Internet and possessed basic computer skills to the deepest rural parts of Namibia where students had never seen a computer. Because each school was different, each WorldTeacher had to adapt lessons to the environment and resources that were available. My first school at Edundja in the north had a lab of 23 computers that students used on a weekly basis. It required a completely different teaching approach than my second school in Uis, which had three computers, only one of which could access the Internet.

Another factor that should have been considered in the implementation of the program was the different schedules of the schools. Toward the end of the summer, the students were studying for their trimester exams. Trying to fit another class into their schedules was a daunting task, especially given only three weeks at each school. I worked closely with the principals and maximized my time at each school by staying after class and working with the teachers in small groups. The age of students was another factor that affected teaching styles. WorldTeachers placed in primary schools could not teach things the same way as WorldTeachers placed in secondary schools taught them. Even in secondary schools, ages ranged from 13 to 21.

Financing the Future

Because money is the single most important factor in bringing the Internet to the African continent, securing funding will be vital to developing technology there. Already, advances are being made at an extraordinary rate. The Africa Bureau of the United Nations Development Programme has already agreed to a US$6 million fund to improve Internet connectivity in Africa with a project called the Internet Initiative for Africa. The United Nations has also announced the beginning of a US$11.5 million program called Harnessing Information Technology for Development, which provides funding for various information and communications technology (ICT) projects throughout Africa.

As a result of this push for ICT, the number of Internet users in Africa will soon rise to four million, according to Mike Jensen, an independent Internet development researcher. With Eritrea finally getting permanent Internet connectivity in 2001, every country on the continent has some degree of Internet access. In 1996, only 11 countries were connected. I personally saw how Internet use grew more common as I moved from rural towns to the Namibian capital of Windhoek. Benefiting from the influx of capital and technology investment by the United Nations, Namibia is poised to make good on the promise of Internet growth.

Jensen's report on African technological advances also discusses the launching of another program to spur technological growth. The United Nations Economic, Social and Cultural Organization (UNESCO) has recently established the Creating Learning Networks for African Teachers project to connect teachers to the Internet and to assist teaching colleges in using ICTs in literacy and other education programs. The project has already been implemented in Zimbabwe, is being initiated in Senegal, and will eventually reach 20 countries with further outside support.

Along with the US$10 million donation to digital-divide initiatives, the private sector is contributing to the technological growth of Africa in many capacities. Hewlett-Packard has begun a global technology outreach program that has spurred the economic growth of countries such as Senegal and Ghana. Its new project, dubbed "HP e-inclusion," is expanding the possibilities for Internet technology in these countries. Other ICT companies, if not directly involved, are advising countries on how to best integrate technology into their development processes. Academics and executives from top ICT companies met with South African President Thabo Mbeki in October 2001 to begin this process. This was the first meeting of Mbeki's International Task Force on Information, Society, and Development, which was set up to advise him on how South Africa can develop its ICT capacity to promote investment, economic growth, and job creation. "The private sector must respond to the goals, objectives, programs, and mission as defined by the government and its people," said Hewlett-Packard CEO Carly Fiorina, who attended the conference. "In responding to the programs we see an opportunity not only to do well, but also an opportunity to do good."

This historic meeting in South Africa has opened the way for the rest of the continent. With more companies looking to diversify into new markets, Africa is looking attractive to investors. As nations that gained independence in recent decades have grown into fledgling democracies, they have solidified property rights laws and done much to improve the business environment. While the overall outlook may appear positive, many issues could thwart the development of this technological revolution. At the level of ideas, concepts, and statistics, the Internet enjoys enthusiastic support, but it is the practical matters that will determine the pace of technological advancement.

Medicine Without Doctors

In Africa, just 2 percent of people with AIDS get the treatment they need. But drugs are cheap, access to them is improving and a new grass-roots effort gives reason to hope.

By Geoffrey Cowley

THE FIRST PART OF NOZUKO MAVUKA'S story is nothing unusual in sub-Saharan Africa. A young woman comes down with aches and diarrhea, and her strong limbs wither into twigs. As she grows too weak to gather firewood for her family, she makes her way to a provincial hospital, where she is promptly diagnosed with tuberculosis and AIDS. Six weeks of treatment will cure the TB, a medical officer explains, but there is little to be done for her HIV infection. It is destroying her immune system and will soon take her life. Mavuka becomes a pariah as word of her condition gets around the community. Reviled by her parents and ridiculed by her neighbors, she flees with her children to a shack in the weeds beyond the village, where she settles down to die.

In the usual version of this tragedy, the young mother perishes at 35, leaving her kids to beg or steal. But Mavuka's story doesn't end that way. While waiting to die last year, she started visiting a two-room clinic in Mpoza, a scruffy village near her home in South Africa's rural Eastern Cape. Health activists were setting up support groups for HIV-positive villagers, and Medecins sans Frontieres (also known as MSF or Doctors Without Borders) was spearheading a plan to bring lifesaving AIDS drugs to a dozen villages around the impoverished Lusikisiki district. Mavuka could hardly swallow water by the time she got her first dose of anti-HIV medicine in late January. But when I met her at the same clinic in May, I couldn't tell she had ever been sick. The clinic itself felt more like a social club than a medical facility. Patients from the surrounding hills had packed the place for an afternoon meeting, and their spirits and voices were soaring. As they stomped and clapped and sang about hope and survival, Mavuka thumbed through her treatment diary to show me how faithfully she'd taken the medicine and how much it had done for her. Her weight had shot from 104 pounds to 124, and her energy was high. "I feel strong," she said, eyes beaming. "I can fetch water, wash clothes—everything. My sons are glad to see me well again. My parents no longer shun me. I would like to find a job."

It would be rash to call Nozuko Mavuka the new face of AIDS in Africa. The disease killed more than 2 million people on the continent last year, and it could kill 20 million more by the end of the decade. The treatments that have made HIV survivable in wealthier parts of the world still reach fewer than 2 percent of the Africans who need them. Yet mass salvation is no longer a fool's dream. The cost of antiretroviral (ARV) drugs has fallen by 98 percent in the past few years, with the result that a life can be saved for less than a dollar a day. The Bush administration and the Geneva-based Global Fund to Fight AIDS, TB and Malaria are financing large international treatment initiatives, and the World Health Organization is orchestrating a global effort to get 3 million people onto ARVs by the end of 2005—an ambition on the scale of smallpox eradication. What will it take to make this hope a reality? Raising more money and buying more drugs are only first steps. The greater challenge is to mobilize millions of people to seek out testing and treatment, and to build health systems capable of delivering it. Those systems don't exist at the moment, and they won't be built in a year. But as I discovered on a recent journey through southern Africa, there's more than one way to get medicine to people who need it. This crisis may require a whole new approach—a grass-roots effort led not by doctors in high-tech hospitals but by nurses and peasants on bicycles.

Until recently, mainstream health experts despaired at the thought of treating AIDS in Africa. The drugs seemed too costly, the regimens too hard to manage. Unlike meningitis or malaria, which can be cured with a short course of strong medicine, HIV stays with you. A three-drug cocktail can suppress the virus and protect the immune system—but only if you take the medicine on schedule, every day, for life. Used haphazardly, the drugs foster less treatable strains of HIV, which can then spread. Strict adherence is a challenge even in rich countries, the experts reasoned, and it might prove impossible in poor ones. In light of the dangers, prevention seemed a more appropriate strategy.

Caregivers working on the front lines resented the idea that anyone should die for having the wrong address. So they set out to prove that treatment could work in tough settings, and by 2001 they'd succeeded. In a project led by Dr. Paul Farmer of Harvard, two physicians and a small army of community outreach workers introduced ARVs into 60 villages near the Haitian town of Cange. Around the same time, MSF teamed up with South Africa's Treatment Action Campaign to make the drugs available in an urban slum called Khayelitsha. The upstarts simplified the drug regimens and dialed back on lab tests, and most of the patients were monitored by nurses or outreach workers instead of physicians. But none of this made treatment less effective. The cocktails worked as well in the slums as they did in San Francisco—and the patients were often *more* steadfast than Americans about taking their pills. The obstacle to treatment was not a lack of infrastructure, the activists proclaimed. It was a lack of political will.

The climate has changed since then. Yesterday's unacceptable risk is today's moral imperative, and the world's highest-ranking health authorities are pushing hard to realize it. "We still believe in prevention," the WHO's director-general, Dr. Jong-wook Lee, told me during an interview in Geneva this spring. "But 25 million HIV-positive Africans are facing certain death. If we fail to help them, it can't be because we didn't try." Since Lee took office last year, staffers in the agency's HIV/AIDS department have worked at a furious pace to devise a global treatment strategy and help besieged countries design programs that the Global Fund will pay for. Proposals are rolling in, and the fund is responding favorably. Grants approved so far could finance treatment for 1.6 million people over the next five years.

THE TROUBLE IS, FEW OF THE COUNTRIES winning those grants are ready to absorb them. Their health systems have withered under austerity plans imposed by foreign creditors. Doctors and nurses have left in droves to take private-sector jobs or work in wealthier countries. And those left behind are overwhelmed and exhausted. While traveling in Zambia, I visited Lusaka's University Teaching Hospital, the 1,600-bed facility at the forefront of the country's two-year-old treatment program. Dr. Peter Mwaba, the hospital's stout, vigorous chief of medicine, detailed the country's strategy for treating 100,000 people (50 times the current number) by the end of next year. Yet his own facility was half abandoned. In 1990 the hospital had 42 nurses for every shift. Today it has 24—and the patients are sicker. "I've been here for 30 years," Violet Nsemiwe, the hospital's grandmotherly head nurse, confided as we walked the dim corridors. "It has never been this bad."

In an ideal world, the clock would stop while countries in this predicament trained tens of thousands of health professionals, quintupled their salaries and dispatched them to underserved areas. But the clock is ticking at a rate of 56,000 deaths a week, so the WHO is embracing a different approach—one rooted in the populism of Cange and Khayelitsha. "AIDS care, as we practice it in the North, is about elite specialists using costly tests to monitor individual patients," says Dr. Charles Gilks, the English physician coordinating the WHO's "3 by 5" treatment initiative. "I've done that and it's great. But it's irrelevant in a place like Uganda, where there is one physician for every 18,000 people and that physician is busy at the moment. If we're going to make a difference in Africa, we've got to simplify the regimens and expand the pool of people who can administer them."

That's precisely the agenda that activists are pursuing in Lusikisiki, the remote South African district where Nozuko Mavuka got her life back. When MSF and the Treatment Action Campaign launched their project there last year, the local hospital was performing the occasional HIV test but had little to offer people who were positive—a population that includes 30 percent of pregnant women. Lusikisiki is the poorest part of the poorest province in South Africa, but the activists used what they found—a struggling hospital and a dozen small day clinics—to start a movement. A small team led by Dr. Hermann Reuter, a veteran of the Khayelitsha project, set up a voluntary testing center at each site, organized support groups for positive people and emboldened them to stand up to stigma. Before long, people like Mavuka were donning HIV-POSITIVE T shirts, singing about the virtue of condoms and quizzing each other on the difference between a nucleoside-analogue reverse transcriptase inhibitor and a *non*-nucleoside-analogue reverse transcriptase inhibitor.

By the time the first drugs arrived last fall, people in the support groups were poised not to receive treatment but to *claim* it. They shared an almost religious commitment to adherence, and some had become counselors and pharmacy assistants. Twenty-eight-year-old Akona Siziwe was as sick as Mavuka when she joined a support group in Lusikisiki last year. Weary of her husband's incessant criticism (he didn't like the way she limped), she had packed up her 7-year-old son and her HIV-positive toddler and gone home to die with her mom. But her health returned quickly when she started treatment in December, and she went to work as a community organizer. She now runs workshops and counsels patients in three villages. "What's a good CD4 count?" she asks. "The nurses don't have time to explain, but people want to know. When I share information that can help them, they're grateful and happy and full of praise. I can't even sleep because they are knocking on my door! They want testing and treatment tonight!"

The Lusikisiki project has only two nurses and two full-time doctors, but it was treating 255 patients when I visited in May, and people from the villages were flocking to the clinics as the good news spread. Many of them show up expecting a quick test and a jar of pills, but as the program's head nurse, Nozie Ntuli, likes to say, "Giving out pills is the final step in the process." First the patient has to join a support group and get treated for secondary infections such as thrush and TB. A counselor then conducts a home study to make sure the person is ready for a long-term commitment. When the supports are all in place, the counselor takes the patient's case to a community-based selection committee. And everyone shares the joy when a patient succeeds. "I see people transformed every day," Ntuli says. "It is a new dispensation."

This isn't the first time village volunteers have launched a successful health initiative. "Home-based care" is a tradition throughout southern Africa, and a cornerstone of countless successful programs. In rural Malawi, minimally trained community volunteers manage everything from pregnancy to cholera. They work with TB patients to ensure adherence, and they supply vitamins, aspirin and antibiotics to people living with HIV/AIDS. When Malawi's Health Ministry starts distributing antiretrovirals through a national program this fall, the volunteers will help administer those, too.

They'll play an especially important role in Thyolo, a desperately poor district surrounded by tall mountains and jade green tea plantations. Roughly 50,000 of Thyolo's half-million residents are HIV-positive, and 8,000 have reached advanced stages of illness. When I visited Thyolo this spring, MSF was treating several hundred of them at the local district hospital, a converted colonial-era country club run by nurses and clinical officers (non-M.D.s with four years of training). But the hospital was in no position to handle thousands more, even if the government provided the drugs. Its two-person AIDS staff was struggling just to keep up with the MSF program. Many of the untreated patients lived too far away to trek in for routine visits anyway.

Dr. Roger Teck, a fiftyish Belgian physician who runs MSF's Thyolo program, described the predicament during a bumpy jeep ride from the hospital to the outlying village of Kapichi, where 20 volunteers were waiting for us in a freshly painted one-room community center. Some were as young as 20, others as old as 70. After an hour of prayers and introductions and soulful choral chants, the group's leader, 49-year-old Kingsley Mathado, peppered us with facts about the 30 villages in his area (3,000 people living with HIV, 500 in need of treatment) and described the volunteers' program for supporting them. When the government drugs reach Thyolo district hospital, the patients will still have to walk a half day to queue up for an exam and an initial two-week supply. They'll also have to return for their first few refills so that a nurse or doctor can see how they're responding. But the volunteers will take over as soon as patients are stable, refilling prescriptions from a village-based pharmacy and charting adherence and side effects.

NEW PATTERNS OF A GLOBAL PLAGUE: Africa remains hardest hit by AIDS. But infections are booming in Asia and Eastern Europe. And most patients go without adequate therapy.

Could this strategy work on a grand scale? Lay health workers are already a mainstay of large-scale TB initiatives, and the Malawian government has assigned them a big role in AIDS treatment as well. The country's nascent ARV program uses a regimen simple enough for anyone to administer after a week of intensive training (three generic drugs in one pill—no substitutions). Physicians from Malawi's Ministry of Health are now traveling the country to conduct training courses for lay health workers. The first drugs should arrive in the fall. "We've taken a radical leap to ensure real access," says Dr. William Aldis, the WHO's Malawi representative and one of the plan's many architects. "We're either going to win a Nobel Prize or get shot."

Malawi's challenge is to foster the kind of engagement that has made treatment so effective in places like Cange, Khayelitsha and Lusikisiki. If 25 years of HIV/AIDS has taught us anything, it's that grass-roots involvement is critical. "One set of characteristics runs through nearly all of the success stories," the London-based Panos Institute concludes in a 2003 report on the pandemic: "ownership, participation and a politicized civil society." No one denies the need for trained experts to manage programs and handle medical emergencies. But people from affected communities are often better than experts at raising awareness, shattering stigmas and motivating people to take charge of their health. Reuter, the Lusikisiki project's director, recalls an experiment in which doctors teamed up with activists to extend a hospital-based ARV program into community clinics in the Cape Town slum of Gugulethu, where access would be easier and peer counselors could play a bigger role. The ghetto-based patients achieved 93 percent adherence during the first year. The hospital's program had never topped 63 percent—a rate Reuter dismisses as "American-style adherence."

With access to treatment, millions of dying people could soon recover as dramatically as Nozuko Mavuka did in Mpoza—and their salvation could revive farms, schools and economies as well as families. But there are hazards, too, and drug resistance isn't the only one. Successful ARV therapy expands the pool of infected people simply by keeping people alive. Unless the survivors can reduce transmission, the epidemic will grow until the demand for treatment is unmanageable. "We can't focus blindly on treatment," Teck mused as our jeep lurched away from Kapichi. "If we don't reduce the infection rate, we're going to end up in a nightmare situation." The patients and counselors in the clinics I visited weren't singing and stomping only about pills. They were celebrating a shared commitment to ending what is already a nightmare. The rest of the world needs to lock arms with them.

Reporting for this story was supported by a grant from the Henry J. Kaiser Family Foundation

Sudan's Ragtag Rebels

The ambitions of Darfur fighters exceed their limited resources

By Emily Wax
Washington Post Foreign Service

FURAWIYA, Sudan—The greasy, stinking Land Cruiser, with its screeching fan belt and goatskin water jug swinging off the back, beat a fast pace across the desert in rebel-held Darfur, until it slammed to a jarring stop.

Riding on the roof was Issac, an energetic sharpshooter from the Sudanese Liberation Army (SLA). He had spotted the target. The Land Cruiser, haphazardly camouflaged with black spray paint, rested quietly for a moment. Issac looked down the sight of his AK-47 rifle. He fired, the crackle of gunshots echoing through the silent desert.

In the distance, a gentle antelope broke for the bush. The famished rebel forces spent the next 45 minutes and nine shots chasing down the swift creature through thorny bushes and thick sand before catching their prey.

Finding food is a matter of survival for thousands of people in this vast area of Sudan, including the warring troops who are fueling a raging humanitarian crisis. Armed conflict in Darfur has left 1.2 million people homeless, 50,000 dead and hundreds of thousands vulnerable to life-threatening diseases.

International condemnation has focused on the government-backed militia known as the Janjaweed, which has terrorized civilians in areas where armed resistance to the state has been strongest. Less attention has been devoted to the SLA rebels, who said they started the conflict to defend the rights of Darfur's African tribes but now preside over corners of acute suffering and desperation on the frontiers of Africa's largest country.

A week spent traveling through rebel-held areas showed the SLA to be an ill-equipped, untrained and disorganized group, with child fighters among its ragtag ranks. Its grand ambitions are not matched by its resources. The only thing the rebels don't seem to be lacking is motivation.

"Give us 500 cars with mounted machine guns and we'll take Khartoum in one month," proclaimed Bahar Ibrahim, a top adviser in the SLA's political wing, referring to Sudan's capital. A graying wisp of man, Ibrahim said over sugary tea at a base camp in the town of Bahai that he admired the ferocity of American action movies and spaghetti westerns. "We can act like that," he said.

Yet on that steamy afternoon, the rebels in the SUV weren't able even to cross a riverbed that had swollen with slow-moving water from seasonal rainfall. At the base camp, they had five cars, all taken in battle from the government. Not one of them would start.

WHILE THE JANJAWEED IS united by ethnic hatred toward African tribes, SLA leaders speak with equal ferocity about the Arab government in Khartoum, which they say has discriminated against generations of black Africans. They see themselves as heroes defending the lives of tribal members who have not fled to disease-ridden camps that the government runs.

As a result, SLA leaders, like the commanders of the Janjaweed, refuse to silence their guns. Peace talks in Nigeria between SLA rebels and the government stalled at the beginning of September, as the sides argued over who scuttled attempts at a cease-fire.

In the first week of this month, 3,000 people have fled more fighting in the town of Zam Zam in North Darfur, a U.N. report said. It was not clear if rebels or Janjaweed sparked the clashes. The United Nations had set an Aug. 30 deadline for the government to improve safety conditions and rein in Janjaweed or face unspecified sanctions.

Last Thursday, Secretary of State Colin L. Powell said teh abuses committed by the militias qualify as genocide against the black African population in teh Darfur region. I remarks prepared for the Senate Foreign Relations Committee, Powell said his conclusion was based on interviews conducted with refugees as well as other evidence.

At the United Nations last Wednesday, the United States circulated a draft resolution that calls for a stronger international force to monitor the situation.

Meanwhile, civilians continue to suffer. The rebels are accused of atrocities, although on a much smaller scale than the Janjaweed and its government sponsors. The rebels control a vast countryside where an estimated 130,000 civilians are beyond the reach of food and medical aid that those in the government-held areas are slowly receiving.

The SLA hadn't planned to gain so much ground so quickly. The group claimed its first major victory last year in the stunning capture of the town of El Fasher. The rebels killed 75 government soldiers, stole weapons and destroyed four helicopter gunships and two Antonov aircraft, government officials says. A second smaller rebel group called the Justice and Equality Movement (JEM) joined the fight against the government.

The government in Khartoum reacted to the defeat by arming the Janjaweed to assist the army. The Sudanese military launced a bombing campaign against hundreds of largely African villages, leaving more than 1 million people as refugees.

The Darfur conflict emerged just as a U.S.-backed peace between the government and separate rebel groups in the south of Sudan promised to end nearly half a century of intermittent warfare there. The SLA asked to be a part of those peace talks, but the government refused. Now, the government faces yet another uprising mounting among the Beja tribe in eastern Sudan.

The SLA leaders are drawn from the elite ranks of African tribes, some of whom say they are fighting the legacy of decades of discrimination for more political power and a share of Sudan' s $1 million-a-day oil revenue. Some leaders say what they really want is to join with other rebellions around Sudan and push for a change of government in Khartoum. However, the groups have little political experience and remain fragmented.

In the town of Faraywaiah, SLA rebels explained their struggle by pointing to a ravine where 12 male bodies lay decomposing. One body was curled up in a fetal position. Another's scalp was rotting in the hot sun. Others had skulls poking through burned hair. The bitter smell of the dead hung in the hot air.

The rebels and a few townspeople who are left said that these were local African men, some of them caught by surprise at the village well and killed by government forces who stormed the town in March. The town, pockmarked with bomb craters, was deserted.

THE REBELS' MOST FAMOUS fighter is known as Kongo. He arrived one afternoon in the shade of a tree where his troops had spread carpets for an afternoon nap amid Belgian assault rifles and ammunition belts. Kongo had a thick swagger. His *nom de guerre* means the man who walks without a stoop.

His men gave him this name because they claim he never cowers from gunfire. That might explain why his left eye is shot out, a bullet is lodged in his jaw and one of his legs is stitched like a pincushion.

Kongo had a pounding headache and asked a reporter for some aspirin and a bottle of *pastis*, an aniseed-flavored, French liqueur, to soothe his aching head. The SLA's star fighter embodies the disparity between the group's determination and its resources.

Last July, Kongo fought in Gourbou Jong, a battle the SLA calls its mini-Stalingrad, after the Soviet defeat of Adolf Hitler's army. Gourbou Jong proved that despite poor training and bad weapons, the rebels could beat one of the most powerful militaries in Africa. A Sudanese Army general was killed in the fighting.

Kongo, whose given name is Kitir Zakariya, is nominally a commander in the SLA. But to his troops he is a god. He removed his sunglasses and revealed a wounded patch of skin where an eye used to be and described his mission to fight the Arabs in Darfur.

"The only language they understand is the gun," Kongo said, fiddling with his charm necklace, dozens of worn, square leather pouches filled with Koranic verses to protect him from gunfire. Then he took a cigarette out of his pocket. "We, the youth of African tribes in Darfur, feel this has been going on for years. Something must be done," he said. "We can take them with just a few cars."

Listening to his every word was baby-faced, curly haired Harry Fadhul, who possessed a large gun and a frown. He's one of many young fighters who have chosen combat over being raised in squalid and humiliating refugee camps. Sipping powdered milk loaded with sugar, he recounted how he joined the force three months ago, after his village was burned. He saw corpses everywhere. His mother fled to a refugee camp, he said.

"I have nowhere to go," said Fadhul, who gave his age as 18 but looked much younger. "The SLA is my family now."

Kongo justified the war by describing the way successive governments in Khartoum discriminated against African tribes, punishing students in school who spoke their tribal language, Zaghawa. The SLA is now renaming towns from Arabic to Zaghawan.

"Whenever the Arabs talked to you it was like they were better than you," he said, a hand resting on the pistol in his belt holster. "No one seems to think they could ever live peacefully with their Arab neighbors again. I don't have any Arab friends. I would never marry even a beautiful Arab woman now. She is not for me."

African tribes in Darfur have festering grievances dating back to the formation of a movement known as the Arab Gathering in the 1980s. Bringing together various Arab tribes, it espoused the supremacy of the "Arab race" and laid plans to help Arab tribes displace African tribes that were granted land under British rule. Heavy clashes began in 1987.

In the early 1990s, the country's southern rebel movement, the Sudan People's Liberation Army, trained rebels from Darfur and deployed forces in the region to fight the government. But the rebellion in Darfur was quickly crushed.

In response to the Arab Gathering, a group of African intellectuals created the Black Book, a fiery catalogue of facts and figures secretly gathered from government records that sought to prove that three Arab tribes dominated positions of power in nearly every sphere of Sudanese society, including hospitals, schools and police forces. Copies were passed out to every government leader, including the president, causing a stir in Khartoum.

It was a sign of what was to come. Today several authors of the Black Book, who called themselves The Seekers of Truth and Justice, are leaders in the SLA and JEM.

"We know there is a well-organized plan to get rid of us," says Ibrahim, who helped advise those who wrote the Black Book. "But we are warriors; that is our culture. The Arabs here don't know how hard we are willing to fight for our land."

The challenge for the SLA is not only to confront the powerful regime in Khartoum, but to somehow administer a territory where a hidden humanitarian crisis is underway. Part of the rebels' popular support will be based on how they feed and protect their own people, analysts say.

In an open field ringed by red mountains, Hawi Bas and her three children were hiding on a recent day under a scrubby tree in Shiga Karo, a rebel-held village about 70 miles east of Sudan's border with Chad.

Her youngest child, 2-year-old Hari, had wilting hair and willowy legs. He had become so thin that he could no longer walk. His name means "strong" in Zaghawa. His family survives on boiling the toxic pea known as *mukheit*, which needs to be soaked in water for three days. But mukheit only fills the stomach; it does not provide real nourishment to children.

Hawi Bas said she could not travel to the safety of refugee camps in Chad. "How are we going to go there?" she asked, crouching in the hot sand and showing a small bowl of mukheit. "We are not feeling fine. I always feel stomach pains. Do you see our condition?"

Aid groups say they believe there are tens of thousands of people struggling to survive in rebel areas, sealed off from aid. The government will not permit aid agencies to travel into rebel-held areas, arguing that the SLA will steal the food. The U.N. food agency has recently been granted access by rebel groups to study the needs in the region, entering through Chad or circumventing government troops.

Refugees say the rebels have not bothered them. Late last month, rebel soldiers distributed sacks of corn flour to some of the displaced, but it was not enough. Hari looked close to death.

Ibrahim, the SLA leader, looked at the child, held him up and then promised to make some calls to aid agencies to work on getting them here to help the refugees.

The rebel group is now suffering from splits within its leadership, and the people can sense a lack of focus. "They seem to have as little as we do," said Ismael Hagar, 25, a teacher who has brought biscuits to the refugees.

ON A RECENT AFTERNOON, the riverbeds were flooded and the only way to cross back into Chad from Sudan was to swim through the dirty water and wade through the mud. Walking alone in the opposite direction was Kongo. He was carrying a burlap sack and his pants were rolled up to his bony knees. He said he was hungry and went to Chad to get meat.

Now after eating, he trudged back across the border.

"To fight," he said. "Always."

Responding to the Special Needs of Children

Educating HIV/AIDS Orphans in Kenya

In Kenya, education is considered the pillar of all development activities. The guiding philosophy for Kenya's education is the belief that every Kenyan, no matter his or her socioeconomic status, has the inalienable right to basic education.

Tata Mbugua

In Africa, as elsewhere in the developing world, education is viewed as the most prominent public policy issue, involving an interplay of national budget allocation and foreign assistance. In Kenya, education is considered the pillar of all development activities (Odiwour, 2000). The guiding philosophy for Kenya's education is the belief that every Kenyan, no matter his or her socioeconomic status, has the inalienable right to basic education (National Development Plan, 1997). Consequently, Kenya spends 40 percent of its official budget on education (United Nations Development Program, 2003).

While Fagerlind and Saha (1989) stressed the complexities affecting the link between education and development, more recent studies have highlighted the long-term positive developmental effects of education (Barnett, 1995; Haveman & Wolfe, 1995; UNICEF, 2003). As enrollments in schools continue to rise, some researchers are predicting an increased demand for education as well as a need to serve children who present increasingly diverse needs (Hernandez, 1995; Odiwour, 2000). Developing countries that are faced with the HIV/AIDS pandemic have additional challenges to address. How does Kenya, a country faced with economic, social, and health challenges, educate and care for those "special needs" children who have been orphaned by the HIV/AIDS pandemic?

The purpose of this article is threefold. First, it will provide a brief overview of facts about the spread of HIV/AIDS. Second, it will characterize HIV/AIDS orphans and discuss indigenous initiatives in Kenya to educate and care for those children. Third, the article will describe the Nyumbani Children's Center, a successful comprehensive model for best practice in responding to the special needs of HIV/AIDS orphans in Kenya.

The Global Context of HIV/AIDS

The HIV/AIDS pandemic continues to claim millions of lives around the world, generating a serious humanitarian crisis that threatens to transcend all other health problems (International Crisis Group, 2001). Those most affected by the pandemic include children orphaned or otherwise burdened by its devastating toll. An estimated 3.2 million children worldwide are infected with HIV/AIDS, with 2,000 new cases each day; by the end of 1999, 13.2 million children under the age of 15 worldwide had lost their mothers or both parents to AIDS (UNAIDS, 2002b). Additional estimates by the Pediatric AIDS Foundation indicate that an estimated 5.6 million children will have died of the epidemic and over 25 million will be orphaned by the year 2010 (Pediatric AIDS Foundation, 2003).

One of the biggest misconceptions about HIV/AIDS is that children born to HIV/AIDS mothers are automatically infected with HIV, the virus that causes AIDS (Pediatric AIDS Foundation, 2003). These children carry the mother's antibodies and thus are not necessarily HIV-positive. Current medical advances confirm that two doses of a drug called Nevirapine administered to an HIV-positive pregnant woman can successfully prevent transmission of the HIV virus to the child and reduces the chance of the infant being born with AIDS to 47 percent (Pediatric AIDS Foundation, 2003). This information is both encouraging and crucial for educators and caregivers, especially in Africa, where one of the largely unarticulated consequences of the HIV/AIDS pandemic is the stigma associated with the disease. Children whose mothers are infected or have died face discrimination even if they are not infected themselves. This discrimination often results in the children being denied the special attention and care that they desperately need.

The Regional Context of HIV/AIDS—Africa

The statistics on HIV/AIDS infections indicate that 26 million adults and 2.6 million children are infected with HIV/AIDS (UNAIDS, 2002a). In addition, about 95 percent of all AIDS orphans live in Africa, where more than one child in every 10 has lost a parent to AIDS (UNAIDS, 2002b). UN Secretary-General Kofi Annan (2000) predicts that sub-Saharan Africa will be home to 40 million orphans by 2010, largely because of AIDS. Consequently, the pandemic is threatening community and social cohesion (Spectar, 2003) by stretching the traditional and critical extended family support systems. There is even an alarming contention that HIV/AIDS is fast reversing some of the economic and social gains made in Africa over the last 40 years (The World Bank, 2002).

The National Context of AIDS—Kenya

Kenya is estimated to have the ninth-highest prevalence of HIV/AIDS in the world; about 14 percent of the adult population is infected with the virus (*World Almanac*, 2002). Currently, 2 million adults and 100,000 children under 15 years are living with AIDS in Kenya. About one million children have been orphaned as a result of HIV/AIDS, and the number is rising (UNAIDS, 2002a).

Taxonomically, the HIV/AIDS pandemic raises three human security issues (Spectar, 2003). On the personal security level, the pandemic damages the individual's ability to sustain him- or herself and the family, as the terminal illness and/or death of a breadwinner reduces or ends income. In terms of economic security for the country, the disease results in loss of jobs, loss of productivity, and loss of expensively trained manpower. From a community security standpoint, the society must absorb the losses and the attendant social and cultural consequences, including the breakdown of extended families and other support systems to which orphans traditionally have turned (Human Rights Watch, 2001).

Kenya has one of the best HIV/AIDS surveillance systems in Africa, with many programs put in place to mitigate the disease (UNAIDS, 2002b). These programs range from volunteer counseling and testing (VCT), the National AIDS/STD Control Program (NASCOP, 2001), training of trainers (TOT), condom usage, adolescent and youth education, and mother to child prevention (MTCP). In addition, Kenya has passed legislation that facilitates the importation of more affordable drugs (Siringi, 2002).

Very few programs, however, address the needs of HIV/AIDS orphans. Most mitigation programs focus on institutional or home-based care for adult patients, adolescent sexuality education, and income-generating activities for caregivers. Therefore, a gap exists in meeting the special needs of affected and infected orphans, especially their education and psychosocial needs.

Gachuhi (1999) defined the pandemic as a war, with the education sector at the frontline. Kelly (2000) conceptualized and summarized HIV/AIDS as having the potential to affect education in 10 different ways, number one of which is meeting the special needs of the increasing number of orphans.

Living As an HIV/AIDS Orphan

There are a number of definitions of HIV/AIDS orphans. According to UNAIDS (2002b), an HIV/AIDS orphan is a child who has lost his or her mother to the disease. However, a more inclusive definition refers to a child who has lost one or both parents to HIV/AIDS (Lusk & Ogara, 2003). Other definitions expand the term to include children abandoned by parents and children heading households (Ayieko, 1998). These children may be infected by HIV or have AIDS, they may be affected by HIV/AIDS through the loss of one or both parents or siblings, or they may be at risk of infection.

These orphans may be vulnerable, isolated, depressed, stigmatized, discriminated against, and uneducated; some live in the streets (Odiwour, 2000; Oywa, 2003). Ayieko (1998) says that these orphans may be resented by wealthier relatives with whom they are sometimes placed. Orphaned children in Kenya usually have only four choices of where to live (Odiwour, 2000). First, they may stay in their parents' house to look after themselves, with a relative a short distance away. Second, they may go to live with grandparents, uncles, or aunts, who "inherit" them. Third, they may go to more distant relatives or to nonrelatives and neighbors. Fourth, they may go into some kind of institutional care. Underscoring the plight of HIV/AIDS orphans, Caldwell and Caldwell (1993) note evidence of orphans being removed from school on the grounds that they must help with their own support.

The health, development, and psychosocial well-being of HIV/AIDS orphans are at risk long before either parent dies (Juma, 2001). The psychological trauma these orphans might undergo includes tending to a dying parent and taking care of siblings.

Indigenous Strategies for Educating HIV/AIDS Orphans

Making inroads against the effects of HIV/AIDS in any country depends on having access to appropriate resources, clear information on prevention, and treatment and care for affected and infected persons, especially young children. Young HIV/AIDS orphans, however, need direct and special attention to health, education, and psychosocial care. While poverty may mitigate many initiatives dealing with HIV/AIDS, Kenyan communities have embraced the daunting task of educating and caring for HIV/AIDS orphans through community mobilization, based on a longstanding philosophy of "harambee," which means "let us pull together." Key to this idea is the recognition that the care and education of society's children is a responsibility shared by the entire community. Families, caregivers, and educators share an ethical and moral responsibility to promote the optimum conditions for the well-being of all children, especially HIV/AIDS orphans.

According to Lo and Mbugua (2000), the concept of special needs is socially constructed; because each society is unique, each will develop their own meaningful concept of special needs, ways to identify gaps in services, and plans for attendant service provisions. A variety of community programs are in place in Kenya that involve the care and education of HIV/AIDS orphans, including village school houses, family care, and community preschools. An important dimension to these community efforts is that of catering to the overall mental health of the orphans. A discussion of preschools will highlight one of these strategies.

Benefits of Schools for HIV/AIDS Orphans

Through education, young children are provided with school readiness skills, stimulation, basic education, and socialization opportunities. The additional benefits of preschools serving HIV/AIDS orphans in Kenya include health care and psychosocial adjustment. Achieving balanced nutrition through an enriched diet is a particularly important benefit derived from enrollment in these preschools. Some of the children enrolled are simply in need of a sense of belonging, acceptance, and appreciation. Essentially, the schools offer the orphans hope for a viable future.

Current Educational Policy

In January 2003, Kenyans reached a notable milestone by voting to make primary education free and compulsory (Integrated Regional Information Network, 2003). With this initiative came an unofficial policy that requires all children under the age of 6 to possess school readiness skills in order to be enrolled in primary school. As a result, numerous private and public preschools catering to all children, including HIV/AIDS orphans, have been established. In some cases, the Kenya government provides preschools within regular primary schools. These schools charge a small fee and also require families to buy school uniforms. Unfortunately, once these fees are added to the costs for books, snacks, and school supplies, many families are unable to send their children to school. Preschools that fall outside of the purview of the government still require fees and school uniforms. On average, the cost of preschool (school fee and provision of a snack) is 100 Kenya shillings (US$ 1.18) per child, per month, or 1,200 Kenya shillings (US$ 14.16) per child, per year (Academy for Educational Development, 2002).

Preschools catering to the needs of HIV/AIDS orphans differ in terms of setup and resources, due to regional disparities. It is not uncommon to find children who are older than 6 attending these preschools. All preschools, however, make every effort to provide a safe, supervised group experience for the children. Some preschools are better equipped than others. Rural preschools tend to be poor compared to urban and suburban preschools (except those in slum areas). In terms of staffing, communities often call on retired teachers, mothers, high school drop-outs, and volunteers, who provide their services free of charge or at a minimal salary.

Preschools for HIV/AIDS orphans are, in many cases, established through community efforts. English is used as the main medium of instruction in all schools. In addition, Ki-Swahili, the national language, is also used. Nongovernmental organizations, church groups, and local/international donor agencies often complement/augment these community efforts by providing financial, social, and medical services. Preschools provide a needed break for caregivers, whose time is now freed to attend to household routines and engage in gainful employment.

At the primary school level, many HIV/AIDS orphans are educated under the auspices of established government schools. Although official policy states that these children should be educated with their peers, there are concerns that the orphans are being discriminated against and denied admission to the regular schools. HIV/AIDS orphans are often enrolled in "preschool" settings, run by community members, that serve their special needs. In this instance, age plays a secondary role compared to the dire need for socialization and provision of the basic necessities for normal growth and stimulation.

Stumbling Blocks and Paths to the Future

As in many developing countries, the barriers to education in Kenya often appear to be overwhelming. Children who are orphans continue to be denied admission to public schools. When they are admitted, they face stigmatization, discrimination, and physical and emotional neglect, which has a negative effect on their education and care (UNESCO, 2003). In addition, overcrowding and the lack of trained staff in community-based preschools are pressing challenges. Efforts to deal with the anxiety, grief, and depression that most of these children undergo are insufficient. Financial support, access to and availability of testing services, and availability and costs associated with medication are additional stumbling blocks. Despite these challenges, a cadre of professionals are working on behalf of young children affected by HIV/AIDS. Some of the progress that they have made can be seen in the Women's Groups and the Children's Homes in Kenya.

Women's Groups. Women's groups are integral to the education and care of young children (Mbugua-Murithi, 1997). These groups are ubiquitous in Kenya and play a significant role in society. They are known by a variety of names, such as "rotating schemes," "merry-go-rounds" (Kabiru, Nienga, & Swadener, 2003), or "katibas." Recently, the women's groups' efforts have been complemented by the work of Women Fighting AIDS in Kenya (WOFAK), an organization that functions at both the community and political levels to advocate for vulnerable women and children.

All the women's groups have one goal in common: raising funds and resources to support community members and community programs that cater to the educational and socioeconomic needs of young children. In effect, these groups help in the education and care of orphans through payment of school fees and purchase of snacks (enriched porridge), school supplies, school bags, and the compulsory school uniforms. In the low-income areas, trash bags often substitute for backpacks.

Nyumbani Children's Home. Nyumbani, which means "home" in Ki-Swahili, epitomizes the indigenous initiatives for the care and education of HIV/AIDS orphans in Kenya. Located in Nairobi's suburbs, Nyumbani Children's Home, a non-profit company, was founded in 1992 by Dr. Angelo D'Agostino, S.J., and recognized as a home by the Ministry of Home Affairs in 1995 (Nyumbani Children's Home, 2004). Currently, Nyumbani houses over 100 children (newborn to age 20) who were orphaned by the HIV/AIDS pandemic. The stigma placed on these children often precludes them from being resettled in their communities.

Recognizing the importance of family in fostering a sense of belonging for children, and keenly aware of the rising number of HIV/AIDS orphans, the Nyumbani Children's Home founders developed a unique and holistic model. They focused initially on providing institutionalized comprehensive care for the orphans, and later utilized a community mobilization approach for home-based care.

The institutional approach to the care and education of HIV/AIDS orphans strives to re-create an environment as close as possible to that of a family in a community. Through its services, Nyumbani works to prevent the dire consequences of neglect, stigmatization, and abandonment that often affects HIV/AIDS orphans. The Children's Home is a village-style set-up of five duplex houses, with two family units and other service buildings. A surrogate mother or uncle lives with four or six children, based on the size of the house.

This model of living helps meet the basic needs of the orphans through the establishment of "homes," a school with a well-equipped play yard, a diagnostic laboratory and clinic to monitor the orphans' health care (especially nutrition and immunizations), a church, a garden, a community hall, and a graveyard. Cadres of volunteers, both local and international, have adopted a multi-disciplinary approach as they give their time and expertise to work with and care for the HIV/AIDS orphans at Nyumbani. Among the volunteers are doctors, nurses, social workers, lab technicians, religious personnel, teachers, and students.

A Typical Day in a Preschool for HIV/AIDS Orphans

TIME	ACTIVITIES
8:30	free activity, outside play (enriched porridge for needy children)
8:45	pledge of allegiance and a song
9:00	letter/number recognition and recitation
9:30	drawing, often on the sand outdoors or drawing on coloring books and papers
10:00	snack (enriched porridge), outdoor play health care and immunizations— mobile vans
10:15	poetry recitation (poems are written by community members)
10:30	puppetry and storytelling to complement reading, due to shortage of books in many areas (storytelling is also a literacy strategy of preference that is culturally sound and encourages sharing of feelings and the development of oratory and public speaking skills)
11:15	puppetry, drama, and riddles
12:00	leave for home

In order to meet the educational needs of the younger HIV/AIDS orphans, Nyumbani Center has its own school home. Older children attend local schools. The school home's curriculum focuses on the physical, cognitive, and psychosocial developmental domains. In line with Vygotsky' s emphasis on social interactions and socially constructed knowledge (as cited in Taylor, 2004), the school emphasizes a play-centered and child-centered approach to education and care. Both indoor and outdoor learning environments reflect these ideas. Early childhood education professionals from Kenyatta University provide early childhood development inservice training for staff and caregivers.

The community mobilization aspect of Nyumbani, the Lea Toto community-based intervention program, was started in 1998 (www.leatoto.com). The goal of the program is to mitigate the impact of the pandemic by mobilizing, equipping, and empowering communities to care for children affected by HIV/AIDS and their families. This is in response to the increasing numbers of HIV/AIDS orphans and the complex demands for service and care placed upon Nyumbani. Research shows that institutionalized care for HIV/AIDS orphans, while still widespread in many parts of Africa, is not a developmentally ideal or financially appropriate option on a large scale (Academy for Educational Development, 2002). A better alternative is devoting resources to creating an enabling environment in which communities can care for and educate the orphans.

This alternative is accomplished through satellite offices in the slums of Nairobi. Orphans affected by or infected with HIV/AIDS are identified by the community and become eligible for quality home-based care and counseling services. Community members also receive these services, in addition to training in caregiving skills. The overarching goal is to bring together the key stakeholders within the commununity to build and sustain the programs and services.

Puppetry As an Educational and Psychosocial Therapeutic Exercise. Puppetry is a unique strategy used widely in preschool and regular school settings for HIV/AIDS education. Community members utilize their talents, locally available materials, and culturally sensitive puppetry to create an educational and clinical tool. This approach provides a creative learning strategy that allows children freedom of expression while stimulating learning through play (Synovitz, 1999). Because young children love to pretend play and engage in dramatic characterizations, using puppets affords them the opportunity to manifest these behaviors.

As a tool for counseling, puppetry helps in reducing stigmatization of HIV/AIDS orphans. More important, the technique aims at catering to the emotional needs of the orphaned children by getting them to express feelings. Through manipulation, the inanimate puppets can be given "life." They provoke emotions, dance, laughter, and reflection as they interact with the audience. Puppets also can offer sympathy, when needed. Community members develop the puppetry themes based on their observations and conversations with the orphans and their families.

Conclusion

In spite of the ravages of the pandemic and its socio-economic impact on society, communities in Kenya have undertaken the invaluable task of responding to the special needs of HIV/AIDS orphans. The essence of the indigenous initiatives to educate and care for the vulnerable orphans in the community are succinctly captured in the words of Fr. D'Agostino, S.J. "*The education and care of children with an as yet incurable disease may not seem a 'profitable enterprise,' but it cannot be surpassed as a humanitarian and spiritually rewarding endeavor. Providing a stress-free family life experience has been proven scientifically to mitigate the ravages of HIV infection*" (Nyumbani Children's Home, 2004).

Indigenous efforts by communities in Kenya need to be acknowledged and validated as unique endeavors that complement long-established local and international institutional efforts to educate and care for HIV/AIDS orphans. This is of particular importance in the wake of renewed and robust global attention to the HIV/AIDS pandemic.

References

Annan, K. (2000). *We the peoples: The role of the United Nations in the twenty-first century.* Retrieved on February 26, 2004, from www.un.org/millenium/sg/report/full.html.

Academy for Educational Development. (2002). Facts about basic education in developing countries. Retrieved January 2, 2004, from http://aed.org.

Ayieko, M. (1998). From single parents to childheaded households: The case of children orphaned by AIDS in Kisumu and Siaya districts. Research Report, Study Paper No.7 and Development Programs, UNDP.education (UNDP, 2003).

Barnett, W. S. (1995). Long-term effects of early childhood programs on cognitive and school outcomes. In R. E. Behrman (Ed.), *The future of children: Long-term outcomes of early childhood programs* (Vol 5, No.3). Los Altos, CA: David and Lucile Packard Foundation.

Caldwell, J. C., & Caldwell, P. (1993). African families and AIDS: Context, reactions, and potential interventions. In *Health Transitions Review*, Vol 3, Supplementary Issue.

Fagerlind, I., & Saha, L. (1989). *Education and national development: A comparative perspective.* Oxford: Pergamon.

Gachuhi, D. (1999). *The impact of HIV/AIDS on education systems in eastern and southern Africa region and the response of education systems to HIV/AIDS: Life skills programmes.* Paper prepared for UNICEF Presentation at the all Sub-Saharan Africa Conference on Education for All 2000, Johannesburg, South Africa.

Haveman, R., & Wolfe, B. (1995). *Succeeding generations: On the effects of investments in children.* New York: Russell Sage Foundation.

Hernandez, D. J. (1995). Changing demographics: Past and future demands for early childhood programs. In R. E. Berman (Ed), *The future of children: Long-term outcomes of early childhood programs* (VoL 5, No.3). Los Altos, CA: David and Lucile Packard Foundation.

Human Rights Watch. (2001). Kenya—In the shadow of death: HIV/AIDS and children's rights in Kenya. *Human Rights Watch,* 13(4).

Integrated Regional Information Network. (2003). *Kenya's challenge of providing free primary education.* New York: UN Office for the Coordination of Humanitarian Affairs.

International Crisis Group. (2001). *International Crisis Group calls AIDS a security threat.* Retrieved May 2004, from www.intl-crisis-group.org.

Juma, M. (2001). *Coping with HIV/AIDS in education: Case studies of Kenya and Tanzania.* London: Commonwealth Secretariat.

Kabiru, M., Nienga, A., &Swadener, B. (2003). Earlychildhood development in Kenya: Empowering young mothers, mobilizing a community. *Childhood Education,* 79, 358-363.

Kelly, M. J. (2000). *Planning for education in the context of HIV/AIDS.* Paris, France: International Institute for Educational Planning.

Lo, D., & Mibugua, T. (2000). Child advocacy and its application to education professionals: International symposium on early childhood education and care for the 21st century. In I. R. Berson, & M. J. Berson (Eds.), *Research in global child advocacy* (pp. 117-137). Washington, DC: American Educational Research Association.

Lusk, D., & Ogara, C. (2003). The two who survive: The impact of HIV/AIDs on young children, their families and communities. In *Coordinators Notebook*, 3(26).

Mbugua-Murithi, T. (1997). Strategies for survival in Kenya: Women, education and self-help groups. *International Journal of Educational Reform*, 6(4), 420-427.

Kenya Government. (1997). *National development plan.* Nairobi, Kenya: Government Printers.

National guidelines for voluntary counseling and testing. (2001). Nairobi, Kenya: Ministry of Health, NASCOP.

Nyumbani Children's Home. (2004). Retrieved on February 20, 2004, from http://nyumbani.org.

Odiwour, W. H. (2000). *The impact of HIV/AIDS on primary education in Kenya: A case study on selected districts of Kenya.* Stockhom University: Institute of International Education.

Oywa, J. (2003, November 10). UNICEF roots for AIDS orphans. *Daily Nation.*

Pediatrics AIDS Foundation. (2003). *Facts about HIV/AIDS.* Retrieved December 19, 2003, from http://pediads.org.

Siringi, S. (2002). African initiative launched to tackle HIV/AIDS in children. *Lancet,* 359(9300), 55.

Spectar, J. (2003). Indiana International & Comparative Law Review. V.13:2.483-542.

Synovitz, L. B. (1999). Using puppetry in a coordinated school health program. *Journal of School Health,* 69(4), 145-147.

Taylor, B. (2004). *A child goes forth: A curriculum guide for preschool children.* Columbus, OH: Merrill/Prentice Hall

UNAIDS. (2002a). *AIDS epidemic update.* Retrieved March 2, 2004, from www.unAIDS.org.

UNAIDS. (2002b). *HIV/AIDS in Africa*. Retrieved February 24, 2004, from www.unAIDS.org.

United Nations Development Program. (2003). *Human development reports: Kenya human development indicators.* New York: Author.

UNESCO. (2003). *HIV/AIDS stigma and discrimination: An anthropological approach.* Proceedings of the Round Table on 29 November, 2002, at UNESCO—Paris, Special Issues #20.

UNICEF. (2003). *The state of the world's children 2004—girls, education and development*. New York: Author.

World Almanac & Book of Facts, The. (2002). Kenya. New York: World Almanac Books.

World Bank, The. (2002). *Kenya—AIDS Disaster Response Project (APL).* Retrieved in March 2003 from http://worldbank.org.

Tata Mbugua is Assistant Professor, Education Department, University of Scranton, Scranton, Pennsylvania.

No cash in this crop

Kenya's tobacco farmers have been harvesting losses for far too long, says Joe Asila.

Quote... The British American Tobacco website offers the following pearls.
'Our approach over many years has been to work through dedicated staff in the field alongside farmers, many of whom are small producers in scattered rural communities. We train, advise and support farmers, providing seed and advice on all aspects of crop production. Our approach benefits the environment and benefits both the farmers and us in improving crop yields and quality.'

Unquote... And this is a testimony submitted to the World Health Organization in 2000 during a public hearing for the Framework Convention on Tobacco Control.
'All you see now are dwellings and tobacco-drying kilns in the compounds. Tobacco, the cash crop, has replaced food crops and livestock, and threatens the food security of every family. Yet tobacco is not yielding enough money for these people to buy food for subsistence and viable livelihoods.'

By Joe Asila

JANE CHACHA, 46, looks at the large green plants growing luxuriantly on her one-hectare farm and shakes her head in anger. She's still waiting for the riches she was assured would accompany growing tobacco.

Ten years after being convinced to convert her maize plantation to tobacco, her life has barely changed. She still lives in the two-room, mud and thatch house that her husband built 15 years ago. 'This is a hopeless dream,' says the mother of four. 'Growing tobacco has been nothing but trouble.'

Chacha is not the only disillusioned farmer in Kuria district, which lies at the heart of Kenya's main tobacco-growing region around Lake Victoria. She is among a growing band who have come to realize that there is no cash in this so-called cash crop and who blame the Government for not warning them about the dangers.

'We were never told that tobacco growing would clear the forest that we relied upon for firewood,' said Peter Masaba, another tobacco farmer. The hills near Masaba's home were once dotted with beautiful trees which provided land cover and helped retain rainwater. Today, the land is bare. The trees were cleared over the years to meet the high demand for wood fuel required in tobacco curing. And a local stream, a major source of water, has gone dry due to the deforestation.

Ties that bind

Worse, working in the tobacco curing barn has affected Masaba's health. His eyes are constantly teary after long years of exposure to tobacco smoke. Masaba has nothing to show by way of savings. The main beneficiaries, says Masaba, are the tobacco companies to whom the farmers are contracted. The Kenyan arm of the transnational British American Tobacco (BAT) rules the roost with 55 per cent of market share, followed by StanCom Company (25 per cent) and Mastermind Tobacco Kenya (20 per cent). 'Each year the companies proudly display figures they have paid in taxes, but those who grow tobacco like us remain poor.'

Legislation introduced in 1994 requires farmers to grow tobacco under contract to only one company and they are forbidden to grow tobacco 'out of season'. BAT Kenya was heavily involved in drafting the Bill, as a fax from its regional director made clear: 'The law was actually drafted by us but the Government is to be congratulated on its wise actions.'

Such wisdom looks very different from the small farmer's perspective (some plots are no larger than a quarter of a hectare). Contracts are signed but 90 per cent of farmers don't understand them. The contracts set the buying price of the tobacco and the points of sale. Inputs such as seeds, fertilizers and pesticides are loaned to farmers—their price, often charged at way above the normal shop price, is deducted by the company when the farmers sell their leaf.

Farmers are basically contract workers who must assume all financial risks. Should a barn burn down or the crop suffer due to a hailstorm, the farmer must shoulder both the losses and the resulting cycle of debt.

A successful tobacco crop takes nine months of labour-intensive, backbreaking work—most of which must be done by hand. The tobacco companies are not known to contribute farm implements or the use of a tractor to help out. The entire family, including children, gets drawn into tending the crop. 'Tobacco farmers can't cost the labour from members of their family,' says civil servant Pauline Mwita. 'Children are withdrawn from school to help in tobacco farm work. I have yet to see a healthy tobacco farmer.'

Local leader Helen Kibwabwa sees other negative social consequences: 'Men marry many wives to have more labour on tobacco farms.

Some women are contracted but at the time of payment men go to collect the money which they then squander while women and children are left to suffer.'

Growing tobacco is chemically intensive and a common complaint is that BAT doesn't provide even the most basic protective gear. 'Tobacco spraying is intoxicating. We have had cases of people becoming unconscious after inhaling the chemicals,' said Samwel Moseti, a farmer in Kuria. Headaches, nausea, constipation, skin and eye irritation and chest pains are common symptoms. Margaret Akinyi from nearby Rangwe added: 'Officials of tobacco companies will never enter tobacco farms during spraying without protection, yet they don't mind us working without protective gear.'

The official line from BAT Kenya has a decidedly different slant. When pressed by British charity Christian Aid, they responded: 'All farmers are provided with protective clothing and training on how to use it. The ultimate responsibility of wearing this clothing lies with the farmer.'

Pesticides also pollute water sources. As local doctor Japeth Opiya puts it: 'We depend on the river for everything—washing, drinking and cooking. All these pesticides are washed into the river; it touches everyone, even if you are not a farmer.'

Come picking time and another problem presents itself. Even discounting pesticide hazards, the tobacco plant itself can be toxic. Picking moist leaves with bare limbs can lead to the direct absorption of large quantities of nicotine through the skin. Green Tobacco Sickness, as it is called, results in nausea, vomiting, dizziness, abdominal cramps, aching joints and an accelerated heartbeat.

When the time comes to sell the crop, farmers allege that BAT graders downgrade the quality of their leaves, thus pushing down the price. It is easy to see how this could happen—independent graders aren't involved and the prices are decided by the company.

When it comes to the crunch, the profits, if any, in growing tobacco are paltry. Christian Aid uncovered BAT Kenya's figures for the Bungoma region and found that farmers were being paid the grand price of seven cents per kilo. They actually received less than half this price once the cost of the agro-chemicals and other inputs sold to them by the company had been deducted.

Rights wronged

A recent study by the Kenyan Health Ministry found 80 per cent of tobacco farmers actually lose money. Because they are usually poorly educated, they lack the skills to organize a budget for their nine months of toil. When the payoff comes they see the cash-in-hand, not the loss they may have incurred. Their only certainty is that tobacco farming does little to improve their hand-to-mouth existence.

The same cannot be said for BAT Kenya. When the going gets tough it tends to play dirty. In 2001, tobacco farmers from Nyanza province produced a bumper crop, much higher than the company had expected. Having already purchased more than it had planned for, BAT Kenya suspended further purchases, in clear breach of contract. In a bid to disempower farmers, BAT Kenya threatened to relocate to Uganda and leave them stranded. The Government did nothing to help.

BAT has also been fighting any meaningful union representation that would give tobacco farmers collective bargaining powers. Previous attempts at forming a union collapsed due to tobacco company interference.

Recently a beast called NEWTFA (the Nyanza, Eastern and Western Tobacco Farmers Association) has emerged. It purports to be a union for tobacco farmers but was in reality sponsored by BAT Kenya to counter the Kenya Tobacco Growers Association. NEWTFA officials can never visit any farmer without an accompanying BAT official.

Picking moist leaves with bare limbs can lead to the direct absorption of large quantities of nicotine through the skin

But NEWTFA is not without opposition. 'NEWTFA cannot champion the rights of the farmers and that is why we changed our organization from a union to an anti-tobacco growing crusade,' says George Kivandah,

Secretary General of the Kenya Anti-Tobacco Growing Association.

The Kenyan Ministry of Health currently has a Tobacco Control Bill before Parliament which calls for far-reaching reforms, including a ban on tobacco advertising and sales to minors. (The Bill was first introduced five years ago; critics say industry meddling has stalled its progress.)

Local NGOs want protection of tobacco farmers to be part of this Bill too. But, more than this, they would like to see other cash and food crops replace tobacco farming.

Joe Asila (socialneeds@hotmail.com) left a career in finance to become a pastor in a Pentecostal church. In 1996 he founded the SocialNEEDS Network, a Kenyan NGO involved in tobacco control advocacy, recycling and clean water projects.

Moving Africa off the Back Burner

Building on the European Model

Challenges for the African Union

BY HENRY OWUOR

AS FOR ANY other international body seeking to stake its place in the international arena, the African Union (AU) still has a long way to go. But as the Chinese say, a journey of a thousand miles starts with a single step. It may be recalled that the European Union (EU), on which the AU is modeled, was initially just a coal-and- steel-trading body. It now boasts the largest common market with 370 million consumers.

The EU's strong point has been the numerous treaties that govern its operations. The first such treaty was signed in Paris in 1951 to formalize the body's operations with an assembly to monitor them.

This is where Africa comes in. Any association of independent states can only function if all its members have a sense of belonging. None should feel left out, as was the case with East and Central African countries that weren't consulted when Nigerian, South African, Senegalese, and Algerian presidents met leaders of the G-8 nations to seek support for the New Partnership for Africa's Development (NEPAD). NEPAD is an initiative of the AU, yet not a single East African country is on its steering committee.

But these are what can be said to be teething problems, as occurred in the early stages of the EU when Britain refused to join and later had a rough time negotiating terms for entry. It was not until 1973 that Britain finally joined the community together with Ireland and Denmark.

Another area where Africa fares worse than the EU is the fact that the European body has only 15 countries to deal with. Africa starts with 53 nations loaded with a myriad of problems and needs. Even if one considers just one task, conflict prevention, a look at all the conflicts on the continent shows that this will not be an easy one to deal with.

But still, the AU comes in well armed. Under its Act, it can intervene in the affairs of member states on issues such as elections. This provision is already in use in Madagascar, where the AU has refused to recognize Marc Ravalomanana, who claims to have won disputed presidential elections, or Didier Ratsiraka, the former president who he claims to have beaten. Instead, it wants a fresh poll.

The AU plans a special summit in six months' time to review some contentious issues, especially under NEPAD, whose peer-review scheme is still being watched suspiciously by long-time dictators not used to outside scrutiny.

The NEPAD Peer Review will be run by retired presidents, judges, and other eminent persons who will police African countries in areas such as the timely holding of elections and the respect for human rights.

Still, some figures are very troubling: The World Bank forecasts that the number of poor people in Africa will rise from just over 300 million out of a total population of 659 million in 2000 to 345 million people by 2015.

Add to this the fact that as the AU was being launched in Durban early this month, more than half the members owed its precursor, the Organization of African Unity, more than US$50 million in membership fees, and you see the challenge it faces.

The task African countries are embarking on has a lot of similarities with that of the EU, says Finn Thilsted, the Danish ambassador to Kenya, whose country currently holds the presidency of the EU. He says both the EU and the AU "are out to promote trade and get rid of borders as they are a hindrance to free movement of goods and services."

Europe, like Africa, says Thilsted, has major differences in the lifestyles of its people. But still, European countries share similarities: For example, England and Denmark were both sea-faring nations and have contacts dating back to the Middle Ages.

The same may not be said for Africa, but still, some cultures did penetrate [the continent]: the Kiswahili language spread from the East African coast all the way to the Congo. Given the debate on adopting a working language for the AU, not even Kiswahili or Arabic can emerge as the main language. Arabic is the main language in northern Africa, while Kiswahili reigns in East Africa. Africa therefore finds itself using three colonial languages— English, French, and Portuguese.

The AU has given itself 10 years in a period dubbed "the age of capacity building" to radically change Africa. What it needs are treaties among African nations on various goals it wants achieved, such as human rights, with the AU setting standards for members to meet. Once this system begins, other treaty areas will be explored. Such power tactics as military coups and rigged elections should be among the first to be tackled in African treaties. The AU is moving in the right direction by creating an Economic and Social Council, through which nongovernmental organizations (NGOs) and trade unions will have the right to participate in the affairs of the continental body. At the moment, most African countries suppress NGO operations, which they see as a threat to their hold on power.

World Press Review, October 2002, p. 7.

A woman's rite

They call it 'something of which you should never speak', though many African countries have now explicitly outlawed the practice. Not so in the Gambia, Nikki van der Gaag discovers.

by Nikki van der Gaag

It always happens on a Monday. On the Sunday, the drumming and dancing go on all night. All the women get dressed in their best and most brightly coloured clothes—purples, lime green, orange, yellow, sky blue—and they dance until they drop, even the grandmothers. They have been preparing for this for a week, pounding vast quantities of rice and groundnuts which is then cooked in an enormous silver pot with three legs. At six in the morning it all starts again, and then a procession wends its way out of the gate and along the road, one woman banging a black jerry can with a stick. The mother of one of the little girls, aged six, cradles her child's head gently on her lap and fans her against the heat. The baby and five girls are spoiled and feted, given new dresses. It is a cause for celebration, and for those who were the centre of attention last year, the first time that they are able to join in properly, to enter the circle of the initiated. A week later there will be another party when the children come home; more food, more dancing, more celebrations.

The mothers tell the children—those who are old enough to enjoy stories—that where they are going there is a tree which has money instead of leaves. The girls dream of sending their father to Mecca, buying new clothes for themselves and all the family.

But at the heart of the party, as at the centre of all good stories, there is pain. For the little girls are going to be 'circumcised'. In her compound, where only the girls and their grandmothers are able to enter, the circumciser, ngaman, will take a razor, and cut off the children's' clitorises and labia minora with a razor.

Around 80 per cent of girls in the Gambia undergo this procedure. Seven of the nine major ethnic groups practise female genital mutilation (FGM) at ages from shortly after birth until 18 years old. Some girls die, others have complications following the operation. Many have difficult childbirths. Though this is also due to malnutrition and to early marriage—the majority of deaths are among women under 20—the scarring of the vaginal area in FGM also makes it less able to stretch for the baby's head to come out. The maternal mortality rate in the Gambia is high, at 1,100 per 100,000 births—107th out of 119 countries in the world—and has, if anything, gone up rather than down. In 1990 it was 1,050 per 100,000 births. Almost everyone you talk to has a sister, mother or other relative who died giving birth. Sometimes the child of a dead mother is called *Malafi* 'the one who was not wanted'.

But it is not difficult to see why mothers in the villages allow their female children to undergo the cutting, especially in the rural areas.

'My husband and I didn't want our daughters to be circumcised,' says one woman, 'but I knew they would be bullied and ostracized if they were not. I felt I had no choice. Everyone would know if they had not had it done, and if they came too near a compound where the circumcision was being performed, they could be brought in and circumcised then and there against their will.'

This is what happened to one girl from the Wolof tribe, who was taken forcibly off the street to be circumcised. As a result of the operation, she had to go to hospital. Her family, unusually, took the case to the police. But that was two years ago and nothing has happened since.

So entrenched is the practice that the *nyansymba*, the village head woman, will not talk about it. 'This is because if you talk about something, it might change, and she doesn't want it to change,' says one observer. But she is not the only one; it is something that it is taboo to discuss. As one Mandinka song goes: '*Jileng nyaka jileng nyaaa ye ku le je ndata foola*—my eyes have seen something about which I should never speak.'[1]

In other African countries—Burkina Faso, Senegal, Côte d'Ivoire, Ghana, Guinea, Central African Republic and Tanzania—things are changing. The governments have outlawed

Life expectancy

FEMALE LIFE EXPECTANCY AT BIRTH 2001

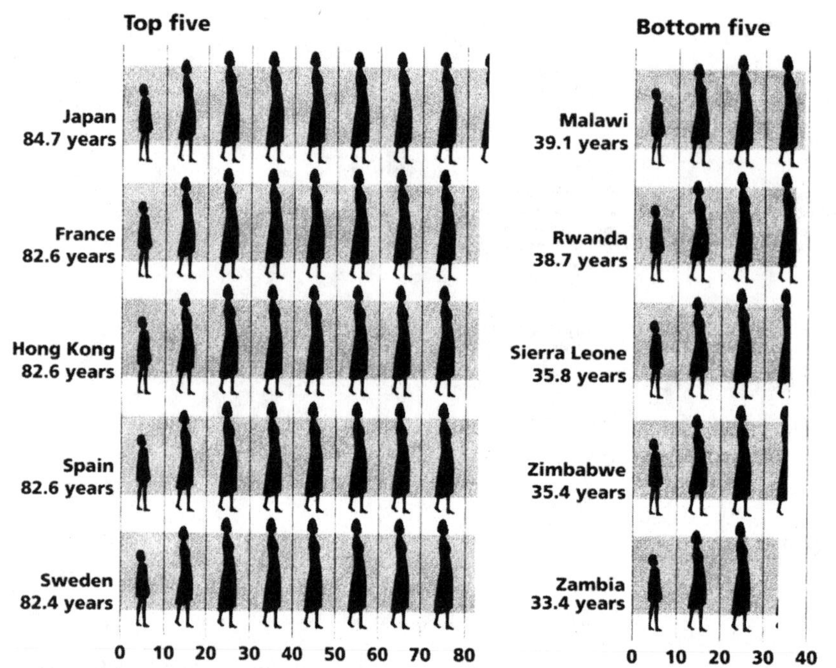

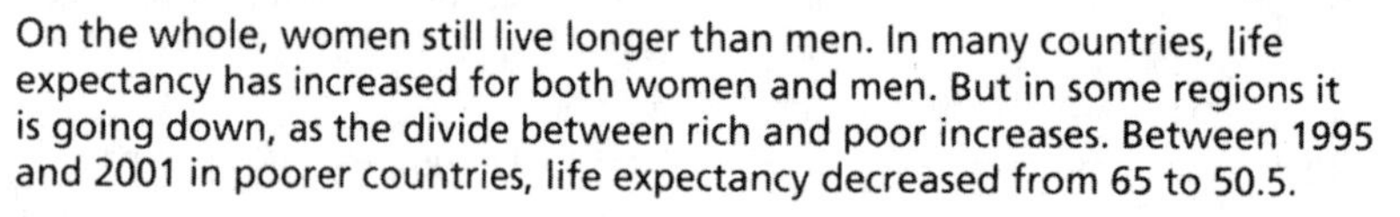

On the whole, women still live longer than men. In many countries, life expectancy has increased for both women and men. But in some regions it is going down, as the divide between rich and poor increases. Between 1995 and 2001 in poorer countries, life expectancy decreased from 65 to 50.5.

Source: Human Development Report, 2003, UNDP

VIOLENCE

Violence against women and girls takes many forms and shows no sign of abating.

- 79 million girls who would otherwise be expected to be alive are 'missing' from various populations, mostly in Asia, as a result of sex-selective abortions, infanticide or neglect.
- Domestic violence is widespread in most societies and is a frequent cause of suicides among women.
- Rape and other forms of sexual violence are increasing. Estimates of the proportion of rapes reported to authorities vary—from less than 3 percent in South Africa to about 16 percent in the US.
- Two million girls between the ages of 5 and 15 are introduced into the commercial sex market each year.
- An estimated 4 million women and girls are bought and sold worldwide each year, either into marriage, prostitution or slavery.
- At least 130 million women have undergone female genital mutilation or cutting; another 2 million are at risk each year.
- So-called 'honour' killings take the lives of thousands of young women every year, mainly in Western Asia, North Africa and parts of South Asia.[1]

FGM and in many cases there are programmes to retrain circumcisers, who are usually older women.

But in the Gambia in 1999 the President, Yahya Jammeh, refused to outlaw the practice, saying that it is part of 'Gambian culture'. Others have argued that it is a Muslim practice, or a Christian one; but in fact where it is practised it is done so by both religions. The President also implied that organizations working against the practice in the Gambia were fair game, saying that anti-FGM campaigners could continue to do their work in rural areas, but that 'there's no guarantee that after they deliver their speeches, they will return to their homes'.

So the four staff and the many volunteers who work at the Gambia Committee on Traditional Practices (GAMCOTRAP), are brave people. People like Sunta Javara, who is serene and determined. 'The President said we should know that our lives are at stake. He said we were not allowed to use public radio or TV, nor would we have access to a lawyer if anything happened. International organizations, which had been funding GAMCOTRAP direct, redirected their funding via the Government.' She looks up. 'We hope that, as time goes by, the Government will change its mind.'

The campaigners work towards changing hearts and minds through workshops, conferences and persuading friends and family, perhaps the most difficult task.

Sunta started as a volunteer after attending one of these workshops: 'I found it hard to hear what they had to say at first. But then I became 100 per cent convinced that this was wrong. After my training I went back and talked to my aunt, who I was living with. You know, in Gambia it is not easy for us to talk to our parents or older relatives about these things. But she also became convinced and we were happy when we found she was pregnant. I asked if I could adopt the baby, and she agreed.

'The President said we should know that our lives are at stake'

When the grandmother came to get the baby for circumcision we said no. My uncle supported us. The child was their last-born and he said: "I don't want anything to harm my baby." That baby is now 12. She still lives with me and she has not been mutilated. I also managed to save two of my cousin's sister's daughters. But it is easier in the town. It is more difficult in the villages and it will take time. But we are patient.'

Meanwhile, on the day Sunta talks to me, the five girls from the village are already in the circumciser's house.

---

1. GAMCOTRAP leaflet

Glossary of Terms and Abbreviations

Acquired Immune Deficiency Syndrome (AIDS) A disease of immune system dysfunction widely believed to be caused by the human immunodeficiency virus (HIV), which allows opportunistic infections to take over the body.

African Development Bank Founded in 1963 under the auspices of the United Nations Economic Commission on Africa, the bank, located in Côte d'Ivoire, makes loans to African countries, although other nations can apply.

African National Congress (ANC) Founded in 1912, the group's goal is to achieve equal rights for blacks in South Africa through nonviolent action. "Spear of the Nation," the ANC wing dedicated to armed struggle, was organized after the Sharpeville massacre in 1960.

African Party for the Independence of Guinea-Bissau and Cape Verde (PAICG) An independence movement that fought during the 1960s and 1970s for the liberation of present-day Guinea-Bissau and Cape Verde from Portuguese rule. The two territories were ruled separately by a united PAIGC until a 1981 coup in Guinea-Bissau caused the party to split along national lines. In 1981 the Cape Verdean PAIGC formally renounced its Guinea links and became the PAICV.

African Socialism A term applied to a variety of ideas (including those of Nkrumah and Senghor) about communal and shared production in Africa's past and present. The concept of African socialism was especially popular in the early 1960s. Adherence to it has not meant governments' exclusion of private-capitalist ventures.

Afrikaners South Africans of European descent who speak Afrikaans and are often referred to as *Boers* (Afrikaans for "farmers").

Algiers Agreement The 1979 peace agreement when Mauritania made peace with the Polisario and abandoned claims to Western Sahara.

Aouzou Strip A barren strip of land between Libya and Chad contested by both countries.

Apartheid Literally, "separatehood," a South African policy that segregated the races socially, legally, and politically.

Arusha Declaration A document issued in 1967 by Tanzanian President Julius Nyerere, committing the country to socialism based on peasant farming, democracy under one party, and self-reliance.

Assimilado The Portuguese term for Africans who became "assimilated" to Western ways. Assimilados enjoyed equal rights under Portuguese law.

Azanian People's Organization (AZAPO) Founded in 1978 at the time of the Black Consciousness Movement and revitalized in the 1980s, the movement works to develop chapters and bring together black organizations in a national forum.

Bantu A major linguistic classification for many Central, Southern, and East African languages.

Bantustans Areas, or "homelands," to which black South Africans were assigned "citizenship" as part of the policy of apartheid.

Basarawa Peoples of Botswana who have historically been hunters and gatherers.

Berber The collective term for the indigenous languages and peoples of North Africa.

Black Consciousness Movement A South African student movement founded by Steve Biko and others in the 1970s to promote pride and empowerment of blacks.

Boers See *Afrikaners*.

Brotherhoods Islamic organizations based on specific religious beliefs and practices. In many areas, brotherhood leaders and their spiritual followers gain political influence.

Cabinda A small, oil-rich portion of Angola separated from the main body of that country by a coastal strip of the Democratic Republic of the Congo.

Caliphate The office or dominion of a caliph, the spiritual head of Islam.

Cassava A tropical plant with a fleshy, edible rootstock; one of the staples of the African diet. Also known as manioc.

Chimurenga A Shona term meaning "fighting in which everyone joins," used to refer to Zimbabwe's fight for independence.

Committee for the Struggle against Drought in the Sahel (CILSS) A grouping of eight West African countries, formed to fight the effects of drought in the region.

Commonwealth of Nations An association of nations and dependencies loosely joined by the common tie of having been part of the British Empire.

Congress of South African Trade Unions (COSATU) Established in 1985 to form a coalition of trade unions to press for workers' rights and an end to apartheid.

Copperbelt A section of Zambia with a high concentration of copper-mining concessions.

Creole A person or language of mixed African and European descent.

Dergue From the Amheric word for "committee," the ruling body of Ethiopia following the Revolution in 1974 to the 1991 Revolution (it was overthrown by the Ethiopian People's Revolutionary Democratic Front).

East African Community (EAC) Established in 1967, this organization grew out of the East African Common Services Organization begun under British rule. The EAC included Kenya, Tanzania, and Uganda in a customs union and involved common currency and development of infrastructure. It was disbanded in 1977, and the final division of assets was completed in 1983.

Economic Commission for Africa (ECA) Founded in 1958 by the Economic and Social Committee of the United Nations to aid African development through regional centers, field agents, and the encouragement of regional efforts, food self-sufficiency, transport, and communications development.

Economic Community of Central African States (CEEAC, also known as ECCA) An organization of all of the Central African states, as well as Rwanda and Burundi, whose goal is to promote economic and social cooperation among its members.

Economic Community of West Africa (CEAO) An economic organization of former French colonies that was formed to promote trade and regional economic cooperation.

Economic Organization of West African States (ECOWAS) Established in 1975 by the Treaty of Lagos, the organization includes all of the West African states except Western Sahara. The organization's goals are to promote trade, cooperation, and self-reliance among its members.

Enclave Industry An industry run by a foreign company that uses imported technology and machinery and exports the product to industrialized countries; often described as a "state within a state."

Eritrean People's Liberation Front (EPLF) The major group fighting the Ethiopian government for the independence of Eritrea.

European Community See *European Union.*

European Union (EU) Known as the European Community until 1994, this is the collective designation of three organizations with common membership—the European Economic Community, the European Coal and Steel Community, and the European Atomic Energy Community. Sometimes also referred to as the Common Market.

Evolués A term used in colonial Zaire (the Democratic Republic of the Congo) to refer to Western-educated Congolese.

Fokonolas Indigenous village management bodies.

Food and Agricultural Organization of the United Nations (FAO) Established in 1945 to oversee good nutrition and agricultural development. Franc Zone (Commonly known as the CFA [*le franc des Colonies Françaises d'Afrique*] franc zone.) This organization includes members of the West African Monetary Union and the monetary organizations of Central Africa that have currencies linked to the French franc. Reserves are managed by the French treasury and guaranteed by the French franc.

Free French Conference A 1944 conference of French-speaking territories, which proposed a union of all the territories in which Africans would be represented and their development furthered.

Freedom Charter Established in 1955, this charter proclaimed equal rights for all South Africans and has been a foundation for almost all groups in the resistance against apartheid.

Frelimo See *Mozambique Liberation Front.*

French Equatorial Africa (FEA) The French colonial federation that included present-day Democratic Republic of the Congo, Central African Republic, Chad, and Gabon.

French West Africa The administrative division of the former French colonial empire that included the current independent countries of Senegal, Côte d'Ivoire, Guinea, Mali, Niger, Burkina Faso, Benin, and Mauritania.

Frontline States A caucus supported by the Organization of African Unity (consisting of Tanzania, Zambia, Mozambique, Botswana, Zimbabwe, and Angola) whose goal is to achieve black majority rule in all of Southern Africa.

Green Revolution Use of Western technology and agricultural practices to increase food production and agricultural yields.

Griots Professional bards of West Africa, some of whom tell history and are accompanied by the playing of the kora or harplute.

Gross Domestic Product (GDP) The value of production attributable to the factors of production in a given country regardless of their ownership. GDP equals GNP minus the product of a country's residents originating in the rest of the world.

Gross National Product (GNP) The sum of the values of all goods and services produced by a country's residents at home and abroad in any given year, less income earned by foreign residents and remitted abroad.

Guerrilla A member of a small force of irregular soldiers. Generally, guerrilla forces are made up of volunteers who make surprise raids against the incumbent military or political force.

Harmattan In West Africa, the dry wind that blows in from the Sahara Desert during January and February, which now reaches many parts of the West African coast. Its dust and haze are a sign of the new year and of new agricultural problems.

Homelands See *Bantustans.*

Horn of Africa A section of northeastern Africa including the countries of Djibouti, Ethiopia, Somalia, and the Sudan.

Hut Tax Instituted by the colonial governments in Africa, this measure required families to pay taxes on each building in the village.

International Monetary Fund (IMF) Established in 1945 to promote international monetary cooperation.

Irredentism An effort to unite certain people and territory in one state with another, on the grounds that they belong together.

Islam A religious faith started in Arabia during the seventh century A.D. by the Prophet Muhammad and spread in Africa through African Muslim leaders, migrations, and wars.

Jihad A struggle, or "holy war," waged as a religious duty on behalf of Islam to rid the world of disbelief and error.

Koran Writings accepted by Muslims as the word of God, as revealed to the Prophet Mohammed.

Lagos Plan of Action Adopted by the Organization of African Unity in 1980, this agreement calls for self-reliance, regional economic cooperation, and the creation of a pan-African economic community and common market by the year 2000.

League of Nations Established at the Paris Peace Conference in 1919, this forerunner of the modern-day United Nations had 52 member nations at its peak (the United States never joined the organization) and mediated in international affairs. The league was dissolved in 1945 after the creation of the United Nations.

Least Developed Countries (LDCs) A term used to refer to the poorest countries of the world, including many African countries.

Maghrib An Arabic term, meaning "land of the setting sun," that is often used to refer to the former French colonies of Morocco, Algeria, and Tunisia.

Mahdi The expected messiah of Islamic tradition; or a Muslim leader who plays a messianic role.

Malinke (Mandinka, or Mandinga) One of the major groups of people speaking Mande languages. The original homeland of the Malinke was Mali, but the people are now found in Mali, Guinea-Bissau, The Gambia, and other areas, where they are sometimes called Mandingoes. Some trading groups are called Dyoula.

Marabout A dervish Muslim in Africa believed to have supernatural power.

Marxist-Leninism Sometimes called "scientific socialism," this doctrine derived from the ideas of Karl Marx as modified by Vladimir Lenin; it was the ideology of the Communist Party of the Soviet Union and has been modified in many ways by other persons and groups who still use the term. In Africa, some political parties or movements have claimed to be Marxist-Leninist but have often followed policies that conflict in practice with the ideology; these governments have usually not stressed Marx's philosophy of class struggle.

Mfecane The movement of people in the nineteenth century in the eastern areas of present-day South Africa to the west and north as the result of wars led by the Zulus.

Movement for the Liberation of Angola (MPLA) A major Angolan liberation movement that has its strongest following among assimilados and Kimbundu speakers, who are predominant in Luanda, the capital, and the interior to the west of the city.

Mozambique Liberation Front (Frelimo) Mozambique's single ruling party following a 10-year struggle against Portuguese colonial rule, which ended in 1974.

Mozambique National Resistance See *Renamo.*

Muslim A follower of the Islamic faith.

Naam A traditional work cooperative in Burkina Faso.

National Front for the Liberation of Angola (FNLA) One of the major Angolan liberation movements; its original focus was limited to the northern Kongo-speaking population.

National Union for the Total Independence of Angola (UNITA) One of three groups that fought the Portuguese during the colonial period in Angola, later backed by South Africa and the U.S. CIA in fighting the independent government of Angola.

National Youth Service Service to the state required of youth after completing education, a common practice in many African countries.

Nkomati Accords An agreement signed in 1984 between South Africa and Mozambique, pledging that both sides would no longer support opponents of the other.

Nonaligned Movement (NAM) A group of nations that chose not to be politically or militarily associated with either the West or the former communist bloc.

Nongovernmental Organizations (NGO) A private voluntary organization or agency working in relief and development programs.

Organization for the Development of the Senegal River (OMVS) A regional grouping of countries bordering the Senegal River that sponsors joint research and projects.

Organization of African Unity (OAU) An association of all the independent states of Africa (except South Africa) whose goal is to promote peace and security as well as economic and social development.

Organization of Petroleum Exporting Countries (OPEC) Established in 1960, this association of some of the world's major oil-producing countries seeks to coordinate the petroleum policies of its members.

Pan Africanist Congress (PAC) A liberation organization of black South Africans that broke away from the ANC in the 1950s.

Parastatals Agencies for production or public service that are established by law and that are, in some measure, government organized and controlled. Private enterprise may be involved, and the management of the parastatal may be in private hands.

Pastoralist A person, usually a nomad, who raises livestock for a living.

Polisario Front Originally a liberation group in Western Sahara seeking independence from Spanish rule. Today, it is battling Morocco, which claims control over the Western Sahara. See Saharawi Arab Democratic Republic (SADR).

Popular Movement for the Liberation of Angola (MPLA) A Marxist liberation movement in Angola during the resistance to Portuguese rule; now the governing party in Angola.

Renamo A South African-backed rebel movement that attacked civilians in an attampt to overthrow the government of Mozambique.

Rinderpest A cattle disease that periodically decimates herds in savanna regions.

Saharawi Arab Democratic Republic (SADR) The Polisario Front name for Western Sahara, declared in 1976 in the struggle for independence from Morocco.

Sahel In West Africa, the borderlands between savanna and desert.

Sanctions Coercive measures, usually economic, adopted by nations acting together against a nation violating international law.

Savanna Tropical or subtropical grassland with scattered trees and undergrowth.

Shari'a The Islamic code of law.

Sharpeville Massacre The 1960 demonstration in South Africa in which 60 people were killed when police fired into the crowd; it became a rallying point for many antiapartheid forces.

Sorghum A tropical grain that is a traditional staple in the savanna regions.

Southern African Development Community (SADC) (Formerly the Southern African Development Coordination Conference. Its name was changed in 1992.) An organization of nine African states (Angola, Zambia, Malawi, Mozambique, Zimbabwe, Lesotho, Botswana, Swaziland, and Tanzania) whose goal is to free themselves from dependence on South Africa and to cooperate on projects of economic development.

South-West Africa People's Organization (SWAPO) Angola-based freedom fighters who had been waging guerrilla warfare against the presence of South Africa in Namibia since the 1960s. The United Nations and the Organization of African Unity now recognize SWAPO as the only authentic representative of the Namibian people.

Structural Adjustment Program (SAP) Economic reforms encouraged by the International Monetary Fund, which include devaluation of currency, cutting government subsidies on commodities, and reducing government expenditures.

Swahili A trade and government Bantu language that covers much of East Africa and Congo region.

Tsetse Fly An insect that transmits sleeping sickness to cattle and humans. It is usually found in the scrub-tree and forest regions of Central Africa.

Ujaama In Swahili, "familyhood"; government-sponsored cooperative villages in Tanzania.

Unicameral A political structure with a single legislative branch.

Unilateral Declaration of Independence (UDI) A declaration of white minority settlers in Rhodesia, claiming independence from the United Kingdom in 1965.

United Democratic Front (UDF) A multiracial, black-led group in South Africa that gained prominence during the 1983 campaign to defeat the government's Constitution, which gave only limited political rights to Asians and Coloureds.

United Nations (UN) An international organization established on June 26, 1945, through official approval of the charter by delegates of 50 nations at a conference in San Francisco, California. The charter went into effect on October 24, 1945.

United Nations Development Program (UNDP) Established to create local organizations for increasing wealth through better use of human and natural resources.

United Nations Educational, Scientific, and Cultural Organization (UNESCO) Established on November 4, 1946, to promote international collaboration in education, science, and culture.

United Nations High Commission for Refugees (UNHCR) Established in 1951 to provide international protection for people with refugee status.

Villagization A policy whereby a government relocates rural dwellers to create newer, more concentrated communities.

West African Monetary Union (WAMU) A regional association of member countries in West Africa (Benin, Burkina Faso, Côte d'Ivoire, Mali, Niger, Senegal, and Togo) that have vested authority to conduct monetary policy in a common central bank.

World Bank A closely integrated group of international institutions providing financial and technical assistance to developing countries.

World Health Organization (WHO) Established by the United Nations in 1948, this organization promotes the highest possible state of health in countries throughout the world.

Bibliography

RESOURCE CENTERS

African Studies Centers provide special services for schools, libraries, and community groups. Contact the center nearest you for further information about resources available.

African Studies Center
Boston University
270 Bay State Road
Boston, MA 02215

African Studies Program
Indiana University
Woodburn Hall 221
Bloomington, IN 47405

African Studies Educational Resource Center
100 International Center
Michigan State University
East Lansing, MI 49923

African Studies Program
630 Dartmouth
Northwestern University
Evanston, IL 60201

Africa Project
Lou Henry Hoover Room 223
Stanford University
Stanford, CA 94305

African Studies Center
University of California
Los Angeles, CA 90024

Center for African Studies
470 Grinter Hall
University of Florida
Gainesville, FL 32611

African Studies Program
University of Illinois
1208 W. California, Room 101
Urbana, IL 61801

African Studies Program
1450 Van Hise Hall
University of Wisconsin
Madison, WI 53706

Council on African Studies
Yale University
New Haven, CT 06520

Foreign Area Studies
The American University
5010 Wisconsin Avenue, NW
Washington, DC 20016

African Studies Program
Center for Strategic and International Studies
Georgetown University
1800 K Street, NW
Washington, DC 20006

REFERENCE WORKS, BIBLIOGRAPHIES, AND OTHER SOURCES

Africa Research Bulletin (Political Series), Africa Research Ltd., Exeter, Devon, England (monthly).

Africa South of the Sahara (updated yearly) (Detroit: Gale Research).

Africa Today: An Atlas of Reproductible Pages, rev. ed. (Wellesley: World Eagle, 1990).

Rosalid Baucham, *African-American Organizations: A Selective Bibliography* (Organizations and Institutional Groups) (New York: Garland, 1997).

Chris Cook and David Killingray, *African Political Facts Since 1945* (New York: Facts on File, 1990).

David E. Gardinier, *Africana Journal* notes, Volume xvii. A Bibliographic Library Guide and Review Forum (New York: Holmes & Meier, 1997).

Colin Legum, ed., *Africa Contemporary Record* (New York: Holmes & Meier) (annual).

MAGAZINES AND PERIODICALS

Africa News, P.O. Box 3851, Durham, NC 27702.

Africa Now, 212 Fifth Avenue, Suite 1409, New York, NY 10010.

Africa Recovery, DPI, Room S-1061, United Nations, New York, NY 10017.

Africa Today, 64 Washburn Avenue, Wellesley, MA 02181.

African Arts, University of California, Los Angeles, CA.

African Concord, 5–15 Cromer Street, London WCIH 8LS, England.

The Economist, 122 E. 42nd Street, 14th Floor, New York, NY 10168.

Newswatch, 62 Oregun Road, P.M.B. 21499, Ikeja, Nigeria.

The UNESCO Courier, 31, Rue François Bonvin, 75732, Paris CEDEX 15, France.

The Weekly Review, P.O. Box 42271, Nairobi, Kenya.

West Africa, Graybourne House, 52/54 Gray Inn Road, London WCIX 8LT, England.

NOVELS AND AUTOBIOGRAPHICAL WRITINGS

Chinua Achebe, *Things Fall Apart* (Portsmouth: Heinemann, 1965).

This is the story of the life and values of residents of a traditional Igbo village in the nineteenth century and of its first contacts with the West.

___, *No Longer at Ease* (Portsmouth: Heinemann, 1963).
The grandson of the major character of *Things Fall Apart* lives an entirely different life in the modern city of Lagos.

Ayi Kwei Armah, *The Beautyful Ones Are Not Yet Born* (London: Heinemann, 1992).

André Brink, *A Dry White Season* (New York: Penguin, 1989).

Syl Cheney-Choker, *The Last Harmattan of Alusine Dunba* (London: Heinemann, 1991).

Tsitsi Dangarembga, *Nervous Conditions* (Seal Press Feminist Publishing, 2002).

Buchi Emecheta, *The Joys of Motherhood* (New York: G. Braziller, 1979).
The story of a Nigerian woman who overcomes great obstacles to raise a large family and then finds that the meaning of motherhood has changed.

Nadine Gordimer, *Burgher's Daughter* (New York: Viking, 1980).

___, *A Soldier's Embrace* (New York: Viking, 1982).
These short stories treat the effects of apartheid on people's relations with each other. Films made from some of these stories are available at the University of Illinois Film Library, Urbana-Champaign, IL and the Boston University Film Library, Boston, MA.

Bessie Head, *Question of Power* (London: Heinemann, 1974).

Cheik Amadou Kane, *Ambiguous Adventure* (Portsmouth: Heinemann, 1972).
This autobiographical novel of a young man coming of age in Senegal, in a Muslim society, and, later, in a French school.

F. Kietseng, *Comrade Fish: Memoirs of a Motswana in the ANC Underground* (Pula Publishing, 1999).

Alex LaGuma, *Time of the Butcherbird* (Portsmouth: Heinemann, 1979).
The people of a long-standing black community in South Africa's countryside are to be removed to a Bantustan.

Camara Laye, *The Dark Child* (Farrar Straus and Giroux, 1954).
This autobiographical novel gives a loving and nostalgic picture of a Malinke family of Guinea.

Nelson Mandela, *Long Walk to Freedom: The Autobiography of Nelson Mandela* (New York: Little, Brown, 1995).

Okot p'Bitek, *Song of Lawino* (Portsmouth: Heinemann, 1983).
A traditional Ugandan wife comments on the practices of her Western-educated husband and reveals her own life-style and values.

Alexander McCall Smith, *The No. 1 Ladies' Detective Agency* (New York: Anchor Books, 2003).
The first in a popular series of novels set in Botswana.

Wole Soyinka, *Ake: The Years of Childhood* (New York: Random House, 1983).
Soyinka's account of his first 11 years is full of the sights, tastes, smells, sounds, and personal encounters of a headmaster's home and a busy Yoruba town.

___, *Death and the King's Horsemen* (New York: W. W. Norton, 2002).

Ngugi wa Thiong'o, *A Grain of Wheat* (Portsmouth: Heinemann, 1968).
A story of how the Mau-Mau movement and the coming of independence affected several individuals after independence as well as during the struggle that preceded it.

Amos Tutuola, *The Palm-Wine Drinkard* (Grove Press, 1994).

Yvonne Vera, *Butterfly Burning* (Baobab Books, 2000).

INTRODUCTORY BOOKS

Philip G. Altbach, *Muse of Modernity: Essays on Culture as Development in Africa* (Lawrenceville, NJ: Africa World, 1997).

Tony Binns, *People and Environment in Africa* (New York: Wiley, 1995).

Raymond Bonner, *At the Hand of Man: Peril and Hope for Africa's Wildlife* (New York: Random House, 1994).

Gwendolen Carter and Patrick O'Meara, eds., *African Independence: The First Twenty-Five Years* (Midland Books, 1986).
Collected essays surrounding issues such as political structures, military rule, and economics.

John Chiasson, *African Journey* (Upland, CT: Bradbury Press, 1987).
An examination into Africa's social life and customs.

Basil Davidson, *Africa in History* (New York: Macmillan, 1991).
A fine discussion of African history.

___, *The African Genius* (Boston: Little, Brown, 1979). Also published as *The Africans.*
Davidson discusses the complex political, social, and economic systems of traditional African societies, translating scholarly works into a popular mode without distorting complex material.

___, *The Black Man's Burden: Africa and the Curse of the Nation State* (New York: Random House, 1992).
A discussion on Africa's government and the status of the nation-state.

___, *A History of Africa,* 2nd ed. (Unwin Hyman, 1989).
A comprehensive look at the historical evolution of Africa.

Christopher Ehret, *An African Classical Age: Eastern and Southern Africa in World History, 1000 B.C. to A.D. 400* (Charlottesville: University of Virginia Press, 2001).

Clementine M. Faik-Nzuji, *Tracing Memory: Glossary of Graphic Signs and Symbols in African Art & Culture* (Seattle, WA: University of Washington Press, 1997).

Timothy J. Keegan, *Colonial South Africa and the Origins of the Racial Order* (Charlottesville: University of Virginia Press, 1997).

Anthony Appiah Kwame, Henry Louis Gates Jr., eds., *Africana: The Encyclopedia of the African and African American Experience* (BasicCivitas Books, 1999).

Paul E. Lovejoy, *Transformations in Slavery: A History of Slavery in Africa* (Cambridge: Cambridge University Press, 2000).

Amina Mama, *Beyond the Mask: Race, Gender and Identity* (London: Routledge, 1995).

John Mbiti, *African Religions and Philosophy* (Portsmouth: Heinemann, 1982).
This work by a Ugandan scholar is the standard introduction to the rich variety of religious beliefs and rituals of African peoples.

V. Y. Mudimbe, *The Invention of Africa* (Bloomington: Indiana University Press, 1988).

Joseph M. Murphy, *Working the Spirit: Ceremonies of the African Diaspora* (Boston: Beacon Press, 1994).

J. H. Kwabena Nketia, *The Music of Africa* (New York: Norton, 1974).
The author, a Ghanaian by birth, is Africa's best-known ethnomusicologist.

Gladson I. Nwanna, *Do's and Don'ts around the World: A Country Guide to Cultural and Social Taboos and Etiquette in Africa* (Baltimore: World Travel Institute, 1998).

Keith R. Richburg, *Out of Africa: A Black Man Confronts Africa* (New York: Basic Books, 1997).

Kevin Shillington, *History of Africa* (New York: Macmillan, 1995).

Bengt Sundkler and Christopher Steed, *A History of the Church in Africa* (Cambridge: Cambridge University Press, 2000).

John Thornton, *Africa and Africans in the Making of the Atlantic World, 1400–1800* (Cambridge: Cambridge University Press, 1998).

J. B. Webster, A. A. Boahen, and M. Tidy, *The Revolutionary Years: West Africa Since 1800* (London: Longman, 1980).
An interesting, enjoyable, and competent introduction.

Frank Willett, *African Art* (New York: Oxford University Press, 1971).
A work to read for both reference and pleasure.

COUNTRY AND REGIONAL STUDIES

Howard Adelman and John Sorenson, eds., *African Refugees* (Boulder: Westview, 1993).

Howard Adelman and Astri Suhrke, eds., *The Path of a Genocide: The Rwanda Crisis From Uganda to Zaire* (Transaction Publishing, 1999).

Allan R. Booth, *Swaziland: Tradition and Change in a Southern African Kingdom* (Boulder: Westview Press, 1984).

Thomas Borstelmann, *Apartheid, Colonialism, and the Cold War: The United States and Southern Africa* (New York: Oxford University Press, 1993).

Louis Brenner, ed., *Muslim Identity and Social Change in Sub-Saharan Africa* (Bloomington: Indiana University Press, 1993).

Mike Brogden and Clifford Shearing, *Policing for a New South Africa* (New York: Routledge, 1993).

Marcia M. Burdette, *Zambia: Between Two Worlds* (Boulder: Westview Press, 1988).

Amilcar Cabral, *Unity and Struggle* (Monthly Review Press, 1981).

Joao M. Cabrit, *Mozambique: The Tortuous Road to Democracy* (Palgrave Macmillan, 2001).

Thomas Callaghy and John Ravenhill, eds., *Hemmed In: Global Responses to Africa's Economic Decline* (New York: Columbia University Press, 1994).

W. Joesph Campbell, *The Emergent Independent Press in Benin and Coté d'Ivoire: From Voice of the State to Advocate of Democracy* (New York: Praeger, 1998).

Robin Cohen and Harry Goulbourne, eds., *Democracy and Socialism in Africa* (Boulder: Westview Press, 1991).

Maureen Covell, *Madagascar: Politics, Economy, and Society* (London and New York: F. Pinter, 1987).

W. A. Edge and M. H. Lekorwe, *Botswana, Politics and Society* (J. L. van Schaik, 1998).

Norman Etherington, *The Great Treks: The Transformation of Southern Africa, 1815–1854* (London: Longman, 2001).

Toyin Falola and Julius Ihonvbere, *The Rise and Fall of Nigeria's Second Republic, 1979–1984* (London: Zed Press, 1985).

Robert Fatton, *The Making of a Liberal Democracy: Senegal's Passive Revolution, 1975–85* (Boulder: L. Rienner, 1987).

Foreign Area Studies (Washington, DC: Government Printing Office).

Pumla Gobodo-Madikizela, *A Human Being Died That Night: A South African Story of Forgiveness* (New York: Houghton-Mifflin, 2003).

April A. Gordon and Donald L. Gordon, *Understanding Contemporary Africa* (Boulder: L. Rienner Publishers, 1996).

Phillip Gourevitch, *We Wish to Inform You That Tomorrow We Will Be Killed With Our Families: Stories From Rwanda* (Picador, 1999).

Joseph Hanlon, *Mozambique: The Revolution Under Fire* (London: Zed Press, 1984).

Angelique Haugerud, *The Culture of Politics in Modern Kenya* (Cambridge: Cambridge University Press, 1997).

Adam Hochschild, *King Leopold's Ghost* (New York: Houghton-Mifflin, 1999).
A fascinating and wrenching account of how Belgium's King Leopold ravaged Congo.

Tony Hodges, *Western Sahara: The Roots of a Desert War* (Westport: Laurence Hill & Co., 1983).

Gaim Kibreab, *Refugees and Development in Africa: The Case of Eritrea* (Trenton: Red Sea Press, 1987).

Gerhard Kraus, *Human Development from an African Ancestry* (London: Karnak House, 1990).

David D. Laitin and Said S. Samatar, *Somalia: Nation in Search of a State* (Boulder: Westview Press, 1987).

Karl Maier, *Angola: Promises and Lies* (Independence Educational Publishers, 2002).

Mahmoud Mamdani, *Citizen and Subject* (Princeton: Princeton University Press, 1996).

David Martin and Phyllis Johnson, *The Struggle for Zimbabwe: The Chimurenga War* (Boston: Faber & Faber, 1981).

Georges Nzongola-Ntalaja, *The Congo: From Leopold to Kabila: A People's History* (Zed Books, 2002).

Adebayo O. Olukoshi and Liisa Laakso, *Challenges to the Nation-State in Africa* (Uppsala: Nordiska Afrikainstitutel, in cooperation with Institute of Development Studies, University of Helsinki, 1996).

Eghosa E. Osaghae, *Crippled Giant: Nigeria Since Independence* (Bloomington: Indiana University Press, 1998).

Thomas O'Toole, *The Central African Republic: The Continent's Hidden Heart* (Boulder: Westview Press, 1986).

Richard Pankhurst, *The Ethiopians: A History* (London: Blackwell, 2001).

Deborah Pellow and Naomi Chazan, *Ghana: Coping With Uncertainty* (Boulder: Westview Press, 1986).

F. Jeffress Ramsay, Barry Morton, and Themba Mgadla, *Building a Nation: A History of Botswana* (Gaborone: Longman Botswana, 1996).

Richard Sandbrook, *The Politics of Africa's Economic Recovery* (Cambridge: Cambridge University Press, 1993).

Wisdom J. Tettey, ed., et al., *Critical Perspectives on Politics and Socio-Economic Development in Ghana* (African Social Studies Series, 6) (Brill Adademic Publishers, 2003).

Teun Voeten, Roz Vatter-Buck, trans., *How de Body? One Man's Terrifying Journey Through an African War* (Thomas Dunne Books, 2002).

C. W. Wigwe, *Language, Culture, and Society in West Africa* (Elms Court, UK: Arthur H. Stockwell, 1990).

Gabriel Williams, *Liberia: The Heart of Darkness* (Trafford, 2002).

Edwin Wilmsen, *Land Filled With Flies, A Political Economy of the Kalahari* (Chicago: University of Chicago Press, 1989).

Index

Y

Z